CONVERSION FACTORS FROM ENGLISH TO SI UNITS

Length:

1 ft	= 0.3048 m
1 ft	= 30.48 cm
1 ft	= 304.8 mm
1 in.	= 0.0254 m
1 in.	= 2.54 cm
1 in.	= 25.4 mm

Area:

1 ft^2	= 929.03 $\times$ 10^{-4} m^2
1 ft^2	= 929.03 cm^2
1 ft^2	= 929.03 $\times$ 10^2 mm^2
1 in^2	= 6.452 $\times$ 10^{-4} m^2
1 in^2	= 6.452 cm^2
1 in^2	= 645.16 mm^2

Volume:

1 ft^3	= 28.317 $\times$ 10^{-3} m^3
1 ft^3	= 28.317 $\times$ 10^3 cm^3
1 in^3	= 16.387 $\times$ 10^{-6} m^3
1 in^3	= 16.387 cm^3

Force:

1 lb	= 4.448 N
1 lb	= 4.448 $\times$ 10^{-3} kN
1 lb	= 0.4536 kgf
1 kip	= 4.448 kN
1 U.S. ton	= 8.896 kN
1 lb	= 0.4536 $\times$ 10^{-3} metric ton
1 lb/ft	= 14.593 N/m

Stress:

1 lb/ft^2	= 47.88 N/m^2
1 lb/ft^2	= 0.04788 kN/m^2
1 U.S. ton/ft^2	= 95.76 kN/m^2
1 kip/ft^2	= 47.88 kN/m^2
1 lb/in^2	= 6.895 kN/m^2

Unit weight:

1 lb/ft^3	= 0.1572 kN/m^3
1 lb/in^3	= 271.43 kN/m^3

Moment:

1 lb-ft	= 1.3558 N$\cdot$m
1 lb-in.	= 0.11298 N$\cdot$m

Energy:

1 ft-lb	= 1.3558 J

Moment of inertia:

1 in^4	= 0.4162 $\times$ 10^6 mm^4
1 in^4	= 0.4162 $\times$ 10^{-6} m^4

Section modulus:

1 in^3	= 0.16387 $\times$ 10^5 mm^3
1 in^3	= 0.16387 $\times$ 10^{-4} m^3

Hydraulic conductivity:

1 ft/min	= 0.3048 m/min
1 ft/min	= 30.48 cm/
1 ft/min	= 304.8 mm/
1 ft/sec	= 0.3048 m/
1 ft/sec	= 304.8 mm/
1 in./min	= 0.0254 m/
1 in./sec	= 2.54 cm/
1 in./sec	= 25.4 mm/

Coefficient of consolidation:

1 in^2/sec	= 6.452 cm
1 in^2/sec	= 20.346 $\rangle$
1 ft^2/sec	= 929.03 c

Principles of Foundation Engineering, SI

Seventh Edition

BRAJA M. DAS

CENGAGE
Learning™

Australia • Brazil • Japan • Korea • Mexico • Singapore • Spain • United Kingdom • United States

CENGAGE
Learning™

Principles of Foundation Engineering, SI
Seventh Edition
Author Braja M. Das

Publisher, Global Engineering:
Christopher M. Shortt

Senior Developmental Editor: Hilda Gowans

Editorial Assistant: Tanya Altieri

Team Assistant: Carly Rizzo

Marketing Manager: Lauren Betsos

Production Manager: Patricia M. Boies

Content Project Manager: Darrell Frye

Production Service: RPK Editorial Services, Inc.

Copyeditor: Shelly Gerger-Knecthl

Proofreader: Martha McMaster

Indexer: Braja M. Das

Compositor: Integra

Senior Art Director: Michelle Kunkler

Internal Designer: Carmela Pereira

Cover Designer: Andrew Adams

Cover Images:
Courtesy of ADSC : The International
Association of Foundation Drillers, Dallas, Texas
D. B. M. Contractors, Inc., Federal Way,
Washington

Image Permissions Researcher: Deanna Ettinger

Text Permissions Researcher: Katie Huha

Text and Image Permissions Researcher:
Kristiina Paul

First Print Buyer: Arethea Thomas

For product information and technology assistance,
contact us at **Cengage Learning Customer &
Sales Support, 1-800-354-9706**

For permission to use material from this text or product,
submit all requests online at **cengage.com/permissions**
Further permissions questions can be emailed to
permissionrequest@cengage.com

Library of Congress Control Number: 2010922634

ISBN-13: 978-0-495-66812-1

ISBN-10: 0-495-66812-5

Cengage Learning
200 First Stamford Place, Suite 400
Stamford, CT 06902
USA

Cengage Learning is a leading provider of customized
learning solutions with office locations around the globe,
including Singapore, the United Kingdom, Australia, Mexico,
Brazil, and Japan. Locate your local office at:
international.cengage.com/region

Cengage Learning products are represented in Canada by
Nelson Education Ltd.

For your course and learning solutions, visit
www.cengage.com/engineering

Purchase any of our products at your local college store or at
our preferred online store **www.CengageBrain.com**

Printed in the United States of America
1 2 3 4 5 6 7 13 12 11 10

To our granddaughter, Elizabeth Madison

Contents

4   Ultimate Bearing Capacity of Shallow Foundations: Special Cases   181

5   Shallow Foundations: Allowable Bearing Capacity and Settlement   223

9 | Sheet Pile Walls   437

14 Soil Improvement and Ground Modification 722

Preface

Soil mechanics and foundation engineering have developed rapidly during the last fifty years. Intensive research and observation in the field and the laboratory have refined and improved the science of foundation design. Originally published in the fall of 1983 with a 1984 copyright, this text on the principles of foundation engineering is now in the seventh edition. The use of this text throughout the world has increased greatly over the years; it also has been translated into several languages. New and improved materials that have been published in various geotechnical engineering journals and conference proceedings have been incorporated into each edition of the text.

Principles of Foundation Engineering is intended primarily for undergraduate civil engineering students. The first chapter, on Geotechnical Properties of Soil, reviews the topics covered in the introductory soil mechanics course, which is a prerequisite for the foundation engineering course. The text is composed of fourteen chapters with examples and problems, and an answer section for selected problems. The chapters are mostly devoted to the geotechnical aspects of foundation design. Both Systéime International (SI) units and English units are used in the text.

Because the text introduces the application of fundamental concepts of foundation analysis and design to civil engineering students, the mathematical derivations are not always presented; instead, just the final form of the equation is given. A list of references for further information and study is included at the end of each chapter.

Each chapter contains many example problems that will help students understand the application of the equations and graphs. For better understanding and visualization of the ideas and field practices, about thirty new photographs have been added in this edition.

A number of practice problems also are given at the end of each chapter. Answers to some of these problems are given at the end of the text.

The following is a brief overview of the changes from the sixth edition.

- In several parts of the text, the presentation has been thoroughly reorganized for better understanding.
- A number of new case studies have been added to familiarize students with the deviations from theory to practice.
- In Chapter 1 on Geotechnical Properties of Soil, new sections on liquidity index and activity have been added. The discussions on hydraulic conductivity of clay, relative density, and the friction angle of granular soils have been expanded.
- Expanded treatment of the weathering process of rocks is given in Chapter 2, Natural Soil Deposits and Subsoil Exploration.

- In Chapter 3 (Shallow Foundations: Ultimate Bearing Capacity), a new case study on bearing capacity failure in soft saturated clay has been added. Also included is the *reduction factor method* for estimating the ultimate bearing capacity of eccentrically loaded strip foundations on granular soil.
- Chapter 4, Ultimate Bearing Capacity of Shallow Foundations: Special Cases, has new sections on the ultimate bearing capacity of weaker soil underlain by a stronger soil, the seismic bearing capacity of foundations at the edge of a granular slope, foundations on rocks, and the stress characteristics solution for foundations located on the top of granular slopes.
- Stress distribution due to a point load and uniformly loaded circular and rectangular areas located on the surface of a Westergaard-type material has been added to Chapter 5 on Allowable Bearing Capacity and Settlement. Also included in this chapter is the procedure to estimate foundation settlement based on Pressuremeter test results.
- Lateral earth pressure due to a surcharge on unyielding retaining structures is now included in Chapter 7 (Lateral Earth Pressure). Also included in this chapter is the solution for passive earth pressure on a retaining wall with inclined back face and horizontal granular backfill using the method of triangular slices.
- Chapter 8 on Retaining Walls has a new case study. A more detailed discussion is provided on the design procedure for geogrid-reinforced retaining walls.
- Chapter 9 on Sheet Pile Walls has an added section on the holding capacity of plate anchors based on the stress characteristics solution.
- Two case studies have been added to the chapter on Braced Cuts (Chapter 10).
- The chapter on Pile Foundations (Chapter 11) has been thoroughly reorganized for better understanding.
- Based on recent publications, new recommendations have been made to estimate the load-bearing capacity of drilled shafts extending to rock (Chapter 12).

As my colleagues in the geotechnical engineering area well know, foundation analysis and design is not just a matter of using theories, equations and graphs from a textbook. Soil profiles found in nature are seldom homogeneous, elastic, and isotropic. The educated judgment needed to properly apply the theories, equations, and graphs to the evaluation of soils, foundations, and foundation design cannot be overemphasized or completely taught in the classroom. Field experience must supplement classroom work.

The following individuals were kind enough to share their photographs which have been included in this new edition.

- Professor A. S. Wayal, K. J. Somayia Polytechnic, Mumbai, India
- Professor Sanjeev Kumar, Southern Illinois University, Carbondale, Illinois
- Mr. Paul J. Koszarek, Professional Service Industries, Inc., Waukesha, Wisconsin
- Professor Khaled Sobhan, Florida Atlantic University, Boca Raton, Florida
- Professor Jean-Louis Briaud, Texas A&M University, College Station, Texas
- Dr. Dharma Shakya, Geotechnical Solutions, Inc., Irvine, California
- Mr. Jon Ridgeway, Tensar International, Atlanta, Georgia
- Professor N. Sivakugan, James Cook University, Townsville, Queensland, Australia
- Professor Anand J. Puppala, University of Texas at Arlington, Arlington, Texas
- Professor Thomas M. Petry, Missouri University of Science and Technology, Rolla, Missouri

Thanks are due to Neill Belk, graduate student at the University of North Carolina at Charlotte, and Jennifer Nicks, graduate student at Texas A&M University, College Station, Texas, for their help during the preparation of this revised edition. I am also grateful for several helpful suggestions of Professor Adel S. Saada of Case Western Reserve University, Cleveland, Ohio.

Thanks are due to Chris Carson, Executive Director, Global Publishing Program; and Hilda Gowans, Senior Developmental Editor, Engineering, Cengage Learning; Lauren Betsos, Marketing Manager; and Rose Kernan of RPK Editorial Services for their interest and patience during the revision and production of the manuscript.

For the past twenty-seven years, my primary source of inspiration has been the immeasurable energy of my wife, Janice. I am grateful for her continual help in the development of the original text and its six subsequent revisions.

Braja M. Das

1 Geotechnical Properties of Soil

1.1 Introduction

The design of foundations of structures such as buildings, bridges, and dams generally requires a knowledge of such factors as (a) the load that will be transmitted by the super-structure to the foundation system, (b) the requirements of the local building code, (c) the behavior and stress-related deformability of soils that will support the foundation system, and (d) the geological conditions of the soil under consideration. To a foundation engineer, the last two factors are extremely important because they concern soil mechanics.

The geotechnical properties of a soil—such as its grain-size distribution, plasticity, compressibility, and shear strength—can be assessed by proper laboratory testing. In addition, recently emphasis has been placed on the *in situ* determination of strength and deformation properties of soil, because this process avoids disturbing samples during field exploration. However, under certain circumstances, not all of the needed parameters can be or are determined, because of economic or other reasons. In such cases, the engineer must make certain assumptions regarding the properties of the soil. To assess the accuracy of soil parameters—whether they were determined in the laboratory and the field or whether they were assumed—the engineer must have a good grasp of the basic principles of soil mechanics. At the same time, he or she must realize that the natural soil deposits on which foundations are constructed are not homogeneous in most cases. Thus, the engineer must have a thorough understanding of the geology of the area—that is, the origin and nature of soil stratification and also the groundwater conditions. Foundation engineering is a clever combination of soil mechanics, engineering geology, and proper judgment derived from past experience. To a certain extent, it may be called an art.

When determining which foundation is the most economical, the engineer must consider the superstructure load, the subsoil conditions, and the desired tolerable settlement. In general, foundations of buildings and bridges may be divided into two major categories: (1) *shallow foundations* and (2) *deep foundations. Spread footings, wall footings,* and *mat foundations* are all shallow foundations. In most shallow foundations, *the depth of embedment can be equal to or less than three to four times the width of the foundation. Pile* and *drilled shaft* foundations are deep foundations. They are used when top layers have poor load-bearing capacity and when the use of shallow foundations will cause considerable structural damage or instability. The problems relating to shallow foundations and mat foundations are considered in Chapters 3, 4, 5, and 6. Chapter 11 discusses pile foundations, and Chapter 12 examines drilled shafts.

This chapter serves primarily as a review of the basic geotechnical properties of soils. It includes topics such as grain-size distribution, plasticity, soil classification, effective stress, consolidation, and shear strength parameters. It is based on the assumption that you have already been exposed to these concepts in a basic soil mechanics course.

1.2 Grain-Size Distribution

In any soil mass, the sizes of the grains vary greatly. To classify a soil properly, you must know its *grain-size distribution*. The grain-size distribution of *coarse-grained* soil is generally determined by means of *sieve analysis*. For a *fine-grained* soil, the grain-size distribution can be obtained by means of *hydrometer analysis*. The fundamental features of these analyses are presented in this section. For detailed descriptions, see any soil mechanics laboratory manual (e.g., Das, 2009).

Sieve Analysis

A sieve analysis is conducted by taking a measured amount of dry, well-pulverized soil and passing it through a stack of progressively finer sieves with a pan at the bottom. The amount of soil retained on each sieve is measured, and the cumulative percentage of soil passing through each is determined. This percentage is generally referred to as *percent finer*. Table 1.1 contains a list of U.S. sieve numbers and the corresponding size of their openings. These sieves are commonly used for the analysis of soil for classification purposes.

Table 1.1 U.S. Standard Sieve Sizes

Sieve No.	Opening (mm)
4	4.750
6	3.350
8	2.360
10	2.000
16	1.180
20	0.850
30	0.600
40	0.425
50	0.300
60	0.250
80	0.180
100	0.150
140	0.106
170	0.088
200	0.075
270	0.053

Figure 1.1 Grain-size distribution curve of a coarse-grained soil obtained from sieve analysis

The percent finer for each sieve, determined by a sieve analysis, is plotted on *semilogarithmic graph paper,* as shown in Figure 1.1. Note that the grain diameter, *D,* is plotted on the *logarithmic scale* and the percent finer is plotted on the *arithmetic scale.*

Two parameters can be determined from the grain-size distribution curves of coarse-grained soils: (1) the *uniformity coefficient* (C_u) and (2) the *coefficient of gradation,* or *coefficient of curvature* (C_c). These coefficients are

$$C_u = \frac{D_{60}}{D_{10}} \tag{1.1}$$

and

$$C_c = \frac{D_{30}^2}{(D_{60})(D_{10})} \tag{1.2}$$

where D_{10}, D_{30}, and D_{60} are the diameters corresponding to percents finer than 10, 30, and 60%, respectively.

For the grain-size distribution curve shown in Figure 1.1, $D_{10} = 0.08$ mm, $D_{30} = 0.17$ mm, and $D_{60} = 0.57$ mm. Thus, the values of C_u and C_c are

$$C_u = \frac{0.57}{0.08} = 7.13$$

and

$$C_c = \frac{0.17^2}{(0.57)(0.08)} = 0.63$$

Parameters C_u and C_c are used in the *Unified Soil Classification System,* which is described later in the chapter.

Hydrometer Analysis

Hydrometer analysis is based on the principle of sedimentation of soil particles in water. This test involves the use of 50 grams of dry, pulverized soil. A *deflocculating agent* is always added to the soil. The most common deflocculating agent used for hydrometer analysis is 125 cc of 4% solution of sodium hexametaphosphate. The soil is allowed to soak for at least 16 hours in the deflocculating agent. After the soaking period, distilled water is added, and the soil–deflocculating agent mixture is thoroughly agitated. The sample is then transferred to a 1000-ml glass cylinder. More distilled water is added to the cylinder to fill it to the 1000-ml mark, and then the mixture is again thoroughly agitated. A hydrometer is placed in the cylinder to measure the specific gravity of the soil–water suspension in the vicinity of the instrument's bulb (Figure 1.2), usually over a 24-hour period. Hydrometers are calibrated to show the amount of soil that is still in suspension at any given time t. The largest diameter of the soil particles still in suspension at time t can be determined by Stokes' law,

$$D = \sqrt{\frac{18\eta}{(G_s - 1)\gamma_w}}\sqrt{\frac{L}{t}} \tag{1.3}$$

where

D = diameter of the soil particle
G_s = specific gravity of soil solids
η = viscosity of water

Figure 1.2 Hydrometer analysis

γ_w = unit weight of water

L = effective length (i.e., length measured from the water surface in the cylinder to the center of gravity of the hydrometer; see Figure 1.2)

t = time

Soil particles having diameters larger than those calculated by Eq. (1.3) would have settled beyond the zone of measurement. In this manner, with hydrometer readings taken at various times, the soil *percent finer* than a given diameter D can be calculated and a grain-size distribution plot prepared. The sieve and hydrometer techniques may be combined for a soil having both coarse-grained and fine-grained soil constituents.

1.3 Size Limits for Soils

Several organizations have attempted to develop the size limits for *gravel, sand, silt,* and *clay* on the basis of the grain sizes present in soils. Table 1.2 presents the size limits recommended by the American Association of State Highway and Transportation Officials (AASHTO) and the Unified Soil Classification systems (Corps of Engineers, Department of the Army, and Bureau of Reclamation). The table shows that soil particles smaller than 0.002 mm have been classified as *clay.* However, clays by nature are cohesive and can be rolled into a thread when moist. This property is caused by the presence of *clay minerals* such as *kaolinite, illite,* and *montmorillonite.* In contrast, some minerals, such as *quartz* and *feldspar,* may be present in a soil in particle sizes as small as clay minerals, but these particles will not have the cohesive property of clay minerals. Hence, they are called *clay-size particles,* not *clay particles.*

1.4 Weight–Volume Relationships

In nature, soils are three-phase systems consisting of solid soil particles, water, and air (or gas). To develop the *weight–volume relationships* for a soil, the three phases can be separated as shown in Figure 1.3a. Based on this separation, the volume relationships can then be defined.

The *void ratio, e,* is the ratio of the volume of voids to the volume of soil solids in a given soil mass, or

$$e = \frac{V_v}{V_s} \tag{1.4}$$

Table 1.2 Soil-Separate Size Limits

Classification system	Grain size (mm)
Unified	Gravel: 75 mm to 4.75 mm Sand: 4.75 mm to 0.075 mm Silt and clay (fines): <0.075 mm
AASHTO	Gravel: 75 mm to 2 mm Sand: 2 mm to 0.05 mm Silt: 0.05 mm to 0.002 mm Clay: <0.002 mm

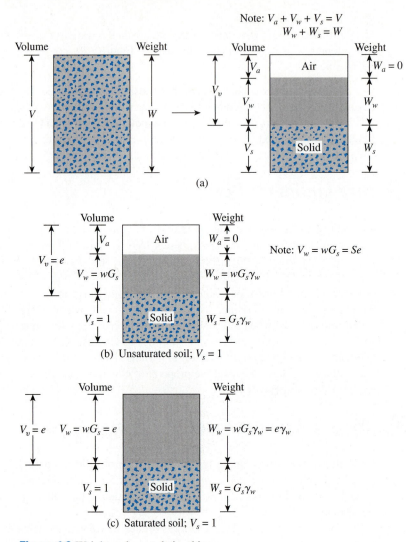

Figure 1.3 Weight–volume relationships

where

V_v = volume of voids
V_s = volume of soil solids

The *porosity, n,* is the ratio of the volume of voids to the volume of the soil specimen, or

$$n = \frac{V_v}{V} \tag{1.5}$$

where

V = total volume of soil

Moreover,

$$n = \frac{V_v}{V} = \frac{V_v}{V_s + V_v} = \frac{\dfrac{V_v}{V_s}}{\dfrac{V_s}{V_s} + \dfrac{V_v}{V_s}} = \frac{e}{1 + e} \tag{1.6}$$

The *degree of saturation, S,* is the ratio of the volume of water in the void spaces to the volume of voids, generally expressed as a percentage, or

$$S(\%) = \frac{V_w}{V_v} \times 100 \tag{1.7}$$

where

V_w = volume of water

Note that, for saturated soils, the degree of saturation is 100%.

The weight relationships are *moisture content, moist unit weight, dry unit weight,* and *saturated unit weight,* often defined as follows:

$$\text{Moisture content} = w(\%) = \frac{W_w}{W_s} \times 100 \tag{1.8}$$

where

W_s = weight of the soil solids
W_w = weight of water

$$\text{Moist unit weight} = \gamma = \frac{W}{V} \tag{1.9}$$

where

W = total weight of the soil specimen = $W_s + W_w$

The weight of air, W_a, in the soil mass is assumed to be negligible.

$$\text{Dry unit weight} = \gamma_d = \frac{W_s}{V} \tag{1.10}$$

When a soil mass is completely saturated (i.e., all the void volume is occupied by water), the moist unit weight of a soil [Eq. (1.9)] becomes equal to the *saturated unit weight* (γ_{sat}). So $\gamma = \gamma_{sat}$ if $V_v = V_w$.

More useful relations can now be developed by considering a representative soil specimen in which the volume of soil solids is equal to *unity,* as shown in Figure 1.3b. Note that if $V_s = 1$, then, from Eq. (1.4), $V_v = e$, and the weight of the soil solids is

$$W_s = G_s \gamma_w$$

where

G_s = specific gravity of soil solids
γ_w = unit weight of water (9.81 kN/m^3)

Also, from Eq. (1.8), the weight of water $W_w = wW_s$. Thus, for the soil specimen under consideration, $W_w = wW_s = wG_s\gamma_w$. Now, for the general relation for moist unit weight given in Eq. (1.9),

$$\gamma = \frac{W}{V} = \frac{W_s + W_w}{V_s + V_v} = \frac{G_s\gamma_w(1 + w)}{1 + e} \qquad (1.11)$$

Similarly, the dry unit weight [Eq. (1.10)] is

$$\gamma_d = \frac{W_s}{V} = \frac{W_s}{V_s + V_v} = \frac{G_s\gamma_w}{1 + e} \qquad (1.12)$$

From Eqs. (1.11) and (1.12), note that

$$\gamma_d = \frac{\gamma}{1 + w} \qquad (1.13)$$

According to Eq. (1.7), degree of saturation is

$$S = \frac{V_w}{V_v}$$

Now, referring to Fig. 1.3(b),

$$V_w = wG_s$$

and

$$V_v = e$$

Thus,

$$S = \frac{V_w}{V_v} = \frac{wG_s}{e} \qquad (1.14)$$

For a saturated soil, $S = 1$. So

$$e = wG_s \qquad (1.15)$$

The saturated unit weight of soil then becomes

$$\gamma_{\text{sat}} = \frac{W_s + W_w}{V_s + V_v} = \frac{G_s\gamma_w + e\gamma_w}{1 + e} \qquad (1.16)$$

In SI units, Newton or kiloNewton is weight and is a derived unit, and g or kg is mass. The relationships given in Eqs. (1.11), (1.12) and (1.16) can be expressed as moist, dry, and saturated densities as follow:

$$\rho = \frac{G_s \rho_w (1 + w)}{1 + e} \tag{1.17}$$

$$\rho_d = \frac{G_s \rho_w}{1 + e} \tag{1.18}$$

$$\rho_{sat} = \frac{\rho_w (G_s + e)}{1 + e} \tag{1.19}$$

where ρ, ρ_d, ρ_{sat} = moist density, dry density, and saturated density, respectively
ρ_w = density of water (= 1000 kg/m^3)

Relationships similar to Eqs. (1.11), (1.12), and (1.16) in terms of porosity can also be obtained by considering a representative soil specimen with a unit volume (Figure 1.3c). These relationships are

$$\gamma = G_s \gamma_w (1 - n)(1 + w) \tag{1.20}$$

$$\gamma_d = (1 - n)G_s \gamma_w \tag{1.21}$$

and

$$\gamma_{sat} = [(1 - n)G_s + n]\gamma_w \tag{1.22}$$

Table 1.3 gives a summary of various forms of relationships that can be obtained for γ, γ_d, and γ_{sat}.

Table 1.3 Various Forms of Relationships for γ, γ_d, and γ_{sat}

Unit-weight relationship	Dry unit weight	Saturated unit weight
$\gamma = \dfrac{(1 + w)G_s \gamma_w}{1 + e}$	$\gamma_d = \dfrac{\gamma}{1 + w}$	$\gamma_{sat} = \dfrac{(G_s + e)\gamma_w}{1 + e}$
$\gamma = \dfrac{(G_s + Se)\gamma_w}{1 + e}$	$\gamma_d = \dfrac{G_s \gamma_w}{1 + e}$	$\gamma_{sat} = [(1 - n)G_s + n]\gamma_w$
$\gamma = \dfrac{(1 + w)G_s \gamma_w}{1 + \dfrac{wG_s}{S}}$	$\gamma_d = G_s \gamma_w (1 - n)$	$\gamma_{sat} = \left(\dfrac{1 + w}{1 + wG_s}\right)G_s \gamma_w$
$\gamma = G_s \gamma_w (1 - n)(1 + w)$	$\gamma_d = \dfrac{G_s}{1 + \dfrac{wG_s}{S}}\gamma_w$	$\gamma_{sat} = \left(\dfrac{e}{w}\right)\left(\dfrac{1 + w}{1 + e}\right)\gamma_w$
	$\gamma_d = \dfrac{eS\gamma_w}{(1 + e)w}$	$\gamma_{sat} = \gamma_d + n\gamma_w$
	$\gamma_d = \gamma_{sat} - n\gamma_w$	$\gamma_{sat} = \gamma_d + \left(\dfrac{e}{1 + e}\right)\gamma_w$
	$\gamma_d = \gamma_{sat} - \left(\dfrac{e}{1 + e}\right)\gamma_w$	

Table 1.4 Specific Gravities of Some Soils

Type of soil	G_s
Quartz sand	2.64–2.66
Silt	2.67–2.73
Clay	2.70–2.9
Chalk	2.60–2.75
Loess	2.65–2.73
Peat	1.30–1.9

Except for peat and highly organic soils, the general range of the values of specific gravity of soil solids (G_s) found in nature is rather small. Table 1.4 gives some representative values. For practical purposes, a reasonable value can be assumed in lieu of running a test.

1.5 Relative Density

In *granular soils,* the degree of compaction in the field can be measured according to the *relative density,* defined as

$$D_r(\%) = \frac{e_{max} - e}{e_{max} - e_{min}} \times 100 \tag{1.23}$$

where

e_{max} = void ratio of the soil in the loosest state
e_{min} = void ratio in the densest state
e = *in situ* void ratio

The relative density can also be expressed in terms of dry unit weight, or

$$D_r(\%) = \left\{ \frac{\gamma_d - \gamma_{d(min)}}{\gamma_{d(max)} - \gamma_{d(min)}} \right\} \frac{\gamma_{d(max)}}{\gamma_d} \times 100 \tag{1.24}$$

where

γ_d = *in situ* dry unit weight
$\gamma_{d(max)}$ = dry unit weight in the *densest* state; that is, when the void ratio is e_{min}
$\gamma_{d(min)}$ = dry unit weight in the *loosest* state; that is, when the void ratio is e_{max}

The denseness of a granular soil is sometimes related to the soil's relative density. Table 1.5 gives a general correlation of the denseness and D_r. For naturally occurring sands, the magnitudes of e_{max} and e_{min} [Eq. (1.23)] may vary widely. The main reasons for such wide variations are the uniformity coefficient, C_u, and the roundness of the particles.

Table 1.5 Denseness of a Granular Soil

Relative density, D_r(%)	Description
0–20	Very loose
20–40	Loose
40–60	Medium
60–80	Dense
80–100	Very dense

Example 1.1

The moist weight of $28.3 \times 10^{-4} \text{ m}^3$ of soil is 54.27 N. If the moisture content is 12% and the specific gravity of soil solids is 2.72, find the following:

a. Moist unit weight (kN/m^3)
b. Dry unit weight (kN/m^3)
c. Void ratio
d. Porosity
e. Degree of saturation (%)
f. Volume occupied by water (m^3)

Solution
Part a
From Eq. (1.9),

$$\gamma = \frac{W}{V} = \frac{54.27 \times 10^{-3}}{0.00283} = \mathbf{19.18 \text{ kN/m}^3}$$

Part b
From Eq. (1.13),

$$\gamma_d = \frac{\gamma}{1 + w} = \frac{19.18}{1 + \frac{12}{100}} = \mathbf{17.13 \text{ kN/m}^3}$$

Part c
From Eq. (1.12),

$$\gamma_d = \frac{G_s \gamma_w}{1 + e}$$

or

$$17.13 = \frac{(2.72)(9.81)}{1 + e}$$

$$e = 0.56$$

Part d
From Eq. (1.6),

$$n = \frac{e}{1+e} = \frac{0.56}{1+0.56} = \textbf{0.359}$$

Part e
From Eq. (1.14),

$$S = \frac{wG_s}{e} = \frac{(0.12)(2.72)}{0.56} = \textbf{0.583}$$

Part f
From Eq. (1.12),

$$W_s = \frac{W}{1+w} = \frac{54.27 \times 10^{-3}}{1.12} = 48.46 \times 10^{-3}\,\text{kN}$$

$$W_w = W - W_s = 54.27 \times 10^{-3} - 48.46 \times 10^{-3} = 5.81 \times 10^{-3}\,\text{kN}$$

$$V_w = \frac{5.81 \times 10^{-3}}{9.81} = \textbf{0.592} \times \textbf{10}^{-3}\,\textbf{m}^3$$ ∎

Example 1.2

The dry density of a sand with a porosity of 0.387 is 1600 kg/m³. Find the void ratio of the soil and the specific gravity of the soil solids.

Solution
Void ratio
Given: $n = 0.387$. From Eq. (1.6),

$$e = \frac{n}{1-n} = \frac{0.387}{1-0.387} = \textbf{0.631}$$

Specific gravity of soil solids
From Eq. (1.18),

$$\rho_d = \frac{G_s \rho_w}{1+e}$$

$$1600 = \frac{G_s(1000)}{1.631}$$

$$G_s = \textbf{2.61}$$ ∎

Example 1.3

The moist unit weight of a soil is 19.2 kN/m³. Given $G_s = 2.69$ and moisture content $w = 9.8\%$, determine

    a. Dry unit weight (kN/m³)
    b. Void ratio
    c. Porosity
    d. Degree of saturation (%)

Solution
Part a
From Eq. (1.13),

$$\gamma_d = \frac{\gamma}{1 + w} = \frac{19.2}{1 + \dfrac{9.8}{100}} = \mathbf{17.49 \ kN/m^3}$$

Part b
From Eq. (1.12),

$$\gamma_d = 17.49 \ \text{kN/m}^3 = \frac{G_s \gamma_w}{1 + e} = \frac{(2.69)(9.81)}{1 + e}$$

$$e = \mathbf{0.509}$$

Part c
From Eq. (1.6),

$$n = \frac{e}{1 + e} = \frac{0.509}{1 + 0.509} = \mathbf{0.337}$$

Part d
From Eq. (1.14),

$$S = \frac{wG_s}{e} = \left[\frac{(0.098)(2.69)}{0.509} \right](100) = \mathbf{51.79\%}$$ ∎

Example 1.4

For a saturated soil, given $w = 40\%$ and $G_s = 2.71$, determine the saturated and dry unit weights in lb/ft³ and kN/m³.

Solution
For saturated soil, from Eq. (1.15),

$$e = wG_s = (0.4)(2.71) = 1.084$$

From Eq. (1.16),

$$\gamma_{\text{sat}} = \frac{(G_s + e)\gamma_w}{1 + e} = \frac{(2.71 + 1.084)9.81}{1 + 1.084} = \mathbf{17.86 \ kN/m^3}$$

From Eq. (1.12),

$$\gamma_d = \frac{G_s \gamma_w}{1 + e} = \frac{(2.71)(9.81)}{1 + 1.084} = 12.76 \ kN/m^3 \qquad \blacksquare$$

Example 1.5

The mass of a moist soil sample collected from the field is 465 grams, and its oven dry mass is 405.76 grams. The specific gravity of the soil solids was determined in the laboratory to be 2.68. If the void ratio of the soil in the natural state is 0.83, find the following:

a. The moist density of the soil in the field (kg/m^3)
b. The dry density of the soil in the field (kg/m^3)
c. The mass of water, in kilograms, to be added per cubic meter of soil in the field for saturation

Solution
Part a
From Eq. (1.8),

$$w = \frac{W_w}{W_s} = \frac{465 - 405.76}{405.76} = \frac{59.24}{405.76} = 14.6\%$$

From Eq. (1.17),

$$\rho = \frac{G_s \rho_w + w G_s \rho_w}{1 + e} = \frac{G_s \rho_w (1 + w)}{1 + e} = \frac{(2.68)(1000)(1.146)}{1.83}$$

$$= \mathbf{1678.3 \ kg/m^3}$$

Part b
From Eq. (1.18),

$$\rho_d = \frac{G_s \rho_w}{1 + e} = \frac{(2.68)(1000)}{1.83} = \mathbf{1468.48 \ kg/m^3}$$

Part c
Mass of water to be added $= \rho_{sat} - \rho$
From Eq. (1.19),

$$\rho_{sat} = \frac{G_s\rho_w + e\rho_w}{1 + e} = \frac{\rho_w(G_s + e)}{1 + e} = \frac{(1000)(2.68 + 0.83)}{1.83} = 1918 \text{ kg/m}^3$$

So, mass of water to be added $= 1918 - 1678.3 = $ **239.7 kg/m³.** ∎

Example 1.6

The maximum and minimum dry unit weights of a sand are 17.1 kN/m³ and 14.2 kN/m³, respectively. The sand in the field has a relative density of 70% with a moisture content of 8%. Determine the moist unit weight of the sand in the field.

Solution
From Eq. (1.24),

$$D_r = \left[\frac{\gamma_d - \gamma_{d(min)}}{\gamma_{d(max)} - \gamma_{d(min)}}\right]\left[\frac{\gamma_{d(max)}}{\gamma_d}\right]$$

$$0.7 = \left[\frac{\gamma_d - 14.2}{17.1 - 14.2}\right]\left[\frac{17.1}{\gamma_d}\right]$$

$$\gamma_d = 16.11 \text{ kN/m}^3$$

$$\gamma = \gamma_d(1 + w) = 16.11\left(1 + \frac{8}{100}\right) = \textbf{17.4 kN/m}^3$$ ∎

1.6 *Atterberg Limits*

When a clayey soil is mixed with an excessive amount of water, it may flow like a *semiliquid*. If the soil is gradually dried, it will behave like a *plastic, semisolid,* or *solid* material, depending on its moisture content. The moisture content, in percent, at which the soil changes from a liquid to a plastic state is defined as the *liquid limit* (LL). Similarly, the moisture content, in percent, at which the soil changes from a plastic to a semisolid state and from a semisolid to a solid state are defined as the *plastic limit* (PL) and the *shrinkage limit* (SL), respectively. These limits are referred to as *Atterberg limits* (Figure 1.4):

- The *liquid limit* of a soil is determined by Casagrande's liquid device (ASTM Test Designation D-4318) and is defined as the moisture content at which a groove closure of 12.7 mm occurs at 25 blows.
- The *plastic limit* is defined as the moisture content at which the soil crumbles when rolled into a thread of 3.18 mm in diameter (ASTM Test Designation D-4318).

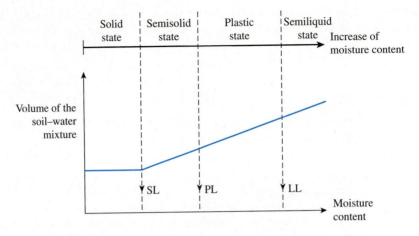

Figure 1.4 Definition of Atterberg limits

- The *shrinkage limit* is defined as the moisture content at which the soil does not undergo any further change in volume with loss of moisture (ASTM Test Designation D-427).

The difference between the liquid limit and the plastic limit of a soil is defined as the *plasticity index* (PI), or

$$PI = LL - PL \tag{1.25}$$

1.7 Liquidity Index

The relative consistency of a cohesive soil in the natural state can be defined by a ratio called the *liquidity index*, which is given by

$$LI = \frac{w - PL}{LL - PL} \tag{1.26}$$

where $w = $ *in situ* moisture content of soil.

The *in situ* moisture content for a sensitive clay may be greater than the liquid limit. In this case,

$$LI > 1$$

These soils, when remolded, can be transformed into a viscous form to flow like a liquid.

Soil deposits that are heavily overconsolidated may have a natural moisture content less than the plastic limit. In this case,

$$LI < 0$$

1.8　Activity

Because the plasticity of soil is caused by the adsorbed water that surrounds the clay particles, we can expect that the type of clay minerals and their proportional amounts in a soil will affect the liquid and plastic limits. Skempton (1953) observed that the plasticity index of a soil increases linearly with the percentage of clay-size fraction (% finer than 2 μm by weight) present. The correlations of PI with the clay-size fractions for different clays plot separate lines. This difference is due to the diverse plasticity characteristics of the various types of clay minerals. On the basis of these results, Skempton defined a quantity called *activity*, which is the slope of the line correlating PI and % finer than 2 μm. This activity may be expressed as

$$A = \frac{PI}{(\% \text{ of clay-size fraction, by weight})} \tag{1.27}$$

Activity is used as an index for identifying the swelling potential of clay soils. Typical values of activities for various clay minerals are given in Table 1.6.

1.9　Soil Classification Systems

Soil classification systems divide soils into groups and subgroups based on common engineering properties such as the *grain-size distribution, liquid limit,* and *plastic limit.* The two major classification systems presently in use are (1) the *American Association of State Highway and Transportation Officials* (AASHTO) *System* and (2) the *Unified Soil Classification System* (also ASTM). The AASHTO system is used mainly for the classification of highway subgrades. It is not used in foundation construction.

Table 1.6 Activities of Clay Minerals

Mineral	Activity (A)
Smectites	1–7
Illite	0.5–1
Kaolinite	0.5
Halloysite ($4H_2O$)	0.5
Halloysite ($2H_2O$)	0.1
Attapulgite	0.5–1.2
Allophane	0.5–1.2

AASHTO System

The AASHTO Soil Classification System was originally proposed by the Highway Research Board's Committee on Classification of Materials for Subgrades and Granular Type Roads (1945). According to the present form of this system, soils can be classified according to eight major groups, A-1 through A-8, based on their grain-size distribution, liquid limit, and plasticity indices. Soils listed in groups A-1, A-2, and A-3 are coarse-grained materials, and those in groups A-4, A-5, A-6, and A-7 are fine-grained materials. Peat, muck, and other highly organic soils are classified under A-8. They are identified by visual inspection.

The AASHTO classification system (for soils A-1 through A-7) is presented in Table 1.7. Note that group A-7 includes two types of soil. For the A-7-5 type, the plasticity

Table 1.7 AASHTO Soil Classification System

General classification	Granular materials (35% or less of total sample passing No. 200 sieve)						
	A-1		A-3	A-2			
Group classification	A-1-a	A-1-b		A-2-4	A-2-5	A-2-6	A-2-7
Sieve analysis (% passing)							
No. 10 sieve	50 max						
No. 40 sieve	30 max	50 max	51 min				
No. 200 sieve	15 max	25 max	10 max	35 max	35 max	35 max	35 max
For fraction passing No. 40 sieve							
Liquid limit (LL)				40 max	41 min	40 max	41 min
Plasticity index (PI)	6 max		Nonplastic	10 max	10 max	11 min	11 min
Usual type of material	Stone fragments, gravel, and sand		Fine sand	Silty or clayey gravel and sand			
Subgrade rating	Excellent to good						

General classification	Silt–clay materials (More than 35% of total sample passing No. 200 sieve)			
Group classification	A-4	A-5	A-6	A-7
				A-7-5[a]
				A-7-6[b]
Sieve analysis (% passing)				
No. 10 sieve				
No. 40 sieve				
No. 200 sieve	36 min	36 min	36 min	36 min
For fraction passing No. 40 sieve				
Liquid limit (LL)	40 max	41 min	40 max	41 min
Plasticity index (PI)	10 max	10 max	11 min	11 min
Usual types of material	Mostly silty soils		Mostly clayey soils	
Subgrade rating	Fair to poor			

[a]If $PI \leq LL - 30$, the classification is A-7-5.
[b]If $PI > LL - 30$, the classification is A-7-6.

index of the soil is less than or equal to the liquid limit minus 30. For the A-7-6 type, the plasticity index is greater than the liquid limit minus 30.

For qualitative evaluation of the desirability of a soil as a highway subgrade material, a number referred to as the *group index* has also been developed. The higher the value of the group index for a given soil, the weaker will be the soil's performance as a subgrade. A group index of 20 or more indicates a very poor subgrade material. The formula for the group index is

$$GI = (F_{200} - 35)[0.2 + 0.005(LL - 40)] + 0.01(F_{200} - 15)(PI - 10) \qquad (1.28)$$

where

F_{200} = percent passing No. 200 sieve, expressed as a whole number
LL = liquid limit
PI = plasticity index

When calculating the group index for a soil belonging to group A-2-6 or A-2-7, use only the partial group-index equation relating to the plasticity index:

$$GI = 0.01(F_{200} - 15)(PI - 10) \qquad (1.29)$$

The group index is rounded to the nearest whole number and written next to the soil group in parentheses; for example, we have

$$\underbrace{\text{A-4}}_{\text{Soil group}} \qquad \underbrace{\text{(5)}}_{\text{Group index}}$$

The group index for soils which fall in groups A-1-a, A-1-b, A-3, A-2-4, and A-2-5 is always zero.

Unified System

The Unified Soil Classification System was originally proposed by A. Casagrande in 1942 and was later revised and adopted by the United States Bureau of Reclamation and the U.S. Army Corps of Engineers. The system is currently used in practically all geotechnical work.

In the Unified System, the following symbols are used for identification:

Symbol	G	S	M	C	O	Pt	H	L	W	P
Description	Gravel	Sand	Silt	Clay	Organic silts and clay	Peat and highly organic soils	High plasticity	Low plasticity	Well graded	Poorly graded

Figure 1.5   Plasticity chart

The plasticity chart (Figure 1.5) and Table 1.8 show the procedure for determining the group symbols for various types of soil. When classifying a soil be sure to provide the group name that generally describes the soil, along with the group symbol. Figures 1.6, 1.7, and 1.8 give flowcharts for obtaining the group names for coarse-grained soil, inorganic fine-grained soil, and organic fine-grained soil, respectively.

Example 1.7

Classify the following soil by the AASHTO classification system.

Percent passing No. 4 sieve = 92
Percent passing No. 10 sieve = 87
Percent passing No. 40 sieve = 65
Percent passing No. 200 sieve = 30
Liquid limit = 22
Plasticity index = 8

Solution

Table 1.7 shows that it is a granular material because less than 35% is passing a No. 200 sieve. With LL = 22 (that is, less than 40) and PI = 8 (that is, less than 10), the soil falls in group A-2-4.

The soil is **A-2-4(0)**. ∎

Table 1.8 Unified Soil Classification Chart (after ASTM, 2009) (ASTM D2487-98: Standard Practice for Classification of Soils for Engineering Purposes (Unified Soil Classification). Copyright ASTM INTERNATIONAL. Reprinted with permission.)

Criteria for assigning group symbols and group names using laboratory tests[a]				Soil classification	
				Group symbol	Group name[b]
Coarse-grained soils More than 50% retained on No. 200 sieve	Gravels More than 50% of coarse fraction retained on No. 4 sieve	Clean Gravels Less than 5% fines[c]	$C_u \geq 4$ and $1 \leq C_c \leq 3^e$	GW	Well-graded gravel[f]
			$C_u < 4$ and/or $1 > C_c > 3^e$	GP	Poorly graded gravel[f]
		Gravels with Fines More than 12% fines[c]	Fines classify as ML or MH	GM	Silty gravel[f,g,h]
			Fines classify as CL or CH	GC	Clayey gravel[f,g,h]
	Sands 50% or more of coarse fraction passes No. 4 sieve	Clean Sands Less than 5% fines[d]	$C_u \geq 6$ and $1 \leq C_c \leq 3^e$	SW	Well-graded sand[i]
			$C_u < 6$ and/or $1 > C_c > 3^e$	SP	Poorly graded sand[i]
		Sand with Fines More than 12% fines[d]	Fines classify as ML or MH	SM	Silty sand[g,h,i]
			Fines classify as CL or CH	SC	Clayey sand[g,h,i]
Fine-grained soils 50% or more passes the No. 200 sieve	Silts and Clays Liquid limit less than 50	Inorganic	PI > 7 and plots on or above "A" line[j]	CL	Lean clay[k,l,m]
			PI < 4 or plots below "A" line[j]	ML	Silt[k,l,m]
		Organic	$\dfrac{\text{Liquid limit—oven dried}}{\text{Liquid limit—not dried}} < 0.75$	OL	Organic clay[k,l,m,n] / Organic silt[k,l,m,o]
	Silts and Clays Liquid limit 50 or more	Inorganic	PI plots on or above "A" line	CH	Fat clay[k,l,m]
			PI plots below "A" line	MH	Elastic silt[k,l,m]
		Organic	$\dfrac{\text{Liquid limit—oven dried}}{\text{Liquid limit—not dried}} < 0.75$	OH	Organic clay[k,l,m,p] / Organic silt[k,l,m,q]
Highly organic soils	Primarily organic matter, dark in color, and organic odor			PT	Peat

[a] Based on the material passing the 75-mm. (3-in) sieve.

[b] If field sample contained cobbles or boulders, or both, add "with cobbles or boulders, or both" to group name.

[c] Gravels with 5 to 12% fines require dual symbols: GW-GM well-graded gravel with silt; GW-GC well-graded gravel with clay; GP-GM poorly graded gravel with silt; GP-GC poorly graded gravel with clay.

[d] Sands with 5 to 12% fines require dual symbols: SW-SM well-graded sand with silt; SW-SC well-graded sand with clay; SP-SM poorly graded sand with silt; SP-SC poorly graded sand with clay.

[e] $C_u = D_{60}/D_{10}$ $C_c = \dfrac{(D_{30})^2}{D_{10} \times D_{60}}$

[f] If soil contains ≥15% sand, add "with sand" to group name.

[g] If fines classify as CL-ML, use dual symbol GC-GM or SC-SM.

[h] If fines are organic, add "with organic fines" to group name.

[i] If soil contains ≥15% gravel, add "with gravel" to group name.

[j] If Atterberg limits plot in hatched area, soil is a CL-ML, silty clay.

[k] If soil contains 15 to 29% plus No. 200, add "with sand" or "with gravel," whichever is predominant.

[l] If soil contains ≥30% plus No. 200, predominantly sand, add "sandy" to group name.

[m] If soil contains ≥30% plus No. 200, predominantly gravel, add "gravelly" to group name.

[n] PI ≥ 4 and plots on or above "A" line.

[o] PI < 4 or plots below "A" line.

[p] PI plots on or above "A" line.

[q] PI plots below "A" line.

Figure 1.6 Flowchart for classifying coarse-grained soils (more than 50% retained on No. 200 Sieve) (After ASTM, 2009) (ASTM D2487-98: Standard Practice for Classification of Soils for Engineering Purposes (Unified Soil Classification). Copyright ASTM INTERNATIONAL. Reprinted with permission.)

Figure 1.7 Flowchart for classifying fine-grained soil (50% or more passes No. 200 Sieve) (After ASTM, 2009)(ASTM D2487-98: Standard Practice for Classification of Soils for Classification of Soils for Engineering Purposes (Unified Soil Classification). Copyright ASTM INTERNATIONAL. Reprinted with permission.)

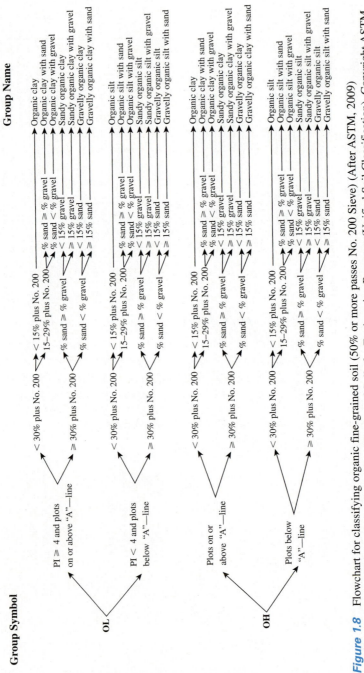

Figure 1.8 Flowchart for classifying organic fine-grained soil (50% or more passes No. 200 Sieve) (After ASTM, 2009) (ASTM D2487-98: Standard Practice for Classification of Soils for Engineering Purposes (Unified Soil Classification). Copyright ASTM INTERNATIONAL. Reprinted with permission.)

Example 1.8

Classify the following soil by the Unified Soil Classification System:

> Percent passing No. 4 sieve = 82
> Percent passing No. 10 sieve = 71
> Percent passing No. 40 sieve = 64
> Percent passing No. 200 sieve = 41
> Liquid limit = 31
> Plasticity index = 12

Solution

We are given that $F_{200} = 41$, LL = 31, and PI = 12. Since 59% of the sample is retained on a No. 200 sieve, the soil is a coarse-grained material. The percentage passing a No. 4 sieve is 82, so 18% is retained on No. 4 sieve (gravel fraction). The coarse fraction passing a No. 4 sieve (sand fraction) is $59 - 18 = 41\%$ (which is more than 50% of the total coarse fraction). Hence, the specimen is a sandy soil.

Now, using Table 1.8 and Figure 1.5, we identify the group symbol of the soil as **SC**.

Again from Figure 1.6, since the gravel fraction is greater than 15%, the group name is **clayey sand with gravel**. ∎

1.10 *Hydraulic Conductivity of Soil*

The void spaces, or pores, between soil grains allow water to flow through them. In soil mechanics and foundation engineering, you must know how much water is flowing through a soil per unit time. This knowledge is required to design earth dams, determine the quantity of seepage under hydraulic structures, and dewater foundations before and during their construction. Darcy (1856) proposed the following equation (Figure 1.9) for calculating the velocity of flow of water through a soil:

$$v = ki \tag{1.30}$$

In this equation,

v = Darcy velocity (unit: cm/sec)
k = hydraulic conductivity of soil (unit: cm/sec)
i = hydraulic gradient

The hydraulic gradient is defined as

$$i = \frac{\Delta h}{L} \tag{1.31}$$

where

Δh = piezometric head difference between the sections at AA and BB
L = distance between the sections at AA and BB

(*Note:* Sections AA and BB are perpendicular to the direction of flow.)

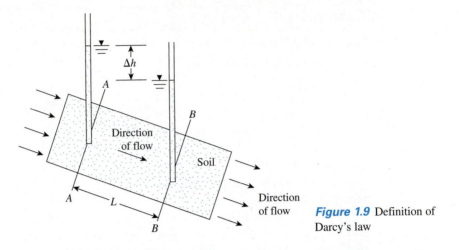

Figure 1.9 Definition of Darcy's law

Darcy's law [Eq. (1.30)] is valid for a wide range of soils. However, with materials like clean gravel and open-graded rockfills, the law breaks down because of the turbulent nature of flow through them.

The value of the hydraulic conductivity of soils varies greatly. In the laboratory, it can be determined by means of *constant-head* or *falling-head* permeability tests. The constant-head test is more suitable for granular soils. Table 1.9 provides the general range for the values of k for various soils. In granular soils, the value depends primarily on the void ratio. In the past, several equations have been proposed to relate the value of k to the void ratio in granular soil. However the author recommends the following equation for use (also see Carrier, 2003):

$$k \propto \frac{e^3}{1 + e} \tag{1.32}$$

where

k = hydraulic conductivity
e = void ratio

Chapuis (2004) proposed an empirical relationship for k in conjunction with Eq. (1.32) as

$$k\,(\text{cm/s}) = 2.4622 \left[D_{10}^2 \frac{e^3}{(1 + e)} \right]^{0.7825} \tag{1.33}$$

where D = effective size (mm).

Table 1.9 Range of the Hydraulic Conductivity for Various Soils

Type of soil	Hydraulic conductivity, k (cm/sec)
Medium to coarse gravel	Greater than 10^{-1}
Coarse to fine sand	10^{-1} to 10^{-3}
Fine sand, silty sand	10^{-3} to 10^{-5}
Silt, clayey silt, silty clay	10^{-4} to 10^{-6}
Clays	10^{-7} or less

The preceding equation is valid for natural, uniform sand and gravel to predict k that is in the range of 10^{-1} to 10^{-3} cm/s. This can be extended to natural, silty sands without plasticity. It is not valid for crushed materials or silty soils with some plasticity.

Based on laboratory experimental results, Amer and Awad (1974) proposed the following relationship for k in granular soil:

$$k = 3.5 \times 10^{-4}\left(\frac{e^3}{1+e}\right)C_u^{0.6}D_{10}^{2.32}\left(\frac{\rho_w}{\eta}\right) \tag{1.34}$$

where

k is in cm/sec
C_u = uniformity coefficient
D_{10} = effective size (mm)
ρ_w = density of water (g/cm^3)
η = viscosity (g·s/cm^2)

At 20°C, $\rho_w = 1$ g/cm^3 and $\eta \approx 0.1 \times 10^{-4}$ g·s/cm^2. So

$$k = 3.5 \times 10^{-4}\left(\frac{e^3}{1+e}\right)C_u^{0.6}D_{10}^{2.32}\left(\frac{1}{0.1 \times 10^{-4}}\right)$$

or

$$k \text{ (cm/sec)} = 35\left(\frac{e^3}{1+e}\right)C_u^{0.6}D_{10}^{2.32} \tag{1.35}$$

Hydraulic Conductivity of Cohesive Soil

According to their experimental observations, Samarasinghe, Huang, and Drnevich (1982) suggested that the hydraulic conductivity of normally consolidated clays could be given by the equation

$$k = C\frac{e^n}{1+e} \tag{1.36}$$

where C and n are constants to be determined experimentally.

Some other empirical relationships for estimating the hydraulic conductivity in clayey soils are given in Table 1.10. One should keep in mind, however, that any empirical

Table 1.10 Empirical Relationships for Estimating Hydraulic Conductivity in Clayey Soil

Type of soil	Source	Relationship[a]
Clay	Mesri and Olson (1971)	$\log k = A' \log e + B'$
	Taylor (1948)	$\log k = \log k_0 - \dfrac{e_0 - e}{C_k}$
		$C_k \approx 0.5e_0$

[a]k_0 = *in situ* hydraulic conductivity at void ratio e_0
k = hydraulic conductivity at void ratio e
C_k = hydraulic conductivity change index

Figure 1.10 Variation of void ratio with hydraulic conductivity of clayey soils (Based on Tavenas, *et al.*, 1983)

relationship of this type is for estimation only, because the magnitude of k is a highly variable parameter and depends on several factors.

Tavenas, *et al.* (1983) also gave a correlation between the void ratio and the hydraulic conductivity of clayey soil. This correlation is shown in Figure 1.10. An important point to note, however, is that in Figure 6.10, PI, the plasticity index, and CF, the clay-size fraction in the soil, are in *fraction* (decimal) form.

1.11 *Steady-State Seepage*

For most cases of seepage under hydraulic structures, the flow path changes direction and is not uniform over the entire area. In such cases, one of the ways of determining the rate of seepage is by a graphical construction referred to as the *flow net,* a concept based on Laplace's theory of continuity. According to this theory, for a steady flow condition, the flow at any point A (Figure 1.11) can be represented by the equation

$$k_x\frac{\partial^2 h}{\partial x^2} + k_y\frac{\partial^2 h}{\partial y^2} + k_z\frac{\partial^2 h}{\partial z^2} = 0 \qquad (1.37)$$

where

k_x, k_y, k_z = hydraulic conductivity of the soil in the x, y, and z directions, respectively

h = hydraulic head at point A (i.e., the head of water that a piezometer placed at A would show with the *downstream water level* as *datum*, as shown in Figure 1.11)

For a two-dimensional flow condition, as shown in Figure 1.11,

$$\frac{\partial^2 h}{\partial^2 y} = 0$$

Figure 1.11 Steady-state seepage

so Eq. (1.37) takes the form

$$k_x\frac{\partial^2 h}{\partial x^2} + k_z\frac{\partial^2 h}{\partial z^2} = 0 \tag{1.38}$$

If the soil is isotropic with respect to hydraulic conductivity, $k_x = k_z = k$, and

$$\frac{\partial^2 h}{\partial x^2} + \frac{\partial^2 h}{\partial z^2} = 0 \tag{1.39}$$

Equation (1.39), which is referred to as Laplace's equation and is valid for confined flow, represents two orthogonal sets of curves known as *flow lines* and *equipotential lines*. A flow net is a combination of numerous equipotential lines and flow lines. A flow line is a path that a water particle would follow in traveling from the upstream side to the downstream side. An equipotential line is a line along which water, in piezometers, would rise to the same elevation. (See Figure 1.11.)

In drawing a flow net, you need to establish the *boundary conditions*. For example, in Figure 1.11, the ground surfaces on the upstream (OO') and downstream (DD') sides are equipotential lines. The base of the dam below the ground surface, $O'BCD$, is a flow line. The top of the rock surface, EF, is also a flow line. Once the boundary conditions are established, a number of flow lines and equipotential lines are drawn by trial and error so that all the flow elements in the net have the same length-to-width ratio (L/B). In most cases, L/B is held to unity, that is, the flow elements are drawn as curvilinear "squares." This method is illustrated by the flow net shown in Figure 1.12. Note that all flow lines must intersect all equipotential lines at *right angles*.

Once the flow net is drawn, the seepage (in unit time per unit length of the structure) can be calculated as

$$q = kh_{max}\frac{N_f}{N_d}n \tag{1.40}$$

Figure 1.12 Flow net

where

N_f = number of flow channels
N_d = number of drops
n = width-to-length ratio of the flow elements in the flow net (B/L)
h_{max} = difference in water level between the upstream and downstream sides

The space between two consecutive flow lines is defined as a *flow channel,* and the space between two consecutive equipotential lines is called a *drop.* In Figure 1.12, $N_f = 2$, $N_d = 7$, and $n = 1$. When square elements are drawn in a flow net,

$$q = kh_{max}\frac{N_f}{N_d} \tag{1.41}$$

1.12   *Effective Stress*

The *total* stress at a given point in a soil mass can be expressed as

$$\sigma = \sigma' + u \tag{1.42}$$

where

σ = total stress
σ' = effective stress
u = pore water pressure

The effective stress, σ', is the vertical component of forces at solid-to-solid contact points over a unit cross-sectional area. Referring to Figure 1.13a, at point A

$$\sigma = \gamma h_1 + \gamma_{sat} h_2$$
$$u = h_2 \gamma_w$$

where

γ_w = unit weight of water
γ_{sat} = saturated unit weight of soil

Figure 1.13 Calculation of effective stress

So

$$\sigma' = (\gamma h_1 + \gamma_{sat} h_2) - (h_2 \gamma_w)$$
$$= \gamma h_1 + h_2(\gamma_{sat} - \gamma_w)$$
$$= \gamma h_1 + \gamma' h_2 \quad (1.43)$$

where γ' = effective or submerged unit weight of soil.

For the problem in Figure 1.13a, there was *no seepage of water* in the soil. Figure 1.13b shows a simple condition in a soil profile in which there is upward seepage. For this case, at point *A*,

$$\sigma = h_1 \gamma_w + h_2 \gamma_{sat}$$

and

$$u = (h_1 + h_2 + h)\gamma_w$$

Thus, from Eq. (1.42),

$$\sigma' = \sigma - u = (h_1\gamma_w + h_2\gamma_{sat}) - (h_1 + h_2 + h)\gamma_w$$
$$= h_2(\gamma_{sat} - \gamma_w) - h\gamma_w = h_2\gamma' - h\gamma_w$$

or

$$\sigma' = h_2\left(\gamma' - \frac{h}{h_2}\gamma_w\right) = h_2(\gamma' - i\gamma_w) \quad (1.44)$$

Note in Eq. (1.44) that h/h_2 is the hydraulic gradient *i*. If the hydraulic gradient is very high, so that $\gamma' - i\gamma_w$ becomes zero, *the effective stress will become zero*. In other words, there is no contact stress between the soil particles, and the soil will break up. This situation is referred to as the *quick condition,* or *failure by heave.* So, for heave,

$$i = i_{cr} = \frac{\gamma'}{\gamma_w} = \frac{G_s - 1}{1 + e} \quad (1.45)$$

where i_{cr} = critical hydraulic gradient.

For most sandy soils, i_{cr} ranges from 0.9 to 1.1, with an average of about unity.

1.13   *Consolidation*

In the field, when the stress on a saturated clay layer is increased—for example, by the construction of a foundation—the pore water pressure in the clay will increase. Because the hydraulic conductivity of clays is very small, some time will be required for the excess pore water pressure to dissipate and the increase in stress to be transferred to the soil skeleton. According to Figure 1.14, if $\Delta\sigma$ is a surcharge at the ground surface over a very large area, the increase in total stress at any depth of the clay layer will be equal to $\Delta\sigma$.

However, at time $t = 0$ (i.e., immediately after the stress is applied), the excess pore water pressure at any depth Δu will equal $\Delta\sigma$, or

$$\Delta u = \Delta h_i \gamma_w = \Delta\sigma \text{ (at time } t = 0)$$

Hence, the increase in effective stress at time $t = 0$ will be

$$\Delta\sigma' = \Delta\sigma - \Delta u = 0$$

Theoretically, at time $t = \infty$, when all the excess pore water pressure in the clay layer has dissipated as a result of drainage into the sand layers,

$$\Delta u = 0 \quad (\text{at time } t = \infty)$$

Then the increase in effective stress in the clay layer is

$$\Delta\sigma' = \Delta\sigma - \Delta u = \Delta\sigma - 0 = \Delta\sigma$$

This gradual increase in the effective stress in the clay layer will cause settlement over a period of time and is referred to as *consolidation*.

Laboratory tests on undisturbed saturated clay specimens can be conducted (ASTM Test Designation D-2435) to determine the consolidation settlement caused by various incremental loadings. The test specimens are usually 63.5 mm in diameter and 25.4 mm in

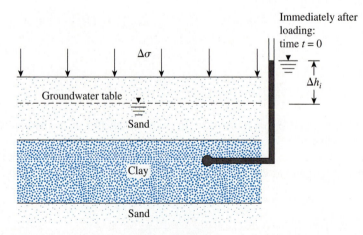

Figure 1.14 Principles of consolidation

height. Specimens are placed inside a ring, with one porous stone at the top and one at the bottom of the specimen (Figure 1.15a). A load on the specimen is then applied so that the total vertical stress is equal to σ. Settlement readings for the specimen are taken periodically for 24 hours. After that, the load on the specimen is doubled and more settlement readings are taken. At all times during the test, the specimen is kept under water. The procedure is continued until the desired limit of stress on the clay specimen is reached.

Based on the laboratory tests, a graph can be plotted showing the variation of the void ratio e at the *end* of consolidation against the corresponding vertical effective stress σ'. (On a semilogarithmic graph, e is plotted on the arithmetic scale and σ' on the log scale.) The nature of the variation of e against $\log \sigma'$ for a clay specimen is shown in Figure 1.15b. After the desired consolidation pressure has been reached, the specimen gradually can be unloaded, which will result in the swelling of the specimen. The figure also shows the variation of the void ratio during the unloading period.

From the e–$\log \sigma'$ curve shown in Figure 1.15b, three parameters necessary for calculating settlement in the field can be determined. They are preconsolidation pressure (σ'_c), compression index (C_c), and the swelling index (C_s). The following are more detailed descriptions for each of the parameters.

Figure 1.15 (a) Schematic diagram of consolidation test arrangement; (b) e–$\log \sigma'$ curve for a soft clay from East St. Louis, Illinois (*Note:* At the end of consolidation, $\sigma = \sigma'$)

Preconsolidation Pressure

The *preconsolidation pressure, σ'_c,* is the *maximum past effective overburden pressure* to which the soil specimen has been subjected. It can be determined by using a simple graphical procedure proposed by Casagrande (1936). The procedure involves five steps (see Figure 1.15b):

a. Determine the point O on the e–log σ' curve that has the sharpest curvature (i.e., the smallest radius of curvature).

b. Draw a horizontal line OA.

c. Draw a line OB that is tangent to the e–log σ' curve at O.

d. Draw a line OC that bisects the angle AOB.

e. Produce the straight-line portion of the e–log σ' curve backwards to intersect OC. This is point D. The pressure that corresponds to point D is the preconsolidation pressure σ'_c.

Natural soil deposits can be *normally consolidated* or *overconsolidated* (or *preconsolidated*). If the present effective overburden pressure $\sigma' = \sigma'_o$ is equal to the preconsolidated pressure σ'_c the soil is *normally consolidated*. However, if $\sigma'_o < \sigma'_c$, the soil is *overconsolidated*.

Stas and Kulhawy (1984) correlated the preconsolidation pressure with liquidity index in the following form:

$$\frac{\sigma'_c}{p_a} = 10^{(1.11 - 1.62 \, \text{LI})} \tag{1.46}$$

where

p_a = atmospheric pressure (≈ 100 kN/m^2)
LI = liquidity index

A similar correlation has also been provided by Kulhawy and Mayne (1990), which is based on the work of Wood (1983) as

$$\sigma'_c = \sigma'_o \left\{ 10^{\left[1 - 2.5\text{LI} - 1.25\log\left(\frac{\sigma'_o}{p_a}\right)\right]} \right\} \tag{1.47}$$

where σ'_o = *in situ* effective overburden pressure.

Nagaraj and Murthy (1985) gave a correlation between σ_c' and the *in situ* effective overburden pressure which can be expressed as

$$\log \sigma'_o (\text{kN/m}^2) = \frac{1.122 - \left(\dfrac{e_o}{e_L}\right) - 0.0463 \log \sigma'_o (\text{kN/m}^2)}{0.188} \tag{1.48}$$

where

σ'_o = *in situ* effective overburden pressure
e_o = *in situ* void ratio
e_L = void ratio at liquid limit $= \left[\dfrac{\text{LL}(\%)}{100}\right] G_s$

G_s = specific gravity of soil solids

Compression Index

The *compression index, C_c*, is the slope of the straight-line portion (the latter part) of the loading curve, or

$$C_c = \frac{e_1 - e_2}{\log \sigma_2' - \log \sigma_1'} = \frac{e_1 - e_2}{\log\left(\dfrac{\sigma_2'}{\sigma_1'}\right)} \tag{1.49}$$

where e_1 and e_2 are the void ratios at the end of consolidation under effective stresses σ_1' and σ_2', respectively.

The *compression index*, as determined from the laboratory e–$\log \sigma'$ curve, will be somewhat different from that encountered in the field. The primary reason is that the soil remolds itself to some degree during the field exploration. The nature of variation of the e–$\log \sigma'$ curve in the field for a normally consolidated clay is shown in Figure 1.16. The curve, generally referred to as the *virgin compression curve*, approximately intersects the laboratory curve at a void ratio of $0.42e_o$ (Terzaghi and Peck, 1967). Note that e_o is the void ratio of the clay in the field. Knowing the values of e_o and σ_c', you can easily construct the virgin curve and calculate its compression index by using Eq. (1.49).

The value of C_c can vary widely, depending on the soil. Skempton (1944) gave an empirical correlation for the compression index in which

$$C_c = 0.009(\text{LL} - 10) \tag{1.50}$$

where LL = liquid limit .

Besides Skempton, several other investigators also have proposed correlations for the compression index. Some of those are given here:

Rendon-Herrero (1983):

$$C_c = 0.141 G_s^{1.2}\left(\frac{1 + e_o}{G_s}\right)^{2.38} \tag{1.51}$$

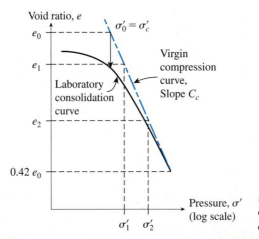

Figure 1.16 Construction of virgin compression curve for normally consolidated clay

Nagaraj and Murty (1985):

$$C_c = 0.2343 \left[\frac{\text{LL}(\%)}{100} \right] G_s \qquad (1.52)$$

Park and Koumoto (2004):

$$C_c = \frac{n_o}{371.747 - 4.275 n_o} \qquad (1.53)$$

where $n_o = $ *in situ* porosity of soil.
Wroth and Wood (1978):

$$C_c = 0.5 G_s \left(\frac{\text{PI}(\%)}{100} \right) \qquad (1.54)$$

If a typical value of $G_s = 2.7$ is used in Eq. (1.54), we obtain (Kulhawy and Mayne, 1990)

$$C_c = \frac{\text{PI}(\%)}{74} \qquad (1.55)$$

Swelling Index

The *swelling index, C_s,* is the slope of the unloading portion of the *e*–log σ' curve. In Figure 1.15b, it is defined as

$$C_s = \frac{e_3 - e_4}{\log\left(\dfrac{\sigma'_4}{\sigma'_3} \right)} \qquad (1.56)$$

In most cases, the value of the swelling index is $\frac{1}{4}$ to $\frac{1}{5}$ of the compression index. Following are some representative values of C_s/C_c for natural soil deposits:

Description of soil	C_s/C_c
Boston Blue clay	0.24–0.33
Chicago clay	0.15–0.3
New Orleans clay	0.15–0.28
St. Lawrence clay	0.05–0.1

The swelling index is also referred to as the *recompression index.*

The determination of the swelling index is important in the estimation of consolidation settlement of *overconsolidated clays.* In the field, depending on the pressure increase, an overconsolidated clay will follow an *e*–log σ' path *abc,* as shown in Figure 1.17. Note that point *a,* with coordinates σ'_o and e_o, corresponds to the field conditions before any increase in pressure. Point *b* corresponds to the preconsolidation pressure (σ'_c) of the clay. Line *ab* is approximately parallel to the laboratory unloading curve *cd* (Schmertmann, 1953). Hence, if you know e_o, σ'_o, σ'_c, C_c, and C_s, you can easily construct the field consolidation curve.

Using the modified Cam clay model and Eq. (1.54), Kulhawy and Mayne (1990) have shown that

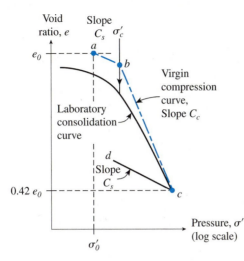

Figure 1.17 Construction of field consolidation curve for overconsolidated clay

$$C_s = \frac{PI(\%)}{370} \tag{1.57}$$

Comparing Eqs. (1.55) and (1.57), we obtain

$$C_s \approx \frac{1}{5}C_c \tag{1.58}$$

1.14 *Calculation of Primary Consolidation Settlement*

The one-dimensional primary consolidation settlement (caused by an additional load) of a clay layer (Figure 1.18) having a thickness H_c may be calculated as

$$S_c = \frac{\Delta e}{1 + e_o}H_c \tag{1.59}$$

Figure 1.18 One-dimensional settlement calculation

where

S_c = primary consolidation settlement
Δe = total change of void ratio caused by the additional load application
e_o = void ratio of the clay before the application of load

For normally consolidated clay (that is, $\sigma'_o = \sigma'_c$)

$$\Delta e = C_c \log \frac{\sigma'_o + \Delta\sigma'}{\sigma'_o} \tag{1.60}$$

where

σ'_o = average effective vertical stress on the clay layer
$\Delta\sigma'$ = $\Delta\sigma$ (that is, added pressure)

Now, combining Eqs. (1.59) and (1.60) yields

$$S_c = \frac{C_c H_c}{1 + e_o} \log \frac{\sigma'_o + \Delta\sigma'}{\sigma'_o} \tag{1.61}$$

For overconsolidated clay with $\sigma'_o + \Delta\sigma' \leq \sigma'_c$,

$$\Delta e = C_s \log \frac{\sigma'_o + \Delta\sigma'}{\sigma'_o} \tag{1.62}$$

Combining Eqs. (1.59) and (1.62) gives

$$S_c = \frac{C_s H_c}{1 + e_o} \log \frac{\sigma'_o + \Delta\sigma'}{\sigma'_o} \tag{1.63}$$

For overconsolidated clay, if $\sigma'_o < \sigma'_c < \sigma'_o + \Delta\sigma'$, then

$$\Delta e = \Delta e_1 + \Delta e_2 = C_s \log \frac{\sigma'_c}{\sigma'_o} + C_c \log \frac{\sigma'_o + \Delta\sigma'}{\sigma'_c} \tag{1.64}$$

Now, combining Eqs. (1.59) and (1.64) yields

$$S_c = \frac{C_s H_c}{1 + e_o} \log \frac{\sigma'_c}{\sigma'_o} + \frac{C_c H_c}{1 + e_o} \log \frac{\sigma'_o + \Delta\sigma'}{\sigma'_c} \tag{1.65}$$

1.15   *Time Rate of Consolidation*

In Section 1.13 (see Figure 1.14), we showed that consolidation is the result of the gradual dissipation of the excess pore water pressure from a clay layer. The dissipation of pore water pressure, in turn, increases the effective stress, which induces settlement. Hence, to estimate the degree of consolidation of a clay layer at some time *t* after the load is applied, you need to know the rate of dissipation of the excess pore water pressure.

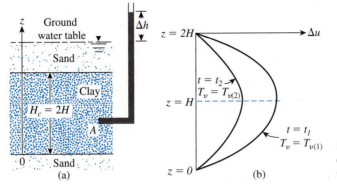

Figure 1.19 (a) Derivation of Eq. (1.68); (b) nature of variation of Δu with time

Figure 1.19 shows a clay layer of thickness H_c that has highly permeable sand layers at its top and bottom. Here, the excess pore water pressure at any point A at any time t after the load is applied is $\Delta u = (\Delta h)\gamma_w$. For a vertical drainage condition (that is, in the direction of z only) from the clay layer, Terzaghi derived the differential equation

$$\frac{\partial(\Delta u)}{\partial t} = C_v \frac{\partial^2(\Delta u)}{\partial z^2} \tag{1.66}$$

where C_v = coefficient of consolidation, defined by

$$C_v = \frac{k}{m_v \gamma_w} = \frac{k}{\dfrac{\Delta e}{\Delta \sigma'(1 + e_{av})}\gamma_w} \tag{1.67}$$

in which

  k = hydraulic conductivity of the clay
Δe = total change of void ratio caused by an effective stress increase of $\Delta \sigma'$
e_{av} = average void ratio during consolidation
m_v = volume coefficient of compressibility = $\Delta e/[\Delta \sigma'(1 + e_{av})]$

Equation (1.66) can be solved to obtain Δu as a function of time t with the following boundary conditions:

1. Because highly permeable sand layers are located at $z = 0$ and $z = H_c$, the excess pore water pressure developed in the clay at those points will be immediately dissipated. Hence,

$$\Delta u = 0 \quad \text{at} \quad z = 0$$

and

$$\Delta u = 0 \quad \text{at} \quad z = H_c = 2H$$

where H = length of maximum drainage path (due to two-way drainage condition—that is, at the top and bottom of the clay).

2. At time $t = 0$, $\Delta u = \Delta u_0 = $ initial excess pore water pressure after the load is applied. With the preceding boundary conditions, Eq. (1.66) yields

$$\Delta u = \sum_{m=0}^{m=\infty}\left[\frac{2(\Delta u_0)}{M}\sin\left(\frac{Mz}{H}\right)\right]e^{-M^2 T_v} \tag{1.68}$$

where

$M = [(2m + 1)\pi]/2$
$m = $ an integer $= 1, 2, \ldots$
$T_v = $ nondimensional time factor $= (C_v t)/H^2$ \hspace{1cm} (1.69)

The value of Δu for various depths (i.e., $z = 0$ to $z = 2H$) at any given time t (and thus T_v) can be calculated from Eq. (1.68). The nature of this variation of Δu is shown in - Figures 1.20a and b. Figure 1.20c shows the variation of $\Delta u/\Delta u_0$ with T_v and H/H_c using Eqs.(1.68) and (1.69).

The *average degree of consolidation* of the clay layer can be defined as

$$U = \frac{S_{c(t)}}{S_{c(\max)}} \tag{1.70}$$

where

$S_{c(t)} = $ settlement of a clay layer at time t after the load is applied
$S_{c(\max)} = $ maximum consolidation settlement that the clay will undergo under a given loading

If the initial pore water pressure (Δu_0) distribution is constant with depth, as shown in Figure 1.20a, the average degree of consolidation also can be expressed as

$$U = \frac{S_{c(t)}}{S_{c(\max)}} = \frac{\displaystyle\int_0^{2H}(\Delta u_0)dz - \int_0^{2H}(\Delta u)dz}{\displaystyle\int_0^{2H}(\Delta u_0)dz} \tag{1.71}$$

or

$$U = \frac{(\Delta u_0)2H - \displaystyle\int_0^{2H}(\Delta u)dz}{(\Delta u_0)2H} = 1 - \frac{\displaystyle\int_0^{2H}(\Delta u)dz}{2H(\Delta u_0)} \tag{1.72}$$

Now, combining Eqs. (1.68) and (1.72), we obtain

$$U = \frac{S_{c(t)}}{S_{c(\max)}} = 1 - \sum_{m=0}^{m=\infty}\left(\frac{2}{M^2}\right)e^{-M^2 T_v} \tag{1.73}$$

The variation of U with T_v can be calculated from Eq. (1.73) and is plotted in Figure 1.21. Note that Eq. (1.73) and thus Figure 1.21 are also valid when an impermeable layer is located at the bottom of the clay layer (Figure 1.20). In that case, the dissipation of

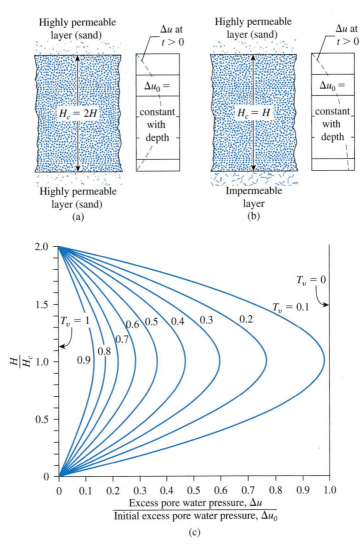

Figure 1.20 Drainage condition for consolidation: (a) two-way drainage; (b) one-way drainage; (c) plot of $\Delta u / \Delta u_0$ with T_v and H / H_c

excess pore water pressure can take place in one direction only. The length of the *maximum drainage path* is then equal to $H = H_c$.

The variation of T_v with U shown in Figure 1.21 can also be approximated by

$$T_v = \frac{\pi}{4}\left(\frac{U\%}{100}\right)^2 \quad \text{(for } U = 0 \text{ to } 60\%) \tag{1.74}$$

and

$$T_v = 1.781 - 0.933 \log\left(100 - U\%\right) \quad \text{(for } U > 60\%) \tag{1.75}$$

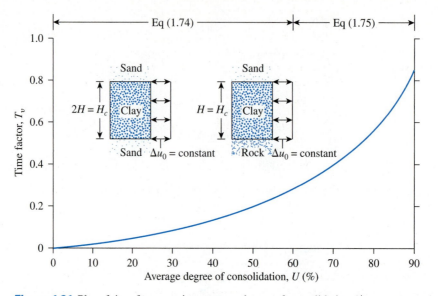

Figure 1.21 Plot of time factor against average degree of consolidation (Δu_0 = constant)

Table 1.11 gives the variation of T_v with U on the basis of Eqs. (1.74) and (1.75)

Sivaram and Swamee (1977) gave the following equation for U varying from 0 to 100%:

$$\frac{U\%}{100} = \frac{(4T_v/\pi)^{0.5}}{[1 + (4T_v/\pi)^{2.8}]^{0.179}} \tag{1.76}$$

or

$$T_v = \frac{(\pi/4)(U\%/100)^2}{[1 - (U\%/100)^{5.6}]^{0.357}} \tag{1.77}$$

Equations (1.76) and (1.77) give an error in T_v of less than 1% for 0% < U < 90% and less than 3% for 90% < U < 100%.

Table 1.11 Variation of T_v with U

U (%)	T_v	U (%)	T_v	U (%)	T_v	U (%)	T_v
0	0	26	0.0531	52	0.212	78	0.529
1	0.00008	27	0.0572	53	0.221	79	0.547
2	0.0003	28	0.0615	54	0.230	80	0.567
3	0.00071	29	0.0660	55	0.239	81	0.588
4	0.00126	30	0.0707	56	0.248	82	0.610
5	0.00196	31	0.0754	57	0.257	83	0.633
6	0.00283	32	0.0803	58	0.267	84	0.658
7	0.00385	33	0.0855	59	0.276	85	0.684
8	0.00502	34	0.0907	60	0.286	86	0.712
9	0.00636	35	0.0962	61	0.297	87	0.742
10	0.00785	36	0.102	62	0.307	88	0.774

Table 1.11 (*Continued*)

U (%)	T_v	U (%)	T_v	U (%)	T_v	U (%)	T_v
11	0.0095	37	0.107	63	0.318	89	0.809
12	0.0113	38	0.113	64	0.329	90	0.848
13	0.0133	39	0.119	65	0.304	91	0.891
14	0.0154	40	0.126	66	0.352	92	0.938
15	0.0177	41	0.132	67	0.364	93	0.993
16	0.0201	42	0.138	68	0.377	94	1.055
17	0.0227	43	0.145	69	0.390	95	1.129
18	0.0254	44	0.152	70	0.403	96	1.219
19	0.0283	45	0.159	71	0.417	97	1.336
20	0.0314	46	0.166	72	0.431	98	1.500
21	0.0346	47	0.173	73	0.446	99	1.781
22	0.0380	48	0.181	74	0.461	100	∞
23	0.0415	49	0.188	75	0.477		
24	0.0452	50	0.197	76	0.493		
25	0.0491	51	0.204	77	0.511		

Example 1.9

A laboratory consolidation test on a normally consolidated clay showed the following results:

Load, $\Delta\sigma'$ (kN/m²)	Void ratio at the end of consolidation, e
140	0.92
212	0.86

The specimen tested was 25.4 mm in thickness and drained on both sides. The time required for the specimen to reach 50% consolidation was 4.5 min.

A similar clay layer in the field 2.8 m thick and drained on both sides, is subjected to a similar increase in average effective pressure (i.e., $\sigma'_0 = 140 \text{ kN/m}^2$ and $\sigma'_0 + \Delta\sigma' = 212 \text{ kN/m}^2$). Determine

a. the expected maximum primary consolidation settlement in the field.
b. the length of time required for the total settlement in the field to reach 40 mm. (Assume a uniform initial increase in excess pore water pressure with depth.)

Solution
Part a
For normally consolidated clay [Eq. (1.49)],

$$C_c = \frac{e_1 - e_2}{\log\left(\dfrac{\sigma'_2}{\sigma'_1}\right)} = \frac{0.92 - 0.86}{\log\left(\dfrac{212}{140}\right)} = 0.333$$

From Eq. (1.61),

$$S_c = \frac{C_c H_c}{1 + e_0} \log \frac{\sigma'_0 + \Delta\sigma'}{\sigma'_0} = \frac{(0.333)\,(2.8)}{1 + 0.92} \log \frac{212}{140} = 0.0875 \text{ m} = \textbf{87.5 mm}$$

Part b

From Eq. (1.70), the average degree of consolidation is

$$U = \frac{S_{c(t)}}{S_{c(\max)}} = \frac{40}{87.5}(100) = 45.7\%$$

The coefficient of consolidation, C_v, can be calculated from the laboratory test. From Eq. (1.69),

$$T_v = \frac{C_v t}{H^2}$$

For 50% consolidation (Figure 1.21), $T_v = 0.197$, $t = 4.5$ min, and $H = H_c/2 = 12.7$ mm, so

$$C_v = T_{50} \frac{H^2}{t} = \frac{(0.197)\,(12.7)^2}{4.5} = 7.061 \text{ mm}^2/\text{min}$$

Again, for field consolidation, $U = 45.7\%$. From Eq. (1.74)

$$T_v = \frac{\pi}{4}\left(\frac{U\%}{100}\right)^2 = \frac{\pi}{4}\left(\frac{45.7}{100}\right)^2 = 0.164$$

But

$$T_v = \frac{C_v t}{H^2}$$

or

$$t = \frac{T_v H^2}{C_v} = \frac{0.164\left(\dfrac{2.8 \times 1000}{2}\right)^2}{7.061} = 45{,}523 \text{ min} = \textbf{31.6 days} \quad \blacksquare$$

1.16   *Degree of Consolidation Under Ramp Loading*

The relationships derived for the average degree of consolidation in Section 1.15 assume that the surcharge load per unit area $(\Delta\sigma)$ is applied instantly at time $t = 0$. However, in most practical situations, $\Delta\sigma$ increases gradually with time to a maximum value and remains constant thereafter. Figure 1.22 shows $\Delta\sigma$ increasing linearly with time (t) up to a maximum at time t_c (a condition called ramp loading). For $t \geq t_c$, the magnitude of $\Delta\sigma$

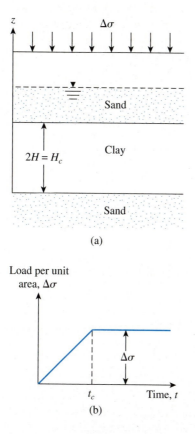

(a)

(b)

Figure 1.22 One-dimensional consolidation due to single ramp loading

remains constant. Olson (1977) considered this phenomenon and presented the average degree of consolidation, U, in the following form:

For $T_v \leq T_c$,

$$U = \frac{T_v}{T_c}\left\{1 - \frac{2}{T_v}\sum_{m=0}^{m=\infty}\frac{1}{M^4}[1 - \exp(-M^2 T_v)]\right\} \tag{1.78}$$

and for $T_v \geq T_c$,

$$U = 1 - \frac{2}{T_c}\sum_{m=0}^{m=\infty}\frac{1}{M^4}[\exp(M^2 T_c) - 1]\exp(-M^2 T_c) \tag{1.79}$$

where m, M, and T_v have the same definition as in Eq. (1.68) and where

$$T_c = \frac{C_v t_c}{H^2} \tag{1.80}$$

Figure 1.23 shows the variation of U with T_v for various values of T_c, based on the solution given by Eqs. (1.78) and (1.79).

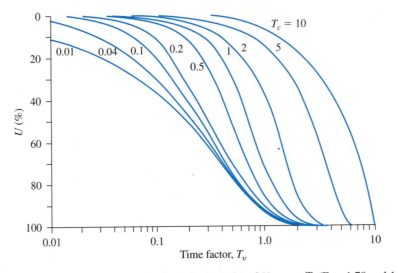

Figure 1.23 Olson's ramp-loading solution: plot of U versus T_v (Eqs. 1.78 and 1.79)

Example 1.10

In Example 1.9, Part (b), if the increase in $\Delta\sigma$ would have been in the manner shown in Figure 1.24, calculate the settlement of the clay layer at time $t = 31.6$ days after the beginning of the surcharge.

Solution

From Part (b) of Example 1.9, $C_v = 7.061$ mm^2/min. From Eq. (1.80),

$$T_c = \frac{C_v t_c}{H^2} = \frac{(7.061 \text{ mm}^2/\text{min}) (15 \times 24 \times 60 \text{ min})}{\left(\dfrac{2.8}{2} \times 1000 \text{ mm}\right)^2} = 0.0778$$

Figure 1.24 Ramp loading

Also,

$$T_v = \frac{C_v t}{H^2} = \frac{(7.061 \text{ mm}^2/\text{min})\,(31.6 \times 24 \times 60 \text{ min})}{\left(\dfrac{2.8}{2} \times 1000 \text{ mm}\right)^2} = 0.164$$

From Figure 1.23, for $T_v = 0.164$ and $T_c = 0.0778$, the value of U is about 36%. Thus,

$$S_{c(t=31.6 \text{ days})} = S_{c(\max)}(0.36) = (87.5)\,(0.36) = \mathbf{31.5 \text{ mm}}$$ ∎

1.17 Shear Strength

The shear strength of a soil, defined in terms of effective stress, is

$$s = c' + \sigma' \tan \phi' \tag{1.81}$$

where

σ' = effective normal stress on plane of shearing
c' = cohesion, or apparent cohesion
ϕ' = effective stress angle of friction

Equation (1.81) is referred to as the *Mohr–Coulomb failure criterion.* The value of c' for sands and normally consolidated clays is equal to zero. For overconsolidated clays, $c' > 0$.

For most day-to-day work, the shear strength parameters of a soil (i.e., c' and ϕ') are determined by two standard laboratory tests: the *direct shear test* and the *triaxial test.*

Direct Shear Test

Dry sand can be conveniently tested by direct shear tests. The sand is placed in a shear box that is split into two halves (Figure 1.25a). First a normal load is applied to the specimen. Then a shear force is applied to the top half of the shear box to cause failure in the sand. The normal and shear stresses at failure are

$$\sigma' = \frac{N}{A}$$

and

$$s = \frac{R}{A}$$

where A = area of the failure plane in soil—that is, the cross-sectional area of the shear box.

Figure 1.25 Direct shear test in sand: (a) schematic diagram of test equipment; (b) plot of test results to obtain the friction angle ϕ'

Several tests of this type can be conducted by varying the normal load. The angle of friction of the sand can be determined by plotting a graph of s against σ' ($= \sigma$ for dry sand), as shown in Figure 1.25b, or

$$\phi' = \tan^{-1}\left(\frac{s}{\sigma'}\right) \tag{1.82}$$

For sands, the angle of friction usually ranges from 26° to 45°, increasing with the relative density of compaction. A general range of the friction angle, ϕ', for sands is given in Table 1.12.

In 1970, Brinch Hansen (see Hansbo, 1975, and Thinh, 2001) gave the following correlation for ϕ' of granular soils.

$$\phi' \text{ (deg)} = 26° + 10D_r + 0.4C_u + 1.6 \log (D_{50}) \tag{1.83}$$

where

D_r = relative density (fraction)
C_u = uniformity coefficient
D_{50} = mean grain size, in mm (i.e., the diameter through which 50% of the soil passes)

Table 1.12 Relationship between Relative Density and Angle of Friction of Cohesionless Soils

State of packing	Relative density (%)	Angle of friction, ϕ' (deg.)
Very loose	<20	<30
Loose	20–40	30–35
Compact	40–60	35–40
Dense	60–80	40–45
Very dense	>80	>45

Teferra (1975) suggested the following empirical correlation based on a large data base.

$$\phi'(\text{deg}) = \tan^{-1}\left(\frac{1}{ae+b}\right) \tag{1.84}$$

where

e = void ratio

$$a = 2.101 + 0.097\left(\frac{D_{85}}{D_{15}}\right) \tag{1.85}$$

$$b = 0.845 - 0.398a \tag{1.86}$$

D_{85} and D_{15} = diameters through which, respectively, 85% and 15% of soil passes

Thinh (2001) suggested that Eq. (1.84) provides as better correlation for ϕ' compared to Eq. (1.83).

Triaxial Tests

Triaxial compression tests can be conducted on sands and clays Figure 1.26a shows a schematic diagram of the triaxial test arrangement. Essentially, the test consists of placing a soil specimen confined by a rubber membrane into a lucite chamber and then applying an all-around confining pressure (σ_3) to the specimen by means of the chamber fluid (generally, water or glycerin). An added stress ($\Delta\sigma$) can also be applied to the specimen in the axial direction to cause failure ($\Delta\sigma = \Delta\sigma_f$ at failure). Drainage from the specimen can be allowed or stopped, depending on the condition being tested. For clays, three main types of tests can be conducted with triaxial equipment (see Figure 1.27):

1. Consolidated-drained test (CD test)
2. Consolidated-undrained test (CU test)
3. Unconsolidated-undrained test (UU test)

Consolidated-Drained Tests:

Step 1. Apply chamber pressure σ_3. Allow complete drainage, so that the pore water pressure ($u = u_0$) developed is zero.

Step 2. Apply a deviator stress $\Delta\sigma$ slowly. Allow drainage, so that the pore water pressure ($u = u_d$) developed through the application of $\Delta\sigma$ is zero. At failure, $\Delta\sigma = \Delta\sigma_f$; the total pore water pressure $u_f = u_0 + u_d = 0$.

So for *consolidated-drained tests,* at failure,

Major principal effective stress = $\sigma_3 + \Delta\sigma_f = \sigma_1 = \sigma_1'$
Minor principal effective stress = $\sigma_3 = \sigma_3'$

Changing σ_3 allows several tests of this type to be conducted on various clay specimens. The shear strength parameters (c' and ϕ') can now be determined by plotting Mohr's circle at failure, as shown in Figure 1.26b, and drawing a common tangent to the Mohr's circles. This is the *Mohr–Coulomb failure envelope.* (*Note:* For normally consolidated clay, $c' \approx 0$.) At failure,

$$\sigma_1' = \sigma_3' \tan^2\left(45 + \frac{\phi'}{2}\right) + 2c' \tan\left(45 + \frac{\phi'}{2}\right) \tag{1.87}$$

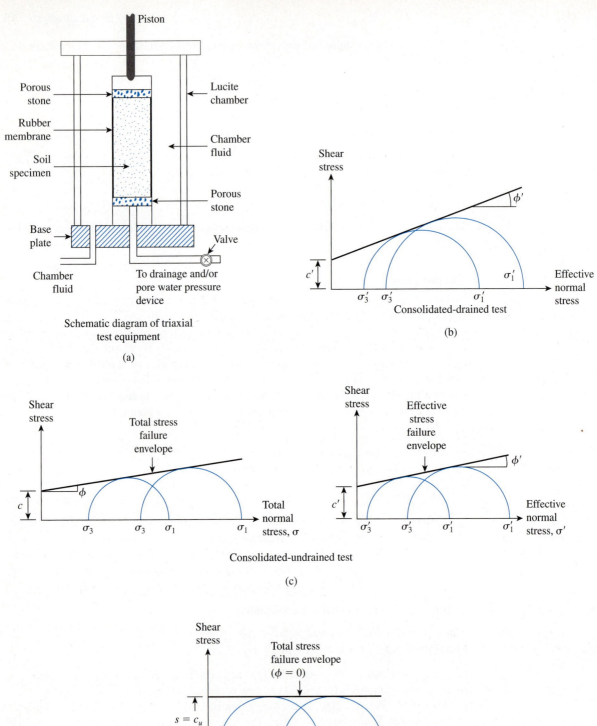

Schematic diagram of triaxial test equipment

(a)

Consolidated-drained test

(b)

Consolidated-undrained test

(c)

Unconsolidated-undrained test

(d)

Figure 1.26 Triaxial test

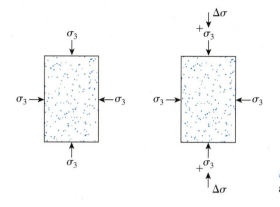

Figure 1.27 Sequence of stress application in triaxial test

Consolidated-Undrained Tests:

Step 1. Apply chamber pressure σ_3. Allow complete drainage, so that the pore water pressure $(u = u_0)$ developed is zero.

Step 2. Apply a deviator stress $\Delta\sigma$. Do not allow drainage, so that the pore water pressure $u = u_d \neq 0$. At failure, $\Delta\sigma = \Delta\sigma_f$; the pore water pressure $u_f = u_0 + u_d = 0 + u_{d(f)}$.

Hence, at failure,

Major principal total stress $= \sigma_3 + \Delta\sigma_f = \sigma_1$
Minor principal total stress $= \sigma_3$
Major principal effective stress $= (\sigma_3 + \Delta\sigma_f) - u_f = \sigma_1'$
Minor principal effective stress $= \sigma_3 - u_f = \sigma_3'$

Changing σ_3 permits multiple tests of this type to be conducted on several soil specimens. The total stress Mohr's circles at failure can now be plotted, as shown in Figure 1.26c, and then a common tangent can be drawn to define the *failure envelope*. This *total stress failure envelope* is defined by the equation

$$s = c + \sigma \tan\phi \tag{1.88}$$

where c and ϕ are the *consolidated-undrained cohesion* and *angle of friction*, respectively. (*Note:* $c \approx 0$ for normally consolidated clays.)

Similarly, effective stress Mohr's circles at failure can be drawn to determine the *effective stress failure envelope* (Figure 1.26c), which satisfy the relation expressed in Eq. (1.81).

Unconsolidated-Undrained Tests:

Step 1. Apply chamber pressure σ_3. Do not allow drainage, so that the pore water pressure $(u = u_0)$ developed through the application of σ_3 is not zero.

Step 2. Apply a deviator stress $\Delta\sigma$. Do not allow drainage $(u = u_d \neq 0)$. At failure, $\Delta\sigma = \Delta\sigma_f$; the pore water pressure $u_f = u_0 + u_{d(f)}$

For *unconsolidated-undrained* triaxial tests,

Major principal total stress $= \sigma_3 + \Delta\sigma_f = \sigma_1$
Minor principal total stress $= \sigma_3$

The total stress Mohr's circle at failure can now be drawn, as shown in Figure 1.26d. For saturated clays, the value of $\sigma_1 - \sigma_3 = \Delta\sigma_f$ is a constant, irrespective of the chamber confining pressure σ_3 (also shown in Figure 1.26d). The tangent to these Mohr's circles will be a horizontal line, called the $\phi = 0$ condition. The shear strength for this condition is

$$s = c_u = \frac{\Delta\sigma_f}{2} \tag{1.89}$$

where c_u = undrained cohesion (or undrained shear strength).

The pore pressure developed in the soil specimen during the unconsolidated-undrained triaxial test is

$$u = u_0 + u_d \tag{1.90}$$

The pore pressure u_0 is the contribution of the hydrostatic chamber pressure σ_3. Hence,

$$u_0 = B\sigma_3 \tag{1.91}$$

where B = Skempton's pore pressure parameter.

Similarly, the pore parameter u_d is the result of the added axial stress $\Delta\sigma$, so

$$u_d = A\Delta\sigma \tag{1.92}$$

where A = Skempton's pore pressure parameter.

However,

$$\Delta\sigma = \sigma_1 - \sigma_3 \tag{1.93}$$

Combining Eqs. (1.90), (1.91), (1.92), and (1.93) gives

$$u = u_0 + u_d = B\sigma_3 + A(\sigma_1 - \sigma_3) \tag{1.94}$$

The pore water pressure parameter B in soft saturated soils is approximately 1, so

$$u = \sigma_3 + A(\sigma_1 - \sigma_3) \tag{1.95}$$

The value of the pore water pressure parameter A at failure will vary with the type of soil. Following is a general range of the values of A at failure for various types of clayey soil encountered in nature:

Type of soil	A at failure
Sandy clays	0.5–0.7
Normally consolidated clays	0.5–1
Overconsolidated clays	−0.5–0

1.18   *Unconfined Compression Test*

The *unconfined compression test* (Figure 1.28a) is a special type of unconsolidated-undrained triaxial test in which the confining pressure $\sigma_3 = 0$, as shown in Figure 1.28b. In this test, an axial stress $\Delta\sigma$ is applied to the specimen to cause failure

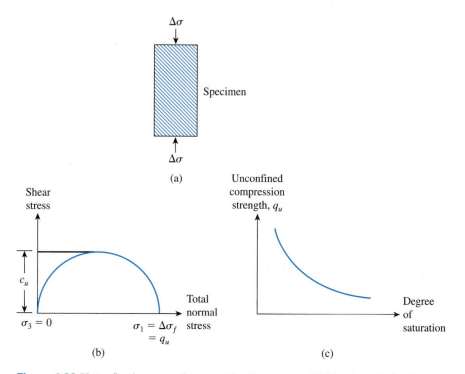

Figure 1.28 Unconfined compression test: (a) soil specimen; (b) Mohr's circle for the test; (c) variation of q_u with the degree of saturation

(i.e., $\Delta\sigma = \Delta\sigma_f$). The corresponding Mohr's circle is shown in Figure 1.28b. Note that, for this case,

Major principal total stress $= \Delta\sigma_f = q_u$
Minor principal total stress $= 0$

The axial stress at failure, $\Delta\sigma_f = q_u$, is generally referred to as the *unconfined compression strength*. The shear strength of saturated clays under this condition ($\phi = 0$), from Eq. (1.81), is

$$s = c_u = \frac{q_u}{2} \tag{1.96}$$

The unconfined compression strength can be used as an indicator of the consistency of clays.

Unconfined compression tests are sometimes conducted on unsaturated soils. With the void ratio of a soil specimen remaining constant, the unconfined compression strength rapidly decreases with the degree of saturation (Figure 1.28c).

1.19 *Comments on Friction Angle, ϕ'*

Effective Stress Friction Angle of Granular Soils

In general, the direct shear test yields a higher angle of friction compared with that obtained by the triaxial test. Also, note that the failure envelope for a given soil is actually curved. The Mohr–Coulomb failure criterion defined by Eq. (1.81) is only an approximation. Because of the curved nature of the failure envelope, a soil tested at higher normal stress will yield a lower value of ϕ'. An example of this relationship is shown in Figure 1.29, which is a plot of ϕ' versus the void ratio e for Chattachoochee River sand near Atlanta, Georgia (Vesic, 1963). The friction angles shown were obtained from triaxial tests. Note that, for a given value of e, the magnitude of ϕ' is about 4° to 5° smaller when the confining pressure σ'_3 is greater than about 70 kN/m², compared with that when $\sigma'_3 < 70$ kN/m².

Effective Stress Friction Angle of Cohesive Soils

Figure 1.30 shows the variation of effective stress friction angle, ϕ', for several normally consolidated clays (Bjerrum and Simons, 1960; Kenney, 1959). It can be seen from the figure that, in general, the friction angle ϕ' decreases with the increase in plasticity index. The value of ϕ' generally decreases from about 37 to 38° with a plasticity index of about 10 to about 25° or less with a plasticity index of about 100. The consolidated undrained friction angle (ϕ) of normally consolidated saturated clays generally ranges from 5 to 20°.

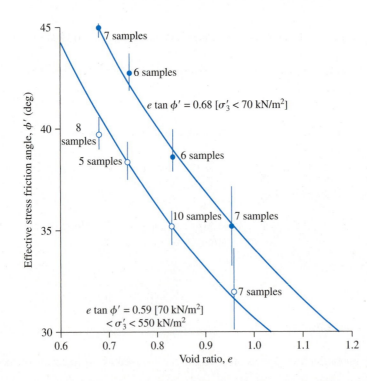

Figure 1.29 Variation of friction angle ϕ' with void ratio for Chattachoochee River sand (After Vesic, 1963) (From Vesic, A. B. Bearing Capacity of Deep Foundations in Sand. In Highway Research Record 39, Highway Research Board, National Research Council, Washington, D.C., 1963, Figure 11, p. 123. Reproduced with permission of the Transportation Research Board.)

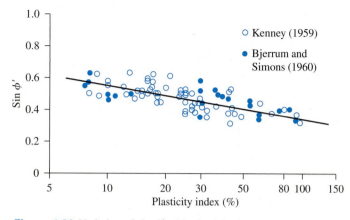

Figure 1.30 Variation of sin ϕ' with plasticity index (PI) for several normally consolidated clays

The consolidated drained triaxial test was described in Section 1.17. Figure 1.31 shows a schematic diagram of a plot of $\Delta\sigma$ versus axial strain in a drained triaxial test for a clay. At failure, for this test, $\Delta\sigma = \Delta\sigma_f$. However, at large axial strain (i.e., the ultimate strength condition), we have the following relationships:

Major principal stress: $\sigma'_{1(ult)} = \sigma_3 + \Delta\sigma_{ult}$
Minor principal stress: $\sigma'_{3(ult)} = \sigma_3$

At failure (i.e., peak strength), the relationship between σ'_1 and σ'_3 is given by Eq. (1.87). However, for ultimate strength, it can be shown that

$$\sigma'_{1(ult)} = \sigma'_3 \tan^2\left(45 + \frac{\phi'_r}{2}\right) \tag{1.97}$$

where ϕ'_r = residual effective stress friction angle.

Figure 1.32 shows the general nature of the failure envelopes at peak strength and ultimate strength (or *residual strength*). The residual shear strength of clays is important in the evaluation of the long-term stability of new and existing slopes and the design of remedial measures. The effective stress residual friction angles ϕ'_r of clays may be substantially smaller than the effective stress peak friction angle ϕ'. Past research has shown that the clay fraction (i.e., the percent finer than 2 microns) present in a given soil, CF, and

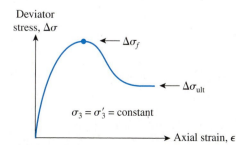

$\sigma_3 = \sigma'_3$ = constant

Figure 1.31 Plot of deviator stress versus axial strain–drained triaxial test

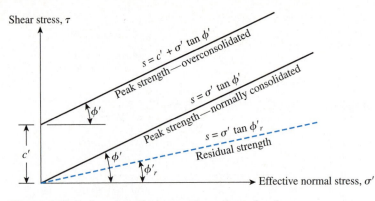

Figure 1.32 Peak- and residual-strength envelopes for clay

the clay mineralogy are the two primary factors that control ϕ_r'. The following is a summary of the effects of CF on ϕ_r'.

1. If CF is less than about 15%, then ϕ_r' is greater than about 25°.
2. For CF > about 50%, ϕ_r' is entirely governed by the sliding of clay minerals and may be in the range of about 10 to 15°.
3. For kaolinite, illite, and montmorillonite, ϕ_r' is about 15°, 10°, and 5°, respectively.

Illustrating these facts, Figure 1.33 shows the variation of ϕ_r' with CF for several soils (Skempton, 1985).

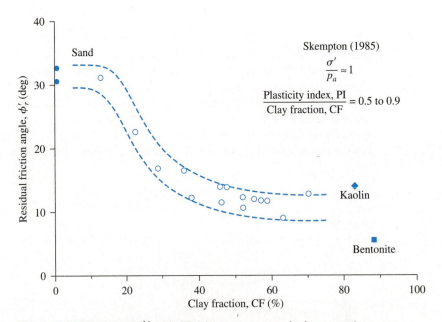

Figure 1.33 Variation of ϕ_r' with CF (*Note:* p_a = atmospheric pressure)

1.20   *Correlations for Undrained Shear Strength, C_u*

Several empirical relationships can be observed between c_u and the effective overburden pressure (σ_0') in the field. Some of these relationships are summarized in Table 1.13.

Table 1.13 Empirical Equations Related to c_u and σ_0'

Reference	Relationship	Remarks
Skempton (1957)	$\dfrac{c_{u(\text{VST})}}{\sigma_0'} = 0.11 + 0.00037\,(\text{PI})$ PI = plasticity index (%) $c_{u(\text{VST})}$ = undrained shear strength from vane shear test	For normally consolidated clay
Chandler (1988)	$\dfrac{c_{u(\text{VST})}}{\sigma_c'} = 0.11 + 0.0037\,(\text{PI})$ σ_c' = preconsolidation pressure	Can be used in overconsolidated soil; accuracy ±25%; not valid for sensitive and fissured clays
Jamiolkowski, *et al.* (1985)	$\dfrac{c_u}{\sigma_c'} = 0.23 \pm 0.04$	For lightly overconsolidated clays
Mesri (1989)	$\dfrac{c_u}{\sigma_0'} = 0.22$	
Bjerrum and Simons (1960)	$\dfrac{c_u}{\sigma_0'} = 0.45\left(\dfrac{\text{PI\%}}{100}\right)^{0.5}$ for PI > 50%	Normally consolidated clay
	$\dfrac{c_u}{\sigma_0'} = 0.118\,(\text{LI})^{0.15}$ for LI = liquidity index > 0.5	Normally consolidated clay
Ladd, *et al.* (1977)	$\dfrac{\left(\dfrac{c_u}{\sigma_0'}\right)_{\text{overconsolidated}}}{\left(\dfrac{c_u}{\sigma_0'}\right)_{\text{normally consolidated}}} = \text{OCR}^{0.8}$ OCR = overconsolidation ratio = σ_c'/σ_0'	

1.21   *Sensitivity*

For many naturally deposited clay soils, the unconfined compression strength is much less when the soils are tested after remolding without any change in the moisture content. This property of clay soil is called *sensitivity*. The degree of sensitivity is the ratio of the unconfined compression strength in an undisturbed state to that in a remolded state, or

$$S_t = \frac{q_{u(\text{undisturbed})}}{q_{u(\text{remolded})}}$$

(1.98)

The sensitivity ratio of most clays ranges from about 1 to 8; however, highly flocculent marine clay deposits may have sensitivity ratios ranging from about 10 to 80. Some clays turn to viscous liquids upon remolding, and these clays are referred to as "quick" clays. The loss of strength of clay soils from remolding is caused primarily by the destruction of the clay particle structure that was developed during the original process of sedimentation.

Problems

1.1 A moist soil has a void ratio of 0.65. The moisture content of the soil is 14% and $G_s = 2.7$. Determine:
a. Porosity
b. Degree of saturation (%)
c. Dry unit weight (kN/m^3)

1.2 For the soil described in Problem 1.1:
a. What would be the saturated unit weight in kN/m^3?
b. How much water, in kN/m^3, needs to be added to the soil for complete saturation?
c. What would be the moist unit weight in kN/m^3 when the degree of saturation is 70%?

1.3 The moist unit weight of a soil is 18.79 kN/m^3. For a moisture content of 12% and $G_s = 2.65$, calculate:
a. Void ratio
b. Porosity
c. Degree of saturation
d. Dry unit weight

1.4 A saturated soil specimen has $w = 36\%$ and $\gamma_d = 13.43$ kN/m^3. Determine:
a. Void ratio
b. Porosity
c. Specific gravity of soil solids
d. Saturated unit weight (kN/m^3)

1.5 The laboratory test results of a sand are $e_{max} = 0.91$, $e_{min} = 0.48$, and $G_s = 2.67$. What would be the dry and moist unit weights of this sand when compacted at a moisture content of 10% to a relative density of 65%?

1.6 The laboratory test results of six soils are given in the following table. Classify the soils by the AASHTO soil classification system and give the group indices.

Sieve Analysis—Percent Passing

Sieve No.	Soil A	Soil B	Soil C	Soil D	Soil E	Soil F
4	92	100	100	95	100	100
10	48	60	98	90	91	82
40	28	41	82	79	80	74
200	13	33	72	64	30	55
Liquid limit	31	38	56	35	43	35
Plastic limit	26	25	31	26	29	21

1.7 Classify the soils in Problem 1.6 using the Unified Soil Classification System. Give group symbols and group names.

1.8 The permeability of a sand was tested in the laboratory at a void ratio of 0.6 and was determined to be 0.14 cm/sec. Use Eq. (1.32) to estimate the hydraulic conductivity of this sand at a void ratio of 0.8.

1.9 A sand has the following: $D_{10} = 0.08$ mm; $D_{60} = 0.37$ mm; void ratio $e = 0.6$.
  a. Determine the hydraulic conductivity using Eq. (1.33).
  b. Determine the hydraulic conductivity using Eq. (1.35).

1.10 The *in situ* hydraulic conductivity of a normally consolidated clay is 5.4×10^{-6} cm/sec at a void ratio of 0.92. What would be its hydraulic conductivity at a void ratio of 0.72? Use Eq. (1.36) and $n = 5.1$.

1.11 Refer to the soil profile shown in Figure P1.11. Determine the total stress, pore water pressure, and effective stress at A, B, C, and D.

1.12 For a normally consolidated clay layer, the following are given:

$$\text{Thickness} = 3 \text{ m}$$
$$\text{Void ratio} = 0.75$$
$$\text{Liquid limit} = 42$$
$$G_s = 2.72$$
$$\text{Average effective stress on the clay layer} = 110 \text{ kN/m}^2$$

How much consolidation settlement would the clay undergo if the average effective stress on the clay layer is increased to 155 kN/m² as the result of the construction of a foundation? Use Eq. (1.51) to estimate the compression index.

1.13 Refer to Problem 1.12. Assume that the clay layer is preconsolidated, $\sigma_c' = 130$ kN/m², and $C_S = \frac{1}{5}C_c$. Estimate the consolidation settlement.

1.14 A saturated clay deposit below the ground water table in the field has liquid limit = 61%, plastic limit = 21%, and moisture content = 38%. Estimate the preconsolidation pressure, σ_c' (kN/m²) using Eq. (1.46).

1.15 A normally consolidated clay layer in the field has a thickness of 3.2 m with an average effective stress of 70 kN/m². A laboratory consolidation test on the clay gave the following results.

Pressure (kN/m²)	Void ratio
100	0.905
200	0.815

  a. Determine the compression index, C_c.
  b. If the average effective stress on the clay layer $(\sigma_o' + \Delta\sigma)$ is increased to 115 kN/m², what would be the total consolidation settlement?

Figure P1.11

1.16 A clay soil specimen, 25.4 mm thick (drained on top and bottom), was tested in the laboratory. For a given load increment, the time for 50% consolidation was 5 min 20 sec. How long will it take for 50% consolidation of a similar clay layer in the field that is 2.5 m thick and drained on one side only?

1.17 A clay soil specimen 25.4 mm thick (drained on top only) was tested in the laboratory. For a given load increment, the time for 60% consolidation was 6 min 20 sec. How long will it take for 50% consolidation for a similar clay layer in the field that is 3.05 m thick and drained on both sides?

1.18 Refer to Figure P1.18. A total of 60-mm consolidation settlement is expected in the two clay layers due to a surcharge $\Delta\sigma$. Find the duration of surcharge application at which 30 mm of total settlement would take place.

1.19 The coefficient of consolidation of a clay for a given pressure range was obtained as 8×10^{-3} mm²/sec on the basis of one-dimensional consolidation test results. In the field, there is a 2-m thick layer of the same clay (Figure P1.19a). Based on the assumption that a uniform surcharge of 70 kN/m² was to be applied instantaneously, the total consolidation settlement was estimated to be 150 mm.

However, during construction, the loading was gradual; the resulting surcharge can be approximated as shown in Figure P1.19b. Estimate the settlement at $t = 30$ days and $t = 120$ days after the beginning of construction.

1.20 A direct shear test was conducted on a 2×2 specimen of dry sand had the following results.

Normal force (N)	Shear force at failure (N)
146.8	91.9
245.4	159.2
294.3	178.8

Draw a graph of shear stress at failure versus normal stress and determine the soil friction angle.

Figure P1.18

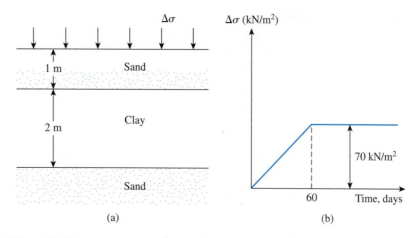

(a)   (b)

Figure P1.19

1.21 For a sand, given:

$D_{85} = 0.21$ mm

$D_{50} = 0.13$ mm

$D_{15} = 0.09$ mm

Uniformity coefficient, $C_u = 2.1$

Void ratio, $e = 0.68$

Relative density = 53%

Estimate the soil friction angle using

a. Eq. (1.83)

b. Eq. (1.84)

1.22 A consolidated-drained triaxial test on a normally consolidated clay yields the following results.

All around confining pressure, $\sigma_3' = 138$ kN/m²

Added axial stress at failure, $\Delta\sigma = 276$ kN/m²

Determine the shear-strength parameters.

1.23 The following are the results of two consolidated-drained triaxial tests on a clay.

Test I: $\sigma_3' = 82.8$ kN/m²; $\sigma_{1(failure)}' = 329.2$ kN/m²

Test II: $\sigma_3' = 165.6$ kN/m²; $\sigma_{1(failure)}' = 558.6$ kN/m²

Determine the shear-strength parameters—that is, c' and ϕ'.

1.24 A consolidated-undrained triaxial test was conducted on a saturated normally consolidated clay. The test results are

$\sigma_3 = 89.7$ kN/m²

$\sigma_{1(failure)} = 220.8$ kN/m²

Pore water pressure at failure = 37.95 kN/m²

Determine c, ϕ, c', and ϕ'.

1.25 A normally consolidated clay soil has $\phi = 20°$ and $\phi' = 28°$. If a consolidated-undrained test is conducted on this clay with an all around pressure of $\sigma_3 = 148.35$ kN/m², what would be the magnitude of the major principal stress, σ_1, and the pore water pressure, u, at failure?

References

AMER, A. M., and AWAD, A. A. (1974). "Permeability of Cohesionless Soils," *Journal of the Geotechnical Engineering Division*, American Society of Civil Engineers, Vol. 100, No. GT12, pp. 1309–1316.

AMERICAN SOCIETY FOR TESTING AND MATERIALS (2009). *Annual Book of ASTM Standards,* Vol. 04.08, West Conshohocken, PA.

BJERRUM, L., and SIMONS, N. E. (1960). "Comparison of Shear Strength Characteristics of Normally Consolidated Clay," *Proceedings, Research Conference on Shear Strength of Cohesive Soils,* ASCE, pp. 711–726.

CARRIER III, W. D. (2003). "Goodbye, Hazen; Hello, Kozeny-Carman," *Journal of Geotechnical and Geoenvironmental Engineering,* ASCE, Vol. 129, No. 11, pp. 1054–1056.

CASAGRANDE, A. (1936). "Determination of the Preconsolidation Load and Its Practical Significance," *Proceedings, First International Conference on Soil Mechanics and Foundation Engineering,* Cambridge, MA, Vol. 3, pp. 60–64.

CHANDLER, R. J. (1988). "The *in situ* Measurement of the Undrained Shear Strength of Clays Using the Field Vane," *STP 1014, Vane Shear Strength Testing in Soils: Field and Laboratory Studies*, American Society for Testing and Materials, pp. 13–44.

CHAPUIS, R. P. (2004). "Predicting the Saturated Hydraulic Conductivity of Sand and Gravel Using Effective Diameter and Void Ratio," *Canadian Geotechnical Journal*, Vol. 41, No. 5, pp. 787–795.

DARCY, H. (1856). *Les Fontaines Publiques de la Ville de Dijon,* Paris.

DAS, B. M. (2009). *Soil Mechanics Laboratory Manual,* 7th ed., Oxford University Press, New York.

HANSBO, S. (1975). *Jordmateriallära*: 211, Stockholm, Awe/Gebers.

HIGHWAY RESEARCH BOARD (1945). *Report of the Committee on Classification of Materials for Subgrades and Granular Type Roads,* Vol. 25, pp. 375–388.

JAMILKOWSKI, M., LADD, C. C., GERMAINE, J. T., and LANCELLOTTA, R. (1985). "New Developments in Field and Laboratory Testing of Soils," *Proceedings, XI International Conference on Soil Mechanics and Foundations Engineering*, San Francisco, Vol. 1, pp. 57–153.

KENNEY, T. C. (1959). "Discussion," *Journal of the Soil Mechanics and Foundations Division,* American Society of Civil Engineers, Vol. 85, No. SM3, pp. 67–69.

KULHAWY, F. H., and MAYNE, P. W. (1990). *Manual of Estimating Soil Properties for Foundation Design*, Electric Power Research Institute, Palo Alto, California.

LADD, C. C., FOOTE, R., ISHIHARA, K., SCHLOSSER, F., and POULOS, H. G. (1977). "Stress Deformation and Strength Characteristics," *Proceedings, Ninth International Conference on Soil Mechanics and Foundation Engineering,* Tokyo, Vol. 2, pp. 421–494.

MESRI, G. (1989). "A Re-evaluation of $s_{u(mob)} \approx 0.22\sigma_p$ Using Laboratory Shear Tests," *Canadian Geotechnical Journal*, Vol. 26, No. 1, pp. 162–164.

MESRI, G., and OLSON, R. E. (1971). "Mechanism Controlling the Permeability of Clays," *Clay and Clay Minerals*, Vol. 19, pp. 151–158.

NAGARAJ, T. S., and MURTHY, B. R. S. (1985). "Prediction of the Preconsolidation Pressure and Recompression Index of Soils," *Geotechnical Testing Journal,* American Society for Testing and Materials, Vol. 8, No. 4, pp. 199–202.

OLSON, R. E. (1977). "Consolidation Under Time-Dependent Loading," *Journal of Geotechnical Engineering,* ASCE, Vol. 103, No. GT1, pp. 55–60.

PARK, J. H., and KOUMOTO, T. (2004). "New Compression Index Equation," *Journal of Geotechnical and Geoenvironmental Engineering*, ASCE, Vol. 130, No. 2, pp. 223–226.

RENDON-HERRERO, O. (1980). "Universal Compression Index Equation," *Journal of the Geotechnical Engineering Division*, American Society of Civil Engineers, Vol. 106, No. GT11, pp. 1178–1200.

SAMARASINGHE, A. M., HUANG, Y. H., and DRNEVICH, V. P. (1982). "Permeability and Consolidation of Normally Consolidated Soils," *Journal of the Geotechnical Engineering Division,* ASCE, Vol. 108, No. GT6, pp. 835–850.

SCHMERTMANN, J. H. (1953). "Undisturbed Consolidation Behavior of Clay," *Transactions, American Society of Civil Engineers,* Vol. 120, p. 1201.

SIVARAM, B., and SWAMEE, A. (1977), "A Computational Method for Consolidation Coefficient," *Soils and Foundations,* Vol. 17, No. 2, pp. 48–52.

SKEMPTON, A. W. (1944). "Notes on the Compressibility of Clays," *Quarterly Journal of Geological Society,* London, Vol. C, pp. 119–135.

SKEMPTON, A. W. (1953). "The Colloidal Activity of Clays," *Proceedings, 3rd International Conference on Soil Mechanics and Foundation Engineering,* London, Vol. 1, pp. 57–61.

SKEMPTON, A. W. (1957). "The Planning and Design of New Hong Kong Airport," *Proceedings, The Institute of Civil Engineers,* London, Vol. 7, pp. 305–307.

SKEMPTON, A. W. (1985). "Residual Strength of Clays in Landslides, Folded Strata, and the Laboratory," *Geotechnique,* Vol. 35, No. 1, pp. 3–18.

STAS, C. V., and KULHAWY, F. H. (1984). "Critical Evaluation of Design Methods for Foundations Under Axial Uplift and Compression Loading, *REPORT EL-3771,* Electric Power Research Institute, Palo Alto, California.

TAVENAS, F., JEAN, P., LEBLOND, F. T. P., and LEROUEIL, S. (1983). "The Permeability of Natural Soft Clays. Part II: Permeability Characteristics," *Canadian Geotechnical Journal,* Vol. 20, No. 4, pp. 645–660.

TAYLOR, D. W. (1948). *Fundamentals of Soil Mechanics,* Wiley, New York.

TEFERRA, A. (1975). *Beziehungen zwischen Reibungswinkel, Lagerungsdichte und Sonderwiderständen nichtbindiger Böden mit verschiedener Kornverteilung.* Ph.D. Thesis, Technical University of Aachen Germany.

TERZAGHI, K., and PECK, R. B. (1967). *Soil Mechanics in Engineering Practice,* Wiley, New York.

THINH, K. D. (2001). "How Reliable is Your Angle of Internal Friction?" *Proceedings, XV International Conference on Soil Mechanics and Geotechnical Engineering,* Istanbul, Turkey, Vol. 1, pp. 73–76.

VESIC, A. S. (1963). "Bearing Capacity of Deep Foundations in Sand," *Highway Research Record No. 39,* National Academy of Sciences, Washington, DC., pp. 112–154.

WOOD, D. M. (1983). "Index Properties and Critical State Soil Mechanics," *Proceedings, Symposium on Recent Developments in Laboratory and Field Tests and Analysis of Geotechnical Problems,* Bangkok, p. 309.

WROTH, C. P., and WOOD, D. M. (1978). "The Correlation of Index Properties with Some Basic Engineering Properties of Soils," *Canadian Geotechnical Journal,* Vol. 15, No. 2, pp. 137–145.

2 Natural Soil Deposits and Subsoil Exploration

2.1 ## *Introduction*

To design a foundation that will support a structure, an engineer must understand the types of soil deposits that will support the foundation. Moreover, foundation engineers must remember that soil at any site frequently is nonhomogeneous; that is, the soil profile may vary. Soil mechanics theories involve idealized conditions, so the application of the theories to foundation engineering problems involves a judicious evaluation of site conditions and soil parameters. To do this requires some knowledge of the geological process by which the soil deposit at the site was formed, supplemented by subsurface exploration. Good professional judgment constitutes an essential part of geotechnical engineering—and it comes only with practice.

This chapter is divided into two parts. The first is a general overview of natural soil deposits generally encountered, and the second describes the general principles of subsoil exploration.

Natural Soil Deposits

2.2 ## *Soil Origin*

Most of the soils that cover the earth are formed by the weathering of various rocks. There are two general types of weathering: (1) mechanical weathering and (2) chemical weathering.

Mechanical weathering is a process by which rocks are broken down into smaller and smaller pieces by physical forces without any change in the chemical composition. Changes in temperature result in *expansion and contraction of rock* due to gain and loss of heat. Continuous expansion and contraction will result in the development of cracks in rocks. Flakes and large fragments of rocks are split. *Frost action* is another source of mechanical weathering of rocks. Water can enter the pores, cracks, and other openings in the rock. When the temperature drops, the water freezes, thereby increasing the volume by about 9%. This results in an outward pressure from inside the rock. Continuous freezing and thawing will result in the breakup of a rock mass. *Exfoliation* is another mechanical weathering process by which rock plates are peeled off from large rocks by physical forces. Mechanical weathering of rocks also takes place due to the action of *running water, glaciers, wind, ocean waves*, and so forth.

Chemical weathering is a process of decomposition or mineral alteration in which the original minerals are changed into something entirely different. For example, the common minerals in igneous rocks are quartz, feldspars, and ferromagnesian minerals. The decomposed products of these minerals due to chemical weathering are listed in Table 2.1.

Table 2.1 Some Decomposed Products of Minerals in Igneous Rock

Mineral	Decomposed Product
Quartz	Quartz (sand grains)
Potassium feldspar ($KAlSi_3O_8$) and Sodium feldspar ($NaAlSi_3O_8$)	Kaolinite (clay) Bauxite Illite (clay) Silica
Calcium feldspar ($CaAl_2Si_2O_8$)	Silica Calcite
Biotite	Clay Limonite Hematite Silica Calcite
Olivine ($(Mg, Fe)_2SiO_4$)	Limonite Serpentine Hematite Silica

Most rock weathering is a combination of mechanical and chemical weathering. Soil produced by the weathering of rocks can be transported by physical processes to other places. The resulting soil deposits are called *transported soils*. In contrast, some soils stay where they were formed and cover the rock surface from which they derive. These soils are referred to as *residual soils*.

Transported soils can be subdivided into five major categories based on the *transporting agent:*

1. *Gravity transported* soil
2. *Lacustrine* (lake) deposits
3. *Alluvial* or *fluvial* soil deposited by running water
4. *Glacial* deposited by glaciers
5. *Aeolian* deposited by the wind

In addition to transported and residual soils, there are *peats* and *organic soils,* which derive from the decomposition of organic materials.

2.3 Residual Soil

Residual soils are found in areas where the rate of weathering is more than the rate at which the weathered materials are carried away by transporting agents. The rate of weathering is higher in warm and humid regions compared to cooler and drier regions and, depending on the climatic conditions, the effect of weathering may vary widely.

Residual soil deposits are common in the tropics, on islands such as the Hawaiian Islands, and in the southeastern United States. The nature of a residual soil deposit will generally depend on the parent rock. When hard rocks such as granite and gneiss undergo weathering, most of the materials are likely to remain in place. These soil deposits generally have a top layer of clayey or silty clay material, below which are silty or sandy soil layers. These layers in turn are generally underlain by a partially weathered rock and then sound bedrock. The depth of the sound bedrock may vary widely, even within a distance of a few meters. Figure 2.1 shows the boring log of a residual soil deposit derived from the weathering of granite.

In contrast to hard rocks, there are some chemical rocks, such as limestone, that are chiefly made up of calcite ($CaCO_3$) mineral. Chalk and dolomite have large concentrations of dolomite minerals [$Ca\,Mg(CO_3)_2$]. These rocks have large amounts of soluble materials,

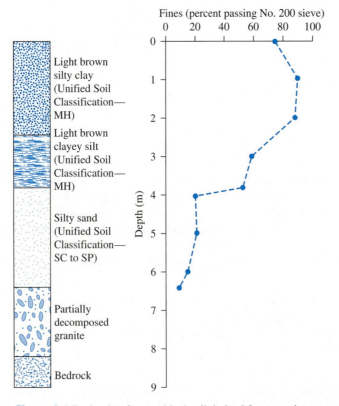

Figure 2.1 Boring log for a residual soil derived from granite

some of which are removed by groundwater, leaving behind the insoluble fraction of the rock. Residual soils that derive from chemical rocks do not possess a gradual transition zone to the bedrock, as seen in Figure 2.1. The residual soils derived from the weathering of limestone-like rocks are mostly red in color. Although uniform in kind, the depth of weathering may vary greatly. The residual soils immediately above the bedrock may be normally consolidated. Large foundations with heavy loads may be susceptible to large consolidation settlements on these soils.

2.4 *Gravity Transported Soil*

Residual soils on a natural slope can move downwards. Cruden and Varnes (1996) proposed a velocity scale for soil movement on a slope, which is summarized in Table 2.2. When residual soils move down a natural slope very slowly, the process is usually referred to as *creep*. When the downward movement of soil is sudden and rapid, it is called a *landslide*. The deposits formed by down-slope creep and landslides are *colluvium*.

Table 2.2 Velocity Scale for Soil Movement on a Slope

Description	Velocity (mm/sec)
Very slow	5×10^{-5} to 5×10^{-7}
Slow	5×10^{-3} to 5×10^{-5}
Moderate	5×10^{-1} to 5×10^{-3}
Rapid	5×10^{1} to 5×10^{-1}

Colluvium is a heterogeneous mixture of soils and rock fragments ranging from clay-sized particles to rocks having diameters of one meter or more. *Mudflows* are one type of gravity-transported soil. Flows are downward movements of earth that resemble a viscous fluid (Figure 2.2) and come to rest in a more dense condition. The soil deposits derived from past mudflows are highly heterogeneous in composition.

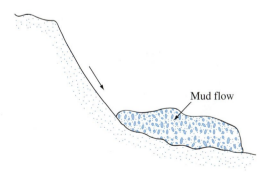

Mud flow

Figure 2.2 Mudflow

2.5   *Alluvial Deposits*

Alluvial soil deposits derive from the action of streams and rivers and can be divided into two major categories: (1) *braided-stream deposits* and (2) deposits caused by the *meandering belt of streams.*

Deposits from Braided Streams

Braided streams are high-gradient, rapidly flowing streams that are highly erosive and carry large amounts of sediment. Because of the high bed load, a minor change in the velocity of flow will cause sediments to deposit. By this process, these streams may build up a complex tangle of converging and diverging channels separated by sandbars and islands.

   The deposits formed from braided streams are highly irregular in stratification and have a wide range of grain sizes. Figure 2.3 shows a cross section of such a deposit. These deposits share several characteristics:

1. The grain sizes usually range from gravel to silt. Clay-sized particles are generally *not* found in deposits from braided streams.
2. Although grain size varies widely, the soil in a given pocket or lens is rather uniform.
3. At any given depth, the void ratio and unit weight may vary over a wide range within a lateral distance of only a few meters. This variation can be observed during soil exploration for the construction of a foundation for a structure. The standard penetration resistance at a given depth obtained from various boreholes will be highly irregular and variable.

   Alluvial deposits are present in several parts of the western United States, such as Southern California, Utah, and the basin and range sections of Nevada. Also, a large amount of sediment originally derived from the Rocky Mountain range was carried eastward to form the alluvial deposits of the Great Plains. On a smaller scale, this type of natural soil deposit, left by braided streams, can be encountered locally.

Meander Belt Deposits

The term *meander* is derived from the Greek word *maiandros,* after the Maiandros (now Menderes) River in Asia, famous for its winding course. Mature streams in a valley curve back and forth. The valley floor in which a river meanders is referred to as the *meander*

Figure 2.3 Cross section of a braided-stream deposit

Figure 2.4 Formation of point bar deposits and oxbow lake in a meandering stream

belt. In a meandering river, the soil from the bank is continually eroded from the points where it is concave in shape and is deposited at points where the bank is convex in shape, as shown in Figure 2.4. These deposits are called *point bar deposits,* and they usually consist of sand and silt-size particles. Sometimes, during the process of erosion and deposition, the river abandons a meander and cuts a shorter path. The abandoned meander, when filled with water, is called an *oxbow lake.* (See Figure 2.4.)

During floods, rivers overflow low-lying areas. The sand and silt-size particles carried by the river are deposited along the banks to form ridges known as *natural levees* (Figure 2.5).

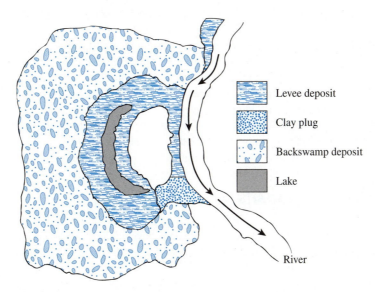

Figure 2.5 Levee and backswamp deposit

Table 2.3 Properties of Deposits within the Mississippi Alluvial Valley

Environment	Soil texture	Natural water content (%)	Liquid limit	Plasticity index
Natural levee	Clay (CL)	25–35	35–45	15–25
	Silt (ML)	15–35	NP–35	NP–5
Point bar	Silt (ML) and silty sand (SM)	25–45	30–55	10–25
Abandoned channel	Clay (CL, CH)	30–95	30–100	10–65
Backswamps	Clay (CH)	25–70	40–115	25–100
Swamp	Organic clay (OH)	100–265	135–300	100–165

(*Note:* NP—Nonplastic)

Finer soil particles consisting of silts and clays are carried by the water farther onto the flood-plains. These particles settle at different rates to form what is referred to as *backswamp deposits* (Figure 2.5), often highly plastic clays.

Table 2.3 gives some properties of soil deposits found in natural levees, point bars, abandoned channels, backswamps and swamps within the alluvial Mississippi Valley (Kolb and Shockley, 1959).

2.6 Lacustrine Deposits

Water from rivers and springs flows into lakes. In arid regions, streams carry large amounts of suspended solids. Where the stream enters the lake, granular particles are deposited in the area forming a delta. Some coarser particles and the finer particles (that is, silt and clay) that are carried into the lake are deposited onto the lake bottom in alternate layers of coarse-grained and fine-grained particles. The deltas formed in humid regions usually have finer grained soil deposits compared to those in arid regions.

Varved clays are alternate layers of silt and silty clay with layer thicknesses rarely exceeding about 13 mm. The silt and silty clay that constitute the layers were carried into fresh water lakes by melt water at the end of the Ice Age. The hydraulic conductivity (Section 1.10) of varved clays exhibits a high degree of anisotropy.

2.7 Glacial Deposits

During the Pleistocene Ice Age, glaciers covered large areas of the earth. The glaciers advanced and retreated with time. During their advance, the glaciers carried large amounts of sand, silt, clay, gravel, and boulders. *Drift* is a general term usually applied to the deposits laid down by glaciers. The drifts can be broadly divided into two major categories: (a) unstratified drifts and (b) stratified drifts. A brief description of each category follows.

Unstratified Drifts

The *unstratified drifts* laid down by melting glaciers are referred to as *till*. The physical characteristics of till may vary from glacier to glacier. Till is called *clay till* because of the presence of the large amount of clay-sized particles in it. In some areas, tills constitute large amounts of boulders, and they are referred to as *boulder till*. The range of grain sizes in a given till varies greatly. The amount of clay-sized fractions present and the plasticity indices of tills also vary widely. During the field exploration program, erratic values of standard penetration resistance (Section 2.13) also may be expected.

The land forms that developed from the till deposits are called *moraines*. A *terminal moraine* (Figure 2.6) is a ridge of till that marks the maximum limit of a glacier's advance. *Recessional moraines* are ridges of till developed behind the terminal moraine at varying distances apart. They are the result of temporary stabilization of the glacier during the recessional period. The till deposited by the glacier between the moraines is referred to as *ground moraine* (Figure 2.6). Ground moraines constitute large areas of the central United States and are called *till plains*.

Figure 2.6 Terminal moraine, ground moraine, and outwash plain

Stratified Drifts

The sand, silt, and gravel that are carried by the melting water from the front of a glacier are called *outwash*. The melted water sorts out the particles by the grain size and forms stratified deposits. In a pattern similar to that of braided-stream deposits, the melted water also deposits the outwash, forming *outwash plains* (Figure 2.6), also called *glaciofluvial deposits*.

2.8 Aeolian Soil Deposits

Wind is also a major transporting agent leading to the formation of soil deposits. When large areas of sand lie exposed, wind can blow the sand away and redeposit it elsewhere. Deposits of windblown sand generally take the shape of *dunes* (Figure 2.7). As dunes are formed, the sand is blown over the crest by the wind. Beyond the crest, the sand particles roll down the slope. The process tends to form a *compact sand deposit* on the *windward side,* and a rather *loose deposit* on the *leeward side,* of the dune.

Dunes exist along the southern and eastern shores of Lake Michigan, the Atlantic Coast, the southern coast of California, and at various places along the coasts of Oregon and

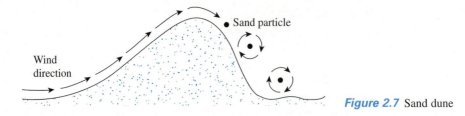

Figure 2.7 Sand dune

Washington. Sand dunes can also be found in the alluvial and rocky plains of the western United States. Following are some of the typical properties of *dune sand:*

1. The grain-size distribution of the sand at any particular location is surprisingly uniform. This uniformity can be attributed to the sorting action of the wind.
2. The general grain size decreases with distance from the source, because the wind carries the small particles farther than the large ones.
3. The relative density of sand deposited on the windward side of dunes may be as high as 50 to 65%, decreasing to about 0 to 15% on the leeward side.

Figure 2.8 shows a sand dune from the Thar Desert, which is a large and arid region located in the northwestern part of India that covers an area of about 200,000 square kilometers.

Loess is an aeolian deposit consisting of silt and silt-sized particles. The grain-size distribution of loess is rather uniform. The cohesion of loess is generally derived from a clay

Figure 2.8 A sand dune from the Thar Desert, India (*Courtesy of A. S. Wayal, K. J. Somaiya Polytechnic, Mumbai, India*)

coating over the silt-sized particles, which contributes to a stable soil structure in an unsaturated state. The cohesion may also be the result of the precipitation of chemicals leached by rainwater. Loess is a *collapsing* soil, because when the soil becomes saturated, it loses its binding strength between particles. Special precautions need to be taken for the construction of foundations over loessial deposits. There are extensive deposits of loess in the United States, mostly in the midwestern states of Iowa, Missouri, Illinois, and Nebraska and for some distance along the Mississippi River in Tennessee and Mississippi.

Volcanic ash (with grain sizes between 0.25 to 4 mm) and volcanic dust (with grain sizes less than 0.25 mm) may be classified as wind-transported soil. Volcanic ash is a lightweight sand or sandy gravel. Decomposition of volcanic ash results in highly plastic and compressible clays.

2.9 Organic Soil

Organic soils are usually found in low-lying areas where the water table is near or above the ground surface. The presence of a high water table helps in the growth of aquatic plants that, when decomposed, form organic soil. This type of soil deposit is usually encountered in coastal areas and in glaciated regions. Organic soils show the following characteristics:

1. Their natural moisture content may range from 200 to 300%.
2. They are highly compressible.
3. Laboratory tests have shown that, under loads, a large amount of settlement is derived from secondary consolidation.

2.10 Some Local Terms for Soils

Soils are sometimes referred to by local terms. The following are a few of these terms with a brief description of each.

1. *Caliche*: a Spanish word derived from the Latin word *calix*, meaning *lime*. It is mostly found in the desert southwest of the United States. It is a mixture of sand, silt, and gravel bonded together by *calcareous deposits*. The calcareous deposits are brought to the surface by a net upward migration of water. The water evaporates in the high local temperature. Because of the sparse rainfall, the carbonates are not washed out of the top layer of soil.
2. *Gumbo*: a highly plastic, clayey soil.
3. *Adobe*: a highly plastic, clayey soil found in the southwestern United States.
4. *Terra Rossa*: residual soil deposits that are red in color and derive from limestone and dolomite.
5. *Muck*: organic soil with a very high moisture content.
6. *Muskeg*: organic soil deposit.
7. *Saprolite*: residual soil deposit derived from mostly insoluble rock.
8. *Loam*: a mixture of soil grains of various sizes, such as sand, silt, and clay.
9. *Laterite*: characterized by the accumulation of iron oxide ($Fe_2 O_3$) and aluminum oxide ($Al_2 O_3$) near the surface, and the leaching of silica. Lateritic soils in Central America contain about 80 to 90% of clay and silt-size particles. In the United States, lateritic soils can be found in the southeastern states, such as Alabama, Georgia, and the Carolinas.

Subsurface Exploration

2.11   Purpose of Subsurface Exploration

The process of identifying the layers of deposits that underlie a proposed structure and their physical characteristics is generally referred to as *subsurface exploration.* The purpose of subsurface exploration is to obtain information that will aid the geotechnical engineer in

1. Selecting the type and depth of foundation suitable for a given structure.
2. Evaluating the load-bearing capacity of the foundation.
3. Estimating the probable settlement of a structure.
4. Determining potential foundation problems (e.g., expansive soil, collapsible soil, sanitary landfill, and so on).
5. Determining the location of the water table.
6. Predicting the lateral earth pressure for structures such as retaining walls, sheet pile bulkheads, and braced cuts.
7. Establishing construction methods for changing subsoil conditions.

Subsurface exploration may also be necessary when additions and alterations to existing structures are contemplated.

2.12   Subsurface Exploration Program

Subsurface exploration comprises several steps, including the collection of preliminary information, reconnaissance, and site investigation.

Collection of Preliminary Information

This step involves obtaining information regarding the type of structure to be built and its general use. For the construction of buildings, the approximate column loads and their spacing and the local building-code and basement requirements should be known. The construction of bridges requires determining the lengths of their spans and the loading on piers and abutments.

A general idea of the topography and the type of soil to be encountered near and around the proposed site can be obtained from the following sources:

1. United States Geological Survey maps.
2. State government geological survey maps.
3. United States Department of Agriculture's Soil Conservation Service county soil reports.
4. Agronomy maps published by the agriculture departments of various states.
5. Hydrological information published by the United States Corps of Engineers, including records of stream flow, information on high flood levels, tidal records, and so on.
6. Highway department soil manuals published by several states.

The information collected from these sources can be extremely helpful in planning a site investigation. In some cases, substantial savings may be realized by anticipating problems that may be encountered later in the exploration program.

Reconnaissance

The engineer should always make a visual inspection of the site to obtain information about

1. The general topography of the site, the possible existence of drainage ditches, abandoned dumps of debris, and other materials present at the site. Also, evidence of creep of slopes and deep, wide shrinkage cracks at regularly spaced intervals may be indicative of expansive soils.
2. Soil stratification from deep cuts, such as those made for the construction of nearby highways and railroads.
3. The type of vegetation at the site, which may indicate the nature of the soil. For example, a mesquite cover in central Texas may indicate the existence of expansive clays that can cause foundation problems.
4. High-water marks on nearby buildings and bridge abutments.
5. Groundwater levels, which can be determined by checking nearby wells.
6. The types of construction nearby and the existence of any cracks in walls or other problems.

The nature of the stratification and physical properties of the soil nearby also can be obtained from any available soil-exploration reports on existing structures.

Site Investigation

The site investigation phase of the exploration program consists of planning, making test boreholes, and collecting soil samples at desired intervals for subsequent observation and laboratory tests. The approximate required minimum depth of the borings should be predetermined. The depth can be changed during the drilling operation, depending on the subsoil encountered. To determine the approximate minimum depth of boring, engineers may use the rules established by the American Society of Civil Engineers (1972):

1. Determine the net increase in the effective stress, $\Delta\sigma'$, under a foundation with depth as shown in Figure 2.9. (The general equations for estimating increases in stress are given in Chapter 5.)
2. Estimate the variation of the vertical effective stress, σ_o', with depth.

Figure 2.9 Determination of the minimum depth of boring

3. Determine the depth, $D = D_1$, at which the effective stress increase $\Delta\sigma'$ is equal to $(\frac{1}{10})q$ (q = estimated net stress on the foundation).
4. Determine the depth, $D = D_2$, at which $\Delta\sigma'/\sigma_o' = 0.05$.
5. Choose the smaller of the two depths, D_1 and D_2, just determined as the approximate minimum depth of boring required, unless bedrock is encountered.

If the preceding rules are used, the depths of boring for a building with a width of 30 m will be approximately the following, according to Sowers and Sowers (1970):

No. of stories	Boring depth
1	3.5 m
2	6 m
3	10 m
4	16 m
5	24 m

To determine the boring depth for hospitals and office buildings, Sowers and Sowers (1970) also used the following rules.

- For light steel or narrow concrete buildings,

$$\frac{D_b}{S^{0.7}} = a \qquad (2.1)$$

where
D_b = depth of boring
S = number of stories
a = 3 if D_b is in meters

- For heavy steel or wide concrete buildings,

$$\frac{D_b}{S^{0.7}} = b \qquad (2.2)$$

where

$b = \begin{array}{l} 6 \text{ if } D_b \text{ is in meters} \\ 20 \text{ if } D_b \text{ is in feet} \end{array}$

When deep excavations are anticipated, the depth of boring should be at least 1.5 times the depth of excavation.

Sometimes, subsoil conditions require that the foundation load be transmitted to bedrock. The minimum depth of core boring into the bedrock is about 3 m. If the bedrock is irregular or weathered, the core borings may have to be deeper.

There are no hard-and-fast rules for borehole spacing. Table 2.4 gives some general guidelines. Spacing can be increased or decreased, depending on the condition of the subsoil. If various soil strata are more or less uniform and predictable, fewer boreholes are needed than in nonhomogeneous soil strata.

Table 2.4 Approximate Spacing of Boreholes

Type of project	Spacing (m)
Multistory building	10–30
One-story industrial plants	20–60
Highways	250–500
Residential subdivision	250–500
Dams and dikes	40–80

The engineer should also take into account the ultimate cost of the structure when making decisions regarding the extent of field exploration. The exploration cost generally should be 0.1 to 0.5% of the cost of the structure. Soil borings can be made by several methods, including auger boring, wash boring, percussion drilling, and rotary drilling.

2.13 *Exploratory Borings in the Field*

Auger boring is the simplest method of making exploratory boreholes. Figure 2.10 shows two types of hand auger: the *posthole auger* and the *helical auger.* Hand augers cannot be used for advancing holes to depths exceeding 3 to 5 m. However, they can be used for soil exploration work on some highways and small structures. *Portable power-driven helical augers* (76 mm to 305 mm in diameter) are available for making deeper boreholes. The soil samples obtained from such borings are highly disturbed. In some noncohesive soils or soils having low cohesion, the walls of the boreholes will not stand unsupported. In such circumstances, a metal pipe is used as a *casing* to prevent the soil from caving in.

(a)   (b)

Figure 2.10 Hand tools: (a) posthole auger; (b) helical auger

When power is available, *continuous-flight augers* are probably the most common method used for advancing a borehole. The power for drilling is delivered by truck- or tractor-mounted drilling rigs. Boreholes up to about 60 to 70 m can easily be made by this method. Continuous-flight augers are available in sections of about 1 to 2 m with either a solid or hollow stem. Some of the commonly used solid-stem augers have outside diameters of 66.68 mm, 82.55 mm, 101.6 mm, and 114.3 mm. Common commercially available hollow-stem augers have dimensions of 63.5 mm ID and 158.75 mm OD, 69.85 mm ID and 177.8 OD, 76.2 mm ID and 203.2 OD, and 82.55 mm ID and 228.6 mm OD.

The tip of the auger is attached to a cutter head (Figure 2.11). During the drilling operation (Figure 2.12), section after section of auger can be added and the hole extended downward. The flights of the augers bring the loose soil from the bottom of the hole to the surface. The driller can detect changes in the type of soil by noting changes in the speed and sound of drilling. When solid-stem augers are used, the auger must be withdrawn at regular intervals to obtain soil samples and also to conduct other operations such as standard penetration tests. Hollow-stem augers have a distinct advantage over solid-stem augers in that they do not have to be removed frequently for

Figure 2.11 Carbide-tipped cutting head on auger flight (*Courtesy of Braja M. Das, Henderson, NV*)

Figure 2.12 Drilling with continuous-flight augers (Danny R. Anderson, PE of Professional Service Industries, Inc, El Paso, Texas.)

sampling or other tests. As shown schematically in Figure 2.13, the outside of the hollow-stem auger acts as a casing.

The hollow-stem auger system includes the following components:

Outer component: (a) hollow auger sections, (b) hollow auger cap, and (c) drive cap

Inner component: (a) pilot assembly, (b) center rod column, and (c) rod-to-cap adapter

The auger head contains replaceable carbide teeth. During drilling, if soil samples are to be collected at a certain depth, the pilot assembly and the center rod are removed. The soil sampler is then inserted through the hollow stem of the auger column.

Wash boring is another method of advancing boreholes. In this method, a casing about 2 to 3 m long is driven into the ground. The soil inside the casing is then removed by means of a chopping bit attached to a drilling rod. Water is forced through the drilling rod and exits at a very high velocity through the holes at the bottom of the chopping bit (Figure 2.14). The water and the chopped soil particles rise in the drill hole and overflow at the top of the casing through a T connection. The washwater is

Drive cap →

Rod-to-cap adapter

Auger connector

Hollow-stem auger section

Center rod

Pilot assembly

Auger connector

Auger head

Center head

Replaceable carbide auger tooth

Figure 2.13 Hollow-stem auger components (After ASTM, 2001) (ASTM D4700-91: Standard Guide for Soil Sampling from the Vadose Zone. Copyright ASTM INTERNATIONAL. Reprinted with permission.)

Rope

Derrick

Pressure water

Tub

Drill rod

Casing

Chopping bit

Driving shoe

Water jet at high velocity

Figure 2.14 Wash boring

collected in a container. The casing can be extended with additional pieces as the borehole progresses; however, that is not required if the borehole will stay open and not cave in. Wash borings are rarely used now in the United States and other developed countries.

Rotary drilling is a procedure by which rapidly rotating drilling bits attached to the bottom of drilling rods cut and grind the soil and advance the borehole. There are several types of drilling bit. Rotary drilling can be used in sand, clay, and rocks (unless they are badly fissured). Water or *drilling mud* is forced down the drilling rods to the bits, and the return flow forces the cuttings to the surface. Boreholes with diameters of 50 to 203 mm can easily be made by this technique. The drilling mud is a slurry of water and bentonite. Generally, it is used when the soil that is encountered is likely to cave in. When soil samples are needed, the drilling rod is raised and the drilling bit is replaced by a sampler. With the environmental drilling applications, rotary drilling with air is becoming more common.

Percussion drilling is an alternative method of advancing a borehole, particularly through hard soil and rock. A heavy drilling bit is raised and lowered to chop the hard soil. The chopped soil particles are brought up by the circulation of water. Percussion drilling may require casing.

2.14 Procedures for Sampling Soil

Two types of soil samples can be obtained during subsurface exploration: *disturbed* and *undisturbed.* Disturbed, but representative, samples can generally be used for the following types of laboratory test:

1. Grain-size analysis
2. Determination of liquid and plastic limits
3. Specific gravity of soil solids
4. Determination of organic content
5. Classification of soil

Disturbed soil samples, however, cannot be used for consolidation, hydraulic conductivity, or shear strength tests. Undisturbed soil samples must be obtained for these types of laboratory tests. Sections 2.15 through 2.18 describe some procedures for obtaining soil samples during field exploration.

2.15 Split-Spoon Sampling

Split-spoon samplers can be used in the field to obtain soil samples that are generally disturbed, but still representative. A section of a *standard split-spoon sampler* is shown in Figure 2.15a. The tool consists of a steel driving shoe, a steel tube that is split longitudinally

Figure 2.15 (a) Standard split-spoon sampler; (b) spring core catcher

in half, and a coupling at the top. The coupling connects the sampler to the drill rod. The standard split tube has an inside diameter of 34.93 mm and an outside diameter of 50.8 mm; however, samplers having inside and outside diameters up to 63.5 mm and 76.2 mm, respectively, are also available. When a borehole is extended to a predetermined depth, the drill tools are removed and the sampler is lowered to the bottom of the hole. The sampler is driven into the soil by hammer blows to the top of the drill rod. The standard weight of the hammer is 622.72 N, and for each blow, the hammer drops a distance of 0.762 m. The number of blows required for a spoon penetration of three 152.4-mm intervals are recorded. The number of blows required for the last two intervals are added to give the *standard penetration number, N,* at that depth. This number is generally referred to as the *N value* (American Society for Testing and Materials, 2001, Designation D-1586-99). The sampler is then withdrawn, and the shoe and coupling are removed. Finally, the soil sample recovered from the tube is placed in a glass bottle and transported to the laboratory. This field test is called the standard penetration test (SPT). Figure 2.16a and b show a split-spoon sampler unassembled before and after sampling.

The degree of disturbance for a soil sample is usually expressed as

$$A_R(\%) = \frac{D_o^2 - D_i^2}{D_i^2}(100) \qquad (2.3)$$

where

A_R = area ratio (ratio of disturbed area to total area of soil)
D_o = outside diameter of the sampling tube
D_i = inside diameter of the sampling tube

When the area ratio is 10% or less, the sample generally is considered to be undisturbed. For a standard split-spoon sampler,

$$A_R(\%) = \frac{(50.8)^2 - (34.93)^2}{(34.93)^2}(100) = 111.5\%$$

Figure 2.16 (a) Unassembled split-spoon sampler; (b) after sampling (*Courtesy of Professional Service Industries, Inc. (PSI), Waukesha, Wisconsin*)

Hence, these samples are highly disturbed. Split-spoon samples generally are taken at intervals of about 1.5 m. When the material encountered in the field is sand (particularly fine sand below the water table), recovery of the sample by a split-spoon sampler may be difficult. In that case, a device such as a *spring core catcher* may have to be placed inside the split spoon (Figure 2.15b).

At this juncture, it is important to point out that several factors contribute to the variation of the standard penetration number N at a given depth for similar soil profiles. Among these factors are the SPT hammer efficiency, borehole diameter, sampling method, and rod length (Skempton, 1986; Seed, et al., 1985). The SPT hammer energy efficiency can be expressed as

$$E_r(\%) = \frac{\text{actual hammer energy to the sampler}}{\text{input energy}} \times 100 \qquad (2.4)$$

$$\text{Theoretical input energy} = Wh \qquad (2.5)$$

where

W = weight of the hammer ≈ 0.623 kN
h = height of drop ≈ 0.76 mm

So,

$$Wh = (0.623)(0.76) = 0.474 \text{ kN-m}$$

In the field, the magnitude of E_r can vary from 30 to 90%. The standard practice now in the U.S. is to express the N-value to an average energy ratio of 60% ($\approx N_{60}$). Thus, correcting for field procedures and on the basis of field observations, it appears reasonable to standardize the field penetration number as a function of the input driving energy and its dissipation around the sampler into the surrounding soil, or

$$N_{60} = \frac{N\eta_H\eta_B\eta_S\eta_R}{60} \qquad (2.6)$$

where

N_{60} = standard penetration number, corrected for field conditions
N = measured penetration number
η_H = hammer efficiency (%)
η_B = correction for borehole diameter
η_S = sampler correction
η_R = correction for rod length

Variations of η_H, η_B, η_S, and η_R, based on recommendations by Seed et al. (1985) and Skempton (1986), are summarized in Table 2.5.

Correlations for N_{60} in Cohesive Soil

Besides compelling the geotechnical engineer to obtain soil samples, standard penetration tests provide several useful correlations. For example, the consistency of clay soils can be estimated from the standard penetration number, N_{60}. In order to achieve that, Szechy and Vargi (1978) calculated the *consistency index* (CI) as

Table 2.5 Variations of $\eta_H, \eta_B, \eta_S,$ and η_R [Eq. (2.6)]

1. Variation of η_H

Country	Hammer type	Hammer release	η_H (%)
Japan	Donut	Free fall	78
	Donut	Rope and pulley	67
United States	Safety	Rope and pulley	60
	Donut	Rope and pulley	45
Argentina	Donut	Rope and pulley	45
China	Donut	Free fall	60
	Donut	Rope and pulley	50

2. Variation of η_B

Diameter mm	η_B
60–120	1
150	1.05
200	1.15

3. Variation of η_S

Variable	η_S
Standard sampler	1.0
With liner for dense sand and clay	0.8
With liner for loose sand	0.9

4. Variation of η_R

Rod length m	η_R
>10	1.0
6–10	0.95
4–6	0.85
0–4	0.75

$$CI = \frac{LL - w}{LL - PL} \tag{2.7}$$

where

w = natural moisture content
LL = liquid limit
PL = plastic limit

The approximate correlation between CI, N_{60}, and the unconfined compression strength (q_u) is given in Table 2.6.

Hara, et al. (1971) also suggested the following correlation between the undrained shear strength of clay (c_u) and N_{60}:

$$\frac{c_u}{p_a} = 0.29 N_{60}^{0.72} \tag{2.8}$$

where p_a = atmospheric pressure (≈ 100 kN/m²; ≈ 2000 lb/in²).

Table 2.6 Approximate Correlation between CI, N_{60}, and q_u

Standard penetration number, N_{60}	Consistency	CI	Unconfined compression strength, q_u (kN/m²)
<2	Very soft	<0.5	<25
2–8	Soft to medium	0.5–0.75	25–80
8–15	Stiff	0.75–1.0	80–150
15–30	Very stiff	1.0–1.5	150–400
>30	Hard	>1.5	>400

The overconsolidation ratio, OCR, of a natural clay deposit can also be correlated with the standard penetration number. On the basis of the regression analysis of 110 data points, Mayne and Kemper (1988) obtained the relationship

$$\text{OCR} = 0.193 \left(\frac{N_{60}}{\sigma'_o} \right)^{0.689} \tag{2.9}$$

where σ'_o = effective vertical stress in MN/m^2.

It is important to point out that any correlation between c_u, OCR, and N_{60} is only approximate.

Correction for N_{60} in Granular Soil

In granular soils, the value of N is affected by the effective overburden pressure, σ'_o. For that reason, the value of N_{60} obtained from field exploration under different effective overburden pressures should be changed to correspond to a standard value of σ'_o. That is,

$$(N_1)_{60} = C_N N_{60} \tag{2.10}$$

where

$(N_1)_{60}$ = value of N_{60} corrected to a standard value of $\sigma'_o [100 \text{ kN/m}^2 \, (2000 \text{ lb/ft}^2)]$
$\quad C_N$ = correction factor
$\quad N_{60}$ = value of N obtained from field exploration [Eq. (2.6)]

In the past, a number of empirical relations were proposed for C_N. Some of the relationships are given next. The most commonly cited relationships are those of Liao and Whitman (1986) and Skempton (1986).

In the following relationships for C_N, note that σ'_o is the effective overburden pressure and p_a = atmospheric pressure $(\approx 100 \text{ kN/m}^2)$
Liao and Whitman's relationship (1986):

$$C_N = \left[\frac{1}{\left(\dfrac{\sigma'_o}{p_a} \right)} \right]^{0.5} \tag{2.11}$$

Skempton's relationship (1986):

$$C_N = \frac{2}{1 + \left(\dfrac{\sigma'_o}{p_a} \right)} \quad \text{(for normally consolidated fine sand)} \tag{2.12}$$

$$C_N = \frac{3}{2 + \left(\dfrac{\sigma'_o}{p_a} \right)} \quad \text{(for normally consolidated coarse sand)} \tag{2.13}$$

$$C_N = \frac{1.7}{0.7 + \left(\dfrac{\sigma'_o}{p_a}\right)} \quad \text{(for overconsolidated sand)} \tag{2.14}$$

Seed et al.'s relationship (1975):

$$C_N = 1 - 1.25 \log\left(\frac{\sigma'_o}{p_a}\right) \tag{2.15}$$

Peck et al.'s relationship (1974):

$$C_N = 0.77 \log\left[\frac{20}{\left(\dfrac{\sigma'_o}{p_a}\right)}\right] \left(\text{for } \frac{\sigma'_o}{p_a} \geq 0.25\right) \tag{2.16}$$

Bazaraa (1967):

$$C_N = \frac{4}{1 + 4\left(\dfrac{\sigma'_o}{p_a}\right)} \left(\text{for } \frac{\sigma'_o}{p_a} \leq 0.75\right) \tag{2.17}$$

$$C_N = \frac{4}{3.25 + \left(\dfrac{\sigma'_o}{p_a}\right)} \left(\text{for } \frac{\sigma'_o}{p_a} > 0.75\right) \tag{2.18}$$

Table 2.7 shows the comparison of C_N derived using various relationships cited above. It can be seen that the magnitude of the correction factor estimated by using any one of the relationships is approximately the same, considering the uncertainties involved in conducting the standard penetration tests. Hence, it is recommended that Eq. (2.11) may be used for all calculations.

Table 2.7 Variation of C_N

$\dfrac{\sigma'_o}{p_a}$	C_N						
	Eq. (2.11)	Eq. (2.12)	Eq. (2.13)	Eq. (2.14)	Eq. (2.15)	Eq. (2.16)	Eqs. (2.17) and (2.18)
0.25	2.00	1.60	1.33	1.78	1.75	1.47	2.00
0.50	1.41	1.33	1.20	1.17	1.38	1.23	1.33
0.75	1.15	1.14	1.09	1.17	1.15	1.10	1.00
1.00	1.00	1.00	1.00	1.00	1.00	1.00	0.94
1.50	0.82	0.80	0.86	0.77	0.78	0.87	0.84
2.00	0.71	0.67	0.75	0.63	0.62	0.77	0.76
3.00	0.58	0.50	0.60	0.46	0.40	0.63	0.65
4.00	0.50	0.40	0.60	0.36	0.25	0.54	0.55

Correlation between N_{60} and Relative Density of Granular Soil

An approximate relationship between the corrected standard penetration number and the relative density of sand is given in Table 2.8. The values are approximate primarily because the effective overburden pressure and the stress history of the soil significantly influence the N_{60} values of sand. Kulhawy and Mayne (1990) modified an empirical relationship for relative density that was given by Marcuson and Bieganousky (1977), which can be expressed as

$$D_r(\%) = 12.2 + 0.75 \left[222N_{60} + 2311 - 711\text{OCR} - 779\left(\frac{\sigma'_o}{p_a}\right) - 50C_u^2 \right]^{0.5} \quad (2.19)$$

where

D_r = relative density
σ'_o = effective overburden pressure
C_u = uniformity coefficient of sand
$\text{OCR} = \dfrac{\text{preconsolidation pressure}, \sigma'_c}{\text{effective overburden pressure}, \sigma'_o}$
p_a = atmospheric pressure

Meyerhof (1957) developed a correlation between D_r and N_{60} as

$$N_{60} = \left[17 + 24\left(\frac{\sigma'_o}{p_a}\right) \right] D_r^2$$

or

$$D_r = \left\{ \frac{N_{60}}{\left[17 + 24\left(\frac{\sigma'_o}{p_a}\right) \right]} \right\}^{0.5} \quad (2.20)$$

Equation (2.20) provides a reasonable estimate only for clean, medium fine sand.

    Cubrinovski and Ishihara (1999) also proposed a correlation between N_{60} and the relative density of sand (D_r) that can be expressed as

$$D_r(\%) = \left[\frac{N_{60}\left(0.23 + \dfrac{0.06}{D_{50}} \right)^{1.7}}{9} \left(\dfrac{1}{\dfrac{\sigma'_o}{p_a}} \right) \right]^{0.5} (100) \quad (2.21)$$

Table 2.8 Relation between the Corrected $(N_1)_{60}$ Values and the Relative Density in Sands

Standard penetration number, $(N_1)_{60}$	Approximate relative density, D_r (%)
0–5	0–5
5–10	5–30
10–30	30–60
30–50	60–95

where

p_a = atmospheric pressure ($\approx 100 \text{ kN/m}^2$)
D_{50} = sieve size through which 50% of the soil will pass (mm)

Kulhawy and Mayne (1990) correlated the corrected standard penetration number and the relative density of sand in the form

$$D_r(\%) = \left[\frac{(N_1)_{60}}{C_p C_A C_{OCR}} \right]^{0.5} (100) \tag{2.22}$$

where

C_P = grain-size correlations factor = $60 + 25 \log D_{50}$ $\tag{2.23}$

C_A = correlation factor for aging = $1.2 + 0.05 \log\left(\dfrac{t}{100}\right)$ $\tag{2.24}$

C_{OCR} = correlation factor for overconsolidation = $OCR^{0.18}$ $\tag{2.25}$
D_{50} = diameter through which 50% soil will pass through (mm)
t = age of soil since deposition (years)
OCR = overconsolidation ratio

Correlation between Angle of Friction and Standard Penetration Number

The peak friction angle, ϕ', of granular soil has also been correlated with N_{60} or $(N_1)_{60}$ by several investigators. Some of these correlations are as follows:

1. Peck, Hanson, and Thornburn (1974) give a correlation between N_{60} and ϕ' in a graphical form, which can be approximated as (Wolff, 1989)

$$\phi'(\text{deg}) = 27.1 + 0.3N_{60} - 0.00054[N_{60}]^2 \tag{2.26}$$

2. Schmertmann (1975) provided the correlation between N_{60}, σ_o', and ϕ'. Mathematically, the correlation can be approximated as (Kulhawy and Mayne, 1990)

$$\phi' = \tan^{-1}\left[\frac{N_{60}}{12.2 + 20.3\left(\dfrac{\sigma_o'}{p_a}\right)} \right]^{0.34} \tag{2.27}$$

where

N_{60} = field standard penetration number
σ_o' = effective overburden pressure
p_a = atmospheric pressure in the same unit as σ_o'
ϕ' = soil friction angle

3. Hatanaka and Uchida (1996) provided a simple correlation between ϕ' and $(N_1)_{60}$ that can be expressed as

$$\phi' = \sqrt{20(N_1)_{60}} + 20 \tag{2.28}$$

The following qualifications should be noted when standard penetration resistance values are used in the preceding correlations to estimate soil parameters:

1. The equations are approximate.
2. Because the soil is not homogeneous, the values of N_{60} obtained from a given borehole vary widely.
3. In soil deposits that contain large boulders and gravel, standard penetration numbers may be erratic and unreliable.

Although approximate, with correct interpretation the standard penetration test provides a good evaluation of soil properties. The primary sources of error in standard penetration tests are inadequate cleaning of the borehole, careless measurement of the blow count, eccentric hammer strikes on the drill rod, and inadequate maintenance of water head in the borehole.

Correlation between Modulus of Elasticity and Standard Penetration Number

The modulus of elasticity of granular soils (E_s) is an important parameter in estimating the elastic settlement of foundations. A first order estimation for E_s was given by Kulhawy and Mayne (1990) as

$$\frac{E_s}{p_a} = \alpha N_{60} \tag{2.29}$$

where

$p_a =$ atmospheric pressure (same unit as E_s)

$$\alpha = \begin{cases} 5 \text{ for sands with fines} \\ 10 \text{ for clean normally consolidated sand} \\ 15 \text{ for clean overconsolidated sand} \end{cases}$$

2.16   *Sampling with a Scraper Bucket*

When the soil deposits are sand mixed with pebbles, obtaining samples by split spoon with a spring core catcher may not be possible because the pebbles may prevent the springs from closing. In such cases, a scraper bucket may be used to obtain disturbed representative samples (Figure 2.17). The scraper bucket has a driving point and can be attached to a drilling rod. The sampler is driven down into the soil and rotated, and the scrapings from the side fall into the bucket.

Figure 2.17 Scraper bucket

2.17 *Sampling with a Thin-Walled Tube*

Thin-walled tubes are sometimes referred to as *Shelby tubes.* They are made of seamless steel and are frequently used to obtain undisturbed clayey soils. The most common thin-walled tube samplers have outside diameters of 50.8 mm and 76.2 mm. The bottom end of the tube is sharpened. The tubes can be attached to drill rods (Figure 2.18). The drill rod with the sampler attached is lowered to the bottom of the borehole, and the sampler is pushed into the soil. The soil sample inside the tube is then pulled out. The two ends are sealed, and the sampler is sent to the laboratory for testing. Figure 2.19 shows the sequence of sampling with a thin-walled tube in the field.

Samples obtained in this manner may be used for consolidation or shear tests. A thin-walled tube with a 50.8-mm (2-in.) outside diameter has an inside diameter of about 47.63 mm ($1\frac{7}{8}$ in.). The area ratio is

$$A_R(\%) = \frac{D_o^2 - D_i^2}{D_i^2}(100) = \frac{(50.8)^2 - (47.63)^2}{(47.63)^2}(100) = 13.75\%$$

Increasing the diameters of samples increases the cost of obtaining them.

Figure 2.18 Thin-walled tube

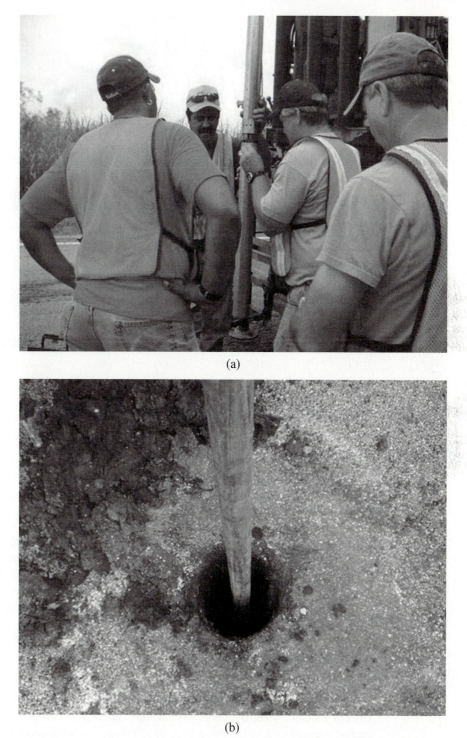

(a)

(b)

Figure 2.19 Sampling with a thin-walled tube: (a) tube being attached to drill rod; (b) tube sampler pushed into soil (*Courtesy of Khaled Sobhan, Florida Atlantic Univetsity, Boca Raton, Florida*)

(c)

Figure 2.19 (*continued*) (c) recovery of soil sample (*Courtesy of Khaled Sobhan, Florida Atlantic University, Boca Raton, Florida*)

2.18   Sampling with a Piston Sampler

When undisturbed soil samples are very soft or larger than 76.2 mm in diameter, they tend to fall out of the sampler. Piston samplers are particularly useful under such conditions. There are several types of piston sampler; however, the sampler proposed by Osterberg (1952) is the most useful. (see Figures 2.20a and 2.20b). It consists of a thin-walled tube with a piston. Initially, the piston closes the end of the tube. The sampler is lowered to the bottom of the borehole (Figure 2.20a), and the tube is pushed into the soil hydraulically, past the piston. Then the pressure is released through a hole in the piston rod (Figure 2.20b). To a large extent, the presence of the piston prevents distortion in the sample by not letting the soil squeeze into the sampling tube very fast and by not admitting excess soil. Consequently, samples obtained in this manner are less disturbed than those obtained by Shelby tubes.

2.19   Observation of Water Tables

The presence of a water table near a foundation significantly affects the foundation's load-bearing capacity and settlement, among other things. The water level will change seasonally. In many cases, establishing the highest and lowest possible levels of water during the life of a project may become necessary.

Figure 2.20 Piston sampler: (a) sampler at the bottom of borehole; (b) tube pushed into the soil hydraulically

If water is encountered in a borehole during a field exploration, that fact should be recorded. In soils with high hydraulic conductivity, the level of water in a borehole will stabilize about 24 hours after completion of the boring. The depth of the water table can then be recorded by lowering a chain or tape into the borehole.

In highly impermeable layers, the water level in a borehole may not stabilize for several weeks. In such cases, if accurate water-level measurements are required, a *piezometer* can be used. A piezometer basically consists of a porous stone or a perforated pipe with a plastic standpipe attached to it. Figure 2.21 shows the general placement of a piezometer in a borehole. This procedure will allow periodic checking until the water level stabilizes.

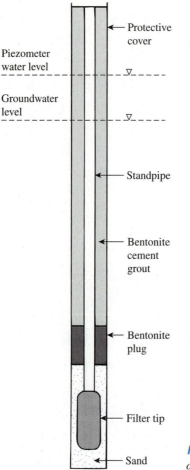

Protective cover

Piezometer water level

Groundwater level

Standpipe

Bentonite cement grout

Bentonite plug

Filter tip

Sand

Figure 2.21 Casagrande-type piezometer (*Courtesy of N. Sivakugan, James Cook University, Australia.*)

2.20   *Vane Shear Test*

The *vane shear test* (ASTM D-2573) may be used during the drilling operation to determine the *in situ* undrained shear strength (c_u) of clay soils—particularly soft clays. The vane shear apparatus consists of four blades on the end of a rod, as shown in Figure 2.22. The height, *H,* of the vane is twice the diameter, *D.* The vane can be either rectangular or tapered (see Figure 2.22). The dimensions of vanes used in the field are given in Table 2.9. The vanes of the apparatus are pushed into the soil at the bottom of a borehole without disturbing the soil appreciably. Torque is applied at the top of the rod to rotate the vanes at a standard rate of 0.1°/sec. This rotation will induce failure in a soil of cylindrical shape surrounding the vanes. The maximum torque, *T,* applied to cause failure is measured. Note that

$$T = f(c_u, H, \text{ and } D) \qquad (2.30)$$

Figure 2.22 Geometry of field vane (After ASTM, 2001) (Annual Book of ASTM Standards, Vol. 04.08. Copyright ASTM INTERNATIONAL. Reprinted with permission.)

or

$$c_u = \frac{T}{K} \qquad (2.31)$$

where

T is in N·m, c_u is in kN/m², and
K = a constant with a magnitude depending on the dimension and shape of the vane

The constant

$$K = \left(\frac{\pi}{10^6}\right)\left(\frac{D^2 H}{2}\right)\left(1 + \frac{D}{3H}\right) \qquad (2.32a)$$

Table 2.9 ASTM Recommended Dimensions of Field Vanes[a] (Annual Book of ASTM Standards, Vol. 04.08. Copyright ASTM INTERNATIONAL. Reprinted with permission.)

Casing size	Diameter, D mm	Height, H mm	Thickness of blade mm	Diameter of rod mm
AX	38.1	76.2	1.6	12.7
BX	50.8	101.6	1.6	12.7
NX	63.5	127.0	3.2	12.7
101.6 mm[b]	92.1	184.1	3.2	12.7

[a]The selection of a vane size is directly related to the consistency of the soil being tested; that is, the softer the soil, the larger the vane diameter should be.
[b]Inside diameter.

where

D = diameter of vane in cm
H = measured height of vane in cm

If $H/D = 2$, Eq. (2.32a) yields

$$K = 366 \times 10^{-8} D^3 \qquad (2.32b)$$
$$\uparrow$$
$$(\text{cm})$$

In English units, if c_u and T in Eq. (2.31) are expressed in lb/ft^2 and lb-ft, respectively, then

$$K = \left(\frac{\pi}{1728}\right)\left(\frac{D^2 H}{2}\right)\left(1 + \frac{D}{3H}\right) \qquad (2.33a)$$

If $H/D = 2$, Eq. (2.33a) yields

$$K = 0.0021 D^3 \qquad (2.33b)$$
$$\uparrow$$
$$(\text{in.})$$

Field vane shear tests are moderately rapid and economical and are used extensively in field soil-exploration programs. The test gives good results in soft and medium-stiff clays and gives excellent results in determining the properties of sensitive clays.

Sources of significant error in the field vane shear test are poor calibration of torque measurement and damaged vanes. Other errors may be introduced if the rate of rotation of the vane is not properly controlled.

For actual design purposes, the undrained shear strength values obtained from field vane shear tests $[c_{u(\text{VST})}]$ are too high, and it is recommended that they be corrected according to the equation

$$c_{u(\text{corrected})} = \lambda c_{u(\text{VST})} \qquad (2.34)$$

where λ = correction factor.

Several correlations have been given previously for the correction factor λ. The most commonly used correlation for λ is that given by Bjerrum (1972), which can be expressed as

$$\lambda = 1.7 - 0.54 \log[\text{PI}(\%)] \tag{2.35a}$$

Morris and Williams (1994) provided the following correlations:

$$\lambda = 1.18e^{-0.08(\text{PI})} + 0.57 \text{ (for PI} > 5) \tag{2.35b}$$

$$\lambda = 7.01e^{-0.08(\text{LL})} + 0.57 \text{ (where LL is in } \%) \tag{2.35c}$$

The field vane shear strength can be correlated with the preconsolidation pressure and the overconsolidation ratio of the clay. Using 343 data points, Mayne and Mitchell (1988) derived the following empirical relationship for estimating the preconsolidation pressure of a natural clay deposit:

$$\sigma_c' = 7.04[c_{u(\text{field})}]^{0.83} \tag{2.36}$$

Here,

$$\sigma_c' = \text{preconsolidation pressure (kN/m}^2)$$
$$c_{u(\text{field})} = \text{field vane shear strength (kN/m}^2)$$

The overconsolidation ratio, OCR, also can be correlated to $c_{u(\text{field})}$ according to the equation

$$\text{OCR} = \beta \frac{c_{u(\text{field})}}{\sigma_o'} \tag{2.37}$$

where $\sigma_o' = $ effective overburden pressure.

The magnitudes of β developed by various investigators are given below.

• Mayne and Mitchell (1988):

$$\beta = 22[\text{PI}(\%)]^{-0.48} \tag{2.38}$$

• Hansbo (1957):

$$\beta = \frac{222}{w(\%)} \tag{2.39}$$

• Larsson (1980):

$$\beta = \frac{1}{0.08 + 0.0055(\text{PI})} \tag{2.40}$$

2.21   *Cone Penetration Test*

The cone penetration test (CPT), originally known as the Dutch cone penetration test, is a versatile sounding method that can be used to determine the materials in a soil profile and estimate their engineering properties. The test is also called the *static penetration test,* and no boreholes are necessary to perform it. In the original version, a 60° cone with a base area of 10 cm² was pushed into the ground at a steady rate of about 20 mm/sec, and the resistance to penetration (called the point resistance) was measured.

The cone penetrometers in use at present measure (a) the *cone resistance* (q_c) to penetration developed by the cone, which is equal to the vertical force applied to the cone, divided by its horizontally projected area; and (b) the *frictional resistance* (f_c), which is the resistance measured by a sleeve located above the cone with the local soil surrounding it. The frictional resistance is equal to the vertical force applied to the sleeve, divided by its surface area—actually, the sum of friction and adhesion.

Generally, two types of penetrometers are used to measure q_c and f_c:

1. *Mechanical friction-cone penetrometer* (Figure 2.23). The tip of this penetrometer is connected to an inner set of rods. The tip is first advanced about 40 mm, giving the

Collapsed   Extended

Figure 2.23 Mechanical friction-cone penetrometer (After ASTM, 2001) (Annual Book of ASTM Standards, Vol. 04.08. Copyright ASTM INTERNATIONAL. Reprinted with permission.)

cone resistance. With further thrusting, the tip engages the friction sleeve. As the inner rod advances, the rod force is equal to the sum of the vertical force on the cone and sleeve. Subtracting the force on the cone gives the side resistance.

2. *Electric friction-cone penetrometer* (Figure 2.24). The tip of this penetrometer is attached to a string of steel rods. The tip is pushed into the ground at the rate of 20 mm/sec. Wires from the transducers are threaded through the center of the rods and continuously measure the cone and side resistances. Figure 2.25 shows a photograph of an electric friction-cone penetrometer.

1 Conical point (10 cm^2)
2 Load cell
3 Strain gauges
4 Friction sleeve (150 cm^2)
5 Adjustment ring
6 Waterproof bushing
7 Cable
8 Connection with rods

Figure 2.24 Electric friction-cone penetrometer (After ASTM, 2001) (Annual Book of ASTM Standards, Vol. 04.08. Copyright ASTM INTERNATIONAL. Reprinted with permission.)

Figure 2.25 Photograph of an electric friction-cone penetrometer (*Courtesy of Sanjeev Kumar, Southern Illinois University, Carbondale, Illinois*)

Figure 2.26 shows the sequence of a cone penetration test in the field. A truck-mounted CPT rig is shown in Figure 2.26a. A hydraulic ram located inside the truck pushes the cone into the ground. Figure 2.26b shows the cone penetrometer in the truck being put in the proper location. Figure 2.26c shows the progress of the CPT.

(a)

(a)

Figure 2.26 Cone penetration test in field: (a) mounted CPT rig; (b) cone penetrometer being set in proper location (*Courtesy of Sanjeev Kumar, Southern Illinois University, Carbondale, Illinois*)

Figure 2.26 (*continued*) (c) test in progress (*Courtesy of Sanjeev Kumar, Southern Illinois University, Carbondale, Illinois*)

Figure 2.27 shows the results of penetrometer test in a soil profile with friction measurement by an electric friction-cone penetrometer.

Several correlations that are useful in estimating the properties of soils encountered during an exploration program have been developed for the point resistance (q_c) and the friction ratio (F_r) obtained from the cone penetration tests. The friction ratio is defined as

$$F_r = \frac{\text{frictional resistance}}{\text{cone resistance}} = \frac{f_c}{q_c} \qquad (2.41)$$

In a more recent study on several soils in Greece, Anagnostopoulos et al. (2003) expressed F_r as

$$F_r(\%) = 1.45 - 1.36 \log D_{50} \text{ (electric cone)} \qquad (2.42)$$

and

$$F_r(\%) = 0.7811 - 1.611 \log D_{50} \text{ (mechanical cone)} \qquad (2.43)$$

where D_{50} = size through which 50% of soil will pass through (mm).

The D_{50} for soils based on which Eqs. (2.42) and (2.43) have been developed ranged from 0.001 mm to about 10 mm.

As in the case of standard penetration tests, several correlations have been developed between q_c and other soil properties. Some of these correlations are presented next.

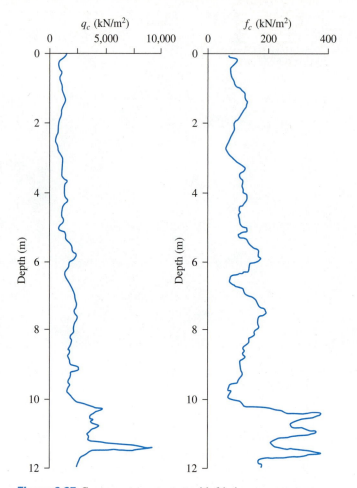

Figure 2.27 Cone penetrometer test with friction measurement

Correlation between Relative Density (D_r) and q_c for Sand

Lancellotta (1983) and Jamiolkowski et al. (1985) showed that the relative density of *normally consolidated sand, D_r,* and q_c can be correlated according to the formula (Figure 2.28).

$$D_r(\%) = A + B \log_{10}\left(\frac{q_c}{\sqrt{\sigma_o'}}\right) \qquad (2.44)$$

The preceding relationship can be rewritten as (Kulhawy and Mayne, 1990)

$$D_r(\%) = 68\left[\log\left(\frac{q_c}{\sqrt{p_a \cdot \sigma_0'}}\right) - 1\right] \qquad (2.45)$$

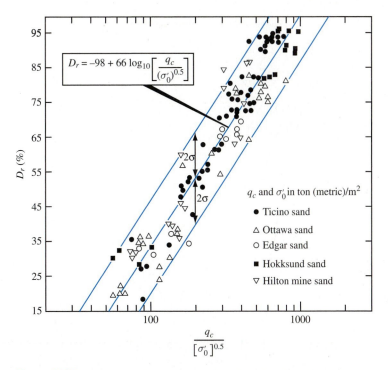

Figure 2.28 Relationship between D_r and q_c (Based on Lancellotta, 1983, and Jamiolski et al., 1985)

where

p_a = atmospheric pressure ($\approx$ 100 kN/m²)
σ'_o = vertical effective stress

Baldi et al. (1982), and Robertson and Campanella (1983) recommended the empirical relationship shown in Figure 2.29 between vertical effective stress (σ'_o), relative density (D_r), and q_c for *normally consolidated sand.*

Kulhawy and Mayne (1990) proposed the following relationship to correlate D_r, q_c, and the vertical effective stress σ'_o:

$$D_r = \sqrt{\left[\frac{1}{305Q_c\text{OCR}^{1.8}}\right]\left[\frac{\dfrac{q_c}{p_a}}{\left(\dfrac{\sigma'_o}{p_a}\right)^{0.5}}\right]} \qquad (2.46)$$

In this equation,

OCR = overconsolidation ratio
p_a = atmospheric pressure
Q_c = compressibility factor

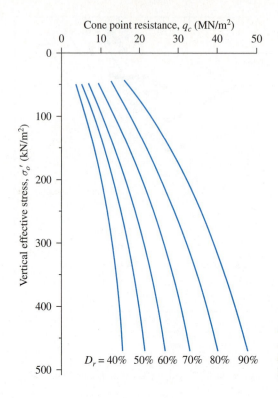

Figure 2.29 Variation of q_c, σ'_o, and D_r for normally consolidated quartz sand (Based on Baldi et al., 1982, and Robertson and Campanella, 1983)

The recommended values of Q_c are as follows:

Highly compressible sand = 0.91
Moderately compressible sand = 1.0
Low compressible sand = 1.09

Correlation between q_c and Drained Friction Angle (ϕ') for Sand

On the basis of experimental results, Robertson and Campanella (1983) suggested the variation of D_r, σ'_o, and ϕ' for normally consolidated quartz sand. This relationship can be expressed as (Kulhawy and Mayne, 1990)

$$\phi' = \tan^{-1}\left[0.1 + 0.38 \log\left(\frac{q_c}{\sigma'_o}\right) \right] \tag{2.47}$$

Based on the cone penetration tests on the soils in the Venice Lagoon (Italy), Ricceri et al. (2002) proposed a similar relationship for soil with classifications of ML and SP-SM as

$$\phi' = \tan^{-1}\left[0.38 + 0.27 \log\left(\frac{q_c}{\sigma'_o}\right) \right] \tag{2.48}$$

In a more recent study, Lee et al. (2004) developed a correlation between ϕ', q_c, and the horizontal effective stress (σ'_h) in the form

$$\phi' = 15.575\left(\frac{q_c}{\sigma'_h}\right)^{0.1714} \tag{2.49}$$

Figure 2.30 General range of variation of q_c/N_{60} for various types of soil

Correlation between q_c and N_{60}

Figure 2.30 shows a plot of q_c $(kN/m^2)/N_{60}$ (N_{60} = standard penetration resistance) against the mean grain size (D_{50} in mm) for various types of soil. This was developed from field test results by Robertson and Campanella (1983).

Anagnostopoulos et al. (2003) provided a similar relationship correlating q_c, N_{60}, and D_{50}. Or

$$\frac{\left(\dfrac{q_c}{p_a}\right)}{N_{60}} = 7.6429 D_{50}^{0.26} \tag{2.50}$$

where p_a = atmospheric pressure (same unit as q_c).

Correlations of Soil Types

Robertson and Campanella (1983) provided the correlations shown in Figure 2.31 between q_c and the friction ratio [Eq. (2.41)] to identify various types of soil encountered in the field.

Correlations for Undrained Shear Strength (c_u), Preconsolidation Pressure (σ'_c), and Overconsolidation Ratio (OCR) for Clays

The undrained shear strength, c_u, can be expressed as

$$c_u = \frac{q_c - \sigma_o}{N_K} \tag{2.51}$$

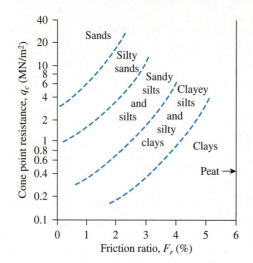

Figure 2.31 Robertson and Campanella's correlation (1983) between q_c, F_r, and the type of soil (Robertson and Campanella, 1983)

where

σ_o = total vertical stress
N_K = bearing capacity factor

The bearing capacity factor, N_K, may vary from 11 to 19 for normally consolidated clays and may approach 25 for overconsolidated clay. According to Mayne and Kemper (1988)

$$N_K = 15 \ (\text{for electric cone})$$

and

$$N_K = 20 \ (\text{for mechanical cone})$$

Based on tests in Greece, Anagnostopoulos et al. (2003) determined

$$N_K = 17.2 \ (\text{for electric cone})$$

and

$$N_K = 18.9 \ (\text{for mechanical cone})$$

These field tests also showed that

$$c_u = \frac{f_c}{1.26} \ (\text{for mechanical cones}) \tag{2.52}$$

and

$$c_u = f_c \ (\text{for electrical cones}) \tag{2.53}$$

Mayne and Kemper (1988) provided correlations for preconsolidation pressure (σ_c') and overconsolidation ratio (OCR) as

$$\sigma_c' = 0.243 (q_c)^{0.96}$$
$$\uparrow \qquad\qquad \uparrow$$
$$\text{MN/m}^2 \qquad \text{MN/m}^2 \tag{2.54}$$

and

$$\text{OCR} = 0.37\left(\frac{q_c - \sigma_o}{\sigma_o'}\right)^{1.01} \tag{2.55}$$

where σ_o and σ_o' = total and effective stress, respectively.

2.22   *Pressuremeter Test (PMT)*

The pressuremeter test is an *in situ* test conducted in a borehole. It was originally developed by Menard (1956) to measure the strength and deformability of soil. It has also been adopted by ASTM as Test Designation 4719. The Menard-type PMT consists essentially of a probe with three cells. The top and bottom ones are *guard cells* and the middle one is the *measuring cell,* as shown schematically in Figure 2.32a. The test is conducted in a prebored hole with a diameter that is between 1.03 and 1.2 times the nominal diameter of the probe. The probe that is most commonly used has a diameter of 58 mm and a length of 420 mm. The probe cells can be expanded by either liquid or gas. The guard cells are expanded to reduce the end-condition effect on the measuring cell, which has a volume (V_o) of 535 cm³. Following are the dimensions for the probe diameter and the diameter of the borehole, as recommended by ASTM:

Probe diameter (mm)	Borehole diameter	
	Nominal (mm)	Maximum (mm)
44	45	53
58	60	70
74	76	89

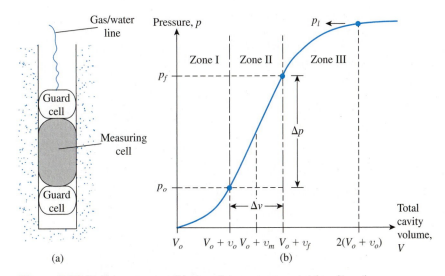

Figure 2.32 (a) Pressuremeter; (b) plot of pressure versus total cavity volume

In order to conduct a test, the measuring cell volume, V_o, is measured and the probe is inserted into the borehole. Pressure is applied in increments and the new volume of the cell is measured. The process is continued until the soil fails or until the pressure limit of the device is reached. The soil is considered to have failed when the total volume of the expanded cavity (V) is about twice the volume of the original cavity. After the completion of the test, the probe is deflated and advanced for testing at another depth.

The results of the pressuremeter test are expressed in the graphical form of pressure versus volume, as shown in Figure 2.32b. In the figure, Zone I represents the reloading portion during which the soil around the borehole is pushed back into the initial state (i.e., the state it was in before drilling). The pressure p_o represents the *in situ* total horizontal stress. Zone II represents a pseudoelastic zone in which the cell volume versus cell pressure is practically linear. The pressure p_f represents the creep, or yield, pressure. The zone marked III is the plastic zone. The pressure p_l represents the limit pressure. Figure 2.33 shows some photographs for a pressuremeter test in the field.

The pressuremeter modulus, E_p, of the soil is determined with the use of the theory of expansion of an infinitely thick cylinder. Thus,

$$E_p = 2(1 + \mu_s)(V_o + v_m)\left(\frac{\Delta p}{\Delta v}\right) \tag{2.56}$$

where

$$v_m = \frac{v_o + v_f}{2}$$

$$\Delta p = p_f - p_o$$

$$\Delta v = v_f - v_o$$

μ_s = Poisson's ratio (which may be assumed to be 0.33)

The limit pressure p_l is usually obtained by extrapolation and not by direct measurement.

In order to overcome the difficulty of preparing the borehole to the proper size, self-boring pressuremeters (SBPMTs) have also been developed. The details concerning SBPMTs can be found in the work of Baguelin et al. (1978).

Correlations between various soil parameters and the results obtained from the pressuremeter tests have been developed by various investigators. Kulhawy and Mayne (1990) proposed that, for clays,

$$\sigma'_c = 0.45 p_l \tag{2.57}$$

where σ'_c = preconsolidation pressure.

On the basis of the cavity expansion theory, Baguelin et al. (1978) proposed that

$$c_u = \frac{(p_l - p_o)}{N_p} \tag{2.58}$$

Figure 2.33 Pressuremeter test in the field: (a) the pressuremeter probe; (b) drilling the bore hole by wet rotary method; (c) pressuremeter control unit with probe in the background; (d) getting ready to insert the pressuremeter probe into the bore hole (*Courtesy of Jean-Louis Briaud, Texas A&M University, College Station, Texas*)

where

c_u = undrained shear strength of a clay

$$N_p = 1 + \ln\left(\frac{E_p}{3c_u}\right)$$

Typical values of N_p vary between 5 and 12, with an average of about 8.5. Ohya et al. (1982) (see also Kulhawy and Mayne, 1990) correlated E_p with field standard penetration numbers (N_{60}) for sand and clay as follows:

$$\text{Clay: } E_p(\text{kN/m}^2) = 1930N_{60}^{0.63} \tag{2.59}$$

$$\text{Sand: } E_p(\text{kN/m}^2) = 908N_{60}^{0.66} \tag{2.60}$$

2.23   *Dilatometer Test*

The use of the flat-plate dilatometer test (DMT) is relatively recent (Marchetti, 1980; Schmertmann, 1986). The equipment essentially consists of a flat plate measuring 220 mm (length) $\times$ 95 mm (width) $\times$ 14 mm (thickness). A thin, flat, circular, expandable steel membrane having a diameter of 60 mm is located flush at the center on one side of the plate (Figure 2.34a). Figure 2.35 shows two flat-plate dilatometers with other instruments for conducting a test in the field. The dilatometer probe is inserted into the ground with a cone penetrometer testing rig (Figure 2.34b). Gas and electric lines extend from the surface control box, through the penetrometer rod, and into the blade. At the required depth, high-pressure nitrogen gas is used to inflate the membrane. Two pressure readings are taken:

1. The pressure A required to "lift off" the membrane.
2. The pressure B at which the membrane expands 1.1 mm into the surrounding soil

Figure 2.34 (a) Schematic diagram of a flat-plate dilatometer; (b) dilatometer probe inserted into ground

Figure 2.35 Dilatometer and other equipment (*Courtesy of N. Sivakugan, James Cook University, Australia*)

The A and B readings are corrected as follows (Schmertmann, 1986):

$$\text{Contact stress, } p_o = 1.05(A + \Delta A - Z_m) - 0.05(B - \Delta B - Z_m) \qquad (2.61)$$
$$\text{Expansion stress, } p_1 = B - Z_m - \Delta B \qquad (2.62)$$

where

ΔA = vacuum pressure required to keep the membrane in contact with its seating
ΔB = air pressure required inside the membrane to deflect it outward to a center expansion of 1.1 mm
Z_m = gauge pressure deviation from zero when vented to atmospheric pressure

The test is normally conducted at depths 200 to 300 mm apart. The result of a given test is used to determine three parameters:

1. Material index, $I_D = \dfrac{p_1 - p_o}{p_o - u_o}$

2. Horizontal stress index, $K_D = \dfrac{p_o - u_o}{\sigma_o'}$

3. Dilatometer modulus, $E_D(\text{kN/m}^2) = 34.7(p_1 \, \text{kN/m}^2 - p_o \, \text{kN/m}^2)$

where

u_o = pore water pressure

σ'_o = *in situ* vertical effective stress

Figure 2.36 shows the results of a dilatometer test conducted in Bangkok soft clay and reported by Shibuya and Hanh (2001). Based on his initial tests, Marchetti (1980) provided the following correlations.

$$K_o = \left(\frac{K_D}{1.5}\right)^{0.47} - 0.6 \tag{2.63}$$

$$OCR = (0.5K_D)^{1.56} \tag{2.64}$$

$$\frac{c_u}{\sigma'_o} = 0.22 \qquad \text{(for normally consolidated clay)} \tag{2.65}$$

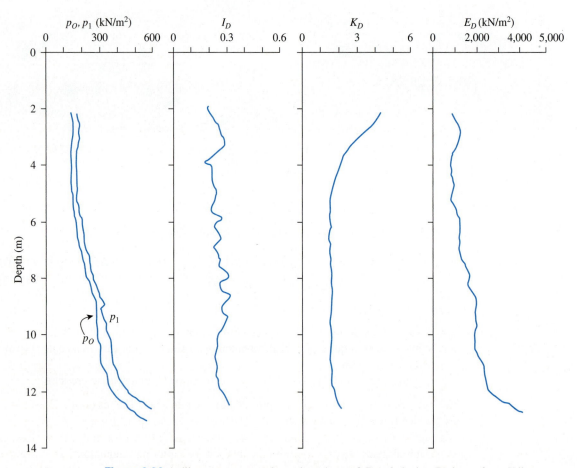

Figure 2.36 A dilatometer test result conducted on soft Bangkok clay (Redrawn from Shibuya and Hanh, 2001)

$$\left(\frac{c_u}{\sigma'_o}\right)_{\text{OC}} = \left(\frac{c_u}{\sigma'_o}\right)_{\text{NC}} (0.5K_D)^{1.25} \tag{2.66}$$

$$E_s = (1 - \mu_s^2)E_D \tag{2.67}$$

where

K_o = coefficient of at-rest earth pressure
OCR = overconsolidation ratio
OC = overconsolidated soil
NC = normally consolidated soil
E_s = modulus of elasticity

Other relevant correlations using the results of dilatometer tests are as follows:

- For undrained cohesion in clay (Kamei and Iwasaki, 1995):

$$c_u = 0.35\,\sigma'_0\,(0.47K_D)^{1.14} \tag{2.68}$$

- For soil friction angle (ML and SP-SM soils) (Ricceri et al., 2002):

$$\phi' = 31 + \frac{K_D}{0.236 + 0.066K_D} \tag{2.69a}$$

$$\phi'_{\text{ult}} = 28 + 14.6\log K_D - 2.1(\log K_D)^2 \tag{2.69b}$$

Schmertmann (1986) also provided a correlation between the material index (I_D) and the dilatometer modulus (E_D) for a determination of the nature of the soil and its unit weight (γ). This relationship is shown in Figure 2.37.

2.24 *Coring of Rocks*

When a rock layer is encountered during a drilling operation, rock coring may be necessary. To core rocks, a *core barrel* is attached to a drilling rod. A *coring bit* is attached to the bottom of the barrel (Fig. 2.38). The cutting elements may be diamond, tungsten, carbide, and so on. Table 2.10 summarizes the various types of core barrel and their sizes, as well as the compatible drill rods commonly used for exploring foundations. The coring is advanced by rotary drilling. Water is circulated through the drilling rod during coring, and the cutting is washed out.

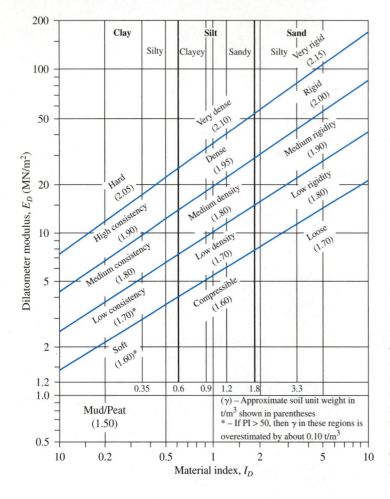

Figure 2.37 Chart for determination of soil description and unit weight (After Schmertmann, 1986) (*Note:* 1 t/m^3 = 9.81 kN/m^3) (Schmertmann, J. H. (1986). "Suggested method for performing the flat dilatometer test," Geotechnical Testing Journal, ASTM, Vol. 9, No. 2, pp. 93-101, Fig. 2. Copyright ASTM INTERNATIONAL. Reprinted with permission.)

Table 2.10 Standard Size and Designation of Casing, Core Barrel, and Compatible Drill Rod

Casing and core barrel designation	Outside diameter of core barrel bit (mm)	Drill rod designation	Outside diameter of drill rod (mm)	Diameter of borehole (mm)	Diameter of core sample (mm)
EX	36.51	E	33.34	38.1	22.23
AX	47.63	A	41.28	50.8	28.58
BX	58.74	B	47.63	63.5	41.28
NX	74.61	N	60.33	76.2	53.98

Two types of core barrel are available: the *single-tube core barrel* (Figure 2.38a) and the *double-tube core barrel* (Figure 2.38b). Rock cores obtained by single-tube core barrels can be highly disturbed and fractured because of torsion. Rock cores smaller than the BX size tend to fracture during the coring process. Figure 2.39 shows

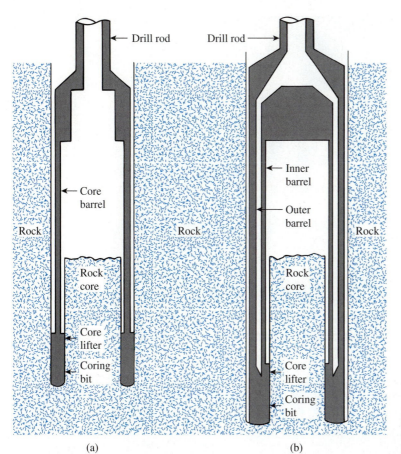

(a)   (b)

Figure 2.38 Rock coring: (a) single-tube core barrel; (b) double-tube core barrel

Figure 2.39 Diamond coring bit (*Courtesy of Braja M. Das, Henderson, NV*)

(a)

(b)

Figure 2.40 Diamond coring bit attached to a double-tube core barrel: (a) end view; (b) side view (*Courtesy of Professional Service Industries, Inc. (PSI), Waukesha, Wisconsin*)

the photograph of a diamond coring bit. Figure 2.40 shows the end and side views of a diamond coring bit attached to a double-tube core barrel.

When the core samples are recovered, the depth of recovery should be properly recorded for further evaluation in the laboratory. Based on the length of the rock core recovered from each run, the following quantities may be calculated for a general evaluation of the rock quality encountered:

$$\text{Recovery ratio} = \frac{\text{length of core recovered}}{\text{theoretical length of rock cored}} \quad (2.70)$$

Rock quality designation (RQD)

$$= \frac{\Sigma \text{ length of recovered pieces equal to or larger than 101.6 mm}}{\text{theoretical length of rock cored}} \quad (2.71)$$

A recovery ratio of unity indicates the presence of intact rock; for highly fractured rocks, the recovery ratio may be 0.5 or smaller. Table 2.11 presents the general relationship (Deere, 1963) between the RQD and the *in situ* rock quality.

Table 2.11 Relation between *in situ* Rock Quality and RQD

RQD	Rock quality
0–0.25	Very poor
0.25–0.5	Poor
0.5–0.75	Fair
0.75–0.9	Good
0.9–1	Excellent

2.25 *Preparation of Boring Logs*

The detailed information gathered from each borehole is presented in a graphical form called the *boring log*. As a borehole is advanced downward, the driller generally should record the following information in a standard log:

1. Name and address of the drilling company
2. Driller's name
3. Job description and number
4. Number, type, and location of boring
5. Date of boring
6. Subsurface stratification, which can be obtained by visual observation of the soil brought out by auger, split-spoon sampler, and thin-walled Shelby tube sampler
7. Elevation of water table and date observed, use of casing and mud losses, and so on
8. Standard penetration resistance and the depth of SPT
9. Number, type, and depth of soil sample collected
10. In case of rock coring, type of core barrel used and, for each run, the actual length of coring, length of core recovery, and RQD

This information should never be left to memory, because doing so often results in erroneous boring logs.

Boring Log

Name of the Project Two-story apartment building

Location Johnson & Olive St.   Date of Boring March 2, 2005

Boring No. 3   Type of Hollow-stem auger   Ground   60.8 m
                Boring   Elevation

Soil description	Depth (m)	Soil sample type and number	N_{60}	w_n (%)	Comments
Light brown clay (fill)					
Silty sand (SM)	1 — 2	SS-1	9	8.2	
°G.W.T. ▼ 3.5 m	3 — 4	SS-2	12	17.6	LL = 38 PI = 11
Light gray clayey silt (ML)	5	ST-1		20.4	LL = 36 q_u = 112 kN/m²
	6	SS-3	11	20.6	
Sand with some gravel (SP)	7				
End of boring @ 8 m	8	SS-4	27	9	

N_{60} = standard penetration number
w_n = natural moisture content
LL = liquid limit; PI = plasticity index
q_u = unconfined compression strength
SS = split-spoon sample; ST = Shelby tube sample

Groundwater table
observed after one
week of drilling

Figure 2.41 A typical boring log

After completion of the necessary laboratory tests, the geotechnical engineer pre-
pares a finished log that includes notes from the driller's field log and the results of tests
conducted in the laboratory. Figure 2.41 shows a typical boring log. These logs have to be
attached to the final soil-exploration report submitted to the client. The figure also lists the
classifications of the soils in the left-hand column, along with the description of each soil
(based on the Unified Soil Classification System).

2.26 *Geophysical Exploration*

Several types of geophysical exploration techniques permit a rapid evaluation of subsoil
characteristics. These methods also allow rapid coverage of large areas and are less expen-
sive than conventional exploration by drilling. However, in many cases, definitive interpretation of the results is difficult. For that reason, such techniques should be used for
preliminary work only. Here, we discuss three types of geophysical exploration technique:
the seismic refraction survey, cross-hole seismic survey, and resistivity survey.

Seismic Refraction Survey

Seismic refraction surveys are useful in obtaining preliminary information about the
thickness of the layering of various soils and the depth to rock or hard soil at a site.

Refraction surveys are conducted by impacting the surface, such as at point *A* in Figure 2.42a, and observing the first arrival of the disturbance (stress waves) at several other points (e.g., *B, C, D, . . .*). The impact can be created by a hammer blow or by a small explosive charge. The first arrival of disturbance waves at various points can be recorded by geophones.

The impact on the ground surface creates two types of *stress wave: P waves* (or *plane waves*) and *S waves* (or *shear waves*). *P* waves travel faster than *S* waves; hence, the first arrival of disturbance waves will be related to the velocities of the *P* waves in various layers. The velocity of *P* waves in a medium is

$$ v = \sqrt{\frac{E_s}{\left(\dfrac{\gamma}{g}\right)} \frac{(1 - \mu_s)}{(1 - 2\mu_s)(1 + \mu_s)}} \tag{2.72} $$

where

E_s = modulus of elasticity of the medium
γ = unit weight of the medium
g = acceleration due to gravity
μ_s = Poisson's ratio

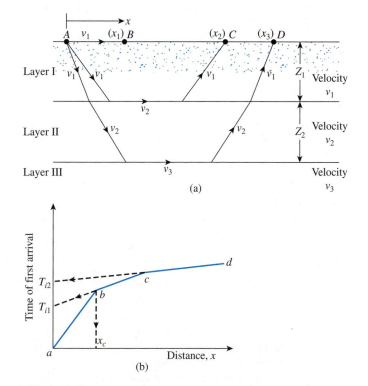

(a)

(b)

Figure 2.42 Seismic refraction survey

To determine the velocity v of P waves in various layers and the thicknesses of those layers, we use the following procedure:

Step 1. Obtain the times of first arrival, $t_1, t_2, t_3, \ldots$, at various distances $x_1, x_2, x_3, \ldots$ from the point of impact.

Step 2. Plot a graph of time t against distance x. The graph will look like the one shown in Figure 2.42b.

Step 3. Determine the slopes of the lines *ab, bc, cd, . . .* :

$$\text{Slope of } ab = \frac{1}{v_1}$$

$$\text{Slope of } bc = \frac{1}{v_2}$$

$$\text{Slope of } cd = \frac{1}{v_3}$$

Here, $v_1, v_2, v_3, \ldots$ are the P-wave velocities in layers I, II, III, $\ldots$, respectively (Figure 2.42a).

Step 4. Determine the thickness of the top layer:

$$Z_1 = \frac{1}{2}\sqrt{\frac{v_2 - v_1}{v_2 + v_1}}\, x_c \tag{2.73}$$

The value of x_c can be obtained from the plot, as shown in Figure 2.42b.

Step 5. Determine the thickness of the second layer:

$$Z_2 = \frac{1}{2}\left[T_{i2} - 2Z_1\frac{\sqrt{v_3^2 - v_1^2}}{v_3 v_1} \right]\frac{v_3 v_2}{\sqrt{v_3^2 - v_2^2}} \tag{2.74}$$

Here, T_{i2} is the time intercept of the line *cd* in Figure 2.42b, extended backwards.

(For detailed derivatives of these equations and other related information, see Dobrin, 1960, and Das, 1992).

The velocities of P waves in various layers indicate the types of soil or rock that are present below the ground surface. The range of the P-wave velocity that is generally encountered in different types of soil and rock at shallow depths is given in Table 2.12.

In analyzing the results of a refraction survey, two limitations need to be kept in mind:

1. The basic equations for the survey—that is, Eqs. (2.73) and (2.74)—are based on the assumption that the P-wave velocity $v_1 < v_2 < v_3 < \cdots$.
2. When a soil is saturated below the water table, the P-wave velocity may be deceptive. P waves can travel with a velocity of about 1500 m/sec through water. For dry, loose soils, the velocity may be well below 1500 m/sec. However, in a saturated condition, the waves will travel through water that is present in the void spaces with a velocity of about 1500 m/sec. If the presence of groundwater has not been detected, the P-wave velocity may be erroneously interpreted to indicate a stronger material (e.g., sandstone) than is actually present *in situ*. In general, geophysical interpretations should always be verified by the results obtained from borings.

Table 2.12 Range of *P*-Wave Velocity in Various Soils and Rocks

Type of soil or rock	P-wave velocity m/sec
Soil	
Sand, dry silt, and fine-grained topsoil	200–1000
Alluvium	500–2000
Compacted clays, clayey gravel, and dense clayey sand	1000–2500
Loess	250–750
Rock	
Slate and shale	2500–5000
Sandstone	1500–5000
Granite	4000–6000
Sound limestone	5000–10,000

Example 2.1

The results of a refraction survey at a site are given in the following table:

Distance of geophone from the source of disturbance (m)	Time of first arrival (sec × 10^3)
2.5	11.2
5	23.3
7.5	33.5
10	42.4
15	50.9
20	57.2
25	64.4
30	68.6
35	71.1
40	72.1
50	75.5

Determine the *P*-wave velocities and the thickness of the material encountered.

Solution

Velocity
In Figure 2.43, the times of first arrival of the *P* waves are plotted against the distance of the geophone from the source of disturbance. The plot has three straight-line segments. The velocity of the top three layers can now be calculated as follows:

$$\text{Slope of segment } 0a = \frac{1}{v_1} = \frac{\text{time}}{\text{distance}} = \frac{23 \times 10^{-3}}{5.25}$$

Figure 2.43 Plot of first arrival time of *P* wave versus distance of geophone from source of disturbance

or

$$v_1 = \frac{5.25 \times 10^3}{23} = \textbf{228 m/sec (top layer)}$$

$$\text{Slope of segment } ab = \frac{1}{v_2} = \frac{13.5 \times 10^{-3}}{11}$$

or

$$v_2 = \frac{11 \times 10^3}{13.5} = \textbf{814.8 m/sec (middle layer)}$$

$$\text{Slope of segment } bc = \frac{1}{v_3} = \frac{3.5 \times 10^{-3}}{14.75}$$

or

$$v_3 = \textbf{4214 m/sec (third layer)}$$

Comparing the velocities obtained here with those given in Table 2.12 indicates that the third layer is a *rock layer.*

Thickness of Layers
From Figure 2.43, $x_c = 10.5$ m, so

$$Z_1 = \frac{1}{2}\sqrt{\frac{v_2 - v_1}{v_2 + v_1}}x_c$$

Thus,

$$Z_1 = \frac{1}{2}\sqrt{\frac{814.8 - 228}{814.8 + 228}} \times 10.5 = \textbf{3.94 m}$$

Again, from Eq. (2.74)

$$Z_2 = \frac{1}{2}\left[T_{i2} - \frac{2Z_1\sqrt{v_3^2 - v_1^2}}{(v_3 v_1)} \right] \frac{(v_3)(v_2)}{\sqrt{v_3^2 - v_2^2}}$$

The value of T_{i2} (from Figure 2.43) is 65×10^{-3} sec. Hence,

$$Z_2 = \frac{1}{2}\left[65 \times 10^{-3} - \frac{2(3.94)\sqrt{(4214)^2 - (228)^2}}{(4214)(228)} \right] \frac{(4214)(814.8)}{\sqrt{(4214)^2 - (814.8)^2}}$$

$$= \frac{1}{2}(0.065 - 0.0345)830.47 = \textbf{12.66 m}$$

Thus, the rock layer lies at a depth of $Z_1 + Z_2 = 3.94 + 12.66 = $ **16.60 m from the surface of the ground**. ∎

Cross-Hole Seismic Survey

The velocity of shear waves created as the result of an impact to a given layer of soil can be effectively determined by the *cross-hole seismic survey* (Stokoe and Woods, 1972). The principle of this technique is illustrated in Figure 2.44, which shows two holes drilled into the ground a distance L apart. A vertical impulse is created at the bottom of one borehole by means of an impulse rod. The shear waves thus generated are recorded by a vertically sensitive transducer. The velocity of shear waves can be calculated as

$$v_s = \frac{L}{t} \tag{2.75}$$

where $t = $ travel time of the waves.

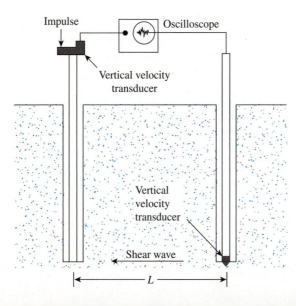

Figure 2.44 Cross-hole method of seismic survey

The shear modulus G_s of the soil at the depth at which the test is taken can be determined from the relation

$$v_s = \sqrt{\frac{G_s}{(\gamma/g)}}$$

or

$$G_s = \frac{v_s^2 \gamma}{g} \qquad (2.76)$$

where

v_s = velocity of shear waves
γ = unit weight of soil
g = acceleration due to gravity

The shear modulus is useful in the design of foundations to support vibrating machinery and the like.

Resistivity Survey

Another geophysical method for subsoil exploration is the *electrical resistivity survey*. The electrical resistivity of any conducting material having a length L and an area of cross section A can be defined as

$$\rho = \frac{RA}{L} \qquad (2.77)$$

where R = electrical resistance.

The unit of resistivity is *ohm-centimeter* or *ohm-meter*. The resistivity of various soils depends primarily on their moisture content and also on the concentration of dissolved ions in them. Saturated clays have a very low resistivity; dry soils and rocks have a high resistivity. The range of resistivity generally encountered in various soils and rocks is given in Table 2.13.

The most common procedure for measuring the electrical resistivity of a soil profile makes use of four electrodes driven into the ground and spaced equally along a straight line. The procedure is generally referred to as the *Wenner method* (Figure 2.45a).

Table 2.13 Representative Values of Resistivity

Material	Resistivity (ohm · m)
Sand	500–1500
Clays, saturated silt	0–100
Clayey sand	200–500
Gravel	1500–4000
Weathered rock	1500–2500
Sound rock	>5000

Figure 2.45 Electrical resistivity survey: (a) Wenner method; (b) empirical method for determining resistivity and thickness of each layer

The two outside electrodes are used to send an electrical current I (usually a dc current with nonpolarizing potential electrodes) into the ground. The current is typically in the range of 50 to 100 milliamperes. The voltage drop, V, is measured between the two inside electrodes. If the soil profile is homogeneous, its electrical resistivity is

$$\rho = \frac{2\pi dV}{I} \tag{2.78}$$

In most cases, the soil profile may consist of various layers with different resistivities, and Eq. (2.78) will yield the *apparent resistivity*. To obtain the *actual resistivity* of various layers and their thicknesses, one may use an empirical method that involves conducting tests at various electrode spacings (i.e., d is changed). The sum of the apparent resistivities, $\Sigma\rho$, is plotted against the spacing d, as shown in Figure 2.45b. The plot thus obtained has relatively straight segments, the slopes of which give the resistivity of individual layers. The thicknesses of various layers can be estimated as shown in Figure 2.45b.

The resistivity survey is particularly useful in locating gravel deposits within a fine-grained soil.

2.27   Subsoil Exploration Report

At the end of all soil exploration programs, the soil and rock specimens collected in the field are subject to visual observation and appropriate laboratory testing. (The basic soil tests were described in Chapter 1.) After all the required information has been compiled, a soil exploration report is prepared for use by the design office and for reference during future construction work. Although the details and sequence of information in such reports may vary to some degree, depending on the structure under consideration and the person compiling the report, each report should include the following items:

1. A description of the scope of the investigation
2. A description of the proposed structure for which the subsoil exploration has been conducted
3. A description of the location of the site, including any structures nearby, drainage conditions, the nature of vegetation on the site and surrounding it, and any other features unique to the site
4. A description of the geological setting of the site
5. Details of the field exploration—that is, number of borings, depths of borings, types of borings involved, and so on
6. A general description of the subsoil conditions, as determined from soil specimens and from related laboratory tests, standard penetration resistance and cone penetration resistance, and so on
7. A description of the water-table conditions
8. Recommendations regarding the foundation, including the type of foundation recommended, the allowable bearing pressure, and any special construction procedure that may be needed; alternative foundation design procedures should also be discussed in this portion of the report
9. Conclusions and limitations of the investigations

The following graphical presentations should be attached to the report:

1. A site location map
2. A plan view of the location of the borings with respect to the proposed structures and those nearby
3. Boring logs
4. Laboratory test results
5. Other special graphical presentations

The exploration reports should be well planned and documented, as they will help in answering questions and solving foundation problems that may arise later during design and construction.

Problems

2.1   For a Shelby tube, given: outside diameter = 76.2 mm and inside diameter 73 mm in. What is the area ratio of the tube?

2.2   A soil profile is shown in Figure P2.2 along with the standard penetration numbers in the clay layer. Use Eqs. (2.8) and (2.9) to determine the variation of c_u and OCR with depth. What is the average value of c_u and OCR?

Figure P2.2

2.3 Following is the variation of the field standard penetration number (N_{60}) in a sand deposit:

Depth (m)	N_{60}
1.5	6
3	8
4.5	9
6	8
7.9	13
9	14

The groundwater table is located at a depth of 6 m. Given: the dry unit weight of sand from 0 to a depth of 6 m is 18 kN/m³, and the saturated unit weight of sand for depth 6 to 12 m is 20.2 kN/m³. Use the relationship of Skempton given in Eq. (2.12) to calculate the corrected penetration numbers.

2.4 For the soil profile described in Problem 2.3, estimate an average peak soil friction angle. Use Eq. (2.28).

2.5 Repeat Problem 2.4 using Eq. (2.27).

2.6 Refer to Problem 2.3. Using Eq. (2.20), determine the average relative density of sand.

2.7 The following table gives the variation of the field standard penetration number (N_{60}) in a sand deposit:

Depth (m)	N_{60}
1.5	5
3.0	11
4.5	14
6.0	18
7.5	16
9.0	21

The groundwater table is located at a depth of 12 m. The dry unit weight of sand from 0 to a depth of 12 m is 17.6 kN/m³. Assume that the mean grain size (D_{50}) of the sand deposit to be about 0.8 mm. Estimate the variation of the relative density with depth for sand. Use Eq. (2.21).

2.8 Following are the standard penetration numbers determined from a sandy soil in the field:

Depth (m)	Unit weight of soil (kN/m³)	N_{60}
3.0	16.66	7
4.5	16.66	9
6.0	16.66	11
7.5	18.55	16
9.0	18.55	18
10.5	18.55	20
12.0	18.55	22

Using Eq. (2.27), determine the variation of the peak soil friction angle, ϕ'. Estimate an average value of ϕ' for the design of a shallow foundation. (*Note:* For depth greater than 6 m, the unit weight of soil is 18.55 kN/m³.)

2.9 Refer to Problem 2.8. Assume that the sand is clean and normally consolidated. Estimate the average value of the modulus of elasticity between depths of 6 m and 9 m.

2.10 Following are the details for a soil deposit in sand:

Depth (m)	Effective overburden pressure (kN/m²)	Field standard penetration number, N_{60}
3.0	55	9
4.5	82	11
6.0	98	12

Assume the uniformity coefficient (C_u) of the sand to be 2.8 and an overconsolidation ratio (OCR) of 2. Estimate the average relative density of the sand between the depth of 3 to 6 m. Use Eq. (2.19).

2.11 Refer to Figure P2.2. Vane shear tests were conducted in the clay layer. The vane dimensions were 63.5 mm $(D) \times 127$ mm (H). For the test at A, the torque required to cause failure was 0.051 N·m. For the clay, given: liquid limit = 46 and plastic limit = 21. Estimate the undrained cohesion of the clay for use in the design by using Bjerrum's λ relationship [Eq. (2.35a)].

2.12 Refer to Problem 2.11. Estimate the overconsolidation ratio of the clay. Use Eqs. (2.37) and (2.38).

2.13 a. A vane shear test was conducted in a saturated clay. The height and diameter of the vane were 101.6 mm and 50.8 mm, respectively. During the test, the maximum torque applied was 23 lb-ft. Determine the undrained shear strength of the clay.

b. The clay soil described in part (a) has a liquid limit of 58 and a plastic limit of 29. What would be the corrected undrained shear strength of the clay for design purposes? Use Bjerrum's relationship for λ [Eq. (2.35a)].

2.14 Refer to Problem 2.13. Determine the overconsolidation ratio for the clay. Use Eqs. (2.37) and (2.40). Use $\sigma_0' = 64.2$ kN/m².

2.15 In a deposit of normally consolidated dry sand, a cone penetration test was conducted. Following are the results:

Depth (m)	Point resistance of cone, q_c (MN/m^2)
1.5	2.06
3.0	4.23
4.5	6.01
6.0	8.18
7.5	9.97
9.0	12.42

Assuming the dry unit weight of sand to be 16 kN/m^3, estimate the average peak friction angle, ϕ', of the sand. Use Eq. (2.48).

2.16 Refer to Problem 2.15. Using Eq. (2.46), determine the variation of the relative density with depth.

2.17 In the soil profile shown in Figure P2.17, if the cone penetration resistance (q_c) at A (as determined by an electric friction-cone penetrometer) is 0.8 MN/m.^{2}, estimate
   a. The undrained cohesion, c_u
   b. The overconsolidation ratio, OCR

2.18 In a pressuremeter test in a soft saturated clay, the measuring cell volume $V_o = 535$ cm^3, $p_o = 42.4$ kN/m^2, $p_f = 326.5$ kN/m^2, $v_o = 46$ cm^3, and $v_f = 180$ cm^3. Assuming Poisson's ratio (μ_s) to be 0.5 and using Figure 2.32, calculate the pressuremeter modulus (E_p).

2.19 A dilatometer test was conducted in a clay deposit. The groundwater table was located at a depth of 3 m below the surface. At a depth of 8 m below the surface, the contact pressure (p_o) was 280 kN/m^2 and the expansion stress (p_1) was 350 kN/m^2. Determine the following:
   a. Coefficient of at-rest earth pressure, K_o
   b. Overconsolidation ratio, OCR
   c. Modulus of elasticity, E_s
   Assume σ'_o at a depth of 8 m to be 95 kN/m^2 and $\mu_s = 0.35$.

Figure P2.17

2.20 A dilatometer test was conducted in a sand deposit at a depth of 6 m. The groundwater table was located at a depth of 2 m below the ground surface. Given, for the sand: $\gamma_d = 14.5$ kN/m^3 and $\gamma_{sat} = 19.8$ kN/m^3. The contact stress during the test was 260 kN/m^2. Estimate the soil friction angle, ϕ'.

2.21 The *P*-wave velocity in a soil is 1900 m/sec. Assuming Poisson's ratio to be 0.32, calculate the modulus of elasticity of the soil. Assume that the unit weight of soil is 18 kN/m^3.

2.22 The results of a refraction survey (Figure 2.42a) at a site are given in the following table. Determine the thickness and the *P*-wave velocity of the materials encountered.

Distance from the source of disturbance (m)	Time of first arrival of *P*-waves (sec $\times$ 10^3)
2.5	5.08
5.0	10.16
7.5	15.24
10.0	17.01
15.0	20.02
20.0	24.2
25.0	27.1
30.0	28.0
40.0	31.1
50.0	33.9

References

AMERICAN SOCIETY FOR TESTING AND MATERIALS (2001). *Annual Book of ASTM Standards,* Vol. 04.08, West Conshohocken, PA.

AMERICAN SOCIETY OF CIVIL ENGINEERS (1972). "Subsurface Investigation for Design and Construction of Foundations of Buildings," *Journal of the Soil Mechanics and Foundations Division,* American Society of Civil Engineers, Vol. 98, No. SM5, pp. 481–490.

ANAGNOSTOPOULOS, A., KOUKIS, G., SABATAKAKIS, N., and TSIAMBAOS, G. (2003). "Empirical Correlations of Soil Parameters Based on Cone Penetration Tests (CPT) for Greek Soils," *Geotechnical and Geological Engineering*, Vol. 21, No. 4, pp. 377–387.

BAGUELIN, F., JÉZÉQUEL, J. F., and SHIELDS, D. H. (1978). *The Pressuremeter and Foundation Engineering,* Trans Tech Publications, Clausthal, Germany.

BALDI, G., BELLOTTI, R., GHIONNA, V., and JAMIOLKOWSKI, M. (1982). "Design Parameters for Sands from CPT," *Proceedings, Second European Symposium on Penetration Testing,* Amsterdam, Vol. 2, pp. 425–438.

BAZARAA, A. (1967). *Use of the Standard Penetration Test for Estimating Settlements of Shallow Foundations on Sand,* Ph.D. Dissertation, Civil Engineering Department, University of Illinois, Champaign-Urbana, Illinois.

BJERRUM, L. (1972). "Embankments on Soft Ground," *Proceedings of the Specialty Conference,* American Society of Civil Engineers, Vol. 2, pp. 1–54.

CRUDEN, D. M., and VARNES, D. J. (1996). "Landslide Types and Processes," *Special Report 247,* Transportation Research Board, pp. 36–75.

CUBRINOVSKI, M., and ISHIHARA, K. (1999). "Empirical Correlations between SPT *N*-Values and Relative Density for Sandy Soils," *Soils and Foundations,* Vol. 39, No. 5, pp. 61–92.

DAS, B. M. (1992). *Principles of Soil Dynamics,* PWS Publishing Company, Boston.

DEERE, D. U. (1963). "Technical Description of Rock Cores for Engineering Purposes," *Felsmechanik und Ingenieurgeologie,* Vol. 1, No. 1, pp. 16–22.

DOBRIN, M. B. (1960). *Introduction to Geophysical Prospecting,* McGraw-Hill, New York.

HANSBO, S. (1957). *A New Approach to the Determination of the Shear Strength of Clay by the Fall Cone Test,* Swedish Geotechnical Institute, Report No. 114.

HARA, A., OHATA, T., and NIWA, M. (1971). "Shear Modulus and Shear Strength of Cohesive Soils," *Soils and Foundations,* Vol. 14, No. 3, pp. 1–12.

HATANAKA, M., and UCHIDA, A. (1996). "Empirical Correlation between Penetration Resistance and Internal Friction Angle of Sandy Soils," *Soils and Foundations,* Vol. 36, No. 4, pp. 1–10.

JAMIOLKOWSKI, M., LADD, C. C., GERMAINE, J. T., and LANCELLOTTA, R. (1985). "New Developments in Field and Laboratory Testing of Soils," *Proceedings, 11th International Conference on Soil Mechanics and Foundation Engineering,* Vol. 1, pp. 57–153.

KAMEI, T., and IWASAKI, K. (1995). "Evaluation of Undrained Shear Strength of Cohesive Soils using a Flat Dilatometer," *Soils and Foundations*, Vol. 35, No. 2, pp. 111–116.

KOLB, C. R., and SHOCKLEY, W. G. (1959). "Mississippi Valley Geology: Its Engineering Significance" *Proceedings*, American Society of Civil Engineers, Vol. 124, pp. 633–656.

KULHAWY, F. H., and MAYNE, P. W. (1990). *Manual on Estimating Soil Properties for Foundation Design,* Electric Power Research Institute, Palo Alto, California.

LANCELLOTTA, R. (1983). *Analisi di Affidabilità in Ingegneria Geotecnica,* Atti Istituto Scienza Construzioni, No. 625, Politecnico di Torino.

LARSSON, R. (1980). "Undrained Shear Strength in Stability Calculation of Embankments and Foundations on Clay," *Canadian Geotechnical Journal,* Vol. 17, pp. 591–602.

LEE, J., SALGADO, R., and CARRARO, A. H. (2004). "Stiffness Degradation and Shear Strength of Silty Sand," *Canadian Geotechnical Journal*, Vol. 41, No. 5, pp. 831–843.

LIAO, S. S. C. AND WHITMAN, R. V. (1986). "Overburden Correction Factors for SPT in Sand," *Journal of Geotechnical Engineering,* American Society of Civil Engineers, Vol. 112, No. 3, pp. 373–377.

MARCHETTI, S. (1980). "*In Situ* Test by Flat Dilatometer," *Journal of Geotechnical Engineering Division,* ASCE, Vol. 106, GT3, pp. 299–321.

MARCUSON, W. F., III, AND BIEGANOUSKY, W. A. (1977). "SPT and Relative Density in Coarse Sands," *Journal of Geotechnical Engineering Division,* American Society of Civil Engineers, Vol. 103, No. 11, pp. 1295–1309.

MAYNE, P. W., AND KEMPER, J. B. (1988). "Profiling OCR in Stiff Clays by CPT and SPT," *Geotechnical Testing Journal,* ASTM, Vol. 11, No. 2, pp. 139–147.

MAYNE, P. W., AND MITCHELL, J. K. (1988). "Profiling of Overconsolidation Ratio in Clays by Field Vane," *Canadian Geotechnical Journal,* Vol. 25, No. 1, pp. 150–158.

MENARD, L. (1956). *An Apparatus for Measuring the Strength of Soils in Place,* master's thesis, University of Illinois, Urbana, Illinois.

MEYERHOF, G. G. (1957). "Discussion on Research on Determining the Density of Sands by Spoon Penetration Testing," *Proceedings, Fourth International Conference on Soil Mechanics and Foundation Engineering*, Vol. 3, p. 110.

MORRIS, P. M., and WILLIAMS, D. T. (1994). "Effective Stress Vane Shear Strength Correction Factor Correlations," *Canadian Geotechnical Journal*, Vol. 31, No. 3, pp. 335–342.

OHYA, S., IMAI, T., AND MATSUBARA, M. (1982). "Relationships between N Value by SPT and LLT Pressuremeter Results," *Proceedings, 2nd European Symposium on Penetration Testing,* Vol. 1, Amsterdam, pp. 125–130.

OSTERBERG, J. O. (1952). "New Piston-Type Soil Sampler," *Engineering News-Record,* April 24.

PECK, R. B., HANSON, W. E., and THORNBURN, T. H. (1974). *Foundation Engineering,* 2d ed., Wiley, New York.

RICCERI, G., SIMONINI, P., and COLA, S. (2002). "Applicability of Piezocone and Dilatometer to Characterize the Soils of the Venice Lagoon" *Geotechnical and Geological Engineering,* Vol. 20, No. 2, pp. 89–121.

ROBERTSON, P. K., and CAMPANELLA, R. G. (1983). "Interpretation of Cone Penetration Tests. Part I: Sand," *Canadian Geotechnical Journal,* Vol. 20, No. 4, pp. 718–733.

SCHMERTMANN, J. H. (1975). "Measurement of *In Situ* Shear Strength," *Proceedings, Specialty Conference on* In Situ *Measurement of Soil Properties,* ASCE, Vol. 2, pp. 57–138.

SCHMERTMANN, J. H. (1986). "Suggested Method for Performing the Flat Dilatometer Test," *Geotechnical Testing Journal,* ASTM, Vol. 9, No. 2, pp. 93–101.

SEED, H. B., ARANGO, I., and CHAN, C. K. (1975). *Evaluation of Soil Liquefaction Potential during Earthquakes,* Report No. EERC 75-28, Earthquake Engineering Research Center, University of California, Berkeley.

SEED, H. B., TOKIMATSU, K., HARDER, L. F., and CHUNG, R. M. (1985). "Influence of SPT Procedures in Soil Liquefaction Resistance Evaluations," *Journal of Geotechnical Engineering,* ASCE, Vol. 111, No. 12, pp. 1425–1445.

SHIBUYA, S., and HANH, L. T. (2001). "Estimating Undrained Shear Strength of Soft Clay Ground Improved by Pre-Loading with PVD—Case History in Bangkok," *Soils and Foundations,* Vol. 41, No. 4, pp. 95–101.

SKEMPTON, A. W. (1986). "Standard Penetration Test Procedures and the Effect in Sands of Overburden Pressure, Relative Density, Particle Size, Aging and Overconsolidation," *Geotechnique,* Vol. 36, No. 3, pp. 425–447.

SOWERS, G. B., and SOWERS, G. F. (1970). *Introductory Soil Mechanics and Foundations,* 3d ed., Macmillan, New York.

STOKOE, K. H., and WOODS, R. D. (1972). "*In Situ* Shear Wave Velocity by Cross-Hole Method," *Journal of Soil Mechanics and Foundations Division,* American Society of Civil Engineers, Vol. 98, No. SM5, pp. 443–460.

SZECHY, K., and VARGA, L. (1978). *Foundation Engineering—Soil Exploration and Spread Foundation,* Akademiai Kiado, Budapest, Hungary.

WOLFF, T. F. (1989). "Pile Capacity Prediction Using Parameter Functions," in *Predicted and Observed Axial Behavior of Piles, Results of a Pile Prediction Symposium,* sponsored by the Geotechnical Engineering Division, ASCE, Evanston, IL, June, 1989, ASCE Geotechnical Special Publication No. 23, pp. 96–106.

3 Shallow Foundations: Ultimate Bearing Capacity

3.1 Introduction

To perform satisfactorily, shallow foundations must have two main characteristics:

1. They have to be safe against overall shear failure in the soil that supports them.
2. They cannot undergo excessive displacement, or settlement. (The term *excessive* is relative, because the degree of settlement allowed for a structure depends on several considerations.)

The load per unit area of the foundation at which shear failure in soil occurs is called the *ultimate bearing capacity*, which is the subject of this chapter.

3.2 General Concept

Consider a strip foundation with a width of B resting on the surface of a dense sand or stiff cohesive soil, as shown in Figure 3.1a. Now, if a load is gradually applied to the foundation, settlement will increase. The variation of the load per unit area on the foundation (q) with the foundation settlement is also shown in Figure 3.1a. At a certain point—when the load per unit area equals q_u—a sudden failure in the soil supporting the foundation will take place, and the failure surface in the soil will extend to the ground surface. This load per unit area, q_u, is usually referred to as the *ultimate bearing capacity of the foundation*. When such sudden failure in soil takes place, it is called *general shear failure*.

If the foundation under consideration rests on sand or clayey soil of medium compaction (Figure 3.1b), an increase in the load on the foundation will also be accompanied by an increase in settlement. However, in this case the failure surface in the soil will gradually extend outward from the foundation, as shown by the solid lines in Figure 3.1b. When the load per unit area on the foundation equals $q_{u(1)}$, movement of the foundation will be accompanied by sudden jerks. A considerable movement of the foundation is then required for the failure surface in soil to extend to the ground surface (as shown by the broken lines in the figure). The load per unit area at which this happens is the *ultimate bearing capacity, q_u*. Beyond that point, an increase in load will be

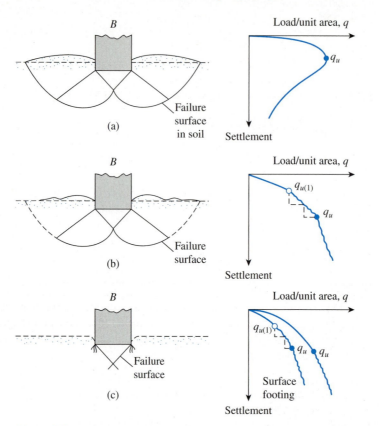

Figure 3.1 Nature of bearing capacity failure in soil: (a) general shear failure: (b) local shear failure; (c) punching shear failure (Redrawn after Vesic, 1973) (Vesic, A. S. (1973). "Analysis of Ultimate Loads of Shallow Foundations," *Journal of Soil Mechanics and Foundations Division*, American Society of Civil Engineers, Vol. 99, No. SM1, pp. 45–73. With permission from ASCE.)

accompanied by a large increase in foundation settlement. The load per unit area of the foundation, $q_{u(1)}$, is referred to as the *first failure load* (Vesic, 1963). Note that a peak value of q is not realized in this type of failure, which is called the *local shear failure* in soil.

If the foundation is supported by a fairly loose soil, the load–settlement plot will be like the one in Figure 3.1c. In this case, the failure surface in soil will not extend to the ground surface. Beyond the ultimate failure load, q_u, the load–settlement plot will be steep and practically linear. This type of failure in soil is called the *punching shear failure.*

Vesic (1963) conducted several laboratory load-bearing tests on circular and rectangular plates supported by a sand at various relative densities of compaction, D_r. The variations of $q_{u(1)}/\frac{1}{2}\gamma B$ and $q_u/\frac{1}{2}\gamma B$ obtained from those tests, where B is the diameter of a circular plate or width of a rectangular plate and γ is a dry unit weight of sand, are shown in Figure 3.2. It is important to note from this figure that, for $D_r \geq$ about 70%, the general shear type of failure in soil occurs.

On the basis of experimental results, Vesic (1973) proposed a relationship for the mode of bearing capacity failure of foundations resting on sands. Figure 3.3

Figure 3.2 Variation of $q_{u(1)}/0.5\gamma B$ and $q_u/0.5\gamma B$ for circular and rectangular plates on the surface of a sand (Adapted from Vesic, 1963) (From Vesic, A. B. Bearing Capacity of Deep Foundations in Sand. In Highway Research Record 39, Highway Research Board, National Research Council, Washington, D.C., 1963, Figure 28, p. 137. Reproduced with permission of the Transportation Research Board.)

shows this relationship, which involves the notation

$$D_r = \text{relative density of sand}$$

$$D_f = \text{depth of foundation measured from the ground surface}$$

$$B^* = \frac{2BL}{B+L} \qquad (3.1)$$

where

B = width of foundation

L = length of foundation

(*Note: L* is always greater than *B*.)

Figure 3.3 Modes of foundation failure in sand (After Vesic, 1973) (Vesic, A. S. (1973). "Analysis of Ultimate Loads of Shallow Foundations," *Journal of Soil Mechanics and Foundations Division*, American Society of Civil Engineers, Vol. 99, No. SM1, pp. 45–73. With permission from ASCE.)

For square foundations, $B = L$; for circular foundations, $B = L =$ diameter, so

$$B^* = B \tag{3.2}$$

Figure 3.4 shows the settlement S of the circular and rectangular plates on the surface of a sand at *ultimate load,* as described in Figure 3.2. The figure indicates a general range of S/B with the relative density of compaction of sand. So, in general, we can say that, for foundations at a shallow depth (i.e., small D_f/B^*), the ultimate load may occur at a foundation settlement of 4 to 10% of B. This condition arises together with general shear failure in soil; however, in the case of local or punching shear failure, the ultimate load may occur at settlements of 15 to 25% of the width of the foundation (B).

3.3 Terzaghi's Bearing Capacity Theory

Terzaghi (1943) was the first to present a comprehensive theory for the evaluation of the ultimate bearing capacity of rough shallow foundations. According to this theory, a foundation is *shallow* if its depth, D_f (Figure 3.5), is less than or equal to its width. Later investigators, however, have suggested that foundations with D_f equal to 3 to 4 times their width may be defined as *shallow foundations.*

Terzaghi suggested that for a *continuous,* or *strip, foundation* (i.e., one whose width-to-length ratio approaches zero), the failure surface in soil at ultimate load may be assumed to be similar to that shown in Figure 3.5. (Note that this is the case of general shear failure, as defined in Figure 3.1a.) The effect of soil above the bottom of the foundation may also be assumed to be replaced by an equivalent surcharge, $q = \gamma D_f$ (where γ is a unit weight of soil). The failure zone under the foundation can be separated into three parts (see Figure 3.5):

Figure 3.4 Range of settlement of circular and rectangular plates at ultimate load ($D_f/B = 0$) in sand (Modified from Vesic, 1963) (From Vesic, A. B. Bearing Capacity of Deep Foundations in Sand. In Highway Research Record 39, Highway Research Board, National Research Council, Washington, D.C., 1963, Figure 29, p. 138. Reproduced with permission of the Transportation Research Board.)

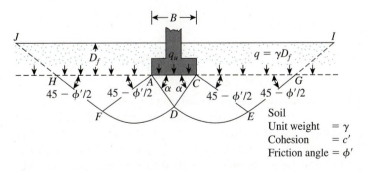

Figure 3.5 Bearing capacity failure in soil under a rough rigid continuous (strip) foundation

1. The *triangular zone ACD* immediately under the foundation
2. The *radial shear zones ADF* and *CDE*, with the curves *DE* and *DF* being arcs of a logarithmic spiral
3. Two triangular *Rankine passive zones AFH* and *CEG*

The angles *CAD* and *ACD* are assumed to be equal to the soil friction angle ϕ'. Note that, with the replacement of the soil above the bottom of the foundation by an equivalent surcharge q, the shear resistance of the soil along the failure surfaces *GI* and *HJ* was neglected.

Using equilibrium analysis, Terzaghi expressed the ultimate bearing capacity in the form

$$q_u = c'N_c + qN_q + \tfrac{1}{2}\gamma B N_\gamma \quad \text{(continuous or strip foundation)} \qquad (3.3)$$

where

$$c' = \text{cohesion of soil}$$
$$\gamma = \text{unit weight of soil}$$
$$q = \gamma D_f$$
$$N_c, N_q, N_\gamma = \text{bearing capacity factors that are nondimensional and are functions only of}$$
$$\text{the soil friction angle } \phi'$$

The bearing capacity factors N_c, N_q, and N_γ are defined by

$$N_c = \cot \phi' \left[\frac{e^{2(3\pi/4 - \phi'/2)\tan \phi'}}{2 \cos^2\left(\dfrac{\pi}{4} + \dfrac{\phi'}{2}\right)} - 1 \right] = \cot \phi' (N_q - 1) \qquad (3.4)$$

$$N_q = \frac{e^{2(3\pi/4 - \phi'/2)\tan \phi'}}{2 \cos^2\left(45 + \dfrac{\phi'}{2}\right)} \qquad (3.5)$$

and

$$N_\gamma = \frac{1}{2}\left(\frac{K_{p\gamma}}{\cos^2 \phi'} - 1\right)\tan \phi' \qquad (3.6)$$

where $K_{p\gamma}$ = passive pressure coefficient.

The variations of the bearing capacity factors defined by Eqs. (3.4), (3.5), and (3.6) are given in Table 3.1.

To estimate the ultimate bearing capacity of *square* and *circular foundations*, Eq. (3.1) may be respectively modified to

$$q_u = 1.3c'N_c + qN_q + 0.4\gamma B N_\gamma \quad \text{(square foundation)} \qquad (3.7)$$

Table 3.1 Terzaghi's Bearing Capacity Factors—Eqs. (3.4), (3.5), and (3.6) a From Kumbhojkar (1993)

ϕ'	N_c	N_q	$N_\gamma{}^a$	ϕ'	N_c	N_q	$N_\gamma{}^a$
0	5.70	1.00	0.00	26	27.09	14.21	9.84
1	6.00	1.10	0.01	27	29.24	15.90	11.60
2	6.30	1.22	0.04	28	31.61	17.81	13.70
3	6.62	1.35	0.06	29	34.24	19.98	16.18
4	6.97	1.49	0.10	30	37.16	22.46	19.13
5	7.34	1.64	0.14	31	40.41	25.28	22.65
6	7.73	1.81	0.20	32	44.04	28.52	26.87
7	8.15	2.00	0.27	33	48.09	32.23	31.94
8	8.60	2.21	0.35	34	52.64	36.50	38.04
9	9.09	2.44	0.44	35	57.75	41.44	45.41
10	9.61	2.69	0.56	36	63.53	47.16	54.36
11	10.16	2.98	0.69	37	70.01	53.80	65.27
12	10.76	3.29	0.85	38	77.50	61.55	78.61
13	11.41	3.63	1.04	39	85.97	70.61	95.03
14	12.11	4.02	1.26	40	95.66	81.27	115.31
15	12.86	4.45	1.52	41	106.81	93.85	140.51
16	13.68	4.92	1.82	42	119.67	108.75	171.99
17	14.60	5.45	2.18	43	134.58	126.50	211.56
18	15.12	6.04	2.59	44	151.95	147.74	261.60
19	16.56	6.70	3.07	45	172.28	173.28	325.34
20	17.69	7.44	3.64	46	196.22	204.19	407.11
21	18.92	8.26	4.31	47	224.55	241.80	512.84
22	20.27	9.19	5.09	48	258.28	287.85	650.67
23	21.75	10.23	6.00	49	298.71	344.63	831.99
24	23.36	11.40	7.08	50	347.50	415.14	1072.80
25	25.13	12.72	8.34				

[a] From Kumbhojkar (1993)

and

$$q_u = 1.3c'N_c + qN_q + 0.3\gamma BN_\gamma \quad \text{(circular foundation)} \tag{3.8}$$

In Eq. (3.7), B equals the dimension of each side of the foundation; in Eq. (3.8), B equals the diameter of the foundation.

For foundations that exhibit the local shear failure mode in soils, Terzaghi suggested the following modifications to Eqs. (3.3), (3.7), and (3.8):

$$q_u = \tfrac{2}{3}c'N_c' + qN_q' + \tfrac{1}{2}\gamma BN_\gamma' \quad \text{(strip foundation)} \tag{3.9}$$

$$q_u = 0.867c'N_c' + qN_q' + 0.4\gamma BN_\gamma' \quad \text{(square foundation)} \tag{3.10}$$

$$q_u = 0.867c'N_c' + qN_q' + 0.3\gamma BN_\gamma' \quad \text{(circular foundation)} \tag{3.11}$$

Table 3.2 Terzaghi's Modified Bearing Capacity Factors N_c', N_q', and N_γ'

ϕ'	N_c'	N_q'	N_γ'	ϕ'	N_c'	N_q'	N_γ'
0	5.70	1.00	0.00	26	15.53	6.05	2.59
1	5.90	1.07	0.005	27	16.30	6.54	2.88
2	6.10	1.14	0.02	28	17.13	7.07	3.29
3	6.30	1.22	0.04	29	18.03	7.66	3.76
4	6.51	1.30	0.055	30	18.99	8.31	4.39
5	6.74	1.39	0.074	31	20.03	9.03	4.83
6	6.97	1.49	0.10	32	21.16	9.82	5.51
7	7.22	1.59	0.128	33	22.39	10.69	6.32
8	7.47	1.70	0.16	34	23.72	11.67	7.22
9	7.74	1.82	0.20	35	25.18	12.75	8.35
10	8.02	1.94	0.24	36	26.77	13.97	9.41
11	8.32	2.08	0.30	37	28.51	15.32	10.90
12	8.63	2.22	0.35	38	30.43	16.85	12.75
13	8.96	2.38	0.42	39	32.53	18.56	14.71
14	9.31	2.55	0.48	40	34.87	20.50	17.22
15	9.67	2.73	0.57	41	37.45	22.70	19.75
16	10.06	2.92	0.67	42	40.33	25.21	22.50
17	10.47	3.13	0.76	43	43.54	28.06	26.25
18	10.90	3.36	0.88	44	47.13	31.34	30.40
19	11.36	3.61	1.03	45	51.17	35.11	36.00
20	11.85	3.88	1.12	46	55.73	39.48	41.70
21	12.37	4.17	1.35	47	60.91	44.45	49.30
22	12.92	4.48	1.55	48	66.80	50.46	59.25
23	13.51	4.82	1.74	49	73.55	57.41	71.45
24	14.14	5.20	1.97	50	81.31	65.60	85.75
25	14.80	5.60	2.25				

N_c', N_q', and N_γ', the *modified bearing capacity factors,* can be calculated by using the bearing capacity factor equations (for N_c, N_q, and N_γ, respectively) by replacing ϕ' by $\overline{\phi}' = \tan^{-1}(\frac{2}{3}\tan\phi')$. The variation of N_c', N_q', and N_γ' with the soil friction angle ϕ' is given in Table 3.2.

Terzaghi's bearing capacity equations have now been modified to take into account the effects of the foundation shape (B/L), depth of embedment (D_f), and the load inclination. This is given in Section 3.6. Many design engineers, however, still use Terzaghi's equation, which provides fairly good results considering the uncertainty of the soil conditions at various sites.

3.4   *Factor of Safety*

Calculating the gross *allowable load-bearing capacity* of shallow foundations requires the application of a factor of safety (FS) to the gross ultimate bearing capacity, or

$$q_{all} = \frac{q_u}{FS} \qquad (3.12)$$

However, some practicing engineers prefer to use a factor of safety such that

$$\text{Net stress increase on soil} = \frac{\text{net ultimate bearing capacity}}{\text{FS}} \qquad (3.13)$$

The net ultimate bearing capacity is defined as the ultimate pressure per unit area of the foundation that can be supported by the soil in excess of the pressure caused by the surrounding soil at the foundation level. If the difference between the unit weight of concrete used in the foundation and the unit weight of soil surrounding is assumed to be negligible, then

$$q_{net(u)} = q_u - q \qquad (3.14)$$

where

$q_{net(u)}$ = net ultimate bearing capacity
$q = \gamma D_f$

So

$$q_{all(net)} = \frac{q_u - q}{\text{FS}} \qquad (3.15)$$

The factor of safety as defined by Eq. (3.15) should be at least 3 in all cases.

Example 3.1

A square foundation is 2 m × 2 m in plan. The soil supporting the foundation has a friction angle of $\phi' = 25°$ and $c' = 20 \text{ kN/m}^2$. The unit weight of soil, γ, is 16.5 kN/m³. Determine the allowable gross load on the foundation with a factor of safety (FS) of 3. Assume that the depth of the foundation (D_f) is 1.5 m and that general shear failure occurs in the soil.

Solution
From Eq. (3.7)

$$q_u = 1.3c'N_c + qN_q + 0.4\gamma BN_\gamma$$

From Table 3.1, for $\phi' = 25°$,

$$N_c = 25.13$$
$$N_q = 12.72$$
$$N_\gamma = 8.34$$

Thus,

$$q_u = (1.3)(20)(25.13) + (1.5 \times 16.5)(12.72) + (0.4)(16.5)(2)(8.34)$$

$$= 653.38 + 314.82 + 110.09 = 1078.29 \text{ kN/m}^2$$

So, the allowable load per unit area of the foundation is

$$q_{all} = \frac{q_u}{FS} = \frac{1078.29}{3} \approx 359.5 \text{ kN/m}^2$$

Thus, the total allowable gross load is

$$Q = (359.5) \, B^2 = (359.5) \, (2 \times 2) = \mathbf{1438 \text{ kN}}$$ ∎

3.5 *Modification of Bearing Capacity Equations for Water Table*

Equations (3.3) and (3.7) through (3.11) give the ultimate bearing capacity, based on the assumption that the water table is located well below the foundation. However, if the water table is close to the foundation, some modifications of the bearing capacity equations will be necessary. (See Figure 3.6.)

Case I. If the water table is located so that $0 \leq D_1 \leq D_f$, the factor q in the bearing capacity equations takes the form

$$q = \text{effective surcharge} = D_1\gamma + D_2(\gamma_{sat} - \gamma_w) \tag{3.16}$$

where

γ_{sat} = saturated unit weight of soil
γ_w = unit weight of water

Also, the value of γ in the last term of the equations has to be replaced by $\gamma' = \gamma_{sat} - \gamma_w$.

Case II. For a water table located so that $0 \leq d \leq B$,

$$q = \gamma D_f \tag{3.17}$$

In this case, the factor γ in the last term of the bearing capacity equations must be replaced by the factor

$$\bar{\gamma} = \gamma' + \frac{d}{B}(\gamma - \gamma') \tag{3.18}$$

Figure 3.6 Modification of bearing capacity equations for water table

The preceding modifications are based on the assumption that there is no seepage force in the soil.

Case III. When the water table is located so that $d \geq B$, the water will have no effect on the ultimate bearing capacity.

3.6 *The General Bearing Capacity Equation*

The ultimate bearing capacity equations (3.3), (3.7), and (3.8) are for continuous, square, and circular foundations only; they do not address the case of rectangular foundations $(0 < B/L < 1)$. Also, the equations do not take into account the shearing resistance along the failure surface in soil above the bottom of the foundation (the portion of the failure surface marked as *GI* and *HJ* in Figure 3.5). In addition, the load on the foundation may be inclined. To account for all these shortcomings, Meyerhof (1963) suggested the following form of the general bearing capacity equation:

$$q_u = c'N_c F_{cs} F_{cd} F_{ci} + qN_q F_{qs} F_{qd} F_{qi} + \tfrac{1}{2}\gamma B N_\gamma F_{\gamma s} F_{\gamma d} F_{\gamma i} \qquad (3.19)$$

In this equation:

$$
\begin{aligned}
c' &= \text{cohesion} \\
q &= \text{effective stress at the level of the bottom of the foundation} \\
\gamma &= \text{unit weight of soil} \\
B &= \text{width of foundation } (= \text{diameter for a circular foundation}) \\
F_{cs}, F_{qs}, F_{\gamma s} &= \text{shape factors} \\
F_{cd}, F_{qd}, F_{\gamma d} &= \text{depth factors} \\
F_{ci}, F_{qi}, F_{\gamma i} &= \text{load inclination factors} \\
N_c, N_q, N_\gamma &= \text{bearing capacity factors}
\end{aligned}
$$

The equations for determining the various factors given in Eq. (3.19) are described briefly in the sections that follow. Note that the original equation for ultimate bearing capacity is derived only for the plane-strain case (i.e., for continuous foundations). The shape, depth, and load inclination factors are empirical factors based on experimental data.

Bearing Capacity Factors

The basic nature of the failure surface in soil suggested by Terzaghi now appears to have been borne out by laboratory and field studies of bearing capacity (Vesic, 1973). However, the angle α shown in Figure 3.5 is closer to $45 + \phi'/2$ than to ϕ'. If this change is accepted, the values of N_c, N_q, and N_γ for a given soil friction angle will also change from those given in Table 3.1. With $\alpha = 45 + \phi'/2$, it can be shown that

$$N_q = \tan^2\left(45 + \frac{\phi'}{2}\right)e^{\pi \tan \phi'} \qquad (3.20)$$

and

$$N_c = (N_q - 1) \cot \phi' \qquad (3.21)$$

Equation (3.21) for N_c was originally derived by Prandtl (1921), and Eq. (3.20) for N_q was presented by Reissner (1924). Caquot and Kerisel (1953) and Vesic (1973) gave the relation for N_γ as

$$N_\gamma = 2(N_q + 1) \tan \phi' \qquad (3.22)$$

Table 3.3 shows the variation of the preceding bearing capacity factors with soil friction angles.

Table 3.3 Bearing Capacity Factors

ϕ'	N_c	N_q	N_γ	ϕ'	N_c	N_q	N_γ
0	5.14	1.00	0.00	26	22.25	11.85	12.54
1	5.38	1.09	0.07	27	23.94	13.20	14.47
2	5.63	1.20	0.15	28	25.80	14.72	16.72
3	5.90	1.31	0.24	29	27.86	16.44	19.34
4	6.19	1.43	0.34	30	30.14	18.40	22.40
5	6.49	1.57	0.45	31	32.67	20.63	25.99
6	6.81	1.72	0.57	32	35.49	23.18	30.22
7	7.16	1.88	0.71	33	38.64	26.09	35.19
8	7.53	2.06	0.86	34	42.16	29.44	41.06
9	7.92	2.25	1.03	35	46.12	33.30	48.03
10	8.35	2.47	1.22	36	50.59	37.75	56.31
11	8.80	2.71	1.44	37	55.63	42.92	66.19
12	9.28	2.97	1.69	38	61.35	48.93	78.03
13	9.81	3.26	1.97	39	67.87	55.96	92.25
14	10.37	3.59	2.29	40	75.31	64.20	109.41
15	10.98	3.94	2.65	41	83.86	73.90	130.22
16	11.63	4.34	3.06	42	93.71	85.38	155.55
17	12.34	4.77	3.53	43	105.11	99.02	186.54
18	13.10	5.26	4.07	44	118.37	115.31	224.64
19	13.93	5.80	4.68	45	133.88	134.88	271.76
20	14.83	6.40	5.39	46	152.10	158.51	330.35
21	15.82	7.07	6.20	47	173.64	187.21	403.67
22	16.88	7.82	7.13	48	199.26	222.31	496.01
23	18.05	8.66	8.20	49	229.93	265.51	613.16
24	19.32	9.60	9.44	50	266.89	319.07	762.89
25	20.72	10.66	10.88				

Shape, Depth, Inclination Factors

Commonly used shape, depth, and inclination factors are given in Table 3.4.

Table 3.4 Shape, Depth and Inclination Factors (DeBeer (1970); Hansen (1970); Meyerhof (1963); Meyerhof and Hanna (1981))

Factor	Relationship	Reference
Shape	$$F_{cs} = 1 + \left(\frac{B}{L}\right)\left(\frac{N_q}{N_c}\right)$$ $$F_{qs} = 1 + \left(\frac{B}{L}\right)\tan\phi'$$ $$F_{\gamma s} = 1 - 0.4\left(\frac{B}{L}\right)$$	DeBeer (1970)
Depth	$$\frac{D_f}{B} \leq 1$$ For $\phi = 0$: $$F_{cd} = 1 + 0.4\left(\frac{D_f}{B}\right)$$ $$F_{qd} = 1$$ $$F_{\gamma d} = 1$$ For $\phi' > 0$: $$F_{cd} = F_{qd} - \frac{1 - F_{qd}}{N_c \tan\phi'}$$ $$F_{qd} = 1 + 2\tan\phi'\,(1 - \sin\phi')^2\left(\frac{D_f}{B}\right)$$ $$F_{\gamma d} = 1$$ $$\frac{D_f}{B} > 1$$ For $\phi = 0$: $$F_{cd} = 1 + 0.4\underbrace{\tan^{-1}\left(\frac{D_f}{B}\right)}_{\text{radians}}$$ $$F_{qd} = 1$$ $$F_{\gamma d} = 1$$ For $\phi' > 0$: $$F_{cd} = F_{qd} - \frac{1 - F_{qd}}{N_c \tan\phi'}$$ $$F_{qd} = 1 + 2\tan\phi'(1 - \sin\phi')^2\underbrace{\tan^{-1}\left(\frac{D_f}{B}\right)}_{\text{radians}}$$ $$F_{\gamma d} = 1$$	Hansen (1970)
Inclination	$$F_{ci} = F_{qi} = \left(1 - \frac{\beta^{\circ}}{90^{\circ}}\right)^2$$ $$F_{\gamma i} = \left(1 - \frac{\beta}{\phi'}\right)$$ $\beta =$ inclination of the load on the foundation with respect to the vertical	Meyerhof (1963); Hanna and Meyerhof (1981)

Example 3.2

Solve Example Problem 3.1 using Eq. (3.19).

Solution
From Eq. (3.19),

$$q_u = c'N_cF_{cs}F_{cd}F_{ci} + qN_qF_{qs}F_{qd}F_{qt} + \frac{1}{2}\gamma BN_\gamma F_{\gamma s}F_{\gamma d}F_{\gamma t}$$

Since the load is vertical, $F_{ci} = F_{qi} = F_{\gamma i} = 1$. From Table 3.3 for $\phi' = 25°$, $N_c = 20.72$, $N_q = 10.66$, and $N_\gamma = 10.88$.

Using Table 3.4,

$$F_{cs} = 1 + \left(\frac{B}{L}\right)\left(\frac{N_q}{N_c}\right) = 1 + \left(\frac{2}{2}\right)\left(\frac{10.66}{20.72}\right) = 1.514$$

$$F_{qs} = 1 + \left(\frac{B}{L}\right)\tan\phi' = 1 + \left(\frac{2}{2}\right)\tan 25 = 1.466$$

$$F_{\gamma s} = 1 - 0.4\left(\frac{B}{L}\right) = 1 - 0.4\left(\frac{2}{2}\right) = 0.6$$

$$F_{qd} = 1 + 2\tan\phi'\,(1 - \sin\phi')^2\left(\frac{D_f}{B}\right)$$

$$= 1 + (2)(\tan 25)(1 - \sin 25)^2\left(\frac{1.5}{2}\right) = 1.233$$

$$F_{cd} = F_{qd} - \frac{1 - F_{qd}}{N_c\tan\phi'} = 1.233 - \left[\frac{1 - 1.233}{(20.72)(\tan 25)}\right] = 1.257$$

$$F_{\gamma d} = 1$$

Hence,

$$q_u = (20)(20.72)(1.514)(1.257)(1)$$
$$+(1.5 \times 16.5)(10.66)(1.466)(1.233)(1)$$
$$+ \frac{1}{2}(16.5)(2)(10.88)(0.6)(1)(1)$$
$$= 788.6 + 476.9 + 107.7 = 1373.2 \text{ kN/m}^2$$

$$q_{all} = \frac{q_u}{FS} = \frac{1373.2}{3} = 457.7 \text{ kN/m}^2$$

$$Q = (457.7)(2 \times 2) = \mathbf{1830.8 \text{ kN}}$$

Example 3.3

A square foundation $(B \times B)$ has to be constructed as shown in Figure 3.7. Assume that $\gamma = 16.5 \text{ kN/m}^3$, $\gamma_{sat} = 18.55 \text{ kN/m}^3$, $\phi' = 34°$, $D_f = 1.22$ m, and $D_1 = 0.61$ m. The gross allowable load, Q_{all}, with FS = 3 is 667.2 kN. Determine the size of the footing. Use Eq. (3.19).

$$B \times B$$

Figure 3.7 A square foundation

Solution
We have

$$q_{all} = \frac{Q_{all}}{B^2} = \frac{667.2}{B^2} \text{ kN/m}^2 \qquad \text{(a)}$$

From Eq. (3.19) (with $c' = 0$), for vertical loading, we obtain

$$q_{all} = \frac{q_u}{\text{FS}} = \frac{1}{3}\left(q N_q F_{qs} F_{qd} + \frac{1}{2}\gamma' B N_\gamma F_{\gamma s} F_{\gamma d} \right)$$

For $\phi' = 34°$, from Table 3.3, $N_q = 29.44$ and $N_\gamma = 41.06$. Hence,

$$F_{qs} = 1 + \frac{B}{L}\tan \phi' = 1 + \tan 34 = 1.67$$

$$F_{\gamma s} = 1 - 0.4\left(\frac{B}{L}\right) = 1 - 0.4 = 0.6$$

$$F_{qd} = 1 + 2\tan\phi'(1 - \sin\phi')^2\frac{D_f}{B} = 1 + 2\tan 34\,(1 - \sin 34)^2\frac{4}{B} = 1 + \frac{1.05}{B}$$

$$F_{\gamma d} = 1$$

and

$$q = (0.61)(16.5) + 0.61(18.55 - 9.81) = 15.4 \text{ kN/m}^2$$

So

$$q_{all} = \frac{1}{3}\left[(15.4)(29.44)(1.67)\left(1 + \frac{1.05}{B}\right) \right.$$

$$\left. + \left(\frac{1}{2}\right)(18.55 - 9.81)(B)(41.06)(0.6)(1) \right] \qquad \text{(b)}$$

$$= 252.38 + \frac{265}{B} + 35.89B$$

Combining Eqs. (a) and (b) results in

$$\frac{667.2}{B^2} = 252.38 + \frac{265}{B} + 35.89B$$

By trial and error, we find that $B \approx \textbf{1.3 m}$. ∎

3.7   *Case Studies on Ultimate Bearing Capacity*

In this section, we will consider two field observations related to the ultimate bearing capacity of foundations on soft clay. The failure loads on the foundations in the field will be compared with those estimated from the theory presented in Section 3.6.

Foundation Failure of a Concrete Silo

An excellent case of bearing capacity failure of a 6-m diameter concrete silo was provided by Bozozuk (1972). The concrete tower silo was 21 m high and was constructed over soft clay on a ring foundation. Figure 3.8 shows the variation of the undrained shear strength (c_u) obtained from field vane shear tests at the site. The groundwater table was located at about 0.6 m below the ground surface.

On September 30, 1970, just after it was filled to capacity for the first time with corn silage, the concrete tower silo suddenly overturned due to bearing capacity failure. Figure 3.9 shows the approximate profile of the failure surface in soil. The failure surface extended to about 7 m below the ground surface. Bozozuk (1972) provided the following average parameters for the soil in the failure zone and the foundation:

- Load per unit area on the foundation when failure occurred $\approx 160 \text{ kN/m}^2$
- Average plasticity index of clay (PI) ≈ 36
- Average undrained shear strength (c_u) from 0.6 to 7 m depth obtained from field vane shear tests $\approx 27.1 \text{ kN/m}^2$
- From Figure 3.9, $B \approx 7.2$ m and $D_f \approx 1.52$ m.

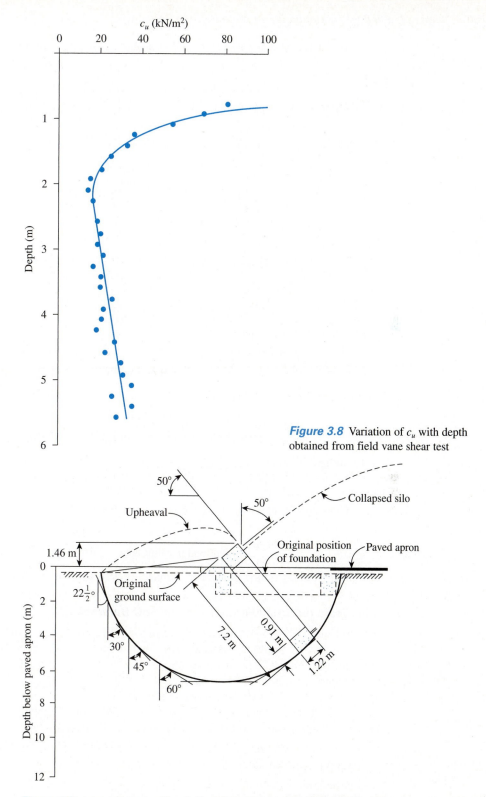

Figure 3.8 Variation of c_u with depth obtained from field vane shear test

Figure 3.9 Approximate profile of silo failure (Adapted from Bozozuk, 1972)

We can now calculate the factor of safety against bearing capacity failure. From Eq. (3.19)

$$q_u = c'N_c F_{cs} F_{cd} F_{ci} + q N_c F_{qs} F_{qd} F_{qi} + \tfrac{1}{2}\gamma B\, N_\gamma F_{\gamma s} F_{\gamma d} F_{\gamma i}$$

For $\phi = 0$ condition and vertical loading, $c' = c_u$, $N_c = 5.14$, $N_q = 1$, $N_\gamma = 0$, and $F_{ci} = F_{qi} = F_{\gamma i} = 0$. Also, from Table 3.4,

$$F_{cs} = 1 + \left(\frac{7.2}{7.2}\right)\left(\frac{1}{5.14}\right) = 1.195$$

$$F_{qs} = 1$$

$$F_{cd} = 1 + (0.4)\left(\frac{1.52}{7.2}\right) = 1.08$$

$$F_{qd} = 1$$

Thus,

$$q_u = (c_u)(5.14)(1.195)(1.08)(1) + (\gamma)(1.52)$$

Assuming $\gamma \approx 18$ kN/m³,

$$q_u = 6.63 c_u + 27.36 \qquad (3.23)$$

According to Eqs. (2.34) and (2.35a),

$$c_{u(\text{corrected})} = \lambda\, c_{u(\text{VST})}$$
$$\lambda = 1.7 - 0.54 \log\left[\text{PI}(\%)\right]$$

For this case, PI ≈ 36 and $c_{u(\text{VST})} = 27.1$ kN/m². So

$$c_{u(\text{corrected})} = \{1.7 - 0.54 \log\left[\text{PI}(\%)\right]\} c_{u(\text{VST})}$$
$$= (1.7 - 0.54 \log 36)(27.1) \approx 23.3 \text{ kN/m}^2$$

Substituting this value of c_u in Eq. (3.23)

$$q_u = (6.63)(23.3) + 27.36 = 181.8 \text{ kN/m}^2$$

The factor of safety against bearing capacity failure

$$\text{FS} = \frac{q_u}{\text{applied load per unit area}} = \frac{181.8}{160} = 1.14$$

This factor of safety is too low and approximately equals one, for which the failure occurred.

Load Tests on Small Foundations in Soft Bangkok Clay

Brand et al. (1972) reported load test results for five small square foundations in soft Bangkok clay in Rangsit, Thailand. The foundations were 0.6 m × 0.6 m, 0.675 m × 0.675 m, 0.75 m × 0.75 m, 0.9 m × 0.9 m, and 1.05 m × 1.05 m. The depth of of the foundations (D_f) was 1.5 m in all cases.

Figure 3.10 shows the vane shear test results for clay. Based on the variation of $c_{u(\text{VST})}$ with depth, it can be approximated that $c_{u(\text{VST})}$ is about 35 kN/m² for depths between zero to 1.5 m measured from the ground surface, and $c_{u(\text{VST})}$ is approximately equal to 24 kN/m² for depths varying from 1.5 to 8 m. Other properties of the clay are

- Liquid limit = 80
- Plastic limit = 40
- Sensitivity ≈ 5

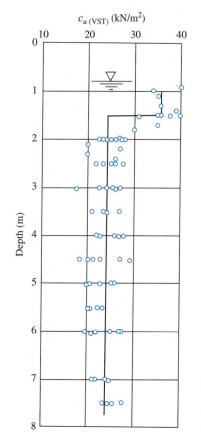

Figure 3.10 Variation of $c_{u(\text{VST})}$ with depth for soft Bangkok clay

Figure 3.11 shows the load-settlement plots obtained from the bearing-capacity tests on all five foundations. The ultimate loads, Q_u, obtained from each test are shown in Figure 3.11 and given in Table 3.5. The ultimate load is defined as the point where the load-settlement plot becomes practically linear.

From Eq. (3.19),

$$q_u = c'N_cF_{cs}F_{cd}F_{ci} + qN_qF_{qs}F_{qd}F_{qi} + \frac{1}{2}\gamma BN_\gamma F_{\gamma s}F_{\gamma d}F_{\gamma i}$$

For undrained condition and vertical loading (that is, $\phi = 0$) from Tables 3.3 and 3.4,

- $F_{ci} = F_{qi} = F_{\gamma i} = 1$
- $c' = c_u$, $N_c = 5.14$, $N_q = 1$, and $N_\gamma = 0$
- $F_{cs} = 1 + \left(\dfrac{B}{L}\right)\left(\dfrac{N_q}{N_c}\right) = 1 + (1)\left(\dfrac{1}{5.14}\right) = 1.195$
- $F_{qs} = 1$
- $F_{qd} = 1$
- $F_{cd} = 1 + 0.4 \tan^{-1}\left(\dfrac{D_f}{B}\right) = 1 + 0.4 \tan^{-1}\left(\dfrac{1.5}{B}\right)$ (3.24)

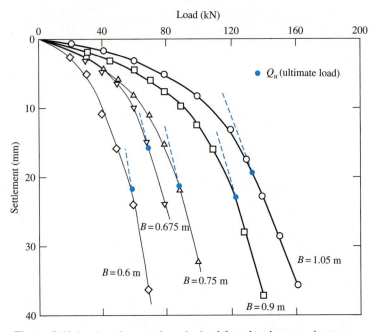

Figure 3.11 Load-settlement plots obtained from bearing capacity tests

(*Note*: $D_f/B > 1$ in all cases)
Thus,

$$q_u = (5.14)(c_u)(1.195)F_{cd} + q \qquad (3.25)$$

The values of $c_{u(VST)}$ need to be corrected for use in Eq. (3.25). From Eq. (2.34),

$$c_u = \lambda c_{u(VST)}$$

From Eq. (2.35b),

$$\lambda = 1.18e^{-0.08(PI)} + 0.57 = 1.18e^{-0.08(80 - 40)} + 0.57 = 0.62$$

From Eq. (2.35c),

$$\lambda = 7.01e^{-0.08(LL)} + 0.57 = 7.01e^{-0.08(80)} + 0.57 = 0.58$$

Table 3.5 Comparison of Ultimate Bearing Capacity—Theory versus Field Test Results

B (m) (1)	D_f (m) (2)	F_{cd}[‡] (3)	$q_{u(theory)}$[‡‡] (kN/m²) (4)	$Q_{u(field)}$ (kN) (5)	$q_{u(field)}$[‡‡‡] (kN/m²) (6)	$\dfrac{q_{u(field)} - q_{u(theory)}}{q_{u(field)}}$(%) (7)
0.600	1.5	1.476	158.3	60	166.6	4.98
0.675	1.5	1.459	156.8	71	155.8	−0.64
0.750	1.5	1.443	155.4	90	160.6	2.87
0.900	1.5	1.412	152.6	124	153.0	0.27
1.050	1.5	1.384	150.16	140	127.0	−18.24

[‡]Eq. (3.24); [‡‡]Eq. (3.26); [‡‡‡]$Q_{u(field)}/B^2 = q_{u(field)}$

So the average value of $\lambda \approx 0.6$. Hence,

$$c_u = \lambda c_{u(\text{VST})} = (0.6)(24) = 14.4 \text{ kN/m}^2$$

Let us assume $\gamma = 18.5 \text{ kN/m}^2$. So

$$q = \gamma D_f = (18.5)(1.5) = 27.75 \text{ kN/m}^2$$

Substituting $c_u = 14.4 \text{ kN/m}^2$ and $q = 27.75 \text{ kN/m}^2$ into Eq. (3.25), we obtain

$$q_u(\text{kN/m}^2) = 88.4 F_{cd} + 27.75 \qquad (3.26)$$

The values of q_u calculated using Eq. (3.26) are given in column 4 of Table 3.5. Also, the q_u determined from the field tests are given in column 6. The theoretical and field values of q_u compare very well. The important lessons learned from this study are

1. The ultimate bearing capacity is a function of c_u. If Eq. (2.35a) would have been used to correct the undrained shear strength, the theoretical values of q_u would have varied between 200 kN/m^2 and 210 kN/m^2. These values are about 25% to 55% more than those obtained from the field and are on the unsafe side.
2. It is important to recognize that empirical correlations like those given in Eqs. (2.35a), (2.35b) and (2.35c) are sometimes site specific. Thus, proper engineering judgment and any record of past studies would be helpful in the evaluation of bearing capacity.

3.8 Effect of Soil Compressibility

In Section 3.3, Eqs. (3.3), (3.7), and (3.8), which apply to the case of general shear failure, were modified to Eqs. (3.9), (3.10), and (3.11) to take into account the change of failure mode in soil (i.e., local shear failure). The change of failure mode is due to soil compressibility, to account for which Vesic (1973) proposed the following modification of Eq. (3.19):

$$q_u = c' N_c F_{cs} F_{cd} F_{cc} + q N_q F_{qs} F_{qd} F_{qc} + \tfrac{1}{2} \gamma B N_\gamma F_{\gamma s} F_{\gamma d} F_{\gamma c} \qquad (3.27)$$

In this equation, F_{cc}, F_{qc}, and $F_{\gamma c}$ are soil compressibility factors.

The soil compressibility factors were derived by Vesic (1973) by analogy to the expansion of cavities. According to that theory, in order to calculate F_{cc}, F_{qc}, and $F_{\gamma c}$, the following steps should be taken:

Step 1. Calculate the *rigidity index, I_r,* of the soil at a depth approximately $B/2$ below the bottom of the foundation, or

$$I_r = \frac{G_s}{c' + q' \tan \phi'} \qquad (3.28)$$

where

G_s = shear modulus of the soil
q = effective overburden pressure at a depth of $D_f + B/2$

Step 2.   The critical rigidity index, $I_{r(cr)}$, can be expressed as

$$I_{r(cr)} = \frac{1}{2}\left\{ \exp\left[\left(3.30 - 0.45\frac{B}{L}\right) \cot\left(45 - \frac{\phi'}{2}\right)\right]\right\} \tag{3.29}$$

The variations of $I_{r(cr)}$ with B/L are given in Table 3.6.

Step 3.   If $I_r \geq I_{r(cr)}$, then

$$F_{cc} = F_{qc} = F_{\gamma c} = 1$$

However, if $I_r < I_{r(cr)}$, then

$$F_{\gamma c} = F_{qc} = \exp\left\{ \left(-4.4 + 0.6\frac{B}{L}\right)\tan\phi' + \left[\frac{(3.07 \sin\phi')(\log 2I_r)}{1 + \sin\phi'}\right]\right\} \tag{3.30}$$

Figure 3.12 shows the variation of $F_{\gamma c} = F_{qc}$ [see Eq. (3.30)] with ϕ' and I_r. For $\phi = 0$,

$$F_{cc} = 0.32 + 0.12\frac{B}{L} + 0.60 \log I_r \tag{3.31}$$

For $\phi' > 0$,

$$F_{cc} = F_{qc} - \frac{1 - F_{qc}}{N_q \tan\phi'} \tag{3.32}$$

Table 3.6 Variation of $I_{r(cr)}$ with ϕ' and B/L

ϕ' (deg)	$I_{r(cr)}$					
	B/L = 0	B/L = 0.2	B/L = 0.4	B/L = 0.6	B/L = 0.8	B/L = 1.0
0	13.56	12.39	11.32	10.35	9.46	8.64
5	18.30	16.59	15.04	13.63	12.36	11.20
10	25.53	22.93	20.60	18.50	16.62	14.93
15	36.85	32.77	29.14	25.92	23.05	20.49
20	55.66	48.95	43.04	37.85	33.29	29.27
25	88.93	77.21	67.04	58.20	50.53	43.88
30	151.78	129.88	111.13	95.09	81.36	69.62
35	283.20	238.24	200.41	168.59	141.82	119.31
40	593.09	488.97	403.13	332.35	274.01	225.90
45	1440.94	1159.56	933.19	750.90	604.26	486.26

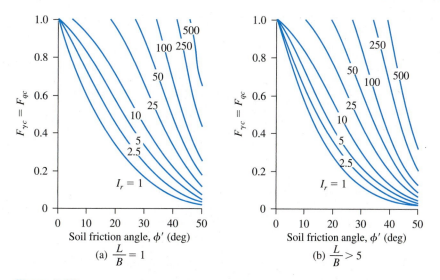

Figure 3.12 Variation of $F_{\gamma c} = F_{qc}$ with I_r and ϕ'

(a) $\dfrac{L}{B} = 1$ (b) $\dfrac{L}{B} > 5$

Example 3.4

For a shallow foundation, $B = 0.6$ m, $L = 1.2$ m, and $D_f = 0.6$ m. The known soil characteristics are as follows:

Soil:

$$\phi' = 25°$$
$$c' = 48 \text{ kN/m}^2$$
$$\gamma = 18 \text{ kN/m}^3$$

Modulus of elasticity, $E_s = 620 \text{ kN/m}^2$

Poisson's ratio, $\mu_s = 0.3$

Calculate the ultimate bearing capacity.

Solution

From Eq. (3.28),

$$I_r = \frac{G_s}{c' + q' \tan \phi'}$$

However,

$$G_s = \frac{E_s}{2(1 + \mu_s)}$$

So

$$I_r = \frac{E_s}{2(1 + \mu_s)[c' + q' \tan \phi']}$$

Now,

$$q' = \gamma \left(D_f + \frac{B}{2} \right) = 18 \left(0.6 + \frac{0.6}{2} \right) = 16.2 \text{ kN/m}^2$$

Thus,

$$I_r = \frac{620}{2(1 + 0.3)[48 + 16.2 \tan 25]} = 4.29$$

From Eq. (3.29),

$$I_{r(\text{cr})} = \frac{1}{2}\left\{\exp\left[\left(3.3 - 0.45\frac{B}{L}\right)\cot\left(45 - \frac{\phi'}{2}\right)\right]\right\}$$

$$= \frac{1}{2}\left\{\exp\left[\left(3.3 - 0.45\frac{0.6}{1.2}\right)\cot\left(45 - \frac{25}{2}\right)\right]\right\} = 62.41$$

Since $I_{r(\text{cr})} > I_r$, we use Eqs. (3.30) and (3.32) to obtain

$$F_{\gamma c} = F_{qc} = \exp\left\{\left(-4.4 + 0.6\frac{B}{L}\right)\tan\phi' + \left[\frac{(3.07 \sin \phi')\log(2I_r)}{1 + \sin\phi'}\right]\right\}$$

$$= \exp\left\{\left(-4.4 + 0.6\frac{0.6}{1.2}\right)\tan 25\right.$$

$$\left. + \left[\frac{(3.07 \sin 25)\log(2 \times 4.29)}{1 + \sin 25}\right]\right\} = 0.347$$

and

$$F_{cc} = F_{qc} - \frac{1 - F_{qc}}{N_c \tan \phi'}$$

For $\phi' = 25°$, $N_c = 20.72$ (see Table 3.3); therefore,

$$F_{cc} = 0.347 - \frac{1 - 0.347}{20.72 \tan 25} = 0.279$$

Now, from Eq. (3.27),

$$q_u = c'N_cF_{cs}F_{cd}F_{cc} + qN_qF_{qs}F_{qd}F_{qc} + \tfrac{1}{2}\gamma BN_\gamma F_{\gamma s}F_{\gamma d}F_{\gamma c}$$

From Table 3.3, for $\phi' = 25°$, $N_c = 20.72$, $N_q = 10.66$, and $N_\gamma = 10.88$. Consequently,

$$F_{cs} = 1 + \left(\frac{N_q}{N_c}\right)\left(\frac{B}{L}\right) = 1 + \left(\frac{10.66}{20.72}\right)\left(\frac{0.6}{1.2}\right) = 1.257$$

$$F_{qs} = 1 + \frac{B}{L}\tan\phi' = 1 + \frac{0.6}{1.2}\tan 25 = 1.233$$

$$F_{\gamma s} = 1 - 0.4\left(\frac{B}{L}\right) = 1 - 0.4\frac{0.6}{1.2} = 0.8$$

$$F_{qd} = 1 + 2\tan\phi'(1 - \sin\phi')^2\left(\frac{D_f}{B}\right)$$

$$= 1 + 2\tan 25(1 - \sin 25)^2\left(\frac{0.6}{0.6}\right) = 1.311$$

$$F_{cd} = F_{qd} - \frac{1 - F_{qd}}{N_c \tan \phi'} = 1.311 - \frac{1 - 1.311}{20.72 \tan 25} = 1.343$$

and

$$F_{\gamma d} = 1$$

Thus,

$$q_u = (48)(20.72)(1.257)(1.343)(0.279) + (0.6 \times 18)(10.66)(1.233)(1.311)$$
$$(0.347) + (\tfrac{1}{2})(18)(0.6)(10.88)(0.8)(1)(0.347) = \mathbf{549.32 \ kN/m^2} \quad \blacksquare$$

3.9 *Eccentrically Loaded Foundations*

In several instances, as with the base of a retaining wall, foundations are subjected to moments in addition to the vertical load, as shown in Figure 3.13a. In such cases, the distribution of pressure by the foundation on the soil is not uniform. The nominal distribution of pressure is

$$q_{max} = \frac{Q}{BL} + \frac{6M}{B^2 L} \qquad (3.33)$$

and

$$q_{min} = \frac{Q}{BL} - \frac{6M}{B^2 L} \qquad (3.34)$$

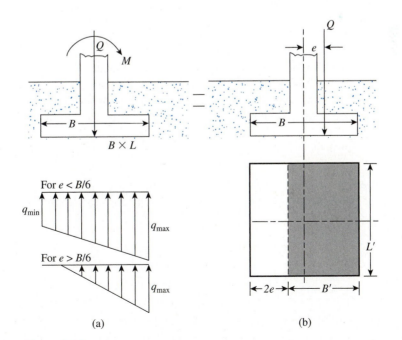

(a)　　　　　　　　　　　　(b)

Figure 3.13 Eccentrically loaded foundations

where

Q = total vertical load
M = moment on the foundation

Figure 3.13b shows a force system equivalent to that shown in Figure 3.13a. The distance

$$e = \frac{M}{Q} \tag{3.35}$$

is the eccentricity. Substituting Eq. (3.35) into Eqs. (3.33) and (3.34) gives

$$q_{max} = \frac{Q}{BL}\left(1 + \frac{6e}{B}\right) \tag{3.36}$$

and

$$q_{min} = \frac{Q}{BL}\left(1 - \frac{6e}{B}\right) \tag{3.37}$$

Note that, in these equations, when the eccentricity e becomes $B/6$, q_{min} is zero. For $e > B/6$, q_{min} will be negative, which means that tension will develop. Because soil cannot take any tension, there will then be a separation between the foundation and the soil underlying it. The nature of the pressure distribution on the soil will be as shown in Figure 3.13a. The value of q_{max} is then

$$q_{max} = \frac{4Q}{3L(B - 2e)} \tag{3.38}$$

The exact distribution of pressure is difficult to estimate.

Figure 3.14 shows the nature of failure surface in soil for a surface strip foundation subjected to an eccentric load. The factor of safety for such type of loading against bearing capacity failure can be evaluated as

$$FS = \frac{Q_{ult}}{Q} \tag{3.39}$$

where Q_{ult} = ultimate load-carrying capacity.

The following sections describe several theories for determining Q_{ult}.

Figure 3.14 Nature of failure surface in soil supporting a strip foundation subjected to eccentric loading (*Note: $D_f = 0$; Q_{ult} is ultimate load per unit length of foundation)

3.10	**Ultimate Bearing Capacity under Eccentric Loading—One-Way Eccentricity**

Effective Area Method (Meyerhoff, 1953)

In 1953, Meyerhof proposed a theory that is generally referred to as the *effective area method.*

 The following is a step-by-step procedure for determining the ultimate load that the soil can support and the factor of safety against bearing capacity failure:

Step 1. Determine the effective dimensions of the foundation (Figure 3.13b):

$$B' = \text{effective width} = B - 2e$$
$$L' = \text{effective length} = L$$

Note that if the eccentricity were in the direction of the length of the foundation, the value of L' would be equal to $L - 2e$. The value of B' would equal B. The smaller of the two dimensions (i.e., L' and B') is the effective width of the foundation.

Step 2. Use Eq. (3.19) for the ultimate bearing capacity:

$$q'_u = c'N_cF_{cs}F_{cd}F_{ci} + qN_qF_{qs}F_{qd}F_{qi} + \tfrac{1}{2}\gamma B'N_\gamma F_{\gamma s}F_{\gamma d}F_{\gamma i} \qquad (3.40)$$

To evaluate F_{cs}, F_{qs}, and $F_{\gamma s}$, use the relationships given in Table 3.4 with *effective length* and *effective width* dimensions instead of L and B, respectively. To determine F_{cd}, F_{qd}, and $F_{\gamma d}$, use the relationships given in Table 3.4. However, do not replace B with B'.

Step 3. The total ultimate load that the foundation can sustain is

$$Q_{\text{ult}} = q'_u \overbrace{(B')\,(L')}^{A'} \qquad (3.41)$$

where $A' = $ effective area.

Step 4. The factor of safety against bearing capacity failure is

$$\text{FS} = \frac{Q_{\text{ult}}}{Q}$$

Prakash and Saran Theory

Prakash and Saran (1971) analyzed the problem of ultimate bearing capacity of eccentrically and vertically loaded continuous (strip) foundations by using the one-sided failure surface in soil, as shown in Figure 3.14. According to this theory, the ultimate load *per unit length of a continuous foundation* can be estimated as

$$Q_{\text{ult}} = B\left[c'N_{c(e)} + qN_{q(e)} + \frac{1}{2}\gamma B N_{\gamma(e)} \right] \qquad (3.42)$$

where $N_{c(e)}, N_{q(e)}, N_{\gamma(e)} = $ bearing capacity factors under eccentric loading.

 The variations of $N_{c(e)}$, $N_{q(e)}$, and $N_{\gamma(e)}$ with soil friction angle ϕ' are given in Figures 3.15, 3.16, and 3.17. For rectangular foundations, the ultimate load can be given as

$$Q_{\text{ult}} = BL\left[c'N_{c(e)}F_{cs(e)} + qN_{q(e)}F_{qs(e)} + \frac{1}{2}\gamma B N_{\gamma(e)}F_{\gamma s(e)} \right]$$
$$(3.43)$$

where $F_{cs(e)}$, $F_{qs(e)}$, and $F_{\gamma s(e)} = $ shape factors.

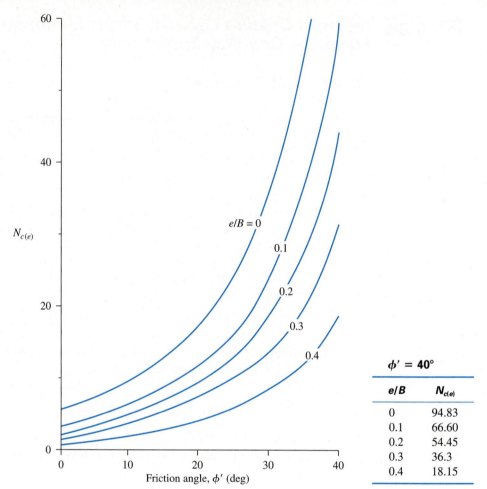

$\phi' = 40°$	
e/B	$N_{c(e)}$
0	94.83
0.1	66.60
0.2	54.45
0.3	36.3
0.4	18.15

Figure 3.15 Variation of $N_{c(e)}$ with ϕ'

Prakash and Saran (1971) also recommended the following for the shape factors:

$$F_{cs(e)} = 1.2 - 0.025\frac{L}{B} \text{ (with a minimum of 1.0)} \tag{3.44}$$

$$F_{qs(e)} = 1 \tag{3.45}$$

and

$$F_{\gamma s(e)} = 1.0 + \left(\frac{2e}{B} - 0.68\right)\frac{B}{L} + \left[0.43 - \left(\frac{3}{2}\right)\left(\frac{e}{B}\right)\right]\left(\frac{B}{L}\right)^2 \tag{3.46}$$

Reduction Factor Method (For Granular Soil)

Purkayastha and Char (1977) carried out stability analysis of eccentrically loaded *continuous foundations supported by a layer of sand* using the method of slices. Based on that analysis, they proposed

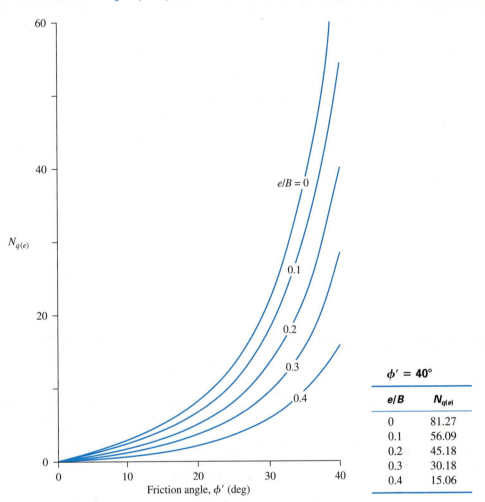

Figure 3.16 Variation of $N_{q(e)}$ with ϕ'

$$R_k = 1 - \frac{q_{u(\text{eccentric})}}{q_{u(\text{centric})}} \qquad (3.47)$$

where

R_k = reduction factor

$q_{u(\text{eccentric})}$ = ultimate bearing capacity of eccentrically loaded continuous foundations

$q_{u(\text{centric})}$ = ultimate bearing capacity of centrally loaded continuous foundations

The magnitude of R_k can be expressed as

$$R_k = a\left(\frac{e}{B}\right)^k \qquad (3.48)$$

where a and k are functions of the embedment ratio D_f/B (Table 3.7).

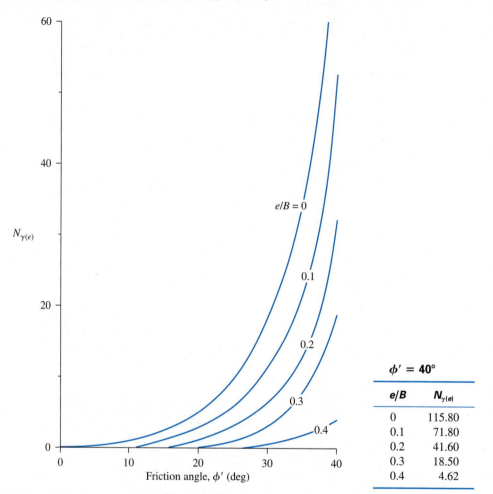

$\phi' = 40°$	
e/B	$N_{\gamma(e)}$
0	115.80
0.1	71.80
0.2	41.60
0.3	18.50
0.4	4.62

Figure 3.17 Variation of $N_{\gamma(e)}$ with ϕ'

Table 3.7 Variations of *a* and *k* [Eq. (3.48)]

D_f/B	*a*	*k*
0.00	1.862	0.73
0.25	1.811	0.785
0.50	1.754	0.80
1.00	1.820	0.888

Hence, combining Eqs. (3.47) and (3.48)

$$q_{u(\text{eccentric})} = q_{u(\text{centric})}(1 - R_k) = q_{u(\text{centric})}\left[1 - a\left(\frac{e}{B}\right)^k\right] \tag{3.49}$$

where

$$q_{u(\text{centric})} = qN_qF_{qd} + \frac{1}{2}\gamma BN_\gamma F_{\gamma d} \tag{3.50}$$

The relationships for F_{qd}, and $F_{\gamma d}$ are given in Table 3.4.

The ultimate load *per unit length* of the foundation can then be given as

$$Q_u = Bq_{u(\text{eccentric})} \tag{3.51}$$

Example 3.5

A continuous foundation is shown in Figure 3.18. If the load eccentricity is 0.2 m, determine the ultimate load, Q_{ult}, per unit length of the foundation. Use Meyerhof's effective area method.

Solution

For $c' = 0$, Eq. (3.40) gives

$$q_u' = qN_qF_{qs}F_{qd}F_{qi} + \frac{1}{2}\gamma'B'N_\gamma F_{\gamma s}F_{\gamma d}F_{\gamma i}$$

where $q = (16.5)(1.5) = 24.75 \text{ kN/m}^2$.

Sand
$\phi' = 40°$
$c' = 0$
$\gamma = 16.5 \text{ kN/m}^3$

1.5 m

2 m

Figure 3.18 A continuous foundation with load eccentricity

For $\phi' = 40°$, from Table 3.3, $N_q = 64.2$ and $N_\gamma = 109.41$. Also,

$$B' = 2 - (2)(0.2) = 1.6 \text{ m}$$

Because the foundation in question is a continuous foundation, B'/L' is zero. Hence, $F_{qs} = 1$, $F_{\gamma s} = 1$. From Table 3.4,

$$F_{qi} = F_{\gamma i} = 1$$

$$F_{qd} = 1 + 2\tan\phi'(1 - \sin\phi')^2\frac{D_f}{B} = 1 + 0.214\left(\frac{1.5}{2}\right) = 1.16$$

$$F_{\gamma d} = 1$$

and

$$q_u' = (24.75)(64.2)(1)(1.16)(1)$$
$$+ \left(\frac{1}{2}\right)(16.5)(1.6)(109.41)(1)(1)(1) = 3287.39 \text{ kN/m}^2$$

Consequently,

$$Q_{\text{ult}} = (B')(1)(q_u') = (1.6)(1)(3287.39) \approx \textbf{5260 kN}$$

Example 3.6

Solve Example 3.5 using Eq. (3.42).

Solution

Since $c' = 0$

$$Q_{ult} = B\left[qN_{q(e)} + \frac{1}{2}\gamma BN_{\gamma(e)}\right]$$

$$\frac{e}{B} = \frac{0.2}{2} = 0.1$$

For $\phi' = 40°$ and $e/B = 0.1$, Figures 3.16 and 3.17 give $N_{q(e)} = 56.09$ and $N_{\gamma(e)} \approx 71.8$. Hence,

$$Q_{ult} = 2[(24.75)(56.09) + (\tfrac{1}{2})(16.5)(2)(71.8)] = \textbf{5146 kN} \qquad \blacksquare$$

Example 3.7

Solve Example 3.5 using Eq. (3.49).

Solution

With $c' = 0$,

$$q_{u(centric)} = qN_qF_{qd} + \frac{1}{2}\gamma BN_\gamma F_{\gamma d}$$

For $\phi' = 40°$, $N_q = 64.2$ and $N_\gamma = 109.41$ (see Table 3.3). Hence,

$$F_{qd} = 1.16 \text{ and } F_{\gamma d} = 1 \text{ (see Example 3.5)}$$

$$q_{u(centric)} = (24.75)(64.2)(1.16) + \frac{1}{2}(16.5)(2)(109.41)(1)$$

$$= 1843.18 + 1805.27 = 3648.45 \text{ kN/m}^2$$

From Eq. (3.48),

$$R_k = a\left(\frac{e}{B}\right)^k$$

For $D_f/B = 1.5/2 = 0.75$, Table 3.7 gives $a \approx 1.75$ and $k \approx 0.85$. Hence,

$$R_k = 1.79\left(\frac{0.2}{2}\right)^{0.85} = 0.253$$

$$Q_u = Bq_{u(eccentric)} = Bq_{u(centric)}(1 - R_k) = (2)(3648.45)(1 - 0.253) \approx \textbf{5451 kN} \qquad \blacksquare$$

Bearing Capacity—Two-way Eccentricity

Consider a situation in which a foundation is subjected to a vertical ultimate load Q_{ult} and a moment M, as shown in Figures 3.19a and b. For this case, the components of the moment M about the x- and y-axes can be determined as M_x and M_y, respectively. (See Figure 3.19.) This condition is equivalent to a load Q_{ult} placed eccentrically on the foundation with $x = e_B$ and $y = e_L$ (Figure 3.19d). Note that

$$e_B = \frac{M_y}{Q_{ult}} \tag{3.52}$$

and

$$e_L = \frac{M_x}{Q_{ult}} \tag{3.53}$$

If Q_{ult} is needed, it can be obtained from Eq. (3.41); that is,

$$Q_{ult} = q'_u A'$$

where, from Eq. (3.40),

$$q'_u = c' N_c F_{cs} F_{cd} F_{ci} + q N_q F_{qs} F_{qd} F_{qi} + \tfrac{1}{2} \gamma B' N_\gamma F_{\gamma s} F_{\gamma d} F_{\gamma i}$$

and

$$A' = \text{effective area} = B'L'$$

As before, to evaluate F_{cs}, F_{qs}, and $F_{\gamma s}$ (Table 3.4), we use the effective length L' and effective width B' instead of L and B, respectively. To calculate F_{cd}, F_{qd}, and $F_{\gamma d}$, we do

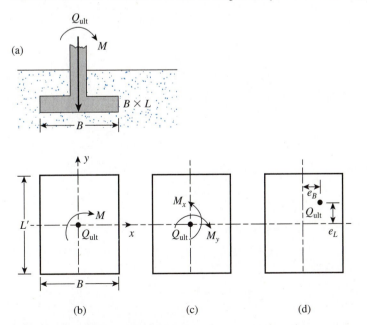

Figure 3.19 Analysis of foundation with two-way eccentricity

not replace B with B'. In determining the effective area A', effective width B', and effective length L', five possible cases may arise (Highter and Anders, 1985).

Case I. $e_L/L \geq \frac{1}{6}$ and $e_B/B \geq \frac{1}{6}$. The effective area for this condition is shown in Figure 3.20, or

$$A' = \tfrac{1}{2}B_1 L_1 \tag{3.54}$$

where

$$B_1 = B\left(1.5 - \frac{3e_B}{B}\right) \tag{3.55}$$

and

$$L_1 = L\left(1.5 - \frac{3e_L}{L}\right) \tag{3.56}$$

The effective length L' is the larger of the two dimensions B_1 and L_1. So the effective width is

$$B' = \frac{A'}{L'} \tag{3.57}$$

Case II. $e_L/L < 0.5$ and $0 < e_B/B < \frac{1}{6}$. The effective area for this case, shown in Figure 3.21a, is

$$A' = \tfrac{1}{2}(L_1 + L_2)B \tag{3.58}$$

The magnitudes of L_1 and L_2 can be determined from Figure 3.21b. The effective width is

$$B' = \frac{A'}{L_1 \text{ or } L_2 \quad (\text{whichever is larger})} \tag{3.59}$$

The effective length is

$$L' = L_1 \text{ or } L_2 \quad (\text{whichever is larger}) \tag{3.60}$$

Case III. $e_L/L < \frac{1}{6}$ and $0 < e_B/B < 0.5$. The effective area, shown in Figure 3.22a, is

$$A' = \tfrac{1}{2}(B_1 + B_2)L \tag{3.61}$$

Figure 3.20 Effective area for the case of $e_L/L \geq \frac{1}{6}$ and $e_B/B \geq \frac{1}{6}$

(a)

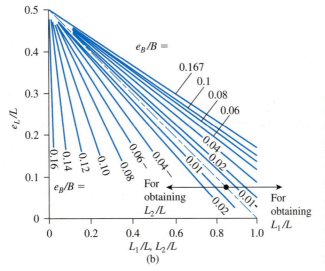

(b)

Figure 3.21 Effective area for the case of $e_L/L < 0.5$ and $0 < e_B/B < \frac{1}{6}$ (After Highter and Anders, 1985) (Highter, W. H. and Anders, J. C. (1985). "Dimensioning Footings Subjected to Eccentric Loads," *Journal of Geotechnical Engineering*, American Society of Civil Engineers, Vol. 111, No. GT5, pp. 659–665. With permission from ASCE.)

The effective width is

$$B' = \frac{A'}{L} \tag{3.62}$$

The effective length is

$$L' = L \tag{3.63}$$

The magnitudes of B_1 and B_2 can be determined from Figure 3.22b.

Case IV. $e_L/L < \frac{1}{6}$ and $e_B/B < \frac{1}{6}$. Figure 3.23a shows the effective area for this case. The ratio B_2/B, and thus B_2, can be determined by using the e_L/L curves that slope upward. Similarly, the ratio L_2/L, and thus L_2, can be determined by using the e_L/L curves that slope downward. The effective area is then

$$A' = L_2 B + \frac{1}{2}(B + B_2)(L - L_2) \tag{3.64}$$

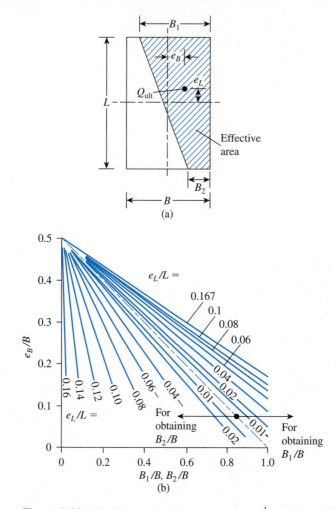

Figure 3.22 Effective area for the case of $e_L/L < \frac{1}{6}$ and $0 < e_B/B < 0.5$ (After Highter and Anders, 1985) Highter, W. H. and Anders, J. C. (1985). "Dimensioning Footings Subjected to Eccentric Loads," *Journal of Geotechnical Engineering*, American Society of Civil Engineers, Vol. 111, No. GT5, pp. 659–665. With permission from ASCE.)

The effective width is

$$B' = \frac{A'}{L} \tag{3.65}$$

The effective length is

$$L' = L \tag{3.66}$$

Case V. (Circular Foundation) In the case of circular foundations under eccentric loading (Figure 3.24a), the eccentricity is always one way. The effective area A' and the effective width B' for a circular foundation are given in a nondimensional form in Table 3.8. Once A' and B' are determined, the effective length can be obtained as

$$L' = \frac{A'}{B'}$$

(a)

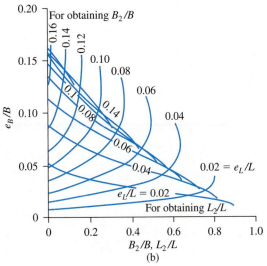

(b)

Figure 3.23 Effective area for the case of $e_L/L < \frac{1}{6}$ and $e_B/B < \frac{1}{6}$ (After Highter and Anders, 1985) (Highter, W. H. and Anders, J. C. (1985). "Dimensioning Footings Subjected to Eccentric Loads," *Journal of Geotechnical Engineering,* American Society of Civil Engineers, Vol. 111, No. GT5, pp. 659–665. With permission from ASCE.)

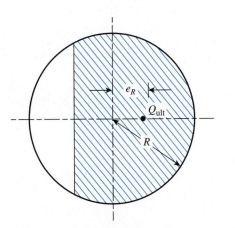

Figure 3.24 Effective area for circular foundation

Table 3.8 Variation of A'/R^2 and B'/R with e_R/R for Circular Foundations

e_R'/R	A'/R^2	B'/R
0.1	2.8	1.85
0.2	2.4	1.32
0.3	2.0	1.2
0.4	1.61	0.80
0.5	1.23	0.67
0.6	0.93	0.50
0.7	0.62	0.37
0.8	0.35	0.23
0.9	0.12	0.12
1.0	0	0

Example 3.8

A square foundation is shown in Figure 3.25, with $e_L = 0.3$ m and $e_B = 0.15$ m. Assume two-way eccentricity, and determine the ultimate load, Q_{ult}.

Solution
We have

$$\frac{e_L}{L} = \frac{0.3}{1.5} = 0.2$$

and

$$\frac{e_B}{B} = \frac{0.15}{1.5} = 0.1$$

This case is similar to that shown in Figure 3.21a. From Figure 3.21b, for $e_L/L = 0.2$ and $e_B/B = 0.1$,

$$\frac{L_1}{L} \approx 0.85; \qquad L_1 = (0.85)(1.5) = 1.275 \text{ m}$$

and

$$\frac{L_2}{L} \approx 0.21; \qquad L_2 = (0.21)(1.5) = 0.315 \text{ m}$$

From Eq. (3.58),

$$A' = \tfrac{1}{2}(L_1 + L_2)B = \tfrac{1}{2}(1.275 + 0.315)(1.5) = 1.193 \text{ m}^2$$

Figure 3.25 An eccentrically loaded foundation

From Eq. (3.60),

$$L' = L_1 = 1.275 \text{ m}$$

From Eq. (3.59),

$$B' = \frac{A'}{L'} = \frac{1.193}{1.275} = 0.936 \text{ m}$$

Note from Eq. (3.40) with $c' = 0$,

$$q'_u = qN_q F_{qs} F_{qd} F_{qi} + \tfrac{1}{2}\gamma B' N_\gamma F_{\gamma s} F_{\gamma d} F_{\gamma i}$$

where $q = (0.7)(18) = 12.6 \text{ kN/m}^2$.

For $\phi' = 30°$, from Table 3.3, $N_q = 18.4$ and $N_\gamma = 22.4$. Thus from Table 3.4,

$$F_{qs} = 1 + \left(\frac{B'}{L'}\right)\tan\phi' = 1 + \left(\frac{0.936}{1.275}\right)\tan 30° = 1.424$$

$$F_{\gamma s} = 1 - 0.4\left(\frac{B'}{L'}\right) = 1 - 0.4\left(\frac{0.936}{1.275}\right) = 0.706$$

$$F_{qd} = 1 + 2\tan\phi'(1 - \sin\phi')^2 \frac{D_f}{B} = 1 + \frac{(0.289)(0.7)}{1.5} = 1.135$$

and

$$F_{\gamma d} = 1$$

So

$$Q_{ult} = A'q'_u = A'(qN_qF_{qs}F_{qd} + \tfrac{1}{2}\gamma B'N_\gamma F_{\gamma s}F_{\gamma d})$$
$$= (1.193)[(12.6)(18.4)(1.424)(1.135)$$
$$+ (0.5)(18)(0.936)(22.4)(0.706)(1)] \approx \textbf{606 kN} \quad\blacksquare$$

Example 3.9

Consider the foundation shown in Figure 3.25 with the following changes:

$$e_L = 0.18 \text{ m}$$
$$e_B = 0.12 \text{ m}$$

For the soil, $\gamma = 16.5$ kN/m^3

$$\phi' = 25°$$
$$c' = 25 \text{ kN/m}^2$$

Determine the ultimate load, Q_{ult}.

Solution

$$\frac{e_L}{L} = \frac{0.18}{1.5} = 0.12; \qquad \frac{e_B}{B} = \frac{0.12}{1.5} = 0.08$$

This is the case shown in Figure 3.23a. From Figure 3.23b,

$$\frac{B_2}{B} \approx 0.1; \qquad \frac{L_2}{L} \approx 0.32$$

So

$$B_2 = (0.1)(1.5) = 0.15 \text{ m}$$
$$L_2 = (0.32)(1.5) = 0.48 \text{ m}$$

From Eq. (3.64),

$$A' = L_2B + \frac{1}{2}(B + B_2)(L - L_2) = (0.48)(1.5) + \frac{1}{2}(1.5 + 0.15)(1.5 - 0.48)$$
$$= 0.72 + 0.8415 = 1.5615 \text{ m}^2$$
$$B' = \frac{A'}{L} = \frac{1.5615}{1.5} = 1.041\text{m}$$
$$L' = 1.5 \text{ m}$$

From Eq. (3.40),

$$q'_u = c'N_cF_{cs}F_{cd} + qN_qF_{qs}F_{qd} + \frac{1}{2}\gamma B'N_\gamma F_{\gamma s}F_{\gamma d}$$

For $\phi' = 25°$, Table 3.3 gives $N_c = 20.72$, $N_q = 10.66$ and $N_\gamma = 10.88$. From Table 3.4,

$$F_{cs} = 1 + \left(\frac{B'}{L'}\right)\left(\frac{N_q}{N_c}\right) = 1 + \left(\frac{1.041}{1.5}\right)\left(\frac{10.66}{20.72}\right) = 1.357$$

$$F_{qs} = 1 + \left(\frac{B'}{L'}\right)\tan\phi' = 1 + \left(\frac{1.041}{1.5}\right)\tan 25 = 1.324$$

$$F_{\gamma s} = 1 - 0.4\left(\frac{B'}{L'}\right) = 1 - 0.4\left(\frac{1.041}{1.5}\right) = 0.722$$

$$F_{qd} = 1 + 2\tan\phi'(1 - \sin\phi')^2\left(\frac{D_f}{B}\right) = 1 + 2\tan 25(1-\sin 25)^2\left(\frac{0.7}{1.5}\right) = 1.145$$

$$F_{cd} = F_{qd} - \frac{1 - F_{qd}}{N_c \tan\phi'} = 1.145 - \frac{1 - 1.145}{20.72 \tan 25} = 1.16$$

$$F_{\gamma d} = 1$$

Hence,

$$q_u' = (25)(20.72)(1.357)(1.16) + (16.5 \times 0.7)(10.66)(1.324)(1.145)$$

$$+ \frac{1}{2}(16.5)(1.041)(10.88)(0.722)(1)$$

$$= 815.39 + 186.65 + 67.46 = 1069.5 \text{ kN/m}^2$$

$$Q_{\text{ult}} = A'q_u' = (1069.5)(1.5615) = \mathbf{1670 \text{ kN}} \quad \blacksquare$$

3.12   *Bearing Capacity of a Continuous Foundation Subjected to Eccentric Inclined Loading*

The problem of ultimate bearing capacity of a *continuous foundation* subjected to an eccentric inclined load was studied by Saran and Agarwal (1991). If a continuous foundation is located at a depth D_f below the ground surface and is subjected to an eccentric load (load eccentricity $= e$) inclined at an angle β to the vertical, the ultimate capacity can be expressed as

$$Q_{\text{ult}} = B\left[c'N_{c(ei)} + qN_{q(ei)} + \frac{1}{2}\gamma B N_{\gamma(ei)}\right] \quad (3.67)$$

where $N_{c(ei)}$, $N_{q(ei)}$, and $N_{\gamma(ei)}$ = bearing capacity factors

$$q = \gamma D_f$$

The variations of the bearing capacity factors with e/B, ϕ', and β derived by Saran and Agarwal are given in Figures 3.26, 3.27, and 3.28.

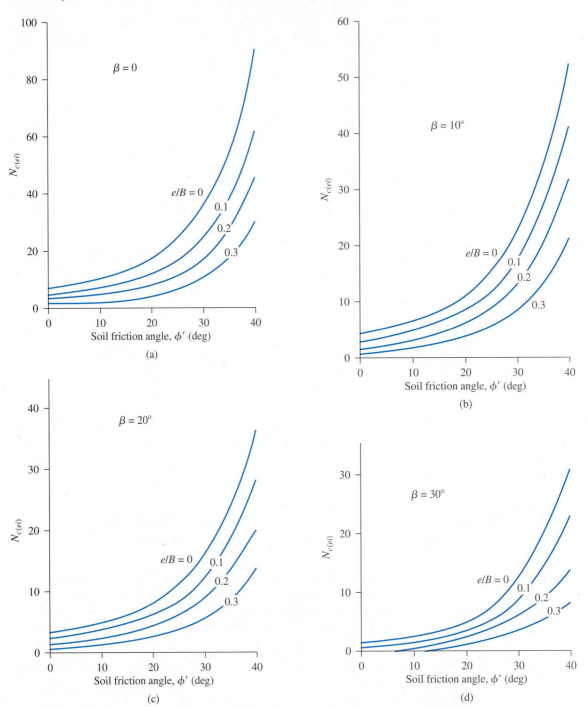

Figure 3.26 Variation of $N_{c(ei)}$ with ϕ', e/B, and β

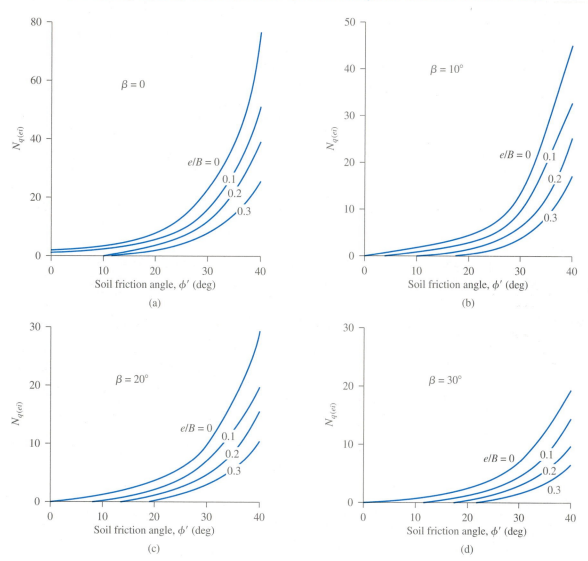

Figure 3.27 Variation of $N_{q(ei)}$ with ϕ', e/B, and β

Example 3.10

A continuous foundation is shown in Figure 3.29. Estimate the ultimate load, Q_{ult} per unit length of the foundation.

Solution

With $c' = 0$, from Eq. (3.67),

$$Q_{ult} = \left[qN_{q(ei)} + \frac{1}{2}\gamma B N_{\gamma(ei)} \right]$$

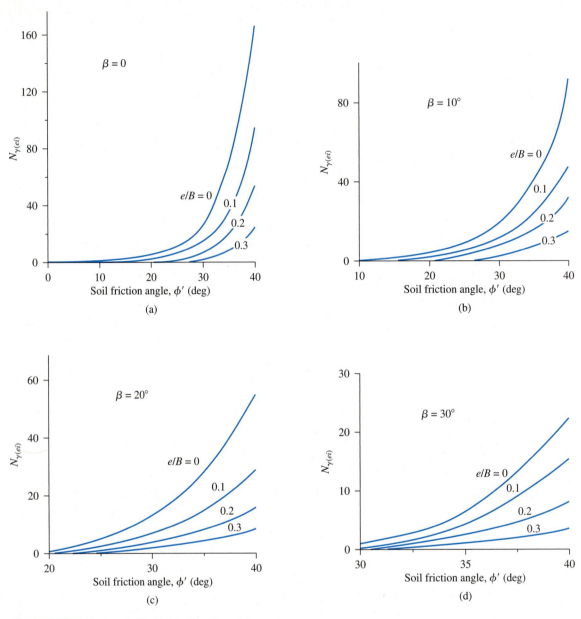

Figure 3.28 Variation of $N_{\gamma(ei)}$ with ϕ', e/B, and β

$B = 1.5$ m, $q = D_f\gamma = (1)(16) = 16$ kN/m^2, $e/B = 0.15/1.5 = 0.1$, and $\beta = 20°$. From Figures 3.27(c) and 3.28(c), $N_{q(ei)} = 14.2$ and $N_{\gamma(ei)} = 20$. Hence,

$$Q_{\text{ult}} = (1.5)[(16)(14.2) + (\tfrac{1}{2})(16)(1.5)(20)] = \mathbf{700.8 \text{ kN/m}}$$

Q_{ult}

20°

$\gamma = 16\ kN/m^3$
$\phi' = 35°$
$c' = 0$

1 m

0.15 m

1.5 m

Figure 3.29

Problems

3.1 For the following cases, determine the allowable gross vertical load-bearing capacity of the foundation. Use Terzaghi's equation and assume general shear failure in soil. Use FS = 4.

Part	B	D_f	ϕ'	c'	γ	Foundation type
a.	1.22 m	0.91 m	25°	28.75 kN/m²	17.29 kN/m³	Continuous
b.	2 m	1 m	30°	0	17 kN/m³	Continuous
c.	3 m	2 m	30°	0	16.5 kN/m³	Square

3.2 A square column foundation has to carry a gross allowable load of 1805 kN (FS = 3). Given: $D_f = 1.5$ m, $\gamma = 15.9\ kN/m^3$, $\phi' = 34°$, and $c' = 0$. Use Terzaghi's equation to determine the size of the foundation (B). Assume general shear failure.

3.3 Use the general bearing capacity equation [Eq. (3.19)] to solve the following:
  a. Problem 3.1a
  b. Problem 3.1b
  c. Problem 3.1c

3.4 The applied load on a shallow square foundation makes an angle of 15° with the vertical. Given: $B = 1.83$ m, $D_f = 0.9$ m, $\gamma = 18.08\ kN/m^3$, $\phi' = 25°$, and $c' = 23.96\ kN/m^2$. Use FS = 4 and determine the gross allowable load. Use Eq. (3.19).

3.5 A column foundation (Figure P3.5) is 3 m × 2 m in plan. Given: $D_f = 1.5$ m, $\phi' = 25°$, $c' = 70\ kN/m^2$. Using Eq. (3.19) and FS = 3, determine the net allowable load [see Eq. (3.15)] the foundation could carry.

3.6 For a square foundation that is $B \times B$ in plan, $D_f = 2$ m; vertical gross allowable load, $Q_{all} = 3330$ kN, $\gamma = 16.5\ kN/m^3$; $\phi' = 30°$; $c' = 0$; and FS = 4. Determine the size of the foundation. Use Eq. (3.19).

Figure P3.5

3.7   For the design of a shallow foundation, given the following:

Soil: $\phi' = 25°$
            $c' = 50 \text{ kN/m}^2$
            Unit weight, $\gamma = 17 \text{ kN/m}^3$
            Modulus of elasticity, $E_s = 1020 \text{ kN/m}^2$
            Poisson's ratio, $\mu_s = 0.35$

Foundation: $L = 1.5 \text{ m}$
             $B = 1 \text{ m}$
             $D_f = 1 \text{ m}$

Calculate the ultimate bearing capacity. Use Eq. (3.27).

3.8   An eccentrically loaded foundation is shown in Figure P3.8. Use FS of 4 and determine the maximum allowable load that the foundation can carry. Use Meyerhof's effective area method.

3.9   Repeat Problem 3.8 using Prakash and Saran's method.

3.10 For an eccentrically loaded continuous foundation on sand, given $B = 1.8 \text{ m}$, $D_f = 0.9 \text{ m}$, $e/B = 0.12$ (one-way eccentricity), $\gamma = 16 \text{ kN/m}^3$, and $\phi' = 35°$. Using the reduction factor method, estimate the ultimate load per unit length of the foundation.

Figure P3.8

3.11 An eccentrically loaded continuous foundation is shown in Figure P3.11. Determine the ultimate load Q_u per unit length that the foundation can carry. Use the reduction factor method.

3.12 A square footing is shown in Figure P3.12. Use FS = 6, and determine the size of the footing. Use Prakash and Saran theory [Eq. (3.43)].

Figure P3.11

Figure P3.12

3.13 The shallow foundation shown in Figure 3.19 measures 1.2 m × 1.8 m and is subjected to a centric load and a moment. If $e_B = 0.12$ m, $e_L = 0.36$ m, and the depth of the foundation is 1 m, determine the allowable load the foundation can carry. Use a factor of safety of 3. For the soil, we are told that unit weight $\gamma = 17$ kN/m³, friction angle $\phi' = 35°$, and cohesion $c' = 0$.

References

BOZOZUK, M. (1972). "Foundation Failure of the Vankleek Hill Tower Site," *Proceedings,* Specialty Conference on Performance of Earth and Earth-Supported Structures, Vol. 1, Part 2, pp. 885–902.

BRAND, E. W., MUKTABHANT, C., and TAECHANTHUMMARAK, A. (1972). "Load Test on Small Foundations in Soft Clay," *Proceedings, Specialty Conference on Performance of Earth and Earth-Supported Structures*, American Society of Civil Engineers, Vol. 1, Part 2, pp. 903–928.

CAQUOT, A., and KERISEL, J. (1953). "Sur le terme de surface dans le calcul des fondations en milieu pulverulent," *Proceedings, Third International Conference on Soil Mechanics and Foundation Engineering,* Zürich, Vol. I, pp. 336–337.

DE BEER, E. E. (1970). "Experimental Determination of the Shape Factors and Bearing Capacity Factors of Sand," *Geotechnique,* Vol. 20, No. 4, pp. 387–411.

HANNA, A. M., and MEYERHOF, G. G. (1981). "Experimental Evaluation of Bearing Capacity of Footings Subjected to Inclined Loads," *Canadian Geotechnical Journal,* Vol. 18, No. 4, pp. 599–603.

HANSEN, J. B. (1970). *A Revised and Extended Formula for Bearing Capacity,* Bulletin 28, Danish Geotechnical Institute, Copenhagen.

HIGHTER, W. H., and ANDERS, J. C. (1985). "Dimensioning Footings Subjected to Eccentric Loads," *Journal of Geotechnical Engineering,* American Society of Civil Engineers, Vol. 111, No. GT5, pp. 659–665.

KUMBHOJKAR, A. S. (1993). "Numerical Evaluation of Terzaghi's N_γ," *Journal of Geotechnical Engineering,* American Society of Civil Engineers, Vol. 119, No. 3, pp. 598–607.

MEYERHOF, G. G. (1953). "The Bearing Capacity of Foundations Under Eccentric and Inclined Loads," *Proceedings, Third International Conference on Soil Mechanics and Foundation Engineering,* Zürich, Vol. 1, pp. 440–445.

MEYERHOF, G. G. (1963). "Some Recent Research on the Bearing Capacity of Foundations," *Canadian Geotechnical Journal,* Vol. 1, No. 1, pp. 16–26.

PRAKASH, S., and SARAN, S. (1971). "Bearing Capacity of Eccentrically Loaded Footings," *Journal of the Soil Mechanics and Foundations Division,* ASCE, Vol. 97, No. SM1, pp. 95–117.

PRANDTL, L. (1921). "Über die Eindringungsfestigkeit (Härte) plastischer Baustoffe und die Festigkeit von Schneiden," *Zeitschrift für angewandte Mathematik und Mechanik,* Vol. 1, No. 1, pp. 15–20.

PURKAYASTHA, R.D., and CHAR, R. A. N. (1977). "Stability Analysis of Eccentrically Loaded Footings," *Journal of Geotechnical Engineering Div.,* ASCE, Vol. 103, No. 6, pp. 647–651.

REISSNER, H. (1924). "Zum Erddruckproblem," *Proceedings, First International Congress of Applied Mechanics,* Delft, pp. 295–311.

SARAN, S., and AGARWAL, R. B. (1991). "Bearing Capacity of Eccentrically Obliquely Loaded Footing," *Journal of Geotechnical Engineering,* ASCE, Vol. 117, No. 11, pp. 1669–1690.

TERZAGHI, K. (1943). *Theoretical Soil Mechanics,* Wiley, New York.

VESIC, A. S. (1963). "Bearing Capacity of Deep Foundations in Sand," *Highway Research Record* No. 39, National Academy of Sciences, pp. 112–153.

VESIC, A. S. (1973). "Analysis of Ultimate Loads of Shallow Foundations," *Journal of the Soil Mechanics and Foundations Division,* American Society of Civil Engineers, Vol. 99, No. SM1, pp. 45–73.

4 Ultimate Bearing Capacity of Shallow Foundations: Special Cases

4.1 Introduction

The ultimate bearing capacity problems described in Chapter 3 assume that the soil supporting the foundation is homogeneous and extends to a great depth below the bottom of the foundation. They also assume that the ground surface is horizontal. However, that is not true in all cases: It is possible to encounter a rigid layer at a shallow depth, or the soil may be layered and have different shear strength parameters. In some instances, it may be necessary to construct foundations on or near a slope, or it may be required to design a foundation subjected to uplifting load.

This chapter discusses bearing capacity problems relating to these special cases.

4.2 Foundation Supported by a Soil with a Rigid Base at Shallow Depth

Figure 4.1(a) shows a shallow, rough *continuous* foundation supported by a soil that extends to a great depth. Neglecting the depth factor, for vertical loading Eq. (3.19) will take the form

$$q_u = c'N_c + qN_q + \frac{1}{2}\gamma BN_\gamma \tag{4.1}$$

The general approach for obtaining expressions for N_c, N_q, and N_γ was outlined in Chapter 3. The extent of the failure zone in soil, D, at ultimate load obtained in the derivation of N_c and N_q by Prandtl (1921) and Reissner (1924) is given in Figure 4.1(b). Similarly, the magnitude of D obtained by Lundgren and Mortensen (1953) in evaluating N_γ is given in the figure.

Now, if a rigid, rough base is located at a depth of $H < D$ below the bottom of the foundation, full development of the failure surface in soil will be restricted. In such a case, the soil failure zone and the development of slip lines at ultimate load will be as shown in Figure 4.2.

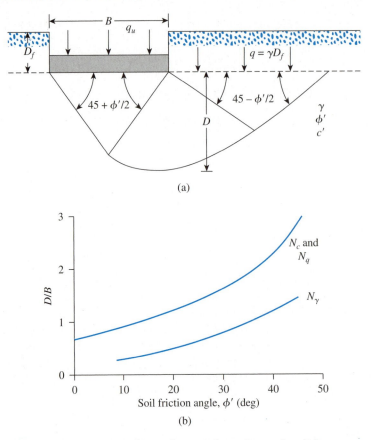

(a)

(b)

Figure 4.1 (a) Failure surface under a rough continuous foundation; (b) variation of D/B with soil friction angle ϕ'

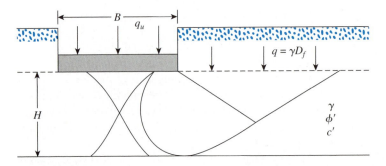

Figure 4.2 Failure surface under a rough, continuous foundation with a rigid, rough base located at a shallow depth

Mandel and Salencon (1972) determined the bearing capacity factors applicable to this case by numerical integration, using the theory of plasticity. According to their theory, the ultimate bearing capacity of a rough continuous foundation with a rigid, rough base located at a shallow depth can be given by the relation

$$q_u = c'N_c^* + qN_q^* + \frac{1}{2}\gamma BN_\gamma^*$$

(4.2)

where

$N_c^*, N_q^*, N_\gamma^* =$ modified bearing capacity factors
$B =$ width of foundation
$\gamma =$ unit weight of soil

Note that, for $H \geqslant D$, $N_c^* = N_c$, $N_q^* = N_q$, and $N_\gamma^* = N_\gamma$ (Lundgren and Mortensen, 1953). The variations of N_c^*, N_q^*, and N_γ^* with H/B and the soil friction angle ϕ' are given in Figures 4.3, 4.4, and 4.5, respectively.

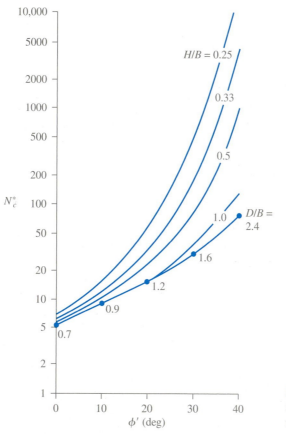

Figure 4.3 Mandel and Salencon's bearing capacity factor N_c^* [Eq. (4.2)]

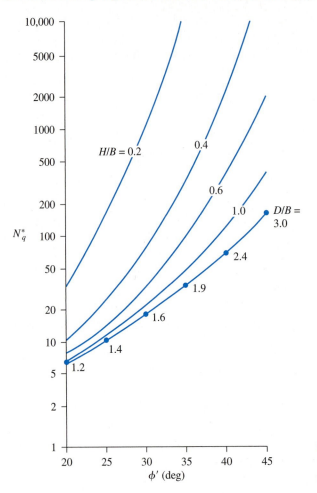

Figure 4.4 Mandel and Salencon's bearing capacity factor N_q^* [Eq. (4.2)]

Rectangular Foundation on Granular Soil

Neglecting the depth factors, the ultimate bearing capacity of rough circular and rectangular foundations on a sand layer ($c' = 0$) with a rough, rigid base located at a shallow depth can be given as

$$q_u = qN_q^*F_{qs}^* + \frac{1}{2}\gamma BN_\gamma^* F_{\gamma s}^* \tag{4.3}$$

where F_{qs}^*, $F_{\gamma s}^*$ = modified shape factors.

The shape factors F_{qs}^* and $F_{\gamma s}^*$ are functions of H/B and ϕ'. On the basis of the work of Meyerhof and Chaplin (1953), and simplifying the assumption that, in radial planes, the stresses and shear zones are identical to those in transverse planes, Meyerhof (1974) proposed that

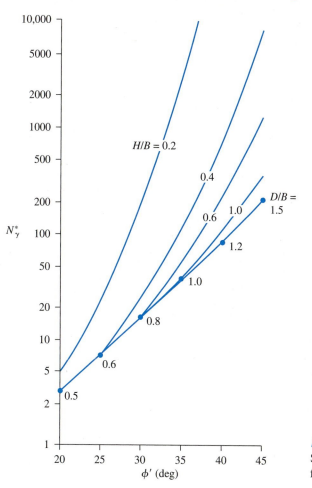

Figure 4.5 Mandel and Salencon's bearing capacity factor N_γ^* [Eq. (4.2)]

$$F_{qs}^* \approx 1 - m_1 \left(\frac{B}{L} \right) \tag{4.4}$$

and

$$F_{\gamma s}^* \approx 1 - m_2 \left(\frac{B}{L} \right) \tag{4.5}$$

where L = length of the foundation. The variations of m_1 and m_2 with H/B and ϕ' are shown in Figure 4.6.

More recently, Cerato and Lutenegger (2006) provided some test results for the bearing capacity factor, N_γ^*. These tests were conducted using *square* and *circular* plates with B varying from 0.152 m (6 in.) to 0.305 m (12 in.). It was assumed that Terzaghi's bearing-capacity equations for *square and circular foundations* can be used. Or, from Eqs. (3.10) and (3.11) with $c' = 0$,

$$q_u = q N_q^* + 0.4 \gamma B N_\gamma^* \quad \text{(square foundation)} \tag{4.6}$$

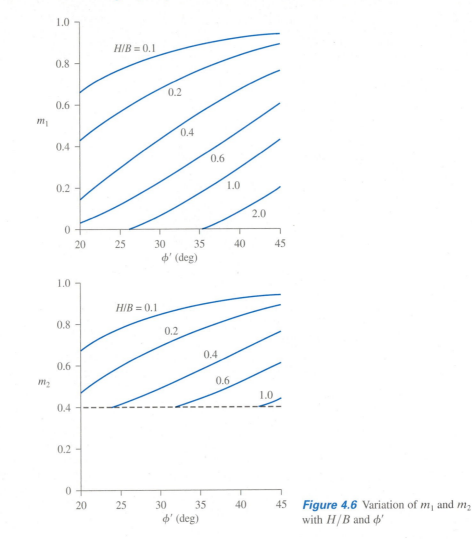

Figure 4.6 Variation of m_1 and m_2 with H/B and ϕ'

and

$$q_u = qN_q^* + 0.3\gamma BN_\gamma^* \text{ (circular foundation)} \tag{4.7}$$

The experimentally determined variation of N_γ^* is shown in Figure 4.7. It also was observed in this study that N_γ^* becomes equal to N_γ at $H/B \approx 3$ instead of D/B, as shown in Figure 4.5. For that reason, Figure 4.7 shows the variation of N_γ^* for $H/B = 0.5$ to 3.0.

Foundation on Saturated Clay

For saturated clay (i.e., under the undrained condition, or $\phi = 0$), Eq. (4.2) will simplify to the form

$$q_u = c_u N_c^* + q \tag{4.8}$$

Figure 4.7 Cerato and Lutenegger's test results for N_γ^*

Mandel and Salencon (1972) performed calculations to evaluate N_c^* for *continuous foundations.* Similarly, Buisman (1940) gave the following relationship for obtaining the ultimate bearing capacity of square foundations:

$$q_{u(\text{square})} = \left(\pi + 2 + \frac{B}{2H} - \frac{\sqrt{2}}{2}\right)c_u + q \qquad \left(\text{for } \frac{B}{2H} - \frac{\sqrt{2}}{2} \geq 0\right) \tag{4.9}$$

In this equation, c_u is the undrained shear strength.

Equation (4.9) can be rewritten as

$$q_{u(\text{square})} = \underbrace{5.14\left(1 + \frac{0.5\dfrac{B}{H} - 0.707}{5.14}\right)}_{N_{c(\text{square})}^*}c_u + q \tag{4.10}$$

Table 4.1 gives the values of N_c^* for continuous and square foundations.

Table 4.1 Values of N_c^* for Continuous and Square Foundations ($\phi = 0$)

$\dfrac{B}{H}$	N_c^*	
	Square[a]	**Continuous**[b]
2	5.43	5.24
3	5.93	5.71
4	6.44	6.22
5	6.94	6.68
6	7.43	7.20
8	8.43	8.17
10	9.43	9.05

[a] Buisman's analysis (1940)
[b] Mandel and Salencon's analysis (1972)

Example 4.1

A square foundation measuring 0.76 m $\times$ 0.76 m is constructed on a layer of sand. We are given that $D_f = 0.61$ m, $\gamma = 17.29$ kN/m^3, $\phi' = 35°$, and $c' = 0$. A rock layer is located at a depth of 0.46 m below the bottom of the foundation. Using a factor of safety of 4, determine the gross allowable load the foundation can carry.

Solution

From Eq. (4.3),

$$q_u = qN_q^*F_{qs}^* + \frac{1}{2}\gamma BN_\gamma^*F_{\gamma s}^*$$

and we also have

$$q = 17.29 \times 0.61 = 10.55 \text{ kN/m}^2$$

For $\phi' = 35°$, $H/B = 0.46/0.76$ m $= 0.6$, $N_q^* \approx 90$ (Figure 4.4), and $N_\gamma^* \approx 50$ (Figure 4.5), and we have

$$F_{qs}^* = 1 - m_1(B/L)$$

From Figure 4.6(a), for $\phi' = 35°$, $H/B = 0.6$, and the value of $m_1 = 0.34$, so

$$F_{qs}^* = 1 - (0.34)(0.76/0.76) = 0.66$$

Similarly,

$$F_{\gamma s}^* = 1 - m_2(B/L)$$

From Figure 4.6(b), $m_2 = 0.45$, so

$$F_{\gamma s}^* = 1 - (0.45)(0.76/0.76) = 0.55$$

Hence,

$$q_u = (10.55)(90)(0.66) + (1/2)(17.29)(0.76)(50)(0.55) = 807.35 \text{ kN/m}^2$$

and

$$Q_{all} = \frac{q_u B^2}{\text{FS}} = \frac{(807.35)(0.76 \times 0.76)}{4} = \textbf{116.58 kN} \qquad \blacksquare$$

Example 4.2

Solve Example 4.1 using Eq. (4.6).

Solution

From Eq. (4.6),

$$q_u = qN_q^* + 0.4\gamma BN_\gamma^*$$

For $\phi' = 35°$ and $H/B = 0.6$, the value of $N_q^* \approx 90$ (Figure 4.4) and $N_\gamma^* \approx 230$ (Figure 4.7).

So

$$q_u = (10.55)(90) + (0.4)(17.29)(0.76)(230) = 949.5 + 1208.9 = 2158.4 \text{ kN/m}^2$$

$$Q_{all} = \frac{q_u B^2}{\text{FS}} \approx \textbf{311.7 kN} \qquad \blacksquare$$

Example 4.3

Consider a square foundation 1 m × 1 m in plan located on a saturated clay layer underlain by a layer of rock. Given:

Clay: $c_u = 72 \text{ kN/m}^2$
Unit weigh*t*: $\gamma = 18 \text{ kN/m}^3$
Distance between the bottom of foundation and the rock layer = 0.25 m
$D_f = 1 \text{ m}$

Estimate the gross allowable bearing capacity of the foundation. Use FS = 3.

Solution

From Eq. (4.10),

$$q_u = 5.14\left(1 + \frac{0.5\dfrac{B}{H} - 0.707}{5.14}\right)c_u + q$$

For $B/H = 1/0.25 = 4$; $c_u = 72$ kN/m²; and $q = \gamma D_f = (18)(1) = 18$ kN/m³.

$$q_u = 5.14\left[1 + \frac{(0.5)(4) - 0.707}{5.14}\right]72 + 18 = 481.2 \text{ kN/m}^2$$

$$q_{\text{all}} = \frac{q_u}{\text{FS}} = \frac{481.2}{3} = \mathbf{160.4 \text{ kN/m}^2} \qquad \blacksquare$$

4.3 Bearing Capacity of Layered Soils: Stronger Soil Underlain by Weaker Soil

The bearing capacity equations presented in Chapter 3 involve cases in which the soil supporting the foundation is homogeneous and extends to a considerable depth. The cohesion, angle of friction, and unit weight of soil were assumed to remain constant for the bearing capacity analysis. However, in practice, layered soil profiles are often encountered. In such instances, the failure surface at ultimate load may extend through two or more soil layers, and a determination of the ultimate bearing capacity in layered soils can be made in only a limited number of cases. This section features the procedure for estimating the bearing capacity for layered soils proposed by Meyerhof and Hanna (1978) and Meyerhof (1974).

Figure 4.8 shows a shallow, continuous foundation supported by a *stronger soil layer,* underlain by a weaker soil that extends to a great depth. For the two soil layers, the physical parameters are as follows:

	Soil properties		
Layer	Unit weight	Friction angle	Cohesion
Top	γ_1	ϕ_1'	c_1'
Bottom	γ_2	ϕ_2'	c_2'

At ultimate load per unit area (q_u), the failure surface in soil will be as shown in the figure. If the depth H is relatively small compared with the foundation width B, a punching shear failure will occur in the top soil layer, followed by a general shear failure in the bottom soil layer. This is shown in Figure 4.8(a). However, if the depth H is relatively large, then the failure surface will be completely located in the top soil layer, which is the upper limit for the ultimate bearing capacity. This is shown in Figure 4.8b.

The ultimate bearing capacity for this problem, as shown in Figure 4.8a, can be given as

$$q_u = q_b + \frac{2(C_a + P_p \sin \delta')}{B} - \gamma_1 H \tag{4.11}$$

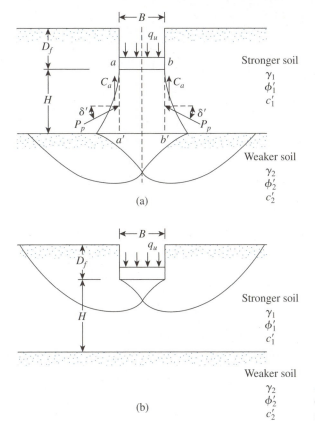

(a)

(b)

Figure 4.8 Bearing capacity of a continuous foundation on layered soil

where

B = width of the foundation
C_a = adhesive force
P_p = passive force per unit length of the faces aa' and bb'
q_b = bearing capacity of the bottom soil layer
δ' = inclination of the passive force P_p with the horizontal

Note that, in Eq. (4.11),

$$C_a = c_a'H$$

where c_a' = adhesion.

Equation (4.11) can be simplified to the form

$$q_u = q_b + \frac{2c_a'H}{B} + \gamma_1 H^2\left(1 + \frac{2D_f}{H}\right)\frac{K_{pH}\tan\delta'}{B} - \gamma_1 H \qquad (4.12)$$

where K_{pH} = horizontal component of passive earth pressure coefficient.
However, let

$$K_{pH}\tan\delta' = K_s\tan\phi_1' \qquad (4.13)$$

where K_s = punching shear coefficient. Then,

$$q_u = q_b + \frac{2c'_a H}{B} + \gamma_1 H^2 \left(1 + \frac{2D_f}{H}\right)\frac{K_s \tan \phi'_1}{B} - \gamma_1 H \qquad (4.14)$$

The punching shear coefficient, K_s, is a function of q_2/q_1 and ϕ'_1, or, specifically,

$$K_s = f\left(\frac{q_2}{q_1}, \phi'_1\right)$$

Note that q_1 and q_2 are the ultimate bearing capacities of a continuous foundation of width B under vertical load on the surfaces of homogeneous thick beds of upper and lower soil, or

$$q_1 = c'_1 N_{c(1)} + \tfrac{1}{2}\gamma_1 B N_{\gamma(1)} \qquad (4.15)$$

and

$$q_2 = c'_2 N_{c(2)} + \tfrac{1}{2}\gamma_2 B N_{\gamma(2)} \qquad (4.16)$$

where

$N_{c(1)}, N_{\gamma(1)}$ = bearing capacity factors for friction angle ϕ'_1 (Table 3.3)

$N_{c(2)}, N_{\gamma(2)}$ = bearing capacity factors for friction angle ϕ'_2 (Table 3.3)

Observe that, for the top layer to be a stronger soil, q_2/q_1 should be less than unity.

The variation of K_s with q_2/q_1 and ϕ'_1 is shown in Figure 4.9. The variation of c'_a/c'_1 with q_2/q_1 is shown in Figure 4.10. If the height H is relatively large, then the

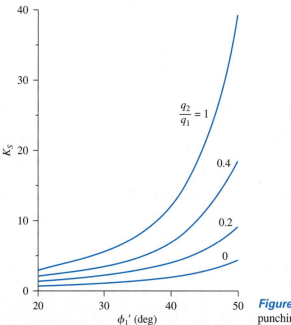

Figure 4.9 Meyerhof and Hanna's punching shear coefficient K_s

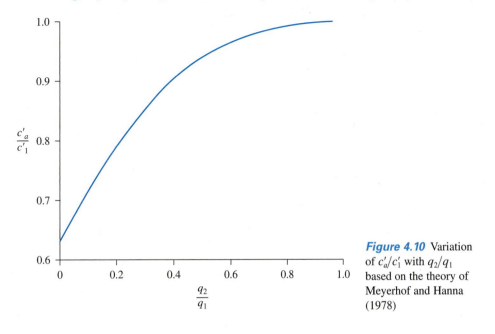

Figure 4.10 Variation of c_a'/c_1' with q_2/q_1 based on the theory of Meyerhof and Hanna (1978)

failure surface in soil will be completely located in the stronger upper-soil layer (Figure 4.8b). For this case,

$$q_u = q_t = c_1'N_{c(1)} + qN_{q(1)} + \tfrac{1}{2}\gamma_1 BN_{\gamma(1)}. \tag{4.17}$$

where $N_{c(1)}$, $N_{q(1)}$, and $N_{\gamma(1)}$ = bearing capacity factors for $\phi' = \phi_1'$ (Table 3.3) and $q = \gamma_1 D_f$.

Combining Eqs. (4.14) and (4.17) yields

$$q_u = q_b + \frac{2c_a'H}{B} + \gamma_1 H^2\left(1 + \frac{2D_f}{H}\right)\frac{K_s \tan \phi_1'}{B} - \gamma_1 H \le q_t \tag{4.18}$$

For rectangular foundations, the preceding equation can be extended to the form

$$q_u = q_b + \left(1 + \frac{B}{L}\right)\left(\frac{2c_a'H}{B}\right)$$
$$+ \gamma_1 H^2\left(1 + \frac{B}{L}\right)\left(1 + \frac{2D_f}{H}\right)\left(\frac{K_s \tan \phi_1'}{B}\right) - \gamma_1 H \le q_t \tag{4.19}$$

where

$$q_b = c_2'N_{c(2)}F_{cs(2)} + \gamma_1(D_f + H)N_{q(2)}F_{qs(2)} + \frac{1}{2}\gamma_2 BN_{\gamma(2)}F_{\gamma s(2)} \tag{4.20}$$

and

$$q_t = c_1' N_{c(1)} F_{cs(1)} + \gamma_1 D_f N_{q(1)} F_{qs(1)} + \frac{1}{2} \gamma_1 B N_{\gamma(1)} F_{\gamma s(1)} \tag{4.21}$$

in which

$F_{cs(1)}, F_{qs(1)}, F_{\gamma s(1)}$ = shape factors with respect to top soil layer (Table 3.4)
$F_{cs(2)}, F_{qs(2)}, F_{\gamma s(2)}$ = shape factors with respect to bottom soil layer (Table 3.4)

Special Cases

1. *Top layer is strong sand and bottom layer is saturated soft clay* $(\phi_2 = 0)$. From Eqs. (4.19), (4.20), and (4.21),

$$q_b = \left(1 + 0.2 \frac{B}{L} \right) 5.14 c_2 + \gamma_1 (D_f + H) \tag{4.22}$$

and

$$q_t = \gamma_1 D_f N_{q(1)} F_{qs(1)} + \frac{1}{2} \gamma_1 B N_{\gamma(1)} F_{\gamma s(1)} \tag{4.23}$$

Hence,

$$q_u = \left(1 + 0.2 \frac{B}{L} \right) 5.14 c_2 + \gamma_1 H^2 \left(1 + \frac{B}{L} \right) \left(1 + \frac{2D_f}{H} \right) \frac{K_s \tan \phi_1'}{B}$$
$$+ \gamma_1 D_f \leq \gamma_1 D_f N_{q(1)} F_{qs(1)} + \frac{1}{2} \gamma_1 B N_{\gamma(1)} F_{\gamma s(1)} \tag{4.24}$$

where c_2 = undrained cohesion.

For a determination of K_s from Figure 4.9,

$$\frac{q_2}{q_1} = \frac{c_2 N_{c(2)}}{\frac{1}{2} \gamma_1 B N_{\gamma(1)}} = \frac{5.14 c_2}{0.5 \gamma_1 B N_{\gamma(1)}} \tag{4.25}$$

2. *Top layer is stronger sand and bottom layer is weaker sand* $(c_1' = 0, c_2' = 0)$. The ultimate bearing capacity can be given as

$$q_u = \left[\gamma_1 (D_f + H) N_{q(2)} F_{qs(2)} + \frac{1}{2} \gamma_2 B N_{\gamma(2)} F_{\gamma s(2)} \right]$$
$$+ \gamma_1 H^2 \left(1 + \frac{B}{L} \right) \left(1 + \frac{2D_f}{H} \right) \frac{K_s \tan \phi_1'}{B} - \gamma_1 H \leq q_t \tag{4.26}$$

where

$$q_t = \gamma_1 D_f N_{q(1)} F_{qs(1)} + \frac{1}{2}\gamma_1 B N_{\gamma(1)} F_{\gamma s(1)} \tag{4.27}$$

Then

$$\frac{q_2}{q_1} = \frac{\frac{1}{2}\gamma_2 B N_{\gamma(2)}}{\frac{1}{2}\gamma_1 B N_{\gamma(1)}} = \frac{\gamma_2 N_{\gamma(2)}}{\gamma_1 N_{\gamma(1)}} \tag{4.28}$$

3. *Top layer is stronger saturated clay* $(\phi_1 = 0)$ *and bottom layer is weaker saturated clay* $(\phi_2 = 0)$. The ultimate bearing capacity can be given as

$$q_u = \left(1 + 0.2\frac{B}{L}\right)5.14c_2 + \left(1 + \frac{B}{L}\right)\left(\frac{2c_aH}{B}\right) + \gamma_1 D_f \leq q_t \tag{4.29}$$

where

$$q_t = \left(1 + 0.2\frac{B}{L}\right)5.14c_1 + \gamma_1 D_f \tag{4.30}$$

and c_1 and c_2 are undrained cohesions. For this case,

$$\frac{q_2}{q_1} = \frac{5.14c_2}{5.14c_1} = \frac{c_2}{c_1} \tag{4.31}$$

Example 4.4

Refer to Figure 4.8(a) and consider the case of a continuous foundation with $B = 2$ m, $D_f = 1.2$ m, and $H = 1.5$ m. The following are given for the two soil layers:

Top sand layer:

Unit weight $\gamma_1 = 17.5$ kN/m^3
$\phi_1' = 40°$
$c_1' = 0$

Bottom clay layer:

Unit weight $\gamma_2 = 16.5$ kN/m^3
$\phi_2' = 0$
$c_2 = 30$ kN/m^2

Determine the gross ultimate load per unit length of the foundation.

Solution

For this case, Eqs. (4.24) and (4.25) apply. For $\phi'_1 = 40°$, from Table 3.3, $N_\gamma = 109.41$ and

$$\frac{q_2}{q_1} = \frac{c_2 N_{c(2)}}{0.5\gamma_1 B N_{\gamma(1)}} = \frac{(30)(5.14)}{(0.5)(17.5)(2)(109.41)} = 0.081$$

From Figure 4.9, for $c_2 N_{c(2)}/0.5\gamma_1 B N_{\gamma(1)} = 0.081$ and $\phi'_1 = 40°$, the value of $K_s \approx 2.5$. Equation (4.24) then gives

$$q_u = \left[1 + (0.2)\left(\frac{B}{L}\right)\right]5.14c_2 + \left(1 + \frac{B}{L}\right)\gamma_1 H^2\left(1 + \frac{2D_f}{H}\right)K_s\frac{\tan\phi'_1}{B} + \gamma_1 D_f$$

$$= [1 + (0.2)(0)](5.14)(30) + (1 + 0)(17.5)(1.5)^2$$

$$\times \left[1 + \frac{(2)(1.2)}{1.5}\right](2.5)\frac{\tan 40}{2.0} + (17.5)(1.2)$$

$$= 154.2 + 107.4 + 21 = 282.6 \text{ kN/m}^2$$

Again, from Eq. (4.27),

$$q_t = \gamma_1 D_f N_{q(1)} F_{qs(1)} + \tfrac{1}{2}\gamma_1 B N_{\gamma(1)} F_{\gamma s(1)}$$

From Table 3.3, for $\phi'_1 = 40°$, $N_\gamma = 109.4$ and $N_q = 64.20$.
From Table 3.4,

$$F_{qs(1)} = 1 + \left(\frac{B}{L}\right)\tan\phi'_1 = 1 + (0)\tan 40 = 1$$

and

$$F_{\gamma s(1)} = 1 - 0.4\frac{B}{L} = 1 - (0.4)(0) = 1$$

so that

$$q_t = (17.5)(1.2)(64.20)(1) + (\tfrac{1}{2})(17.5)(2)(109.4)(1) = 3262.7 \text{ kN/m}^2$$

Hence,

$$q_u = 282.6 \text{ kN/m}^2$$

$$Q_u = (282.6)(B) = (282.6)(2) = \textbf{565.2 kN/m}$$ ■

Example 4.5

A foundation 1.5 m $\times$ 1 m is located at a depth D_f of 1 m in a stronger clay. A softer clay layer is located at a depth H of 3 ft, measured from the bottom of the foundation. For the top clay layer,

Undrained shear strength = 120 kN/m^2
Unit weight = 16.8 kN/m^3

and for the bottom clay layer,

Undrained shear strength = 48 kN/m^2
Unit weight = 16.2 kN/m^3

Determine the gross allowable load for the foundation with an FS of 3.

Solution

For this problem, Eqs. (4.29), (4.30), and (4.31) will apply, or

$$q_u = \left(1 + 0.2\frac{B}{L}\right)5.14c_2 + \left(1 + \frac{B}{L}\right)\left(\frac{2c_aH}{B}\right) + \gamma_1D_f$$

$$\leq \left(1 + 0.2\frac{B}{L}\right)5.14c_1 + \gamma_1D_f$$

We are given the following data:

$$B = 1\text{ m} \qquad H = 1\text{ m} \qquad D_f = 1\text{ m}$$
$$L = 1.5\text{ m} \qquad \gamma_1 = 16.8\text{ kN/m}^3$$

From Figure 4.10 for $c_2/c_1 = 48/120 = 0.4$, the value of $c_a/c_1 \approx 0.9$, so

$$c_a = (0.9)(2500) = 108\text{ kN/m}^2$$

and

$$q_u = \left[1 + (0.2)\left(\frac{1}{1.5}\right)\right](5.14)(48) + \left(1 + \frac{1}{1.5}\right)\left[\frac{(2)(108)(1)}{1}\right] + (16.8)(1)$$

$$= 279.6 + 360 + 16.8 = 656.4\text{ kN/m}^2$$

As a check, we have, from Eq. (4.30),

$$q_t = \left[1 + (0.2)\left(\frac{1}{1.5}\right)\right](5.14)(120) + (16.8)(1)$$

$$= 699 + 16.8 = 715.8\text{ kN/m}^2$$

Thus, $q_u = 656.4$ kN/m^2 (i.e., the smaller of the two values just calculated), and

$$q_{all} = \frac{q_u}{\text{FS}} = \frac{656.4}{3} = 218.8\text{ kN/m}^2$$

The total allowable load is therefore

$$(q_{all})(1 \times 1.5) = \textbf{328.2 kN}$$

4.4 ## *Bearing Capacity of Layered Soil: Weaker Soil Underlain by Stronger Soil*

When a foundation is supported by a weaker soil layer underlain by a stronger layer (Figure 4.11a), the ratio of q_2/q_1 defined by Eqs. (4.15) and (4.16) will be greater than one. Also, if H/B is relatively small, as shown in the left-hand half of Figure 4.11a, the failure surface in soil at ultimate load will pass through both soil layers. However, for larger H/B ratios, the failure surface will be fully located in the top, weaker soil layer, as shown in the

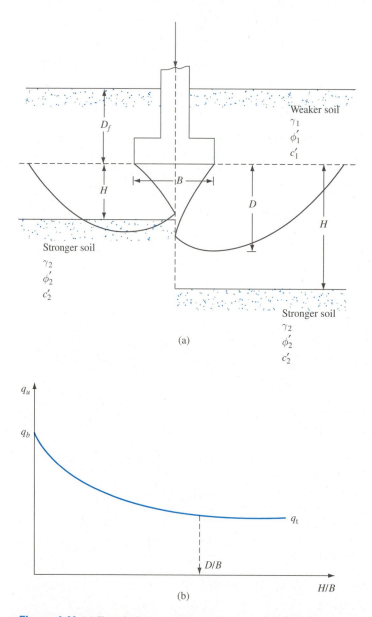

(a)

(b)

Figure 4.11 (a) Foundation on weaker soil layer underlain by stronger sand layer, (b) Nature of variation of q_u with H/B

right-hand half of Figure 4.11a. For this condition, the ultimate bearing capacity (Meyerhof, 1974; Meyerhof and Hanna, 1978) can be given by the empirical equation

$$q_u = q_t + (q_b - q_t)\left(\frac{H}{D}\right)^2 \geq q_t \tag{4.32}$$

where

D = depth of failure surface beneath the foundation in the thick bed of the upper weaker soil layer

q_t = ultimate bearing capacity in a thick bed of the upper soil layer

q_b = ultimate bearing capacity in a thick bed of the lower soil layer

So

$$q_t = c_1 N_{c(1)} F_{cs(1)} + \gamma_1 D_f N_{q(1)} F_{qs(1)} + \frac{1}{2}\gamma_1 B N_{\gamma(1)} F_{\gamma s(1)} \tag{4.33}$$

and

$$q_t = c_2 N_{c(2)} F_{cs(2)} + \gamma_2 D_f N_{q(2)} F_{qs(2)} + \frac{1}{2}\gamma_2 B N_{\gamma(2)} F_{\gamma s(2)} \tag{4.34}$$

where

$N_{c(1)}, N_{q(1)}, N_{\gamma(1)}$ = bearing capacity factors corresponding to the soil friction angle ϕ_1'

$N_{c(2)}, N_{q(2)}, N_{\gamma(2)}$ = bearing capacity factors corresponding to the soil friction angle ϕ_2'

$F_{cs(1)}, F_{qs(1)}, F_{\gamma s(1)}$ = shape factors corresponding to the soil friction angle ϕ_1'

$F_{cs(2)}, F_{qs(2)}, F_{\gamma s(2)}$ = shape factors corresponding to the soil friction angle ϕ_2'

Meyerhof and Hanna (1978) suggested that

- $D \approx B$ for loose sand and clay
- $D \approx 2B$ for dense sand

Equations (4.32), (4.33), and (4.34) imply that the maximum and minimum values of q_u will be q_b and q_t, respectively, as shown in Figure 4.11b.

Example 4.6

Refer to Figure 4.11a. For a layered saturated-clay profile, given: $L = 1.83$ m, $B = 1.22$ m, $D_f = 0.91$ m, $H = 0.61$ m, $\gamma_1 = 17.29$ kN/m³, $\phi_1 = 0$, $c_1 = 57.5$ kN/m², $\gamma_2 = 19.65$ kN/m³, $\phi_2 = 0$, and $c_2 = 119.79$ kN/m². Determine the ultimate bearing capacity of the foundation.

Solution

From Eqs. (4.15) and (4.16),

$$\frac{q_2}{q_1} = \frac{c_2 N_c}{c_1 N_c} = \frac{c_2}{c_1} = \frac{119.79}{57.5} = 2.08 > 1$$

So, Eq. (4.32) will apply.
From Eqs. (4.33) and (4.34) with $\phi_1 = \phi_2 = 0$,

$$q_t = \left(1 + 0.2\frac{B}{L}\right)N_c c_1 + \gamma_1 D_f$$

$$= \left[1 + (0.2)\left(\frac{1.22}{1.83}\right)\right](5.14)(57.5) + (0.91)(17.29) = 334.96 + 15.73$$

$$= 350.69 \text{ kN/m}^2$$

and

$$q_b = \left(1 + 0.2\frac{B}{L}\right)N_c c_2 + \gamma_2 D_f$$

$$= \left[1 + (0.2)\left(\frac{1.22}{1.83}\right)\right](5.14)(119.79) + (0.91)(19.65)$$

$$= 697.82 + 17.88 = 715.7 \text{ kN/m}^2$$

From Eq. (4.32),

$$q_u = q_t + (q_b - q_t)\left(\frac{H}{D}\right)^2$$

$$D \approx B$$

$$q_u = 350.69 + (715.7 - 350.69)\left(\frac{0.61}{1.22}\right)^2 \approx 442 \text{ kN/m}^2 > q_t$$

Hence,

$$q_u = \mathbf{442 \text{ kN/m}^2}$$ ∎

4.5 *Closely Spaced Foundations—Effect on Ultimate Bearing Capacity*

In Chapter 3, theories relating to the ultimate bearing capacity of single rough continuous foundations supported by a homogeneous soil extending to a great depth were discussed. However, if foundations are placed close to each other with similar soil conditions, the ultimate bearing capacity of each foundation may change due to the interference effect of the failure surface in the soil. This was theoretically investigated by Stuart (1962) for *granular soils*. It was assumed that the geometry of the rupture surface in the soil mass would be the same as that assumed by Terzaghi (Figure 3.5). According to Stuart, the following conditions may arise (Figure 4.12).

Case I. (Figure 4.12a) If the center-to-center spacing of the two foundations is $x \geq x_1$, the rupture surface in the soil under each foundation will not overlap. So the ultimate bearing capacity of each continuous foundation can be given by Terzaghi's equation [Eq. (3.3)]. For ($c' = 0$)

$$q_u = qN_q + \tfrac{1}{2}\gamma BN_\gamma \tag{4.35}$$

where N_q, N_γ = Terzaghi's bearing capacity factors (Table 3.1).

Case II. (Figure 4.12b) If the center-to-center spacing of the two foundations ($x = x_2 < x_1$) are such that the Rankine passive zones just overlap, then the magnitude of q_u will still be given by Eq. (4.35). However, the foundation settlement at ultimate load will change (compared to the case of an isolated foundation).

Case III. (Figure 4.12c) This is the case where the center-to-center spacing of the two continuous foundations is $x = x_3 < x_2$. Note that the triangular wedges in the soil

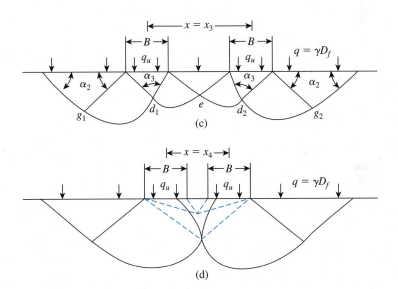

Figure 4.12 Assumptions for the failure surface in granular soil under two closely spaced rough continuous foundations
(*Note:* $\alpha_1 = \phi'$, $\alpha_2 = 45 - \phi'/2$, $\alpha_3 = 180 - 2\phi'$)

under the foundations make angles of $180° - 2\phi'$ at points d_1 and d_2. The arcs of the logarithmic spirals d_1g_1 and d_1e are tangent to each other at d_1. Similarly, the arcs of the logarithmic spirals d_2g_2 and d_2e are tangent to each other at d_2. For this case, the ultimate bearing capacity of each foundation can be given as ($c' = 0$)

$$q_u = qN_q\zeta_q + \tfrac{1}{2}\gamma BN_\gamma\,\zeta_\gamma \qquad (4.36)$$

where ζ_q, ζ_γ = efficiency ratios.

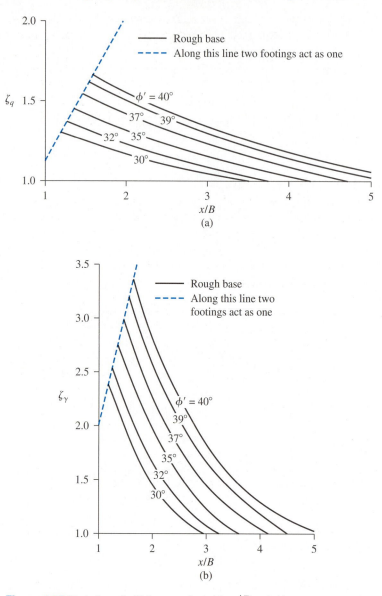

Figure 4.13 Variation of efficiency ratios with x/B and ϕ'

The efficiency ratios are functions of x/B and soil friction angle ϕ'. The theoretical variations of ζ_q and ζ_γ are given in Figure 4.13.

Case IV. (Figure 4.12d): If the spacing of the foundation is further reduced such that $x = x_4 < x_3$, blocking will occur and the pair of foundations will act as a single foundation. The soil between the individual units will form an inverted arch which travels down with the foundation as the load is applied. When the two foundations touch, the zone of arching disappears and the system behaves as a single foundation with a width equal to $2B$. The ultimate bearing capacity for this case can be given by Eq. (4.35), with B being replaced by $2B$ in the second term.

The ultimate bearing capacity of two continuous foundations spaced close to each other may increase since the efficiency ratios are greater than one. However, when the closely spaced foundations are subjected to a similar load per unit area, the settlement S_e will be larger when compared to that for an isolated foundation.

4.6 Bearing Capacity of Foundations on Top of a Slope

In some instances, shallow foundations need to be constructed on top of a slope. In Figure 4.14, the height of the slope is H, and the slope makes an angle β with the horizontal. The edge of the foundation is located at a distance b from the top of the slope. At ultimate load, q_u, the failure surface will be as shown in the figure.

Meyerhof (1957) developed the following theoretical relation for the ultimate bearing capacity for *continuous foundations:*

$$q_u = c'N_{cq} + \tfrac{1}{2}\gamma BN_{\gamma q} \tag{4.37}$$

For purely granular soil, $c' = 0$, thus,

$$q_u = \tfrac{1}{2}\gamma BN_{\gamma q} \tag{4.38}$$

Again, for purely cohesive soil, $\phi = 0$ (the undrained condition); hence,

$$q_u = cN_{cq} \tag{4.39}$$

where c = undrained cohesion.

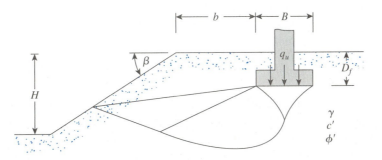

Figure 4.14 Shallow foundation on top of a slope

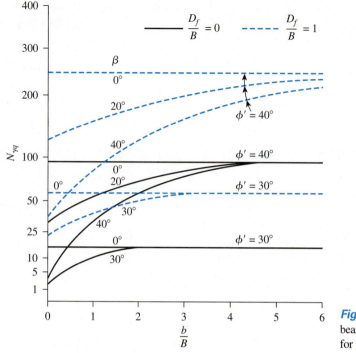

Figure 4.15 Meyerhof's bearing capacity factor $N_{\gamma q}$ for granular soil $(c' = 0)$

The variations of $N_{\gamma q}$ and N_{cq} defined by Eqs. (4.38) and (4.39) are shown in Figures 4.15 and 4.16, respectively. In using N_{cq} in Eq. (4.39) as given in Figure 4.16, the following points need to be kept in mind:

1. The term

$$N_s = \frac{\gamma H}{c} \tag{4.40}$$

is defined as the stability number.
2. If $B < H$, use the curves for $N_s = 0$.
3. If $B \geq H$, use the curves for the calculated stability number N_s.

Stress Characteristics Solution for Granular Soil Slopes

For slopes in granular soils, the ultimate bearing capacity of a continuous foundation can be given by Eq. (4.38), or

$$q_u = \frac{1}{2}\gamma B N_{\gamma q}$$

On the basis of the method of stress characteristics, Graham, Andrews, and Shields (1988) provided a solution for the bearing capacity factor $N_{\gamma q}$ for a shallow continuous foundation on the top of a slope in *granular soil*. Figure 4.17 shows the schematics of the failure zone in the soil for embedment (D_f/B) and setback (b/B) assumed for those authors' analysis. The variations of $N_{\gamma q}$ obtained by this method are shown in Figures 4.18, 4.19, and 4.20.

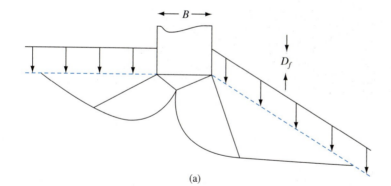

Figure 4.16 Meyerhof's bearing capacity factor N_{cq} for purely cohesive soil

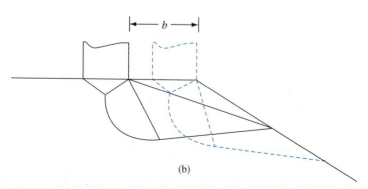

Figure 4.17 Schematic diagram of failure zones for embedment and setback: (a) $D_f/B > 0$; (b) $b/B > 0$

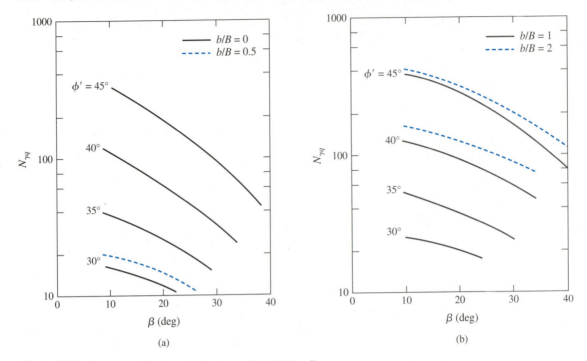

Figure 4.18 Graham et al.'s theoretical values of $N_{\gamma q}(D_f/B = 0)$

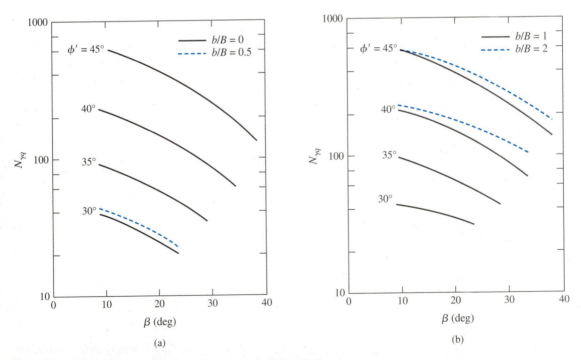

Figure 4.19 Graham et al.'s theoretical values of $N_{\gamma q}(D_f/B = 0.5)$

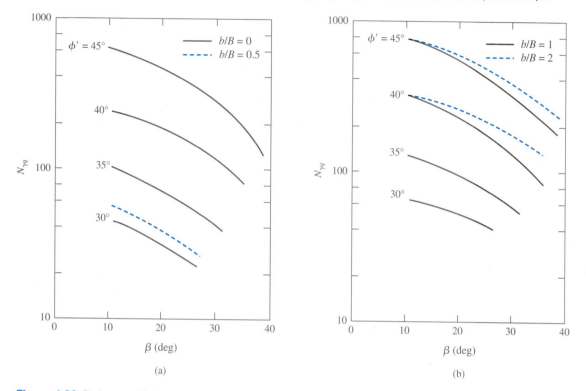

Figure 4.20 Graham et al.'s theoretical values of $N_{\gamma q}(D_f/B = 1)$

Example 4.7

In Figure 4.14, for a shallow continuous foundation in a clay, the following data are given: $B = 1.2$ m; $D_f = 1.2$ m; $b = 0.8$ m; $H = 6.2$ m; $\beta = 30°$; unit weight of soil $= 17.5$ kN/m³; $\phi = 0$; and $c = 50$ kN/m². Determine the gross allowable bearing capacity with a factor of safety FS $= 4$.

Solution

Since $B < H$, we will assume the stability number $N_s = 0$. From Eq. (4.39),

$$q_u = cN_{cq}$$

We are given that

$$\frac{D_f}{B} = \frac{1.2}{1.2} = 1$$

and

$$\frac{b}{B} = \frac{0.8}{1.2} = 0.67$$

For $\beta = 30°$, $D_f/B = 1$ and $b/B = 0.67$, Figure 4.16 gives $N_{cq} = 6.3$. Hence,

$$q_u = (50)(6.3) = 315 \text{ kN/m}^2$$

and

$$q_{all} = \frac{q_u}{\text{FS}} = \frac{315}{4} = \textbf{78.8 kN/m}^2 \quad \blacksquare$$

Example 4.8

Figure 4.21 shows a continuous foundation on a slope of a granular soil. Estimate the ultimate bearing capacity.

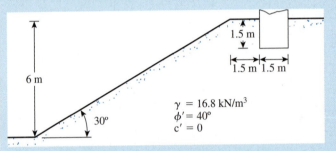

Figure 4.21 Foundation on a granular slope

Solution

For granular soil $(c' = 0)$, from Eq. (4.38),

$$q_u = \tfrac{1}{2}\gamma B N_{\gamma q}$$

We are given that $b/B = 1.5/1.5 = 1$, $D_f/B = 1.5/1.5 = 1$, $\phi' = 40°$, and $\beta = 30°$.

From Figure 4.15, $N_{\gamma q} \approx 120$. So,

$$q_u = \tfrac{1}{2}(16.8)(1.5)(120) = \textbf{1512 kN/m}^2 \quad \blacksquare$$

Example 4.9

Solve Example 4.8 using the stress characteristics solution method.

Solution

$$q_u = \frac{1}{2}\gamma B N_{\gamma q}$$

From Figure 4.20b, $N_{\gamma q} \approx 110$. Hence,

$$q_u = \frac{1}{2}(16.8)(1.5)(110) = \textbf{1386 kN/m}^2 \quad \blacksquare$$

Seismic Bearing Capacity of a Foundation at the Edge of a Granular Soil Slope

Figure 4.22 shows a continuous surface foundation ($B/L = 0$, $D_f/B = 0$) at the edge of a granular slope. The foundation is subjected to a loading inclined at an angle α to the vertical. Let the foundation be subjected to seismic loading with a horizontal coefficient of acceleration, k_h. Based on their analysis of method of slices, Huang and Kang (2008) expressed the ultimate bearing capacity as

$$q_u = \frac{1}{2}\gamma B N_\gamma F_{\gamma i} F_{\gamma \beta} F_{\gamma e}$$ (4.41)

where

N_γ = bearing capacity factor (Table 3.3)
$F_{\gamma i}$ = load inclination factor
$F_{\gamma \beta}$ = slope inclination factor
$F_{\gamma e}$ = correction factor for the inertia force induced by seismic loading

The relationships for $F_{\gamma i}$, $F_{\gamma \beta}$, and $F_{\gamma e}$ are as follow:

$$F_{\gamma i} = \left[1 - \left(\frac{\alpha^\circ}{\phi'^\circ}\right)\right]^{(0.1\phi'-1.21)}$$ (4.42)

$$F_{\gamma \beta} = [1 - (1.062 - 0.014\phi')\tan\phi']^{\left(\frac{\beta^\circ}{10^\circ}\right)}$$ (4.43)

and

$$F_{\gamma e} = 1 - [(2.57 - 0.043\phi')e^{1.45\,\tan\,\beta}]k_h$$ (4.44)

In Eqs. (4.42) through (4.44), ϕ' is in degrees.

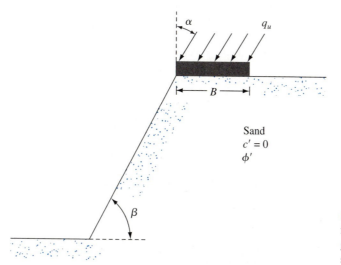

Figure 4.22 Continuous foundation at the edge of a granular slope subjected to seismic loading

Example 4.10

Consider a continuous surface foundation on a granular soil slope subjected to a seismic loading, as shown in Figure 4.22. Given: $B = 1.5$ m, $\gamma = 17.5$ kN/m³, $\phi' = 35°$, $c' = 0$, $\beta = 30°$, $\alpha = 10°$, and $k_h = 0.2$. Calculate the ultimate bearing capacity, q_u.

Solution
From Eq. (4.41),

$$q_u = \frac{1}{2}\gamma B N_\gamma F_{\gamma i} F_{\gamma\beta} F_{\gamma e}$$

For $\phi' = 35°$, $N_\gamma = 48.03$ (Table 3.3). Thus,

$$F_{\gamma t} = \left[1 - \left(\frac{\alpha°}{\phi'°}\right)\right]^{(0.1\phi'-1.21)} = \left[1 - \left(\frac{10}{35}\right)\right]^{[(0.1\times35)-1.21]} = 0.463$$

$$F_{\gamma\beta} = [1 - (1.062 - 0.014\phi')\tan\phi']^{\left(\frac{\beta°}{10°}\right)}$$

$$= [1 - (1.062 - 0.014 \times 35)\tan 35]^{\left(\frac{30}{10}\right)} = 0.215$$

$$F_{\gamma e} = 1 - [(2.57 - 0.043\phi')e^{1.45\tan\beta}]k_h$$

$$= 1 - [(2.57 - 0.043 \times 35)e^{1.45\tan 30}](0.2) = 0.508$$

So

$$q_u = \frac{1}{2}(17.5)(1.5)(48.03)(0.463)(0.215)(0.508) = \mathbf{31.9\ kN/m^2}\quad\blacksquare$$

4.8   *Bearing Capacity of Foundations on a Slope*

A theoretical solution for the ultimate bearing capacity of a shallow foundation located on the face of a slope was developed by Meyerhof (1957). Figure 4.23 shows the nature of the plastic zone developed under a rough continuous foundation of width B. In Figure 4.23, *abc* is an elastic zone, *acd* is a radial shear zone, and *ade* is a mixed shear zone. Based on this solution, the ultimate bearing capacity can be expressed as

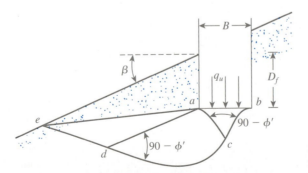

Figure 4.23 Nature of plastic zone under a rough continuous foundation on the face of a slope

$$q_u = cN_{cqs} \text{ (for purely cohesive soil, that is, } \phi = 0) \tag{4.45}$$

and

$$q_u = \tfrac{1}{2}\gamma B N_{\gamma qs} \text{ (for granular soil, that is } c' = 0) \tag{4.46}$$

The variations of N_{cqs} and $N_{\gamma qs}$ with slope angle β are given in Figures 4.24 and 4.25.

Figure 4.24 Variation of N_{cqs} with β.
(*Note:* $N_s = \gamma H/c$)

Figure 4.25 Variation of $N_{\gamma qs}$ with β

4.9   *Foundations on Rock*

On some occasions, shallow foundations may have to be built on rocks, as shown in Figure 4.26. For estimation of the ultimate bearing capacity of shallow foundations on rock, we may use Terzaghi's bearing capacity equations [Eqs. (3.3), (3.7) and (3.8)] with the bearing capacity factors given here (Stagg and Zienkiewicz, 1968; Bowles, 1996):

$$N_c = 5 \tan^4\left(45 + \frac{\phi'}{2}\right) \tag{4.47}$$

$$N_q = \tan^6\left(45 + \frac{\phi'}{2}\right) \tag{4.48}$$

$$N_\gamma = N_q + 1 \tag{4.49}$$

For rocks, the magnitude of the cohesion intercept, c', can be expressed as

$$q_{uc} = 2c'\tan\left(45 + \frac{\phi'}{2}\right) \tag{4.50}$$

where
q_{uc} = unconfined compression strength of rock
ϕ' = angle of friction

The unconfined compression strength and the friction angle of rocks can vary widely. Table 4.2 gives a general range of q_{uc} for various types of rocks. It is important to keep in mind that the magnitude of q_{uc} and ϕ' (hence c') reported from laboratory tests are for intact rock specimens. It does not account for the effect of discontinuities. To account for discontinuities, Bowles (1996) suggested that the ultimate bearing capacity q_u should be modified as

$$q_{u(modified)} = q_u(RQD)^2 \tag{4.51}$$

where RQD = rock quality designation (see Chapter 2).

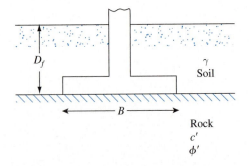

Figure 4.26 Foundation on rock

Table 4.2 Range of the Unconfined Compression Strength of Various Types of Rocks

	q_{uc}		ϕ'
Rock type	**MN/m²**	**kip/m²**	**(deg)**
Granite	65–250	9.5–36	45–55
Limestone	30–150	4–22	35–45
Sandstone	25–130	3.5–19	30–45
Shale	5–40	0.75–6	15–30

In any case, the upper limit of the allowable bearing capacity should not exceed f_c' (28-day compressive strength of concrete).

Example 4.11

Refer to Figure 4.26. A square column foundation is to be constructed over siltstone. Given:

Foundation: $B \times B = 2.5 \text{ m} \times 2.5 \text{ m}$
$D_f = 2 \text{ m}$
Soil: $\gamma = 17 \text{ kN/m}^3$
Siltstone: $c' = 32 \text{ MN/m}^2$
$\phi' = 31°$
$\gamma = 25 \text{ kN/m}^3$
$RDQ = 50\%$

Estimate the allowable load-bearing capacity. Use FS = 4. Also, for concrete, use $f_c' = 30 \text{ MN/m}^2$.

Solution
From Eq. (3.7),

$$q_u = 1.3c'N_c + qN_q + 0.4\gamma BN_\gamma$$

$$N_c = 5 \tan^4\left(45 + \frac{\phi'}{2}\right) = 5 \tan^4\left(45 + \frac{31}{2}\right) = 48.8$$

$$N_q = \tan^6\left(45 + \frac{\phi'}{2}\right) = \tan^6\left(45 + \frac{31}{2}\right) = 30.5$$

$$N_\gamma = N_q + 1 = 30.5 + 1 = 31.5$$

Hence,

$$q_u = (1.3)(32 \times 10^3 \text{ kN/m}^2)(48.8) + (17 \times 2)(30.5) + (0.4)(25)(2.5)(31.5)$$
$$= 2030.08 \times 10^3 + 1.037 \times 10^3 + 0.788 \times 10^3$$
$$= 2031.9 \times 10^3 \text{ kN/m}^2 \approx 2032 \text{ MN/m}^2$$

$$q_{u(\text{modified})} = q_u(\text{RQD})^2 = (2032)(0.5)^2 = 508 \text{ MN/m}^2$$

$$q_{\text{all}} = \frac{508}{4} = 127 \text{ MN/m}^2$$

Since 127 MN/m² is greater than f_c', use $q_{\text{all}} =$ **30 MN/m²**. ∎

4.10   *Uplift Capacity of Foundations*

Foundations may be subjected to uplift forces under special circumstances. During the design process for those foundations, it is desirable to provide a sufficient factor of safety against failure by uplift. This section will provide the relationships for the uplift capacity of foundations in granular and cohesive soils.

Foundations in Granular Soil ($c' = 0$)

Figure 4.27 shows a shallow continuous foundation that is being subjected to an uplift force. At ultimate load, Q_u, the failure surface in soil will be as shown in the figure. The ultimate load can be expressed in the form of a nondimensional breakout factor, F_q. Or

$$F_q = \frac{Q_u}{A\gamma D_f} \tag{4.52}$$

where A = area of the foundation.

The breakout factor is a function of the soil friction angle ϕ' and D_f/B. For a given soil friction angle, F_q increases with D_f/B to a maximum at $D_f/B = (D_f/B)_{cr}$ and remains constant thereafter. For foundations subjected to uplift, $D_f/B \leq (D_f/B)_{cr}$ is considered a shallow foundation condition. When a foundation has an embedment ratio of $D_f/B > (D_f/B)_{cr}$, it is referred to as a deep foundation. Meyerhof and Adams (1968) provided relationships to estimate the ultimate uplifting load Q_u for shallow [that is, $D_f/B \leq (D_f/B)_{cr}$], circular, and rectangular foundations. Using these relationships and Eq. (4.52), Das and Seeley (1975) expressed the breakout factor F_q in the following form

$$F_q = 1 + 2\left[1 + m\left(\frac{D_f}{B}\right)\right]\left(\frac{D_f}{B}\right)K_u \tan\phi' \tag{4.53}$$

(for shallow circular and square foundations)

$$F_q = 1 + \left\{\left[1 + 2m\left(\frac{D_f}{B}\right)\right]\left(\frac{B}{L}\right) + 1\right\}\left(\frac{D_f}{B}\right)K_u \tan\phi' \tag{4.54}$$

(for shallow rectangular foundations)

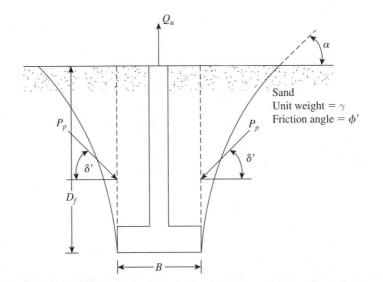

Figure 4.27 Shallow continuous foundation subjected to uplift

where

m = a coefficient which is a function of ϕ'
K_u = nominal uplift coefficient

The variations of K_u, m, and $(D_f/B)_{cr}$ for square and circular foundations are given in Table 4.3 (Meyerhof and Adams, 1968).

For rectangular foundations, Das and Jones (1982) recommended that

$$\left(\frac{D_f}{B}\right)_{cr\text{-rectangular}} = \left(\frac{D_f}{B}\right)_{cr\text{-square}}\left[0.133\left(\frac{L}{B}\right) + 0.867\right] \leq 1.4\left(\frac{D_f}{B}\right)_{cr\text{-square}} \tag{4.55}$$

Using the values of K_u, m, and $(D_f/B)_{cr}$ in Eq. (4.53), the variations of F_q for square and circular foundations have been calculated and are shown in Figure 4.28. A step-by-step procedure to estimate the uplift capacity of foundations in granular soil follows.

Step 1. Determine, D_f, B, L, and ϕ'.
Step 2. Calculate D_f/B.
Step 3. Using Table 4.3 and Eq. (4.55), calculate $(D_f/B)_{cr}$.
Step 4. If D_f/B is less than or equal to $(D_f/B)_{cr}$, it is a shallow foundation.
Step 5. If $D_f/B > (D_f/B)_{cr}$, it is a deep foundation.
Step 6. For shallow foundations, use D_f/B calculated in Step 2 in Eq. (4.53) or (4.54) to estimate F_q. Thus, $Q_u = F_q A\gamma D_f$.
Step 7. For deep foundations, substitute $(D_f/B)_{cr}$ for D_f/B in Eq. (4.53) or (4.54) to obtain F_q, from which the ultimate load Q_u may be obtained.

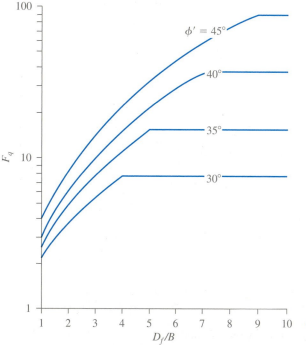

Figure 4.28 Variation of F_q with D_f/B and ϕ'

Table 4.3 Variation of K_u, m, and $(D_f/B)_{cr}$

Soil friction angle, ϕ' (deg)	K_u	m	$(D_f/B)_{cr}$ for square and circular foundations
20	0.856	0.05	2.5
25	0.888	0.10	3
30	0.920	0.15	4
35	0.936	0.25	5
40	0.960	0.35	7
45	0.960	0.50	9

Foundations in Cohesive Soil ($\phi' = 0$)

The ultimate uplift capacity, Q_u, of a foundation in a purely cohesive soil can be expressed as

$$Q_u = A(\gamma D_f + c_u F_c) \tag{4.56}$$

where

A = area of the foundation
c_u = undrained shear strength of soil
F_c = breakout factor

As in the case of foundations in granular soil, the breakout factor F_c increases with embedment ratio and reaches a maximum value of $F_c = F_c^*$ at $D_f/B = (D_f/B)_{cr}$ and remains constant thereafter.

Das (1978) also reported some model test results with square and rectangular foundations. Based on these test results, it was proposed that

$$\left(\frac{D_f}{B}\right)_{cr\text{-square}} = 0.107c_u + 2.5 \leq 7 \tag{4.57}$$

where

$\left(\dfrac{D_f}{B}\right)_{cr\text{-square}}$ = critical embedment ratio of square (or circular) foundations

c_u = undrained cohesion, in kN/m^2

It was also observed by Das (1980) that

$$\left(\frac{D_f}{B}\right)_{cr\text{-rectangular}} = \left(\frac{D_f}{B}\right)_{cr\text{-square}}\left[0.73 + 0.27\left(\frac{L}{B}\right)\right] \leq 1.55\left(\frac{D_f}{B}\right)_{cr\text{-square}} \tag{4.58}$$

where

$\left(\dfrac{D_f}{B}\right)_{cr\text{-rectangular}}$ = critical embedment ratio of rectangular foundations

L = length of foundation

Based on these findings, Das (1980) proposed an empirical procedure to obtain the breakout factors for shallow and deep foundations. According to this procedure, α' and β' are two nondimensional factors defined as

$$\alpha' = \frac{\dfrac{D_f}{B}}{\left(\dfrac{D_f}{B}\right)_{cr}} \tag{4.59}$$

and

$$\beta' = \frac{F_c}{F_c^*} \tag{4.60}$$

For a given foundation, the critical embedment ratio can be calculated using Eqs. (4.57) and (4.58). The magnitude of F_c^* can be given by the following empirical relationship

$$F_{c\text{-rectangular}}^* = 7.56 + 1.44\left(\frac{B}{L}\right) \tag{4.61}$$

where $F_{c\text{-rectangular}}^* = $ breakout factor for deep rectangular foundations

Figure 4.29 shows the experimentally derived plots (upper limit, lower limit, and average of β' and α'. Following is a step-by-step procedure to estimate the ultimate uplift capacity.

Step 1. Determine the representative value of the undrained cohesion, c_u.
Step 2. Determine the critical embedment ratio using Eqs. (4.57) and (4.58).
Step 3. Determine the D_f/B ratio for the foundation.
Step 4. If $D_f/B > (D_f/B)_{cr}$, as determined in Step 2, it is a deep foundation. However, if $D_f/B \leq (D_f/B)_{cr}$, it is a shallow foundation.

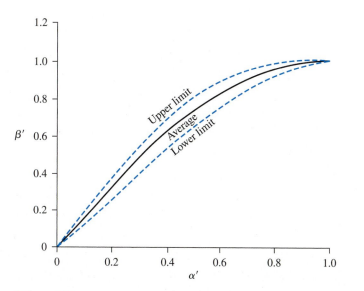

Figure 4.29 Plot of β' versus α'

Step 5.   For $D_f/B > (D_f/B)_{cr}$

$$F_c = F_c^* = 7.56 + 1.44\left(\frac{B}{L}\right)$$

Thus,

$$Q_u = A\left\{\left[7.56 + 1.44\left(\frac{B}{L}\right)\right]c_u + \gamma D_f\right\} \qquad (4.62)$$

where A = area of the foundation.

Step 6.   For $D_f/B \le (D_f/B)_{cr}$

$$Q_u = A(\beta' F_c^* c_u + \gamma D_f) = A\left\{\beta'\left[7.56 + 1.44\left(\frac{B}{L}\right)\right]c_u + \gamma D_f\right\} \quad (4.63)$$

The value of β' can be obtained from the average curve of Figure 4.29. The procedure outlined above gives fairly good results for estimating the net ultimate uplift capacity of foundations and agrees reasonably well with the theoretical solution of Merifield et al. (2003).

Example 4.12

Consider a circular foundation in sand. Given for the foundation: diameter, $B = 1.5$ m and depth of embedment, $D_f = 1.5$ m. Given for the sand: unit weight, $\gamma = 17.4$ kN/m^3, and friction angle, $\phi' = 35°$. Calculate the ultimate bearing capacity.

Solution
$D_f/B = 1.5/1.5 = 1$ and $\phi' = 35°$. For circular foundation, $(D_f/B)_{cr} = 5$. Hence, it is a shallow foundation. From Eq. (4.53)

$$F_q = 1 + 2\left[1 + m\left(\frac{D_f}{B}\right)\right]\left(\frac{D_f}{B}\right)K_u \tan\phi'$$

For $\phi' = 35°$, $m = 0.25$, and $K_u = 0.936$ (Table 4.3). So

$$F_q = 1 + 2[1 + (0.25)(1)](1)(0.936)(\tan 35) = 2.638$$

So

$$Q_u = F_q \gamma A D_f = (2.638)(17.4)\left[\left(\frac{\pi}{4}\right)(1.5)^2\right](1.5) = \textbf{121.7 kN} \qquad \blacksquare$$

Example 4.13

A rectangular foundation in a saturated clay measures 1.5 m × 3 m. Given: $D_f = 1.8$ m, $c_u = 52$ kN/m^2, and $\gamma = 18.9$ kN/m^3. Estimate the ultimate uplift capacity.

Solution
From Eq. (4.57)

$$\left(\frac{D_f}{B}\right)_{\text{cr-square}} = 0.107c_u + 2.5 = (0.107)(52) + 2.5 = 8.06$$

So use $(D_f/B)_{\text{cr-square}} = 7$. Again from Eq. (4.58),

$$\left(\frac{D_f}{B}\right)_{\text{cr-rectangular}} = \left(\frac{D_f}{B}\right)_{\text{cr-square}}\left[0.73 + 0.27\left(\frac{L}{B}\right)\right]$$

$$= 7\left[0.73 + 0.27\left(\frac{3}{1.5}\right)\right] = 8.89$$

Check:
$$1.55\left(\frac{D_f}{B}\right)_{\text{cr-square}} = (1.55)(7) = 10.85$$

So use $(D_f/B)_{\text{cr-rectangular}} = 8.89$. The actual embedment ratio is $D_f/B = 1.8/1.5 = 1.2$. Hence, this is a shallow foundation.

$$\alpha' = \frac{\dfrac{D_f}{B}}{\left(\dfrac{D_f}{B}\right)_{\text{cr}}} = \frac{1.2}{8.89} = 0.135$$

Referring to the average curve of Figure 4.29, for $\alpha' = 0.135$, the magnitude of $\beta' = 0.2$. From Eq. (4.63),

$$Q_u = A\left\{\beta'\left[7.56 + 1.44\left(\frac{B}{L}\right)\right]c_u + \gamma D_f\right\}$$

$$= (1.5)(3)\left\{(0.2)\left[7.56 + 1.44\left(\frac{1.5}{3}\right)\right](52) + (18.9)(1.8)\right\} = \textbf{540.6 kN} \quad\blacksquare$$

Problems

4.1 Refer to Figure 4.2 and consider a rectangular foundation. Given: $B = 0.91$ m, $L = 1.83$ m, $D_f = 0.91$ m, $H = 0.61$ m, $\phi' = 40°$, $c' = 0$, and $\gamma = 18.08$ kN/m^3. Using a factor of safety of 4, determine the gross allowable load the foundation can carry. Use Eq. (4.3).

4.2   Repeat Problem 4.1 with the following data: $B = 1.5$ m, $L = 1.5$ m, $D_f = 1$ m, $H = 0.6$ m, $\phi' = 35°$, $c' = 0$, and $\gamma = 15$ kN/m³. Use FS $= 3$.

4.3   Refer to Figure 4.2. Given: $B = L = 1.75$ m, $D_f = 1$ m, $H = 1.75$ m, $\gamma = 17$ kN/m³, $c' = 0$, and $\phi' = 30°$. Using Eq. (4.6) and FS $= 4$, determine the gross allowable load the foundation can carry.

4.4   Refer to Figure 4.2. A square foundation measuring 1.22 m $\times$ 1.22 m is supported by a saturated clay layer of limited depth underlain by a rock layer. Given that $D_f = 0.91$ m, $H = 0.61$ m, $c_u = 115$ kN/m², and $\gamma = 18.87$ kN/m³, estimate the ultimate bearing capacity of the foundation.

4.5   Refer to Figure 4.8. For a strip foundation in two-layered clay, given:
- $\gamma_1 = 18.08$ kN/m³, $c_1 = 57.5$ kN/m², $\phi_1 = 0$
- $\gamma_2 = 17.29$ kN/m³, $c_2 = 28.75$ kN/m², $\phi_2 = 0$
- $B = 0.91$ m, $D_f = 0.61$ m, $H = 0.61$ m

Find the gross allowable bearing capacity. Use a factor of safety of 3.

4.6   Refer to Figure 4.8. For a strip foundation in two-layered clay, given:
- $B = 0.92$ m, $L = 1.22$ m, $D_f = 0.92$ m, $H = 0.76$ m
- $\gamma_1 = 17$ kN/m³, $\phi_1 = 0$, $c_1 = 72$ kN/m²
- $\gamma_2 = 17$ kN/m³, $\phi_2 = 0$, $c_2 = 43$ kN/m²

Determine the gross ultimate bearing capacity.

4.7   Refer to Figure 4.8. For a square foundation on layered sand, given:
- $B = 1.5$ m, $D_f = 1.5$ m, $H = 1$ m
- $\gamma_1 = 18$ kN/m³, $\phi_1' = 40°$, $c_1' = 0$
- $\gamma_2 = 16.7$ kN/m³, $\phi_2' = 32°$, $c_2' = 0$

Determine the net allowable load that the foundations can support. Use FS $= 4$.

4.8   Refer to Figure 4.11. For a rectangular foundation on layered sand, given:
- $B = 1.22$ m, $L = 1.83$ m, $H = 0.61$ m, $D_f = 0.91$ m
- $\gamma_1 = 15.41$ kN/m³, $\phi_1' = 30°$, $c_1' = 0$
- $\gamma_2 = 16.98$ kN/m³, $\phi_2' = 38°$, $c_2' = 0$

Using a factor of safety of 4, determine the gross allowable load the foundation can carry.

4.9   Two continuous shallow foundations are constructed alongside each other in a granular soil. Given, for the foundation: $B = 1.2$ m, $D_f = 1$ m, and center-to-center spacing $= 2$ m. The soil friction angle, $\phi' = 35°$. Estimate the net allowable bearing capacity of the foundations. Use a factor of safety of FS $= 4$ and a unit weight of soil $\gamma = 16.8$ kN/m³.

4.10   A continuous foundation with a width of 1 m is located on a slope made of clay soil. Refer to Figure 4.14 and let $D_f = 1$ m, $H = 4$ m, $b = 2$ m, $\gamma = 16.8$ kN/m³, $c = 68$ kN/m², $\phi = 0$, and $\beta = 60°$.
a. Determine the allowable bearing capacity of the foundation. Let FS $= 3$.
b. Plot a graph of the ultimate bearing capacity q_u if b is changed from 0 to 6 m.

4.11   A continuous foundation is to be constructed near a slope made of granular soil (see Figure 4.14). If $B = 1.22$ m, $b = 1.83$ m, $H = 4.57$ m, $D_f = 1.22$ m, $\beta = 30°$, $\phi' = 40°$, and $\gamma = 17.29$ kN/m³, estimate the ultimate bearing capacity of the foundation. Use Meyerhof's solution.

4.12   A square foundation in a sand deposit measures 1.22 m $\times$ 1.22 m in plan. Given: $D_f = 1.52$ m, soil friction angle $= 35°$, and unit weight of soil $= 17.6$ kN/m³. Estimate the ultimate uplift capacity of the foundation.

4.13 A foundation measuring 1.2 m × 2.4 m in plan is constructed in a saturated clay. Given: depth of embedment of the foundation = 2 m, unit weight of soil = 18 kN/m^3, and undrained cohesion of clay = 74 kN/m^2. Estimate the ultimate uplift capacity of the foundation.

References

BOWLES, J. E. (1996). *Foundation Analysis and Design*, 5th. edition, McGraw-Hill, New York.

BUISMAN, A. S. K. (1940). *Grondmechanica,* Waltman, Delft, the Netherlands.

CERATO, A. B., and LUTENEGGER, A. J. (2006). "Bearing Capacity of Square and Circular Footings on a Finite Layer of Granular Soil Underlain by a Rigid Base," *Journal of Geotechnical and Geoenvironmental Engineering*, American Society of Civil Engineers, Vol. 132, No. 11, pp. 1496–1501.

DAS, B. M. (1978). "Model Tests for Uplift Capacity of Foundations in Clay," *Soils and Foundations,* Vol. 18, No. 2, pp. 17–24.

DAS, B. M. (1980). "A Procedure for Estimation of Ultimate Uplift Capacity of Foundations in Clay," *Soils and Foundations,* Vol. 20, No. 1, pp. 77–82.

DAS, B. M., and JONES, A. D. (1982). "Uplift Capacity of Rectangular Foundations in Sand," *Transportation Research Record 884,* National Research Council, Washington, D.C., pp. 54–58.

DAS, B. M., and SEELEY, G. R. (1975). "Breakout Resistance of Horizontal Anchors," *Journal of Geotechnical Engineering Division,* ASCE, Vol. 101, No. 9, pp. 999–1003.

GRAHAM, J., ANDREWS, M., and SHIELDS, D. H. (1988). "Stress Characteristics for Shallow Footing on Cohesionless Slopes," *Canadian Geotechnical Journal*, Vol. 25, No. 2, pp. 238–249.

HUANG, C. C., and KANG, W. W. (2008). "Seismic Bearing Capacity of a Rigid Footing Adjacent to a Cohesionless Slope," *Soils and Foundations*, Vol. 48, No. 5, pp. 641–651.

LUNDGREN, H., and MORTENSEN, K. (1953). "Determination by the Theory of Plasticity on the Bearing Capacity of Continuous Footings on Sand," *Proceedings, Third International Conference on Soil Mechanics and Foundation Engineering,* Zurich, Vol. 1, pp. 409–412.

MANDEL, J., and SALENCON, J. (1972). "Force portante d'un sol sur une assise rigide (étude théorique)," *Geotechnique,* Vol. 22, No. 1, pp. 79–93.

MERIFIELD, R. S., LYAMIN, A., and SLOAN, S. W. (2003). "Three Dimensional Lower Bound Solutions for the Stability of Plate Anchors in Clay," *Journal of Geotechnical and Geoenvironmental Engineering,* ASCE, Vo. 129, No. 3, pp. 243–253.

MEYERHOF, G. G. (1957). "The Ultimate Bearing Capacity of Foundations on Slopes," *Proceedings, Fourth International Conference on Soil Mechanics and Foundation Engineering,* London, Vol. 1, pp. 384–387.

MEYERHOF, G. G. (1974). "Ultimate Bearing Capacity of Footings on Sand Layer Overlying Clay," *Canadian Geotechnical Journal,* Vol. 11, No. 2, pp. 224–229.

MEYERHOF, G. G., and ADAMS, J. I. (1968). "The Ultimate Uplift Capacity of Foundations," *Canadian Geotechnical Journal,* Vol. 5, No. 4, pp. 225–244.

MEYERHOF, G. G., and CHAPLIN, T. K. (1953). "The Compression and Bearing Capacity of Cohesive Soils," *British Journal of Applied Physics,* Vol. 4, pp. 20–26.

MEYERHOF, G. G., and HANNA, A. M. (1978). "Ultimate Bearing Capacity of Foundations on Layered Soil under Inclined Load," *Canadian Geotechnical Journal,* Vol. 15, No. 4, pp. 565–572.

PRANDTL, L. (1921). "Über die Eindringungsfestigkeit (Härte) plastischer Baustoffe und die Festigkeit von Schneiden," *Zeitschrift für angewandte Mathematik und Mechanik,* Vol. 1, No. 1, pp. 15–20.

REISSNER, H. (1924). "Zum Erddruckproblem," *Proceedings, First International Congress of Applied Mechanics,* Delft, the Netherlands, pp. 295–311.

STAGG, K. G., and ZIENKIEWICZ, O. C. (1968). *Rock Mechanics in Engineering Practice*, John Wiley & Sons, New York.

5 Shallow Foundations: Allowable Bearing Capacity and Settlement

5.1 Introduction

It was mentioned in Chapter 3 that, in many cases, the allowable settlement of a shallow foundation may control the allowable bearing capacity. The allowable settlement itself may be controlled by local building codes. Thus, the allowable bearing capacity will be the smaller of the following two conditions:

$$q_{\text{all}} = \begin{cases} \dfrac{q_u}{\text{FS}} \\ \text{or} \\ q_{\text{allowable settlement}} \end{cases}$$

The settlement of a foundation can be divided into two major categories: (a) elastic, or immediate, settlement and (b) consolidation settlement. Immediate, or elastic, settlement of a foundation takes place during or immediately after the construction of the structure. Consolidation settlement occurs over time. Pore water is extruded from the void spaces of saturated clayey soils submerged in water. The total settlement of a foundation is the sum of the elastic settlement and the consolidation settlement.

Consolidation settlement comprises two phases: *primary* and *secondary*. The fundamentals of primary consolidation settlement were explained in detail in Chapter 1. Secondary consolidation settlement occurs after the completion of primary consolidation caused by slippage and reorientation of soil particles under a sustained load. Primary consolidation settlement is more significant than secondary settlement in inorganic clays and silty soils. However, in organic soils, secondary consolidation settlement is more significant.

For the calculation of foundation settlement (both elastic and consolidation), it is required that we estimate the vertical stress increase in the soil mass due to the net load applied on the foundation. Hence, this chapter is divided into the following three parts:

1. Procedure for calculation of vertical stress increase
2. Elastic settlement calculation
3. Consolidation settlement calculation

Vertical Stress Increase in a Soil Mass Caused by Foundation Load

5.2 Stress Due to a Concentrated Load

In 1885, Boussinesq developed the mathematical relationships for determining the normal and shear stresses at any point inside *homogeneous, elastic,* and *isotropic* mediums due to a *concentrated point load* located at the surface, as shown in Figure 5.1. According to his analysis, the *vertical stress increase* at point *A* caused by a point load of magnitude *P* is given by

$$\Delta \sigma = \frac{3P}{2\pi z^2 \left[1 + \left(\dfrac{r}{z} \right)^2 \right]^{5/2}} \tag{5.1}$$

where

$$r = \sqrt{x^2 + y^2}$$

x, y, z = coordinates of the point *A*

Note that Eq. (5.1) is not a function of Poisson's ratio of the soil.

5.3 Stress Due to a Circularly Loaded Area

The Boussinesq equation (5.1) can also be used to determine the vertical stress below the center of a flexible circularly loaded area, as shown in Figure 5.2. Let the radius of the loaded area be $B/2$, and let q_o be the uniformly distributed load per unit area. To determine

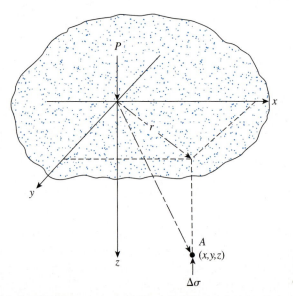

Figure 5.1 Vertical stress at a point *A* caused by a point load on the surface

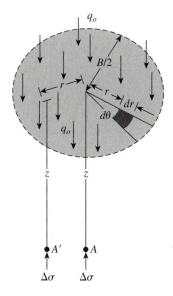

Figure 5.2 Increase in pressure under a uniformly loaded flexible circular area

the stress increase at a point *A,* located at a depth *z* below the center of the circular area, consider an elemental area on the circle. The load on this elemental area may be taken to be a point load and expressed as $q_o r \, d\theta \, dr$. The stress increase at *A* caused by this load can be determined from Eq. (5.1) as

$$d\sigma = \frac{3(q_o r \, d\theta \, dr)}{2\pi z^2 \left[1 + \left(\dfrac{r}{z}\right)^2\right]^{5/2}} \tag{5.2}$$

The total increase in stress caused by the entire loaded area may be obtained by integrating Eq. (5.2), or

$$\Delta\sigma = \int d\sigma = \int_{\theta=0}^{\theta=2\pi} \int_{r=0}^{r=B/2} \frac{3(q_o r \, d\theta \, dr)}{2\pi z^2 \left[1 + \left(\dfrac{r}{z}\right)^2\right]^{5/2}}$$

$$= q_o \left\{ 1 - \frac{1}{\left[1 + \left(\dfrac{B}{2z}\right)^2\right]^{3/2}} \right\} \tag{5.3}$$

Similar integrations could be performed to obtain the vertical stress increase at *A′,* located a distance *r* from the center of the loaded area at a depth *z* (Ahlvin and Ulery, 1962). Table 5.1 gives the variation of $\Delta\sigma/q_o$ with $r/(B/2)$ and $z/(B/2)$ [for $0 \le r/(B/2) \le 1$]. Note that the variation of $\Delta\sigma/q_o$ with depth at $r/(B/2) = 0$ can be obtained from Eq. (5.3).

Table 5.1 Variation of $\Delta\sigma/q_o$ for a Uniformly Loaded Flexible Circular Area

z/(B/2)	r/(B/2)					
	0	0.2	0.4	0.6	0.8	1.0
0	1.000	1.000	1.000	1.000	1.000	1.000
0.1	0.999	0.999	0.998	0.996	0.976	0.484
0.2	0.992	0.991	0.987	0.970	0.890	0.468
0.3	0.976	0.973	0.963	0.922	0.793	0.451
0.4	0.949	0.943	0.920	0.860	0.712	0.435
0.5	0.911	0.902	0.869	0.796	0.646	0.417
0.6	0.864	0.852	0.814	0.732	0.591	0.400
0.7	0.811	0.798	0.756	0.674	0.545	0.367
0.8	0.756	0.743	0.699	0.619	0.504	0.366
0.9	0.701	0.688	0.644	0.570	0.467	0.348
1.0	0.646	0.633	0.591	0.525	0.434	0.332
1.2	0.546	0.535	0.501	0.447	0.377	0.300
1.5	0.424	0.416	0.392	0.355	0.308	0.256
2.0	0.286	0.286	0.268	0.248	0.224	0.196
2.5	0.200	0.197	0.191	0.180	0.167	0.151
3.0	0.146	0.145	0.141	0.135	0.127	0.118
4.0	0.087	0.086	0.085	0.082	0.080	0.075

5.4   *Stress below a Rectangular Area*

The integration technique of Boussinesq's equation also allows the vertical stress at any point A below the corner of a flexible rectangular loaded area to be evaluated. (See Figure 5.3.) To do so, consider an elementary area $dA = dx\,dy$ on the flexible loaded area. If the load per unit area is q_o, the total load on the elemental area is

$$dP = q_o\,dx\,dy \tag{5.4}$$

This elemental load, dP, may be treated as a point load. The increase in vertical stress at point A caused by dP may be evaluated by using Eq. (5.1). Note, however, the need to substitute $dP = q_o\,dx\,dy$ for P and $x^2 + y^2$ for r^2 in that equation. Thus,

$$\text{The stress increase at } A \text{ caused by } dP = \frac{3q_o\,(dx\,dy)z^3}{2\pi(x^2 + y^2 + z^2)^{5/2}}$$

The total stress increase $\Delta\sigma$ caused by the entire loaded area at point A may now be obtained by integrating the preceding equation:

$$\Delta\sigma = \int_{y=0}^{L}\int_{x=0}^{B}\frac{3q_o\,(dx\,dy)z^3}{2\pi(x^2 + y^2 + z^2)^{5/2}} = q_o I \tag{5.5}$$

Figure 5.3 Determination of stress below the corner of a flexible rectangular loaded area

Here,

$$I = \text{influence factor} = \frac{1}{4\pi}\left(\frac{2mn\sqrt{m^2+n^2+1}}{m^2+n^2+m^2n^2+1}\cdot\frac{m^2+n^2+2}{m^2+n^2+1}\right.$$

$$\left. + \tan^{-1}\frac{2mn\sqrt{m^2+n^2+1}}{m^2+n^2+1-m^2n^2}\right) \tag{5.6}$$

where

$$m = \frac{B}{z} \tag{5.7}$$

and

$$n = \frac{L}{z} \tag{5.8}$$

The variations of the influence values with m and n are given in Table 5.2.

The stress increase at any point below a rectangular loaded area can also be found by using Eq. (5.5) in conjunction with Figure 5.4. To determine the stress at a depth z below point O, divide the loaded area into four rectangles, with O the corner common to each. Then use Eq. (5.5) to calculate the increase in stress at a depth z below O caused by each rectangular area. The total stress increase caused by the entire loaded area may now be expressed as

$$\Delta\sigma = q_o\left(I_1 + I_2 + I_3 + I_4\right) \tag{5.9}$$

where I_1, I_2, I_3, and I_4 = the influence values of rectangles 1, 2, 3, and 4, respectively.

In most cases, the vertical stress below the center of a rectangular area is of importance. This can be given by the relationship

$$\Delta\sigma = q_o I_c \tag{5.10}$$

Table 5.2 Variation of Influence Value I [Eq. (5.6)][a]

m	n											
	0.1	0.2	0.3	0.4	0.5	0.6	0.7	0.8	0.9	1.0	1.2	1.4
0.1	0.00470	0.00917	0.01323	0.01678	0.01978	0.02223	0.02420	0.02576	0.02698	0.02794	0.02926	0.03007
0.2	0.00917	0.01790	0.02585	0.03280	0.03866	0.04348	0.04735	0.05042	0.05283	0.05471	0.05733	0.05894
0.3	0.01323	0.02585	0.03735	0.04742	0.05593	0.06294	0.06858	0.07308	0.07661	0.07938	0.08323	0.08561
0.4	0.01678	0.03280	0.04742	0.06024	0.07111	0.08009	0.08734	0.09314	0.09770	0.10129	0.10631	0.10941
0.5	0.01978	0.03866	0.05593	0.07111	0.08403	0.09473	0.10340	0.11035	0.11584	0.12018	0.12626	0.13003
0.6	0.02223	0.04348	0.06294	0.08009	0.09473	0.10688	0.11679	0.12474	0.13105	0.13605	0.14309	0.14749
0.7	0.02420	0.04735	0.06858	0.08734	0.10340	0.11679	0.12772	0.13653	0.14356	0.14914	0.15703	0.16199
0.8	0.02576	0.05042	0.07308	0.09314	0.11035	0.12474	0.13653	0.14607	0.15371	0.15978	0.16843	0.17389
0.9	0.02698	0.05283	0.07661	0.09770	0.11584	0.13105	0.14356	0.15371	0.16185	0.16835	0.17766	0.18357
1.0	0.02794	0.05471	0.07938	0.10129	0.12018	0.13605	0.14914	0.15978	0.16835	0.17522	0.18508	0.19139
1.2	0.02926	0.05733	0.08323	0.10631	0.12626	0.14309	0.15703	0.16843	0.17766	0.18508	0.19584	0.20278
1.4	0.03007	0.05894	0.08561	0.10941	0.13003	0.14749	0.16199	0.17389	0.18357	0.19139	0.20278	0.21020
1.6	0.03058	0.05994	0.08709	0.11135	0.13241	0.15028	0.16515	0.17739	0.18737	0.19546	0.20731	0.21510
1.8	0.03090	0.06058	0.08804	0.11260	0.13395	0.15207	0.16720	0.17967	0.18986	0.19814	0.21032	0.21836
2.0	0.03111	0.06100	0.08867	0.11342	0.13496	0.15326	0.16856	0.18119	0.19152	0.19994	0.21235	0.22058
2.5	0.03138	0.06155	0.08948	0.11450	0.13628	0.15483	0.17036	0.18321	0.19375	0.20236	0.21512	0.22364
3.0	0.03150	0.06178	0.08982	0.11495	0.13684	0.15550	0.17113	0.18407	0.19470	0.20341	0.21633	0.22499
4.0	0.03158	0.06194	0.09007	0.11527	0.13724	0.15598	0.17168	0.18469	0.19540	0.20417	0.21722	0.22600
5.0	0.03160	0.06199	0.09014	0.11537	0.13737	0.15612	0.17185	0.18488	0.19561	0.20440	0.21749	0.22632
6.0	0.03161	0.06201	0.09017	0.11541	0.13741	0.15617	0.17191	0.18496	0.19569	0.20449	0.21760	0.22644
8.0	0.03162	0.06202	0.09018	0.11543	0.13744	0.15621	0.17195	0.18500	0.19574	0.20455	0.21767	0.22652
10.0	0.03162	0.06202	0.09019	0.11544	0.13745	0.15622	0.17196	0.18502	0.19576	0.20457	0.21769	0.22654
∞	0.03162	0.06202	0.09019	0.11544	0.13745	0.15623	0.17197	0.18502	0.19577	0.20458	0.21770	0.22656

Table 5.2 (*Continued*)

m	\(n\) 1.6	1.8	2.0	2.5	3.0	4.0	5.0	6.0	8.0	10.0	∞
0.1	0.03058	0.03090	0.03111	0.03138	0.03150	0.03158	0.03160	0.03161	0.03162	0.03162	0.03162
0.2	0.05994	0.06058	0.06100	0.06155	0.06178	0.06194	0.06199	0.06201	0.06202	0.06202	0.06202
0.3	0.08709	0.08804	0.08867	0.08948	0.08982	0.09007	0.09014	0.09017	0.09018	0.09019	0.09019
0.4	0.11135	0.11260	0.11342	0.11450	0.11495	0.11527	0.11537	0.11541	0.11543	0.11544	0.11544
0.5	0.13241	0.13395	0.13496	0.13628	0.13684	0.13724	0.13737	0.13741	0.13744	0.13745	0.13745
0.6	0.15028	0.15207	0.15326	0.15483	0.15550	0.15598	0.15612	0.15617	0.15621	0.15622	0.15623
0.7	0.16515	0.16720	0.16856	0.17036	0.17113	0.17168	0.17185	0.17191	0.17195	0.17196	0.17197
0.8	0.17739	0.17967	0.18119	0.18321	0.18407	0.18469	0.18488	0.18496	0.18500	0.18502	0.18502
0.9	0.18737	0.18986	0.19152	0.19375	0.19470	0.19540	0.19561	0.19569	0.19574	0.19576	0.19577
1.0	0.19546	0.19814	0.19994	0.20236	0.20341	0.20417	0.20440	0.20449	0.20455	0.20457	0.20458
1.2	0.20731	0.21032	0.21235	0.21512	0.21633	0.21722	0.21749	0.21760	0.21767	0.21769	0.21770
1.4	0.21510	0.21836	0.22058	0.22364	0.22499	0.22600	0.22632	0.22644	0.22652	0.22654	0.22656
1.6	0.22025	0.22372	0.22610	0.22940	0.23088	0.23200	0.23236	0.23249	0.23258	0.23261	0.23263
1.8	0.22372	0.22736	0.22986	0.23334	0.23495	0.23617	0.23656	0.23671	0.23681	0.23684	0.23686
2.0	0.22610	0.22986	0.23247	0.23614	0.23782	0.23912	0.23954	0.23970	0.23981	0.23985	0.23987
2.5	0.22940	0.23334	0.23614	0.24010	0.24196	0.24344	0.24392	0.24412	0.24425	0.24429	0.24432
3.0	0.23088	0.23495	0.23782	0.24196	0.24394	0.24554	0.24608	0.24630	0.24646	0.24650	0.24654
4.0	0.23200	0.23617	0.23912	0.24344	0.24554	0.24729	0.24791	0.24817	0.24836	0.24842	0.24846
5.0	0.23236	0.23656	0.23954	0.24392	0.24608	0.24791	0.24857	0.24885	0.24907	0.24914	0.24919
6.0	0.23249	0.23671	0.23970	0.24412	0.24630	0.24817	0.24885	0.24916	0.24939	0.24946	0.24952
8.0	0.23258	0.23681	0.23981	0.24425	0.24646	0.24836	0.24907	0.24939	0.24964	0.24973	0.24980
10.0	0.23261	0.23684	0.23985	0.24429	0.24650	0.24842	0.24914	0.24946	0.24973	0.24981	0.24989
∞	0.23263	0.23686	0.23987	0.24432	0.24654	0.24846	0.24919	0.24952	0.24980	0.24989	0.25000

[a] After Newmark, 1935.

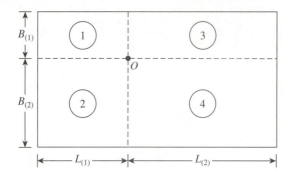

Figure 5.4 Stress below any point of a loaded flexible rectangular area

where

$$I_c = \frac{2}{\pi} \left[\frac{m_1 n_1}{\sqrt{1 + m_1^2 + n_1^2}} \frac{1 + m_1^2 + 2n_1^2}{(1 + n_1^2)(m_1^2 + n_1^2)} \right.$$

$$\left. + \sin^{-1} \frac{m_1}{\sqrt{m_1^2 + n_1^2}\sqrt{1 + n_1^2}} \right] \tag{5.11}$$

$$m_1 = \frac{L}{B} \tag{5.12}$$

$$n_1 = \frac{z}{\left(\dfrac{B}{2}\right)} \tag{5.13}$$

The variation of I_c with m_1 and n_1 is given in Table 5.3.

Table 5.3 Variation of I_c with m_1 and n_1

	m_1									
n_i	1	2	3	4	5	6	7	8	9	10
0.20	0.994	0.997	0.997	0.997	0.997	0.997	0.997	0.997	0.997	0.997
0.40	0.960	0.976	0.977	0.977	0.977	0.977	0.977	0.977	0.977	0.977
0.60	0.892	0.932	0.936	0.936	0.937	0.937	0.937	0.937	0.937	0.937
0.80	0.800	0.870	0.878	0.880	0.881	0.881	0.881	0.881	0.881	0.881
1.00	0.701	0.800	0.814	0.817	0.818	0.818	0.818	0.818	0.818	0.818
1.20	0.606	0.727	0.748	0.753	0.754	0.755	0.755	0.755	0.755	0.755
1.40	0.522	0.658	0.685	0.692	0.694	0.695	0.695	0.696	0.696	0.696
1.60	0.449	0.593	0.627	0.636	0.639	0.640	0.641	0.641	0.641	0.642
1.80	0.388	0.534	0.573	0.585	0.590	0.591	0.592	0.592	0.593	0.593
2.00	0.336	0.481	0.525	0.540	0.545	0.547	0.548	0.549	0.549	0.549
3.00	0.179	0.293	0.348	0.373	0.384	0.389	0.392	0.393	0.394	0.395
4.00	0.108	0.190	0.241	0.269	0.285	0.293	0.298	0.301	0.302	0.303
5.00	0.072	0.131	0.174	0.202	0.219	0.229	0.236	0.240	0.242	0.244
6.00	0.051	0.095	0.130	0.155	0.172	0.184	0.192	0.197	0.200	0.202
7.00	0.038	0.072	0.100	0.122	0.139	0.150	0.158	0.164	0.168	0.171
8.00	0.029	0.056	0.079	0.098	0.113	0.125	0.133	0.139	0.144	0.147
9.00	0.023	0.045	0.064	0.081	0.094	0.105	0.113	0.119	0.124	0.128
10.00	0.019	0.037	0.053	0.067	0.079	0.089	0.097	0.103	0.108	0.112

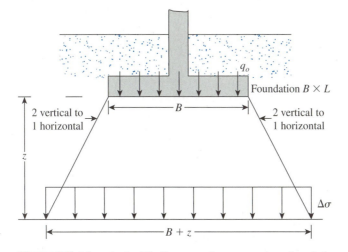

Figure 5.5 2:1 method of finding stress increase under a foundation

Foundation engineers often use an approximate method to determine the increase in stress with depth caused by the construction of a foundation. The method is referred to as the *2:1 method.* (See Figure 5.5.) According to this method, the increase in stress at depth z is

$$\Delta\sigma = \frac{q_o \times B \times L}{(B + z)(L + z)} \qquad (5.14)$$

Note that Eq. (5.14) is based on the assumption that the stress from the foundation spreads out along lines with a *vertical-to-horizontal slope of 2:1.*

Example 5.1

A flexible rectangular area measures 2.5 m $\times$ 5 m in plan. It supports a load of 150 kN/m^2.

Determine the vertical stress increase due to the load at a depth of 6.25 m below the center of the rectangular area.

Solution
Refer to Figure 5.4. For this case,

$$B_1 = B_2 = \frac{2.5}{2} = 1.25 \text{ m}$$

$$L_1 = L_2 = \frac{5}{2} = 2.5 \text{ m}$$

From Eqs. (5.7) and (5.8),

$$m = \frac{B_1}{z} = \frac{B_2}{z} = \frac{1.25}{6.25} = 0.2$$

$$n = \frac{L_1}{z} = \frac{L_2}{z} = \frac{2.5}{6.25} = 0.4$$

From Table 5.2, for $m = 0.2$ and $n = 0.4$, the value of $I = 0.0328$. Thus,

$$\Delta\sigma = q_o(4I) = (150)(4)(0.0328) = \mathbf{19.68\ kN/m^2}$$

Alternate Solution

From Eq. (5.10),

$$\Delta\sigma = q_o I_c$$

$$m_1 = \frac{L}{B} = \frac{5}{2.5} = 2$$

$$n_1 = \frac{z}{\left(\dfrac{B}{2}\right)} = \frac{6.25}{\left(\dfrac{2.5}{2}\right)} = 5$$

From Table 5.3, for $m_1 = 2$ and $n_1 = 5$, the value of $I_c = 0.131$. Thus,

$$\Delta\sigma = (150)(0.131) = \mathbf{19.65\ kN/m^2} \qquad \blacksquare$$

5.5 Average Vertical Stress Increase Due to a Rectangularly Loaded Area

In Section 5.4, the vertical stress increase below the corner of a uniformly loaded rectangular area was given as

$$\Delta\sigma = q_o I$$

In many cases, one must find the average stress increase, $\Delta\sigma_{av}$, below the corner of a uniformly loaded rectangular area with limits of $z = 0$ to $z = H$, as shown in Figure 5.6. This can be evaluated as

$$\Delta\sigma_{av} = \frac{1}{H}\int_0^H (q_o I)\, dz = q_o I_a \qquad (5.15)$$

where

$$I_a = f(m_2, n_2) \qquad (5.16)$$

$$m_2 = \frac{B}{H} \qquad (5.17)$$

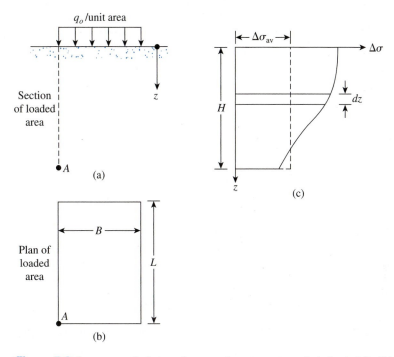

Figure 5.6 Average vertical stress increase due to a rectangularly loaded flexible area

and

$$n_2 = \frac{L}{H} \tag{5.18}$$

The variation of I_a with m_2 and n_2 is shown in Figure 5.7, as proposed by Griffiths (1984).

In estimating the consolidation settlement under a foundation, it may be required to determine the average vertical stress increase in only a given layer—that is, between $z = H_1$ and $z = H_2$, as shown in Figure 5.8. This can be done as (Griffiths, 1984)

$$\Delta\sigma_{av(H_2/H_1)} = q_o \left[\frac{H_2 I_{a(H_2)} - H_1 I_{a(H_1)}}{H_2 - H_1} \right] \tag{5.19}$$

where

$\Delta\sigma_{av(H_2/H_1)}$ = average stress increase immediately below the corner of a uniformly loaded rectangular area between depths $z = H_1$ and $z = H_2$

$$I_{a(H_2)} = I_a \text{ for } z = 0 \text{ to } z = H_2 = f\left(m_2 = \frac{B}{H_2}, n_2 = \frac{L}{H_2} \right)$$

$$I_{a(H_1)} = I_a \text{ for } z = 0 \text{ to } z = H_1 = f\left(m_2 = \frac{B}{H_1}, n_2 = \frac{L}{H_1} \right)$$

Figure 5.7 Griffiths' influence factor I_a

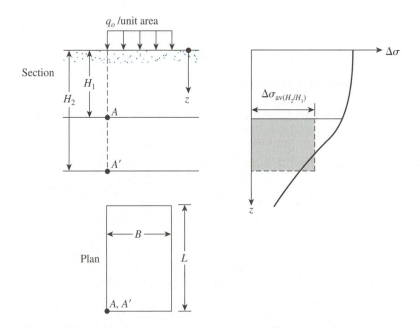

Figure 5.8 Average pressure increase between $z = H_1$ and $z = H_2$ below the corner of a uniformly loaded rectangular area

Example 5.2

Refer to Figure 5.9. Determine the *average* stress increase below the center of the loaded area between $z = 3$ m to $z = 5$ m (that is, between points A and A').

Solution

Refer to Figure 5.9. The loaded area can be divided into four rectangular areas, each measuring 1.5 m $\times$ 1.5 m ($L \times B$). Using Eq. (5.19), the average stress increase (between the required depths) below the corner of each rectangular area can be given as

$$\Delta\sigma_{\text{av} (H_2/H_1)} = q_o \left[\frac{H_2 I_{a(H_2)} - H_1 I_{a(H_1)}}{H_2 - H_1} \right] = 100 \left[\frac{(5) I_{a(H_2)} - (3) I_{a(H_1)}}{5 - 3} \right]$$

For $I_{a(H_2)}$:

$$m_2 = \frac{B}{H_2} = \frac{1.5}{5} = 0.3$$

$$n_2 = \frac{L}{H_2} = \frac{1.5}{5} = 0.3$$

Referring to Figure 5.7, for $m_2 = 0.3$ and $n_2 = 0.3$, $I_{a(H_2)} = 0.126$. For $I_{a(H_1)}$:

$$m_2 = \frac{B}{H_1} = \frac{1.5}{3} = 0.5$$

$$n_2 = \frac{L}{H_1} = \frac{1.5}{3} = 0.5$$

Figure 5.9 Determination of average increase in stress below a rectangular area

Referring to Figure 5.7, $I_{a(H_1)} = 0.175$, so

$$\Delta\sigma_{\text{av}\,(H_2/H_1)} = 100 \left[\frac{(5)(0.126) - (3)(0.175)}{5 - 3} \right] = \textbf{5.25 kN/m}^2$$

The stress increase between $z = 3$ m to $z = 5$ m below the center of the loaded area is equal to

$$4\Delta\sigma_{\text{av}\,(H_2/H_1)} = (4)(5.25) = \textbf{21 kN/m}^2 \qquad\blacksquare$$

5.6   *Stress Increase under an Embankment*

Figure 5.10 shows the cross section of an embankment of height H. For this two-dimensional loading condition, the vertical stress increase may be expressed as

$$\Delta\sigma = \frac{q_o}{\pi} \left[\left(\frac{B_1 + B_2}{B_2} \right)(\alpha_1 + \alpha_2) - \frac{B_1}{B_2}(\alpha_2) \right] \qquad (5.20)$$

where

$q_o = \gamma H$
γ = unit weight of the embankment soil
H = height of the embankment

$$\alpha_1 = \tan^{-1}\left(\frac{B_1 + B_2}{z} \right) - \tan^{-1}\left(\frac{B_1}{z} \right) \qquad (5.21)$$

$$\alpha_2 = \tan^{-1}\left(\frac{B_1}{z} \right) \qquad (5.22)$$

(Note that α_1 and α_2 are in radians.)

Figure 5.10 Embankment loading

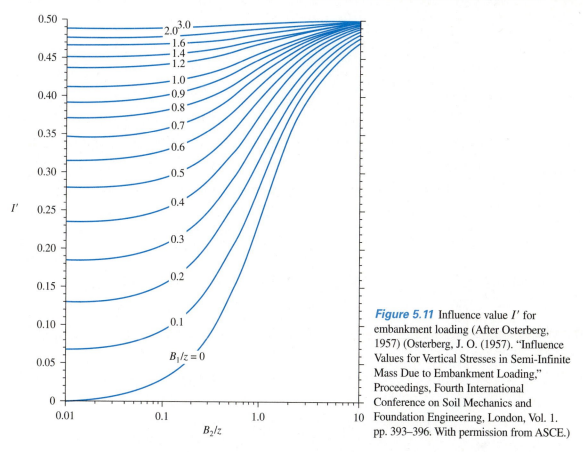

Figure 5.11 Influence value I' for embankment loading (After Osterberg, 1957) (Osterberg, J. O. (1957). "Influence Values for Vertical Stresses in Semi-Infinite Mass Due to Embankment Loading," Proceedings, Fourth International Conference on Soil Mechanics and Foundation Engineering, London, Vol. 1. pp. 393–396. With permission from ASCE.)

For a detailed derivation of Eq. (5.20), see Das (2008). A simplified form of the equation is

$$\Delta\sigma = q_o I' \tag{5.23}$$

where $I' = $ a function of B_1/z and B_2/z.

The variation of I' with B_1/z and B_2/z is shown in Figure 5.11. An application of this diagram is given in Example 5.3.

Example 5.3

An embankment is shown in Figure 5.12a. Determine the stress increase under the embankment at points A_1 and A_2.

Solution

We have

$$\gamma H = (17.5)(7) = 122.5 \text{ kN/m}^2$$

Stress Increase at A_1

The left side of Figure 5.12b indicates that $B_1 = 2.5$ m and $B_2 = 14$ m, so

$$\frac{B_1}{z} = \frac{2.5}{5} = 0.5$$

and

$$\frac{B_2}{z} = \frac{14}{5} = 2.8$$

Figure 5.12 Stress increase due to embankment loading

According to Figure 5.11, in this case $I' = 0.445$. Because the two sides in Figure 5.12b are symmetrical, the value of I' for the right side will also be 0.445, so

$$\Delta\sigma = \Delta\sigma_{(1)} + \Delta\sigma_{(2)} = q_o[I'_{(\text{left side})} + I'_{(\text{right side})}]$$

$$= 122.5[0.445 + 0.445] = \textbf{109.03 kN/m}^2$$

Stress Increase at A_2

In Figure 5.12c, for the left side, $B_2 = 5$ m and $B_1 = 0$, so

$$\frac{B_2}{z} = \frac{5}{5} = 1$$

and

$$\frac{B_1}{z} = \frac{0}{5} = 0$$

According to Figure 5.11, for these values of B_2/z and B_1/z, $I' = 0.24$; hence,

$$\Delta\sigma_{(1)} = 43.75(0.24) = 10.5 \text{ kN/m}^2$$

For the middle section,

$$\frac{B_2}{z} = \frac{14}{5} = 2.8$$

and

$$\frac{B_1}{z} = \frac{14}{5} = 2.8$$

Thus, $I' = 0.495$, so

$$\Delta\sigma_{(2)} = 0.495(122.5) = 60.64 \text{ kN/m}^2$$

For the right side,

$$\frac{B_2}{z} = \frac{9}{5} = 1.8$$

$$\frac{B_1}{z} = \frac{0}{5} = 0$$

and $I' = 0.335$, so

$$\Delta\sigma_{(3)} = (78.75)(0.335) = 26.38 \text{ kN/m}^2$$

The total stress increase at point A_2 is

$$\Delta\sigma = \Delta\sigma_{(1)} + \Delta\sigma_{(2)} - \Delta\sigma_{(3)} = 10.5 + 60.64 - 26.38 = \textbf{44.76 kN/m}^2 \quad \blacksquare$$

| 5.7 | *Westergaard's Solution for Vertical Stress Due to a Point Load* |

Boussinesq's solution for stress distribution due to a point load was presented in Section 5.2. The stress distribution due to various types of loading discussed in Sections 5.3 through 5.6 is based on integration of Boussinesq's solution.

Westergaard (1938) has proposed a solution for the determination of the vertical stress due to a point load P in an elastic solid medium in which there exist alternating layers with thin rigid reinforcements (Figure 5.13a). This type of assumption may be an idealization of a clay layer with thin seams of sand. For such an assumption, the vertical stress increase at a point A (Figure 5.13b) can be given as

$$\Delta\sigma = \frac{P\eta}{2\pi z^2}\left[\frac{1}{\eta^2 + (r/z)^2}\right]^{3/2} \tag{5.24}$$

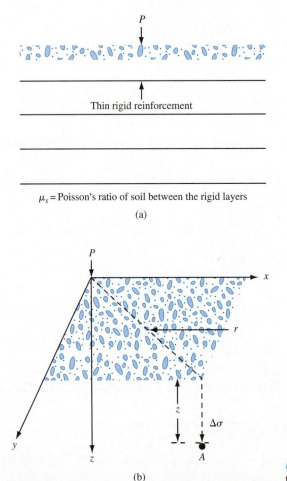

μ_s = Poisson's ratio of soil between the rigid layers

(a)

(b)

Figure 5.13 Westergaard's solution for vertical stress due to a point load

Table 5.4 Variation of I_1 [Eq. (5.27)].

r/z	$\mu_s = 0$	$\mu_s = 0.2$	$\mu_s = 0.4$
		I_1	
0	0.3183	0.4244	0.9550
0.1	0.3090	0.4080	0.8750
0.2	0.2836	0.3646	0.6916
0.3	0.2483	0.3074	0.4997
0.4	0.2099	0.2491	0.3480
0.5	0.1733	0.1973	0.2416
0.6	0.1411	0.1547	0.1700
0.7	0.1143	0.1212	0.1221
0.8	0.0925	0.0953	0.0897
0.9	0.0751	0.0756	0.0673
1.0	0.0613	0.0605	0.0516
1.5	0.0247	0.0229	0.0173
2.0	0.0118	0.0107	0.0076
2.5	0.0064	0.0057	0.0040
3.0	0.0038	0.0034	0.0023
4.0	0.0017	0.0015	0.0010
5.0	0.0009	0.0008	0.0005

where

$$\eta = \sqrt{\frac{1 - 2\mu_s}{2 - 2\mu_s}} \tag{5.25}$$

μ_s = Poisson's ratio of the solid between the rigid reinforcements
$r = \sqrt{x^2 + y^2}$

Equation (5.24) can be rewritten as

$$\Delta\sigma = \left(\frac{P}{z^2}\right)I_1 \tag{5.26}$$

where

$$I_1 = \frac{1}{2\pi\eta^2}\left[\left(\frac{r}{\eta z}\right)^2 + 1\right]^{-3/2} \tag{5.27}$$

Table 5.4 gives the variation of I_1 with μ_s.

5.8   *Stress Distribution for Westergaard Material*

Stress Due to a Circularly Loaded Area

Referring to Figure 5.2, if the circular area is located on a Westergaard-type material, the increase in vertical stress, $\Delta\sigma$, at a point located at a depth z immediately below the center of the area can be given as

$$\Delta\sigma = q_o\left\{ 1 - \frac{\eta}{\left[\eta^2 + \left(\dfrac{B}{2z}\right)^2\right]^{1/2}} \right\} \qquad (5.28)$$

The term η has been defined in Eq. (5.25). The variations of $\Delta\sigma/q_o$ with $B/2z$ and $\mu_s = 0$ are given in Table 5.5.

Stress Due to a Uniformly Loaded Flexible Rectangular Area

Refer to Figure 5.3. If the flexible rectangular area is located on a Westergaard-type material, the stress increase at point A can be given as

$$\Delta\sigma = \frac{q_o}{2\pi}\left[\cot^{-1}\sqrt{\eta^2\left(\frac{1}{m^2} + \frac{1}{n^2}\right) + \eta^4\left(\frac{1}{m^2 n^2}\right)} \right] \qquad (5.29a)$$

where

$$m = \frac{B}{z}$$

$$n = \frac{L}{z}$$

Table 5.5 Variation of $\Delta\sigma/q_o$ with $B/2z$ and $\mu_s = 0$ [Eq. (5.28)]

B/2z	$\Delta\sigma/q_o$
0.00	0.0
0.25	0.0572
0.33	0.0938
0.50	0.1835
0.75	0.3140
1.00	0.4227
1.25	0.5076
1.50	0.5736
1.75	0.6254
2.00	0.6667
2.25	0.7002
2.50	0.7278
2.75	0.7510
3.00	0.7706
4.00	0.8259
5.00	0.8600
6.00	0.8830
7.00	0.8995
8.00	0.9120
9.00	0.9217
10.00	0.9295

Table 5.6. Variation of I_w with m and n ($\mu_s = 0$).

m	n								
	0.1	0.2	0.4	0.5	0.6	1.0	2.0	5.0	10.0
0.1	0.0031	0.0061	0.0110	0.0129	0.0144	0.0182	0.0211	0.0211	0.0223
0.2	0.0061	0.0118	0.0214	0.0251	0.0282	0.0357	0.0413	0.0434	0.0438
0.4	0.0110	0.0214	0.0390	0.0459	0.0516	0.0658	0.0768	0.0811	0.0847
0.5	0.0129	0.0251	0.0459	0.0541	0.0610	0.0781	0.0916	0.0969	0.0977
0.6	0.0144	0.0282	0.0516	0.0610	0.0687	0.0886	0.1044	0.1107	0.1117
1.0	0.0183	0.0357	0.0658	0.0781	0.0886	0.1161	0.1398	0.1491	0.1515
2.0	0.0211	0.0413	0.0768	0.0916	0.1044	0.1398	0.1743	0.1916	0.1948
5.0	0.0221	0.0435	0.0811	0.0969	0.1107	0.1499	0.1916	0.2184	0.2250
10.0	0.0223	0.0438	0.0817	0.0977	0.1117	0.1515	0.1948	0.2250	0.2341

or

$$\frac{\Delta\sigma}{q_o} = \frac{1}{2\pi}\left[\cot^{-1}\sqrt{\eta^2\left(\frac{1}{m^2} + \frac{1}{n^2}\right) + \eta^4\left(\frac{1}{m^2 n^2}\right)}\right] = I_w \qquad (5.29\text{b})$$

Table 5.6 gives the variation of I_w with m and n (for $\mu_s = 0$).

Example 5.4

Solve Example 5.1 using Eq. (5.29). Assume $\mu_s = 0$.

Solution

From Example 5.1

$$m = 0.2$$
$$n = 0.4$$
$$\Delta\sigma = q_o(4I_w)$$

From Table 5.6, for $m = 0.2$ and $n = 0.4$, the value of $I_w \approx 0.0214$. So

$$\Delta\sigma = (150)(4 \times 0.0214) = \textbf{12.84 kN/m}^2 \qquad \blacksquare$$

Elastic Settlement

5.9   Elastic Settlement of Foundations on Saturated Clay ($\mu_s = 0.5$)

Janbu et al. (1956) proposed an equation for evaluating the average settlement of flexible foundations on saturated clay soils (Poisson's ratio, $\mu_s = 0.5$). For the notation used in Figure 5.14, this equation is

$$S_e = A_1 A_2 \frac{q_o B}{E_s} \tag{5.30}$$

where A_1 is a function of H/B and L/B and A_2 is a function of D_f/B.

Christian and Carrier (1978) modified the values of A_1 and A_2 to some extent as presented in Figure 5.14.

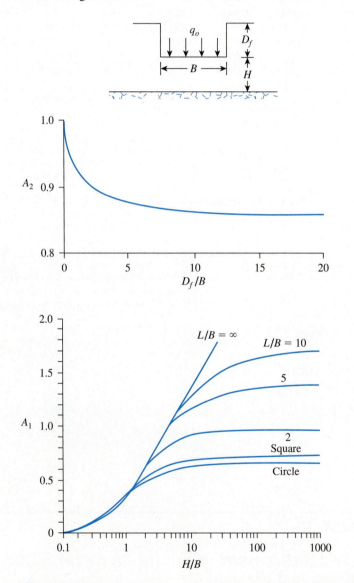

Figure 5.14 Values of A_1 and A_2 for elastic settlement calculation—Eq. (5.30) (After Christian and Carrier, 1978) (Christian, J. T. and and Carrier, W. D. (1978). "Janbu, Bjerrum and Kjaernsli's chart reinterpreted," *Canadian Geotechnical Journal*, Vol. 15, pp. 123–128. © 2008 NRC Canada or its licensors. Reproduced with permission.)

Table 5.7 Range of β for Clay [Eq. (5.31)][a]

Plasticity Index	β				
	OCR = 1	OCR = 2	OCR = 3	OCR = 4	OCR = 5
<30	1500–600	1380–500	1200–580	950–380	730–300
30 to 50	600–300	550–270	580–220	380–180	300–150
>50	300–150	270–120	220–100	180–90	150–75

[a]Interpolated from Duncan and Buchignani, (1976)

The modulus of elasticity (E_s) for clays can, in general, be given as

$$E_s = \beta c_u \tag{5.31}$$

where c_u = undrained shear strength.

The parameter β is primarily a function of the plasticity index and overconsolidation ratio. Table 5.7 provides a general range for β based on that proposed by Duncan and Buchignani (1976). In any case, proper judgment should be used in selecting the magnitude of β.

5.10 Settlement Based on the Theory of Elasticity

The elastic settlement of a shallow foundation can be estimated by using the theory of elasticity. From Hooke's law, as applied to Figure 5.15, we obtain

$$S_e = \int_0^H \varepsilon_z dz = \frac{1}{E_s} \int_0^H (\Delta\sigma_z - \mu_s\Delta\sigma_x - \mu_s\Delta\sigma_y)dz \tag{5.32}$$

Figure 5.15 Elastic settlement of shallow foundation

where

S_e = elastic settlement
E_s = modulus of elasticity of soil
H = thickness of the soil layer
μ_s = Poisson's ratio of the soil
$\Delta\sigma_x, \Delta\sigma_y, \Delta\sigma_z$ = stress increase due to the net applied foundation load in the x, y, and z directions, respectively

Theoretically, if the foundation is perfectly flexible (see Figure 5.16 and Bowles, 1987), the settlement may be expressed as

$$S_e = q_o(\alpha B')\frac{1 - \mu_s^2}{E_s}I_sI_f \tag{5.33}$$

where

q_o = net applied pressure on the foundation
μ_s = Poisson's ratio of soil
E_s = average modulus of elasticity of the soil under the foundation, measured from $z = 0$ to about $z = 5B$
B' = $B/2$ for center of foundation
   = B for corner of foundation
I_s = shape factor (Steinbrenner, 1934)

$$= F_1 + \frac{1 - 2\mu_s}{1 - \mu_s}F_2 \tag{5.34}$$

$$F_1 = \frac{1}{\pi}(A_0 + A_1) \tag{5.35}$$

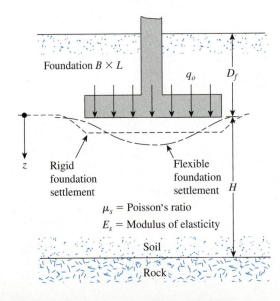

Foundation $B \times L$

q_o D_f

z Rigid foundation settlement   Flexible foundation settlement H

μ_s = Poisson's ratio
E_s = Modulus of elasticity

Soil

Rock

Figure 5.16 Elastic settlement of flexible and rigid foundations

$$F_2 = \frac{n'}{2\pi}\tan^{-1}A_2 \tag{5.36}$$

$$A_0 = m'\ln\frac{\left(1 + \sqrt{m'^2 + 1}\right)\sqrt{m'^2 + n'^2}}{m'\left(1 + \sqrt{m'^2 + n'^2 + 1}\right)} \tag{5.37}$$

$$A_1 = \ln\frac{\left(m' + \sqrt{m'^2 + 1}\right)\sqrt{1 + n'^2}}{m' + \sqrt{m'^2 + n'^2 + 1}} \tag{5.38}$$

$$A_2 = \frac{m'}{n'\sqrt{m'^2 + n'^2 + 1}} \tag{5.39}$$

$$I_f = \text{depth factor (Fox, 1948)} = f\left(\frac{D_f}{B}, \mu_s, \text{ and } \frac{L}{B}\right) \tag{5.40}$$

α = a factor that depends on the location on the foundation where settlement is being calculated

To calculate settlement at the *center* of the foundation, we use

$$\alpha = 4$$

$$m' = \frac{L}{B}$$

and

$$n' = \frac{H}{\left(\dfrac{B}{2}\right)}$$

To calculate settlement at a *corner* of the foundation,

$$\alpha = 1$$

$$m' = \frac{L}{B}$$

and

$$n' = \frac{H}{B}$$

The variations of F_1 and F_2 [see Eqs. (5.35) and (5.36)] with m' and n' are given in Tables 5.8 and 5.9. Also, the variation of I_f with D_f/B (for $\mu_s = 0.3, 0.4,$ and 0.5) is given in Table 5.10. These values are also given in more detailed form by Bowles (1987).

The elastic settlement of a *rigid foundation* can be estimated as

$$S_{e(\text{rigid})} \approx 0.93 S_{e(\text{flexible, center})} \tag{5.41}$$

Table 5.8 Variation of F_1 with m' and n'

n'	m'									
	1.0	1.2	1.4	1.6	1.8	2.0	2.5	3.0	3.5	4.0
0.25	0.014	0.013	0.012	0.011	0.011	0.011	0.010	0.010	0.010	0.010
0.50	0.049	0.046	0.044	0.042	0.041	0.040	0.038	0.038	0.037	0.037
0.75	0.095	0.090	0.087	0.084	0.082	0.080	0.077	0.076	0.074	0.074
1.00	0.142	0.138	0.134	0.130	0.127	0.125	0.121	0.118	0.116	0.115
1.25	0.186	0.183	0.179	0.176	0.173	0.170	0.165	0.161	0.158	0.157
1.50	0.224	0.224	0.222	0.219	0.216	0.213	0.207	0.203	0.199	0.197
1.75	0.257	0.259	0.259	0.258	0.255	0.253	0.247	0.242	0.238	0.235
2.00	0.285	0.290	0.292	0.292	0.291	0.289	0.284	0.279	0.275	0.271
2.25	0.309	0.317	0.321	0.323	0.323	0.322	0.317	0.313	0.308	0.305
2.50	0.330	0.341	0.347	0.350	0.351	0.351	0.348	0.344	0.340	0.336
2.75	0.348	0.361	0.369	0.374	0.377	0.378	0.377	0.373	0.369	0.365
3.00	0.363	0.379	0.389	0.396	0.400	0.402	0.402	0.400	0.396	0.392
3.25	0.376	0.394	0.406	0.415	0.420	0.423	0.426	0.424	0.421	0.418
3.50	0.388	0.408	0.422	0.431	0.438	0.442	0.447	0.447	0.444	0.441
3.75	0.399	0.420	0.436	0.447	0.454	0.460	0.467	0.458	0.466	0.464
4.00	0.408	0.431	0.448	0.460	0.469	0.476	0.484	0.487	0.486	0.484
4.25	0.417	0.440	0.458	0.472	0.481	0.484	0.495	0.514	0.515	0.515
4.50	0.424	0.450	0.469	0.484	0.495	0.503	0.516	0.521	0.522	0.522
4.75	0.431	0.458	0.478	0.494	0.506	0.515	0.530	0.536	0.539	0.539
5.00	0.437	0.465	0.487	0.503	0.516	0.526	0.543	0.551	0.554	0.554
5.25	0.443	0.472	0.494	0.512	0.526	0.537	0.555	0.564	0.568	0.569
5.50	0.448	0.478	0.501	0.520	0.534	0.546	0.566	0.576	0.581	0.584
5.75	0.453	0.483	0.508	0.527	0.542	0.555	0.576	0.588	0.594	0.597
6.00	0.457	0.489	0.514	0.534	0.550	0.563	0.585	0.598	0.606	0.609
6.25	0.461	0.493	0.519	0.540	0.557	0.570	0.594	0.609	0.617	0.621
6.50	0.465	0.498	0.524	0.546	0.563	0.577	0.603	0.618	0.627	0.632
6.75	0.468	0.502	0.529	0.551	0.569	0.584	0.610	0.627	0.637	0.643
7.00	0.471	0.506	0.533	0.556	0.575	0.590	0.618	0.635	0.646	0.653
7.25	0.474	0.509	0.538	0.561	0.580	0.596	0.625	0.643	0.655	0.662
7.50	0.477	0.513	0.541	0.565	0.585	0.601	0.631	0.650	0.663	0.671
7.75	0.480	0.516	0.545	0.569	0.589	0.606	0.637	0.658	0.671	0.680
8.00	0.482	0.519	0.549	0.573	0.594	0.611	0.643	0.664	0.678	0.688
8.25	0.485	0.522	0.552	0.577	0.598	0.615	0.648	0.670	0.685	0.695
8.50	0.487	0.524	0.555	0.580	0.601	0.619	0.653	0.676	0.692	0.703
8.75	0.489	0.527	0.558	0.583	0.605	0.623	0.658	0.682	0.698	0.710
9.00	0.491	0.529	0.560	0.587	0.609	0.627	0.663	0.687	0.705	0.716
9.25	0.493	0.531	0.563	0.589	0.612	0.631	0.667	0.693	0.710	0.723
9.50	0.495	0.533	0.565	0.592	0.615	0.634	0.671	0.697	0.716	0.719
9.75	0.496	0.536	0.568	0.595	0.618	0.638	0.675	0.702	0.721	0.735
10.00	0.498	0.537	0.570	0.597	0.621	0.641	0.679	0.707	0.726	0.740
20.00	0.529	0.575	0.614	0.647	0.677	0.702	0.756	0.797	0.830	0.858
50.00	0.548	0.598	0.640	0.678	0.711	0.740	0.803	0.853	0.895	0.931
100.00	0.555	0.605	0.649	0.688	0.722	0.753	0.819	0.872	0.918	0.956

Table 5.8 (*Continued*)

n'	m'									
	4.5	5.0	6.0	7.0	8.0	9.0	10.0	25.0	50.0	100.0
0.25	0.010	0.010	0.010	0.010	0.010	0.010	0.010	0.010	0.010	0.010
0.50	0.036	0.036	0.036	0.036	0.036	0.036	0.036	0.036	0.036	0.036
0.75	0.073	0.073	0.072	0.072	0.072	0.072	0.071	0.071	0.071	0.071
1.00	0.114	0.113	0.112	0.112	0.112	0.111	0.111	0.110	0.110	0.110
1.25	0.155	0.154	0.153	0.152	0.152	0.151	0.151	0.150	0.150	0.150
1.50	0.195	0.194	0.192	0.191	0.190	0.190	0.189	0.188	0.188	0.188
1.75	0.233	0.232	0.229	0.228	0.227	0.226	0.225	0.223	0.223	0.223
2.00	0.269	0.267	0.264	0.262	0.261	0.260	0.259	0.257	0.256	0.256
2.25	0.302	0.300	0.296	0.294	0.293	0.291	0.291	0.287	0.287	0.287
2.50	0.333	0.331	0.327	0.324	0.322	0.321	0.320	0.316	0.315	0.315
2.75	0.362	0.359	0.355	0.352	0.350	0.348	0.347	0.343	0.342	0.342
3.00	0.389	0.386	0.382	0.378	0.376	0.374	0.373	0.368	0.367	0.367
3.25	0.415	0.412	0.407	0.403	0.401	0.399	0.397	0.391	0.390	0.390
3.50	0.438	0.435	0.430	0.427	0.424	0.421	0.420	0.413	0.412	0.411
3.75	0.461	0.458	0.453	0.449	0.446	0.443	0.441	0.433	0.432	0.432
4.00	0.482	0.479	0.474	0.470	0.466	0.464	0.462	0.453	0.451	0.451
4.25	0.516	0.496	0.484	0.473	0.471	0.471	0.470	0.468	0.462	0.460
4.50	0.520	0.517	0.513	0.508	0.505	0.502	0.499	0.489	0.487	0.487
4.75	0.537	0.535	0.530	0.526	0.523	0.519	0.517	0.506	0.504	0.503
5.00	0.554	0.552	0.548	0.543	0.540	0.536	0.534	0.522	0.519	0.519
5.25	0.569	0.568	0.564	0.560	0.556	0.553	0.550	0.537	0.534	0.534
5.50	0.584	0.583	0.579	0.575	0.571	0.568	0.585	0.551	0.549	0.548
5.75	0.597	0.597	0.594	0.590	0.586	0.583	0.580	0.565	0.583	0.562
6.00	0.611	0.610	0.608	0.604	0.601	0.598	0.595	0.579	0.576	0.575
6.25	0.623	0.623	0.621	0.618	0.615	0.611	0.608	0.592	0.589	0.588
6.50	0.635	0.635	0.634	0.631	0.628	0.625	0.622	0.605	0.601	0.600
6.75	0.646	0.647	0.646	0.644	0.641	0.637	0.634	0.617	0.613	0.612
7.00	0.656	0.658	0.658	0.656	0.653	0.650	0.647	0.628	0.624	0.623
7.25	0.666	0.669	0.669	0.668	0.665	0.662	0.659	0.640	0.635	0.634
7.50	0.676	0.679	0.680	0.679	0.676	0.673	0.670	0.651	0.646	0.645
7.75	0.685	0.688	0.690	0.689	0.687	0.684	0.681	0.661	0.656	0.655
8.00	0.694	0.697	0.700	0.700	0.698	0.695	0.692	0.672	0.666	0.665
8.25	0.702	0.706	0.710	0.710	0.708	0.705	0.703	0.682	0.676	0.675
8.50	0.710	0.714	0.719	0.719	0.718	0.715	0.713	0.692	0.686	0.684
8.75	0.717	0.722	0.727	0.728	0.727	0.725	0.723	0.701	0.695	0.693
9.00	0.725	0.730	0.736	0.737	0.736	0.735	0.732	0.710	0.704	0.702
9.25	0.731	0.737	0.744	0.746	0.745	0.744	0.742	0.719	0.713	0.711
9.50	0.738	0.744	0.752	0.754	0.754	0.753	0.751	0.728	0.721	0.719
9.75	0.744	0.751	0.759	0.762	0.762	0.761	0.759	0.737	0.729	0.727
10.00	0.750	0.758	0.766	0.770	0.770	0.770	0.768	0.745	0.738	0.735
20.00	0.878	0.896	0.925	0.945	0.959	0.969	0.977	0.982	0.965	0.957
50.00	0.962	0.989	1.034	1.070	1.100	1.125	1.146	1.265	1.279	1.261
100.00	0.990	1.020	1.072	1.114	1.150	1.182	1.209	1.408	1.489	1.499

Table 5.9 Variation of F_2 with m' and n'

					m'					
n'	1.0	1.2	1.4	1.6	1.8	2.0	2.5	3.0	3.5	4.0
0.25	0.049	0.050	0.051	0.051	0.051	0.052	0.052	0.052	0.052	0.052
0.50	0.074	0.077	0.080	0.081	0.083	0.084	0.086	0.086	0.0878	0.087
0.75	0.083	0.089	0.093	0.097	0.099	0.101	0.104	0.106	0.107	0.108
1.00	0.083	0.091	0.098	0.102	0.106	0.109	0.114	0.117	0.119	0.120
1.25	0.080	0.089	0.096	0.102	0.107	0.111	0.118	0.122	0.125	0.127
1.50	0.075	0.084	0.093	0.099	0.105	0.110	0.118	0.124	0.128	0.130
1.75	0.069	0.079	0.088	0.095	0.101	0.107	0.117	0.123	0.128	0.131
2.00	0.064	0.074	0.083	0.090	0.097	0.102	0.114	0.121	0.127	0.131
2.25	0.059	0.069	0.077	0.085	0.092	0.098	0.110	0.119	0.125	0.130
2.50	0.055	0.064	0.073	0.080	0.087	0.093	0.106	0.115	0.122	0.127
2.75	0.051	0.060	0.068	0.076	0.082	0.089	0.102	0.111	0.119	0.125
3.00	0.048	0.056	0.064	0.071	0.078	0.084	0.097	0.108	0.116	0.122
3.25	0.045	0.053	0.060	0.067	0.074	0.080	0.093	0.104	0.112	0.119
3.50	0.042	0.050	0.057	0.064	0.070	0.076	0.089	0.100	0.109	0.116
3.75	0.040	0.047	0.054	0.060	0.067	0.073	0.086	0.096	0.105	0.113
4.00	0.037	0.044	0.051	0.057	0.063	0.069	0.082	0.093	0.102	0.110
4.25	0.036	0.042	0.049	0.055	0.061	0.066	0.079	0.090	0.099	0.107
4.50	0.034	0.040	0.046	0.052	0.058	0.063	0.076	0.086	0.096	0.104
4.75	0.032	0.038	0.044	0.050	0.055	0.061	0.073	0.083	0.093	0.101
5.00	0.031	0.036	0.042	0.048	0.053	0.058	0.070	0.080	0.090	0.098
5.25	0.029	0.035	0.040	0.046	0.051	0.056	0.067	0.078	0.087	0.095
5.50	0.028	0.033	0.039	0.044	0.049	0.054	0.065	0.075	0.084	0.092
5.75	0.027	0.032	0.037	0.042	0.047	0.052	0.063	0.073	0.082	0.090
6.00	0.026	0.031	0.036	0.040	0.045	0.050	0.060	0.070	0.079	0.087
6.25	0.025	0.030	0.034	0.039	0.044	0.048	0.058	0.068	0.077	0.085
6.50	0.024	0.029	0.033	0.038	0.042	0.046	0.056	0.066	0.075	0.083
6.75	0.023	0.028	0.032	0.036	0.041	0.045	0.055	0.064	0.073	0.080
7.00	0.022	0.027	0.031	0.035	0.039	0.043	0.053	0.062	0.071	0.078
7.25	0.022	0.026	0.030	0.034	0.038	0.042	0.051	0.060	0.069	0.076
7.50	0.021	0.025	0.029	0.033	0.037	0.041	0.050	0.059	0.067	0.074
7.75	0.020	0.024	0.028	0.032	0.036	0.039	0.048	0.057	0.065	0.072
8.00	0.020	0.023	0.027	0.031	0.035	0.038	0.047	0.055	0.063	0.071
8.25	0.019	0.023	0.026	0.030	0.034	0.037	0.046	0.054	0.062	0.069
8.50	0.018	0.022	0.026	0.029	0.033	0.036	0.045	0.053	0.060	0.067
8.75	0.018	0.021	0.025	0.028	0.032	0.035	0.043	0.051	0.059	0.066
9.00	0.017	0.021	0.024	0.028	0.031	0.034	0.042	0.050	0.057	0.064
9.25	0.017	0.020	0.024	0.027	0.030	0.033	0.041	0.049	0.056	0.063
9.50	0.017	0.020	0.023	0.026	0.029	0.033	0.040	0.048	0.055	0.061
9.75	0.016	0.019	0.023	0.026	0.029	0.032	0.039	0.047	0.054	0.060
10.00	0.016	0.019	0.022	0.025	0.028	0.031	0.038	0.046	0.052	0.059
20.00	0.008	0.010	0.011	0.013	0.014	0.016	0.020	0.024	0.027	0.031
50.00	0.003	0.004	0.004	0.005	0.006	0.006	0.008	0.010	0.011	0.013
100.00	0.002	0.002	0.002	0.003	0.003	0.003	0.004	0.005	0.006	0.006

Table 5.9 (*Continued*)

n'	m'									
	4.5	5.0	6.0	7.0	8.0	9.0	10.0	25.0	50.0	100.0
0.25	0.053	0.053	0.053	0.053	0.053	0.053	0.053	0.053	0.053	0.053
0.50	0.087	0.087	0.088	0.088	0.088	0.088	0.088	0.088	0.088	0.088
0.75	0.109	0.109	0.109	0.110	0.110	0.110	0.110	0.111	0.111	0.111
1.00	0.121	0.122	0.123	0.123	0.124	0.124	0.124	0.125	0.125	0.125
1.25	0.128	0.130	0.131	0.132	0.132	0.133	0.133	0.134	0.134	0.134
1.50	0.132	0.134	0.136	0.137	0.138	0.138	0.139	0.140	0.140	0.140
1.75	0.134	0.136	0.138	0.140	0.141	0.142	0.142	0.144	0.144	0.145
2.00	0.134	0.136	0.139	0.141	0.143	0.144	0.145	0.147	0.147	0.148
2.25	0.133	0.136	0.140	0.142	0.144	0.145	0.146	0.149	0.150	0.150
2.50	0.132	0.135	0.139	0.142	0.144	0.146	0.147	0.151	0.151	0.151
2.75	0.130	0.133	0.138	0.142	0.144	0.146	0.147	0.152	0.152	0.153
3.00	0.127	0.131	0.137	0.141	0.144	0.145	0.147	0.152	0.153	0.154
3.25	0.125	0.129	0.135	0.140	0.143	0.145	0.147	0.153	0.154	0.154
3.50	0.122	0.126	0.133	0.138	0.142	0.144	0.146	0.153	0.155	0.155
3.75	0.119	0.124	0.131	0.137	0.141	0.143	0.145	0.154	0.155	0.155
4.00	0.116	0.121	0.129	0.135	0.139	0.142	0.145	0.154	0.155	0.156
4.25	0.113	0.119	0.127	0.133	0.138	0.141	0.144	0.154	0.156	0.156
4.50	0.110	0.116	0.125	0.131	0.136	0.140	0.143	0.154	0.156	0.156
4.75	0.107	0.113	0.123	0.130	0.135	0.139	0.142	0.154	0.156	0.157
5.00	0.105	0.111	0.120	0.128	0.133	0.137	0.140	0.154	0.156	0.157
5.25	0.102	0.108	0.118	0.126	0.131	0.136	0.139	0.154	0.156	0.157
5.50	0.099	0.106	0.116	0.124	0.130	0.134	0.138	0.154	0.156	0.157
5.75	0.097	0.103	0.113	0.122	0.128	0.133	0.136	0.154	0.157	0.157
6.00	0.094	0.101	0.111	0.120	0.126	0.131	0.135	0.153	0.157	0.157
6.25	0.092	0.098	0.109	0.118	0.124	0.129	0.134	0.153	0.157	0.158
6.50	0.090	0.096	0.107	0.116	0.122	0.128	0.132	0.153	0.157	0.158
6.75	0.087	0.094	0.105	0.114	0.121	0.126	0.131	0.153	0.157	0.158
7.00	0.085	0.092	0.103	0.112	0.119	0.125	0.129	0.152	0.157	0.158
7.25	0.083	0.090	0.101	0.110	0.117	0.123	0.128	0.152	0.157	0.158
7.50	0.081	0.088	0.099	0.108	0.115	0.121	0.126	0.152	0.156	0.158
7.75	0.079	0.086	0.097	0.106	0.114	0.120	0.125	0.151	0.156	0.158
8.00	0.077	0.084	0.095	0.104	0.112	0.118	0.124	0.151	0.156	0.158
8.25	0.076	0.082	0.093	0.102	0.110	0.117	0.122	0.150	0.156	0.158
8.50	0.074	0.080	0.091	0.101	0.108	0.115	0.121	0.150	0.156	0.158
8.75	0.072	0.078	0.089	0.099	0.107	0.114	0.119	0.150	0.156	0.158
9.00	0.071	0.077	0.088	0.097	0.105	0.112	0.118	0.149	0.156	0.158
9.25	0.069	0.075	0.086	0.096	0.104	0.110	0.116	0.149	0.156	0.158
9.50	0.068	0.074	0.085	0.094	0.102	0.109	0.115	0.148	0.156	0.158
9.75	0.066	0.072	0.083	0.092	0.100	0.107	0.113	0.148	0.156	0.158
10.00	0.065	0.071	0.082	0.091	0.099	0.106	0.112	0.147	0.156	0.158
20.00	0.035	0.039	0.046	0.053	0.059	0.065	0.071	0.124	0.148	0.156
50.00	0.014	0.016	0.019	0.022	0.025	0.028	0.031	0.071	0.113	0.142
100.00	0.007	0.008	0.010	0.011	0.013	0.014	0.016	0.039	0.071	0.113

Table 5.10 Variation of I_f with D_f/B, B/L, and μ_s

		B/L		
μ_s	D_f/B	0.2	0.5	1.0
0.3	0.2	0.95	0.93	0.90
	0.4	0.90	0.86	0.81
	0.6	0.85	0.80	0.74
	1.0	0.78	0.71	0.65
0.4	0.2	0.97	0.96	0.93
	0.4	0.93	0.89	0.85
	0.6	0.89	0.84	0.78
	1.0	0.82	0.75	0.69
0.5	0.2	0.99	0.98	0.96
	0.4	0.95	0.93	0.89
	0.6	0.92	0.87	0.82
	1.0	0.85	0.79	0.72

Due to the nonhomogeneous nature of soil deposits, the magnitude of E_s may vary with depth. For that reason, Bowles (1987) recommended using a weighted average of E_s in Eq. (5.33), or

$$E_s = \frac{\sum E_{s(i)} \Delta z}{\bar{z}} \tag{5.42}$$

where

$E_{s(i)}$ = soil modulus of elasticity within a depth Δz
$\bar{z}$ = H or $5B$, whichever is smaller

Example 5.5

Refer to Figure 5.16 and consider a rigid square foundation 2.44 m $\times$ 2.44 m in plan ($D_f = 1.22$ m) on a layer of normally consolidated sand. A rock layer is located at $z = 10.98$ m. The following is an approximation of the standard penetration number (N_{60}) with z.

z (m)	N_{60}
0–2.44	7
2.44–21	6.4
6.4–36	10.98

Given: $\mu_s = 0.3$ and $q_o = 167.7$ kN/m^2. Estimate the elastic settlement of the foundation. Use Eq. (2.29).

Solution

Calculation of Average E_s

From Eq. (2.29),

$$E_s \approx p_a \alpha N_{60}$$

$B = 2.44$ m, $p_a \approx 100$ kN/m^2, and $\alpha \approx 10$ (since it is a normally consolidated clean sand).

Given: $H = 10.98$ m $< 5B$.

The approximate variations of E_s using Eq. (2.29) are given below.

z (m)	Δz (m)	N_{60}	E_s (kN/m^2)
0–2.44	2.44	7	7000
2.44–6.4	3.96	11	11,000
6.4–10.98	4.58	14	14,000

Using Eq. (5.42),

$$E_s = \frac{\Sigma E_{s(i)} \Delta z}{\overline{z}} = \frac{(7,000)(2.44) + (11,000)(3.96) + (14,000)(4.58)}{10.98}$$

$$= 11,362 \text{ kN/m}^2$$

Calculation of S_e below the Center of the Foundation [Eq. (5.33)]

$$S_e = q_o(\alpha B') \frac{1 - \mu_s^2}{E_s} I_s I_f$$

$$B' = \frac{2.44}{2} = 1.22 \text{ m}$$

$$\alpha = 4$$

$$m' = \frac{L}{B} = 1$$

$$n' = \frac{H}{\left(\dfrac{B}{2}\right)} = \frac{10.98}{\left(\dfrac{2.44}{2}\right)} = 9$$

$$I_s = F_1 + \frac{1 - 2\mu_s}{1 - \mu_s} F_2$$

From Table 5.8, $F_1 = 0.491$. From Table 5.9, $F_2 = 0.017$.

$$I_s = 0.491 + \left[\frac{1 - (2)(0.3)}{1 - 0.3}\right](0.017) = 0.5007$$

For $\mu_s = 3$, $D_f/B = 1.22/2.44 = 0.5$, and $B/L = 1$, the value of I_f is about 0.78 (Table 5.10).

$$S_e = (167.7)(4 \times 1.22)\left(\frac{1 - 0.3^2}{11,362}\right)(0.5007)(0.78) = 0.0256 \text{ m} \approx 25.6 \text{ mm}$$

Calculation of S_e for Rigid Foundation
From Eq. (5.41),

$$S_{e(\text{rigid})} \approx 0.93 S_{e(\text{flexible, center})} = (25.6)(0.93) = 23.81 \text{ mm} \approx \textbf{24 mm} \quad \blacksquare$$

5.11 Improved Equation for Elastic Settlement

In 1999, Mayne and Poulos presented an improved formula for calculating the elastic settlement of foundations. The formula takes into account the rigidity of the foundation, the depth of embedment of the foundation, the increase in the modulus of elasticity of the soil with depth, and the location of rigid layers at a limited depth. To use Mayne and Poulos's equation, one needs to determine the equivalent diameter B_e of a rectangular foundation, or

$$B_e = \sqrt{\frac{4BL}{\pi}} \tag{5.43}$$

where

B = width of foundation
L = length of foundation

For circular foundations,

$$B_e = B \tag{5.44}$$

where B = diameter of foundation.

Figure 5.17 shows a foundation with an equivalent diameter B_e located at a depth D_f below the ground surface. Let the thickness of the foundation be t and the modulus of elasticity of the foundation material be E_f. A rigid layer is located at a depth H below the bottom of the foundation. The modulus of elasticity of the compressible soil layer can be given as

$$E_s = E_o + kz \tag{5.45}$$

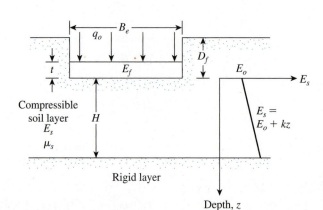

Figure 5.17 Improved equation for calculating elastic settlement: general parameters

With the preceding parameters defined, the elastic settlement below the center of the foundation is

$$S_e = \frac{q_o B_e I_G I_F I_E}{E_o}\left(1 - \mu_s^2\right) \tag{5.46}$$

where

I_G = influence factor for the variation of E_s with depth

$$= f\left(\beta = \frac{E_o}{kB_e}, \frac{H}{B_e}\right)$$

I_F = foundation rigidity correction factor
I_E = foundation embedment correction factor

Figure 5.18 shows the variation of I_G with $\beta = E_o/kB_e$ and H/B_e. The foundation rigidity correction factor can be expressed as

$$I_F = \frac{\pi}{4} + \frac{1}{4.6 + 10\left(\dfrac{E_f}{E_o + \dfrac{B_e}{2}k}\right)\left(\dfrac{2t}{B_e}\right)^3} \tag{5.47}$$

Similarly, the embedment correction factor is

$$I_E = 1 - \frac{1}{3.5\exp(1.22\mu_s - 0.4)\left(\dfrac{B_e}{D_f} + 1.6\right)} \tag{5.48}$$

Figures 5.19 and 5.20 show the variation of I_F and I_E with terms expressed in Eqs. (5.47) and (5.48).

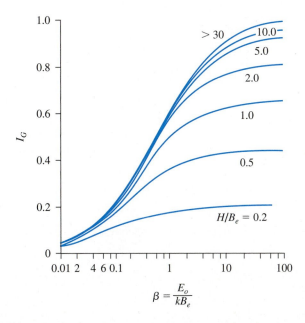

$$\beta = \frac{E_o}{kB_e}$$

Figure 5.18 Variation of I_G with β

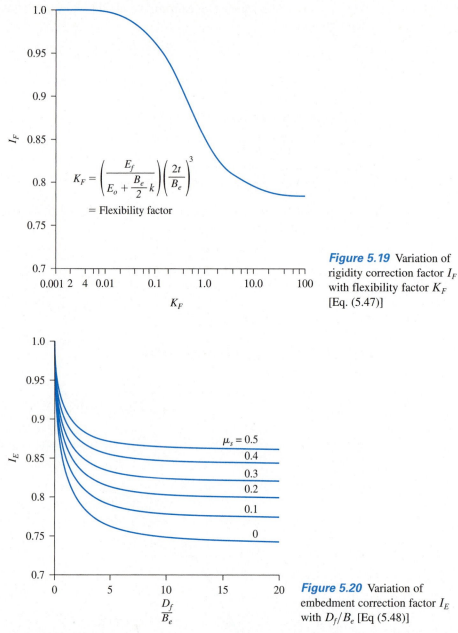

$$K_F = \left(\frac{E_f}{E_o + \frac{B_e}{2}k} \right)\left(\frac{2t}{B_e} \right)^3$$

= Flexibility factor

Figure 5.19 Variation of rigidity correction factor I_F with flexibility factor K_F [Eq. (5.47)]

$\mu_s = 0.5$
0.4
0.3
0.2
0.1
0

Figure 5.20 Variation of embedment correction factor I_E with D_f/B_e [Eq (5.48)]

Example 5.6

For a shallow foundation supported by a silty clay, as shown in Figure 5.17,

Length = L = 3.05 m
Width = B = 1.52 m
Depth of foundation = D_f = 1.52 m

Thickness of foundation $= t = 0.305$ m

Load per unit area $= q_o = 239.6$ kN/m²

$E_f = 15.87 \times 10^6$ kN/m²

The silty clay soil has the following properties:

$H = 3.66$ m

$\mu_s = 0.3$

$E_o = 9660$ kN/m²

$k = 565.6$ kN/m²/m

Estimate the elastic settlement of the foundation.

Solution

From Eq. (5.43), the equivalent diameter is

$$B_e = \sqrt{\frac{4BL}{\pi}} = \sqrt{\frac{(4)(1.52)(3.05)}{\pi}} = 2.43 \text{ m}$$

so

$$\beta = \frac{E_o}{kB_e} = \frac{9660}{(565.6)(2.43)} = 7.02$$

and

$$\frac{H}{B_e} = \frac{3.66}{2.43} = 1.5$$

From Figure 5.18, for $\beta = 7.02$ and $H/B_e = 1.5$, the value of $I_G \approx 0.69$. From Eq. (5.47),

$$I_F = \frac{\pi}{4} + \cfrac{1}{4.6 + 10\left(\cfrac{E_f}{E_o + \dfrac{B_e}{2}k}\right)\left(\dfrac{2t}{B_e}\right)^3}$$

$$= \frac{\pi}{4} + \cfrac{1}{4.6 + 10\left[\cfrac{15.87 \times 10^6}{9660 + \left(\dfrac{2.43}{2}\right)(565.6)}\right]\left[\dfrac{(2)(0.305)}{2.43}\right]^3} = 0.785$$

From Eq. (5.48),

$$I_E = 1 - \cfrac{1}{3.5\exp(1.22\mu_s - 0.4)\left(\dfrac{B_e}{D_f} + 1.6\right)}$$

$$= 1 - \frac{1}{3.5 \exp[(1.22)(0.3) - 0.4]\left(\dfrac{2.43}{1.52} + 1.6\right)} = 0.908$$

From Eq. (5.46),

$$S_e = \frac{q_o B_e I_G I_F I_E}{E_o}(1 - \mu_s^2)$$

so, with $q_o = 239.6 \ \text{kN/m}^2$, it follows that

$$S_e = \frac{(239.6)(2.43)(0.69)(0.785)(0.908)}{(9660)}(1 - 0.3^2) = 0.027 \ \text{m} = \textbf{27 mm} \quad \blacksquare$$

5.12   *Settlement of Sandy Soil: Use of Strain Influence Factor*

The settlement of granular soils can also be evaluated by the use of a semiempirical *strain influence factor* proposed by Schmertmann et al. (1978). According to this method (Figure 5.21), the settlement is

$$S_e = C_1 C_2 (\bar{q} - q) \sum_0^{z_2} \frac{I_z}{E_s} \Delta z \qquad (5.49)$$

where

I_z = strain influence factor
C_1 = a correction factor for the depth of foundation embedment = $1 - 0.5$
    $[q/(\bar{q} - q)]$
C_2 = a correction factor to account for creep in soil
    = $1 + 0.2 \log$ (time in years/0.1)
$\bar{q}$ = stress at the level of the foundation
$q = \gamma D_f$ = effective stress at the base of the foundation

E_s = modulus of elasticity of soil

The recommended variation of the strain influence factor I_z for square ($L/B = 1$) or circular foundations and for foundations with $L/B \geq 10$ is shown in Figure 5.21. The I_z diagrams for $1 < L/B < 10$ can be interpolated.

Note that the maximum value of I_z [that is, $I_{z(m)}$] occurs at $z = z_1$ and then reduces to zero at $z = z_2$. The maximum value of I_z can be calculated as

$$I_{z(m)} = 0.5 + 0.1 \sqrt{\frac{\bar{q} - q}{q'_{z(1)}}} \qquad (5.50)$$

where

$q'_{z(1)}$ = effective stress at a depth of z_1 before construction of the foundation

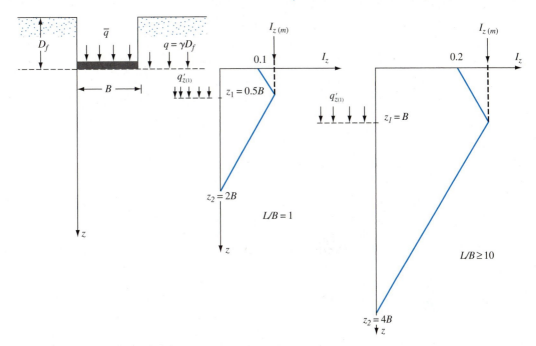

Figure 5.21 Variation of strain influence factor with depth and L/B

The following relations are suggested by Salgado (2008) for interpolation of I_z at $z = 0$, z_1/B, and z_2/B for rectangular foundations.

- I_z at $z = 0$

$$I_z = 0.1 + 0.0111\left(\frac{L}{B} - 1\right) \leq 0.2 \tag{5.51}$$

- Variation of z_1/B for $I_{z(m)}$

$$\frac{z_1}{B} = 0.5 + 0.0555\left(\frac{L}{B} - 1\right) \leq 1 \tag{5.52}$$

- Variation of z_2/B

$$\frac{z_2}{B} = 2 + 0.222\left(\frac{L}{B} - 1\right) \leq 4 \tag{5.53}$$

Schmertmann et al. (1978) suggested that

$$E_s = 2.5q_c \text{ (for square foundation)} \tag{5.54}$$

and

$$E_s = 3.5q_c \text{ (for } L/B \geq 10) \tag{5.55}$$

where q_c = cone penetration resistance.

It appears reasonable to write (Terzaghi et al., 1996)

$$E_{s(\text{rectangle})} = \left(1 + 0.4 \log\frac{L}{B}\right) E_{s(\text{square})} \tag{5.56}$$

The procedure for calculating elastic settlement using Eq. (5.49) is given here (Figure 5.22).

Step 1. Plot the foundation and the variation of I_z with depth to scale (Figure 5.22a).

Step 2. Using the correlation from standard penetration resistance (N_{60}) or cone penetration resistance (q_c), plot the actual variation of E_s with depth (Figure 5.22b).

Step 3. Approximate the actual variation of E_s into a number of layers of soil having a constant E_s, such as $E_{s(1)}, E_{s(2)}, \ldots, E_{s(i)}, \ldots E_{s(n)}$ (Figure 5.22b).

Step 4. Divide the soil layer from $z = 0$ to $z = z_2$ into a number of layers by drawing horizontal lines. The number of layers will depend on the break in continuity in the I_z and E_s diagrams.

Step 5. Prepare a table (such as Table 5.11) to obtain $\Sigma\dfrac{I_z}{E_s}\Delta z$.

Step 6. Calculate C_1 and C_2.

Step 7. Calculate S_e from Eq. (5.49).

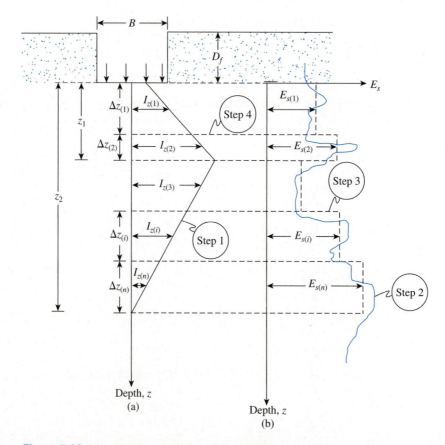

Figure 5.22 Procedure for calculation of S_e using the strain influence factor

Table 5.11 Calculation of $\Sigma \dfrac{I_z}{E_s} \Delta z$

Layer no.	Δz	E_s	I_z at the middle of the layer	$\dfrac{I_z}{E_s} \Delta z$
1	$\Delta z_{(1)}$	$E_{s(1)}$	$I_{z(1)}$	$\dfrac{I_{z(1)}}{E_{s(1)}} \Delta z_1$
2	$\Delta z_{(2)}$	$E_{s(2)}$	$I_{z(2)}$	
$\vdots$	$\vdots$	$\vdots$	$\vdots$	
i	$\Delta z_{(i)}$	$E_{s(i)}$	$I_{z(i)}$	$\dfrac{I_{z(i)}}{E_{s(i)}} \Delta z_i$
$\vdots$	$\vdots$	$\vdots$	$\vdots$	$\vdots$
n	$\Delta z_{(n)}$	$E_{s(n)}$	$I_{z(n)}$	$\dfrac{I_{z(n)}}{E_{s(n)}} \Delta z_n$
				$\Sigma \dfrac{I_z}{E_s} \Delta z$

Example 5.7

Consider a rectangular foundation 2 m × 4 m in plan at a depth of 1.2 m in a sand deposit, as shown in Figure 5.23a. Given: $\gamma = 17.5$ kN/m³; $\bar{q} = 145$ kN/m², and the following approximated variation of q_c with z:

z (m)	q_c (kN/m²)
0–0.5	2250
0.5–2.5	3430
2.5–5.0	2950

Estimate the elastic settlement of the foundation using the strain influence factor method.

Solution

From Eq. (5.52),

$$\frac{z_1}{B} = 0.5 + 0.0555\left(\frac{L}{B} - 1\right) = 0.5 + 0.0555\left(\frac{4}{2} - 1\right) \approx 0.56$$

$$z_1 = (0.56)(2) = 1.12 \text{ m}$$

From Eq. (5.53),

$$\frac{z_2}{B} = 2 + 0.222\left(\frac{L}{B} - 1\right) = 2 + 0.222(2 - 1) = 2.22$$

$$z_2 = (2.22)(2) = 4.44 \text{ m}$$

From Eq. (5.51), at $z = 0$,

$$I_z = 0.1 + 0.0111\left(\frac{L}{B} - 1\right) = 0.1 + 0.0111\left(\frac{4}{2} - 1\right) \approx 0.11$$

From Eq. (5.50),

$$I_{z(m)} = 0.5 + 0.1\sqrt{\frac{\bar{q} - q}{q'_{z(1)}}} = 0.5 + 0.1\left[\frac{145 - (1.2 \times 17.5)}{(1.2 + 1.12)(17.5)}\right]^{0.5} = 0.675$$

The plot of I_z versus z is shown in Figure 5.23c. Again, from Eq. (5.56)

$$E_{s(\text{rectangle})} = \left(1 + 0.4\log\frac{L}{B}\right)E_{s(\text{square})} = \left[1 + 0.4\log\left(\frac{4}{2}\right)\right](2.5 \times q_c) = 2.8q_c$$

Hence, the approximated variation of E_s with z is as follows:

z (m)	q_c (kN/m²)	E_s (kN/m²)
0–0.5	2250	6300
0.5–2.5	3430	9604
2.5–5.0	2950	8260

The plot of E_s versus z is shown in Figure 5.23b.

Figure 5.23

The soil layer is divided into four layers as shown in Figures 5.23b and 5.23c. Now the following table can be prepared.

Layer no.	Δz (m)	E_s (kN/m²)	I_z at middle of layer	$\dfrac{I_z}{E_s}\Delta z$ (m³/ kN)
1	0.50	6300	0.236	1.87×10^{-5}
2	0.62	9604	0.519	3.35×10^{-5}
3	1.38	9604	0.535	7.68×10^{-5}
4	1.94	8260	0.197	4.62×10^{-5}
				$\Sigma 17.52 \times 10^{-5}$

$$S_e = C_1 C_2 (\overline{q} - q) \sum \frac{I_z}{E_s}\Delta z$$

$$C_1 = 1 - 0.5\left(\frac{q}{\overline{q} - q}\right) = 1 - 0.5\left(\frac{21}{145 - 21}\right) = 0.915$$

Assume the time for creep is 10 years. So,

$$C_2 = 1 + 0.2\log\left(\frac{10}{0.1}\right) = 1.4$$

Hence,

$$S_e = (0.915)(1.4)(145 - 21)(17.52 \times 10^{-5}) = 2783 \times 10^{-5} \text{ m} = \textbf{27.83 mm} \quad \blacksquare$$

5.13 Settlement of Foundation on Sand Based on Standard Penetration Resistance

Meyerhof's Method

Meyerhof (1956) proposed a correlation for the *net bearing pressure* for foundations with the standard penetration resistance, N_{60}. The net pressure has been defined as

$$q_{net} = \overline{q} - \gamma D_f$$

where $\overline{q}$ = stress at the level of the foundation.

According to Meyerhof's theory, for 25 mm (1 in.) of estimated maximum settlement,

$$q_{net}(\text{kN/m}^2) = \frac{N_{60}}{0.08} \quad (\text{for } B \leq 1.22 \text{ m}) \tag{5.57}$$

and

$$q_{net}(\text{kN/m}^2) = \frac{N_{60}}{0.125}\left(\frac{B + 0.3}{B}\right)^2 \quad (\text{for } B > 1.22 \text{ m}) \tag{5.58}$$

Since the time that Meyerhof proposed his original correlations, researchers have observed that its results are rather conservative. Later, Meyerhof (1965) suggested that the net allowable bearing pressure should be increased by about 50%. Bowles (1977) proposed that the modified form of the bearing equations be expressed as

$$q_{net}(kN/m^2) = \frac{N_{60}}{2.5} F_d \left(\frac{S_e}{25}\right) \quad \text{(for } B \leq 1.22 \text{ m)} \tag{5.59}$$

and

$$q_{net}(kN/m^2) = \frac{N_{60}}{0.08} \left(\frac{B + 0.3}{B}\right)^2 F_d \left(\frac{S_e}{25}\right) \quad \text{(for } B > 1.22 \text{ m)} \tag{5.60}$$

where

F_d = depth factor = $1 + 0.33(D_f/B)$
B = foundation width, in meters
S_e = settlement, in mm

Hence,

$$S_e(mm) = \frac{1.25 q_{net}(kN/m^2)}{N_{60} F_d} \quad \text{(for } B \leq 1.22 \text{ m)} \tag{5.61}$$

and

$$S_e(mm) = \frac{2 q_{net}(kN/m^2)}{N_{60} F_d} \left(\frac{B}{B + 0.3}\right)^2 \quad \text{(for } B > 1.22 \text{ m)} \tag{5.62}$$

The N_{60} referred to in the preceding equations is the standard penetration resistance between the bottom of the foundation and $2B$ below the bottom.

Burland and Burbidge's Method

Burland and Burbidge (1985) proposed a method of calculating the elastic settlement of sandy soil using the *field standard penetration number, N_{60}.* (See Chapter 2.) The method can be summarized as follows:

1. **Variation of Standard Penetration Number with Depth**
   Obtain the field penetration numbers (N_{60}) with depth at the location of the foundation. The following adjustments of N_{60} may be necessary, depending on the field conditions:
   For gravel or sandy gravel,

   $$N_{60(a)} \approx 1.25 N_{60} \tag{5.63}$$

   For fine sand or silty sand below the groundwater table and $N_{60} > 15$,

   $$N_{60(a)} \approx 15 + 0.5(N_{60} - 15) \tag{5.64}$$

   where $N_{60(a)}$ = adjusted N_{60} value.

2. **Determination of Depth of Stress Influence (z')**
   In determining the depth of stress influence, the following three cases may arise:

Case I. If N_{60} [or $N_{60(a)}$] is approximately constant with depth, calculate z' from

$$\frac{z'}{B_R} = 1.4 \left(\frac{B}{B_R} \right)^{0.75} \tag{5.65}$$

where

B_R = reference width = 0.3 m (if B is in m)

B = width of the actual foundation

Case II. If N_{60} [or $N_{60(a)}$] is increasing with depth, use Eq. (5.65) to calculate z'.

Case III. If N_{60} [or $N_{60(a)}$] is decreasing with depth, $z' = 2B$ or to the bottom of soft soil layer measured from the bottom of the foundation (whichever is smaller).

3. **Calculation of Elastic Settlement S_e**

The elastic settlement of the foundation, S_e, can be calculated from

$$\frac{S_e}{B_R} = \alpha_1 \alpha_2 \alpha_3 \left[\frac{1.25 \left(\dfrac{L}{B} \right)}{0.25 + \left(\dfrac{L}{B} \right)} \right]^2 \left(\frac{B}{B_R} \right)^{0.7} \left(\frac{q'}{p_a} \right) \tag{5.66}$$

where

α_1 = a constant

α_2 = compressibility index

α_3 = correction for the depth of influence

p_a = atmospheric pressure = 100 kN/m^2

L = length of the foundation

Table 5.12 summarizes the values of q', α_1, α_2, and α_3 to be used in Eq. (5.70) for various types of soils. Note that, in this table, $\bar{N}_{60}$ or $\bar{N}_{60(a)}$ = average value of N_{60} or $N_{60(a)}$ in the depth of stress influence.

Table 5.12 Summary of α_1, α_2, and α_3

Soil type	q'	α_1	α_2	α_3
Normally consolidated sand	q_{net}	0.14	$\dfrac{1.71}{[\bar{N}_{60} \text{ or } \bar{N}_{60(a)}]^{1.4}}$	$\alpha_3 = \dfrac{H}{z'} \left(2 - \dfrac{H}{z'} \right)$ (if $H \leq z'$)
Overconsolidated sand ($q_{net} \leq \sigma_c'$) where σ_c' = preconsolidation pressure	q_{net}	0.047	$\dfrac{0.57}{[\bar{N}_{60} \text{ or } \bar{N}_{60(a)}]^{1.4}}$	or $\alpha_3 = 1$ (if $H > z'$) where H = depth of compressible layer
Overconsolidated sand ($q_{net} > \sigma_c'$)	$q_{net} - 0.67\sigma_c'$	0.14	$\dfrac{0.57}{[\bar{N}_{60} \text{ or } \bar{N}_{60(a)}]^{1.4}}$	

Example 5.8

A shallow foundation measuring 1.75 m × 1.75 m is to be constructed over a layer of sand. Given $D_f = 1$ m; N_{60} is generally increasing with depth; $\overline{N}_{60}$ in the depth of stress influence = 10, $q_{net} = 120$ kN/m². The sand is normally consolidated. Estimate the elastic settlement of the foundation. Use the Burland and Burbidge method.

Solution
From Eq. (5.69),

$$\frac{z'}{B_R} = 1.4\left(\frac{B}{B_R}\right)^{0.75}$$

Depth of stress influence,

$$z' = 1.4\left(\frac{B}{B_R}\right)^{0.75} B_R = (1.4)(0.3)\left(\frac{1.75}{0.3}\right)^{0.75} \approx 1.58 \text{ m}$$

From Eq. (5.70),

$$\frac{S_e}{B_R} = \alpha_1\alpha_2\alpha_3\left[\frac{1.25\left(\frac{L}{B}\right)}{0.25 + \left(\frac{L}{B}\right)}\right]^2\left(\frac{B}{B_R}\right)^{0.7}\left(\frac{q'}{p_a}\right)$$

For normally consolidated sand (Table 5.12),

$$\alpha_1 = 0.14$$
$$\alpha_2 = \frac{1.71}{(\overline{N}_{60})^{1.4}} = \frac{1.71}{(10)^{1.4}} = 0.068$$
$$\alpha_3 = 1$$
$$q' = q_{net} = 120 \text{ kN/m}^2$$

So,

$$\frac{S_e}{0.3} = (0.14)(0.068)(1)\left[\frac{(1.25)\left(\frac{1.75}{1.75}\right)}{0.25 + \left(\frac{1.75}{1.75}\right)}\right]^2\left(\frac{1.75}{0.3}\right)^{0.7}\left(\frac{120}{100}\right)$$

$$S_e \approx 0.0118 \text{ m} = \textbf{11.8 mm}$$

Example 5.9

Solve Example 5.8 using Meyerhof's method.

Solution

From Eq. (5.66),

$$S_e = \frac{2q_{net}}{(N_{60})(F_d)}\left(\frac{B}{B + 0.3}\right)^2$$

$$F_d = 1 + 0.33(D_f/B) = 1 + 0.33(1/1.75) = 1.19$$

$$S_e = \frac{(2)(120)}{(10)(1.19)}\left(\frac{1.75}{1.75 + 0.3}\right)^2 = \textbf{14.7 mm}$$

■

5.14 *Settlement in Granular Soil Based on Pressuremeter Test (PMT)*

Briaud (2007) proposed a method based on Pressuremeter tests (Section 2.22) from which the load-settlement diagrams of foundations can be derived. The following is a step-by-step procedure for performing the analysis.

Step 1. Conduct Pressuremeter tests at varying depths at the desired location and obtain plots of p_p (pressure in the measuring cell for cavity expansion; see Figure 2.32) versus $\Delta R/R_o$ (R_o = initial radius of the PMT cavity, and ΔR = increase in the cavity radius), as shown in Figure 5.24a.

Step 2. Extend the straight line part of the PMT curve to zero pressure and shift the vertical axis, as shown in Figure 5.24a. Re-zero the $\Delta R/R_o$ axis.

Step 3. Draw a strain influence factor diagram for the desired foundation (Section 5.12). Using all Pressuremeter test curves within the depth of influence, develop a mean PMT curve. Referring to Figure 5.24b, this can be done as follows:

For each value of $\Delta R/R_o$, let the p_p values be $p_{p(1)}, p_{p(2)}, p_{p(3)}, \ldots$. The mean value of p_p can be obtained as

$$p_{p(m)} = \frac{A_1}{A}p_{p(1)} + \frac{A_2}{A}p_{p(2)} + \frac{A_3}{A}p_{p(3)} + \cdots \qquad (5.67)$$

where

A_1, A_2, A_3 = areas tributary to each test under the strain influence factor diagram

$$A = A_1 + A_2 + A_3 + \cdots \qquad (5.68)$$

Step 4. Based on the results of Step 3, develop a mean $p_{p(m)}$ versus $\Delta R/R_o$ plot (Figure 5.24c).

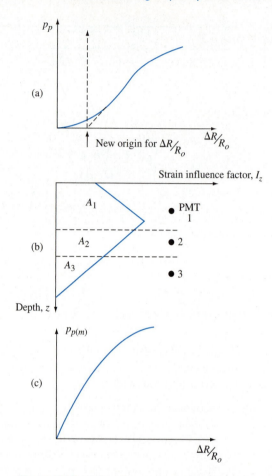

Figure 5.24 (a) Plot of p_p versus $\Delta R/R_o$; (b) averaging the pressuremeter curves within the foundation zone of influence; (c) plot of $p_{p(m)}$ versus $\Delta R/R_o$

Step 5. The mean PMT curve now can be used to develop the load-settlement plot for the foundation via the following equations.

$$\frac{S_e}{B} = 0.24\frac{\Delta R}{R_o} \tag{5.69}$$

and

$$q_o = f_{L/B}f_ef_\delta f_{\beta,d}\Gamma\, p_{p(m)} \tag{5.70}$$

where

S_e = elastic settlement of the foundation
B = width of foundation
L = length of foundation
q_o = net load per unit area on the foundation
Γ = gamma function linking q_o and $p_{p(m)}$

$$f_{L/B} = \text{shape factor} = 0.8 + 0.2\left(\frac{B}{L}\right) \tag{5.71}$$

$$f_e = \text{eccentricity factor} = 1 - 0.33\left(\frac{e}{B}\right)(\text{center}) \tag{5.72}$$

$$f_e = \text{eccentricity factor} = 1 - \left(\frac{e}{B}\right)^{0.5}(\text{edge}) \tag{5.73}$$

$$f_\delta = \text{load inclination factor} = 1 - \left[\frac{\delta(\text{deg})}{90}\right]^2(\text{center}) \tag{5.74}$$

$$f_\delta = \text{load inclination factor} = 1 - \left[\frac{\delta(\text{deg})}{360}\right]^{0.5}(\text{edge}) \tag{5.75}$$

$$f_{\beta,d} = \text{slope factor} = 0.8\left(1 + \frac{d}{B}\right)^{0.1}(\text{3H:1V slope}) \tag{5.76}$$

$$f_{\beta,d} = \text{slope factor} = 0.7\left(1 + \frac{d}{B}\right)^{0.15}(\text{2H:1V slope}) \tag{5.77}$$

δ = inclination of load with respect to the vertical
β = inclination of a slope with the horizontal if the foundation is located on top of a slope
d = distance of the edge of the foundation from the edge of the slope

The parameters δ, β, d, and e are defined in Figure 5.25. Figure 5.26 shows the design plot for Γ with S_e/B or $0.24\frac{\Delta R}{R_o}$.

Figure 5.25 Definition of parameters—β, L, d, δ, β, e, and β

Figure 5.26 Variation of Γ with $S_e/B = 0.24\ \Delta R/R_o$

Step 6. Based on the values of B/L, e/B, δ, and d/B, calculate the values of $f_{L/B}$, f_e, f_δ, and $f_{\beta,d}$ as needed. Let

$$f = (f_{L/B})(f_e)(f_\delta)(f_{\beta,d}) \tag{5.78}$$

Thus,

$$q_o = f\Gamma p_{p(m)} \tag{5.79}$$

Step 7. Now prepare a table, as shown in Table 5.13.
Step 8. Complete Table 5.13 as follows:
   a. Column 1—Assume several values of $\Delta R/R_o$.
   b. Column 2—For given values of $\Delta R/R_o$, obtain $p_{p(m)}$ from Figure 5.24c.
   c. Column 3—From Eq. (5.73), calculate the values of S_e/B from values of $\Delta R/R_o$ given in Column 1.
   d. Column 4—With known values of B, calculate the values of S_e.
   e. Column 5—From Figure 5.26, obtain the desired values of Γ.
   f. Column 6—Use Eq. (5.83) to obtain q_o.
   g. Now plot a graph of S_e (Column 4) versus q_o (Column 6) from which the magnitude of S_e for a given q_o can be determined.

Table 5.13 Calculations to Obtain the Load-Settlement Plot

$\Delta R/R_o$ (1)	$p_{p(m)}$ (2)	S_e/B (3)	S_e (4)	Γ (5)	q_o (6)

Example 5.10

A spread footing, shown in Figure 5.27a with a width of 4 m and a length of 20 m, serves as a bridge abutment foundation. The soil is medium dense sand. A 16,000 kN vertical load acts on the footing. The active pressure on the abutment wall develops a 1,600 kN horizontal load. The resultant reaction force due to the vertical and horizontal load is applied at an eccentricity of 0.13 m. PMT testing at the site produced a mean Pressuremeter curve characterizing the soil and is shown in Figure 5.27b. What is the settlement at the current loading?

Figure 5.27

Solution

Given: $B = 4$ m, $L = 20$ m, $d = 3$ m, and slope $= 3H{:}1V$.
So

$$f_{L/B} = 0.8 + 0.2\left(\frac{B}{L}\right) = 0.8 + 0.2\left(\frac{4}{20}\right) = 0.84$$

$$f_{e(\text{center})} = 1 - 0.33\left(\frac{e}{B}\right) = 1 - 0.33\left(\frac{0.13}{4}\right) = 0.99$$

$$f_{\delta(\text{center})} = 1 - \left(\frac{\delta}{90}\right)^2$$

$$\delta = \tan^{-1}\left(\frac{H}{V}\right) = \tan^{-1}\left(\frac{1600}{16,000}\right) = 5.71°$$

$$f_{\delta} = 1 - \left(\frac{5.71}{90}\right)^2 = 0.996$$

$$f_{\beta,d} = 0.8\left(1 + \frac{d}{B}\right)^{0.1} = 0.8\left(1 + \frac{3}{4}\right)^{0.1} = 0.846$$

$$f = f_{L/B}f_{e}f_{\delta}f_{\beta,d} = (0.84)(0.99)(0.996)(0.845) = 0.7$$

Now the following table can be prepared.

$\Delta R/R_o$ (1)	$p_{p(m)}$ (kN/m²) (2)	S_e/B (3)	S_e (mm) (4)	Γ (5)	q_o (kN/m²) (6)	Q_o (MN) (7)
0.002	50	0.0005	2.0	2.27	79.45	6.36
0.005	150	0.0012	4.8	2.17	227.85	18.23
0.01	250	0.0024	9.6	2.07	362.25	28.98
0.02	450	0.0048	19.2	1.83	576.45	46.12
0.04	800	0.0096	38.4	1.40	784.00	62.72
0.07	1200	0.0168	67.2	1.17	982.8	78.62
0.10	1400	0.024	96.0	1.07	1048.6	83.89
0.20	1700	0.048	192.0	0.90	1071.0	85.68

Note: Columns 1 and 2: From Figure 5.27b
Column 3: (Column 1)(0.24) = S_e/B
Column 4: (Column 3)(B = 4000 mm) = S_e
Column 5: From Figure 5.26
Column 6: $f\Gamma p_{p(m)} = (0.7)(\Gamma)p_{p(m)} = q_o$
Column 7: (Column 6)($B \times L$) = Q_o

Figure 5.28 shows the plot of Q_o versus S_e. From this plot it can be seen that, for a vertical loading of 16,000 kN (16 MN), the value of $S_e \approx$ **4.2 mm**.

Figure 5.28

Consolidation Settlement

5.15 *Primary Consolidation Settlement Relationships*

As mentioned before, consolidation settlement occurs over time in saturated clayey soils subjected to an increased load caused by construction of the foundation. (See Figure 5.29.) On the basis of the one-dimensional consolidation settlement equations given in Chapter 1, we write

$$S_{c(p)} = \int \varepsilon_z dz \tag{5.80}$$

where

ε_z = vertical strain

$\quad = \dfrac{\Delta e}{1 + e_o}$

Δe = change of void ratio

$\quad = f(\sigma'_o, \sigma'_c, \text{ and } \Delta\sigma')$

So,

$$S_{c(p)} = \frac{C_c H_c}{1 + e_o} \log \frac{\sigma'_o + \Delta\sigma'_{av}}{\sigma'_o} \qquad \text{(for normally consolidated clays)} \tag{5.81}$$

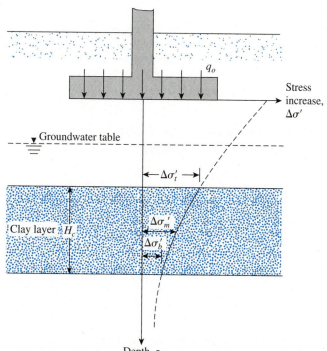

Figure 5.29 Consolidation settlement calculation

$$S_{c(p)} = \frac{C_s H_c}{1 + e_o} \log \frac{\sigma'_o + \Delta\sigma'_{av}}{\sigma'_o} \qquad \begin{array}{l} \text{(for overconsolidated clays} \\ \text{with } \sigma'_o + \Delta\sigma'_{av} < \sigma'_c) \end{array} \qquad (5.82)$$

$$S_{c(p)} = \frac{C_s H_c}{1 + e_o} \log \frac{\sigma'_c}{\sigma'_o} + \frac{C_c H_c}{1 + e_o} \log \frac{\sigma'_o + \Delta\sigma'_{av}}{\sigma'_c} \qquad \begin{array}{l} \text{(for overconsolidated clays} \\ \text{with } \sigma'_o < \sigma'_c < \sigma'_o + \Delta\sigma'_{av}) \end{array} \qquad (5.83)$$

where

σ'_o = average effective pressure on the clay layer before the construction of the foundation
$\Delta\sigma'_{av}$ = average increase in effective pressure on the clay layer caused by the
        construction of the foundation
σ'_c = preconsolidation pressure
e_o = initial void ratio of the clay layer
C_c = compression index
C_s = swelling index
H_c = thickness of the clay layer

The procedures for determining the compression and swelling indexes were discussed in Chapter 1.

Note that the increase in effective pressure, $\Delta\sigma'$, on the clay layer is not constant with depth: The magnitude of $\Delta\sigma'$ will decrease with the increase in depth measured from the bottom of the foundation. However, the average increase in pressure may be approximated by

$$\Delta\sigma'_{av} = \tfrac{1}{6}(\Delta\sigma'_t + 4\Delta\sigma'_m + \Delta\sigma'_b) \qquad (5.84)$$

where $\Delta\sigma'_t$, $\Delta\sigma'_m$, and $\Delta\sigma'_b$ are, respectively, the effective pressure increases at the *top, middle,* and *bottom* of the clay layer that are caused by the construction of the foundation.

The method of determining the pressure increase caused by various types of foundation load using Boussinesq's solution is discussed in Sections 5.2 through 5.6. $\Delta\sigma'_{av}$ can also be directly obtained from the method presented in Section 5.5.

5.16 *Three-Dimensional Effect on Primary Consolidation Settlement*

The consolidation settlement calculation presented in the preceding section is based on Eqs. (1.61), (1.63), and (1.65). These equations, as shown in Chapter 1, are in turn based on one-dimensional laboratory consolidation tests. The underlying assumption is that the increase in pore water pressure, Δu, immediately after application of the load equals the increase in stress, $\Delta\sigma$, at any depth. In this case,

$$S_{c(p)-\text{oed}} = \int \frac{\Delta e}{1 + e_o} dz = \int m_v \Delta\sigma'_{(1)} dz$$

where

$S_{c(p)-\text{oed}}$ = consolidation settlement calculated by using Eqs. (1.61), (1.63), and (1.65)
$\Delta\sigma'_{(1)}$ = effective vertical stress increase
m_v = volume coefficient of compressibility (see Chapter 1)

Flexible
circular load

Clay

$\Delta\sigma_{(1)}$

H_c

$\Delta\sigma_{(3)}$

$\Delta\sigma_{(3)}$

z

Figure 5.30 Circular foundation on a clay layer

In the field, however, when a load is applied over a limited area on the ground surface, such an assumption will not be correct. Consider the case of a circular foundation on a clay layer, as shown in Figure 5.30. The vertical and the horizontal stress increases at a point in the layer immediately below the center of the foundation are $\Delta\sigma_{(1)}$ and $\Delta\sigma_{(3)}$, respectively. For a saturated clay, the pore water pressure increase at that depth (see Chapter 1) is

$$\Delta u = \Delta\sigma_{(3)} + A[\Delta\sigma_{(1)} - \Delta\sigma_{(3)}] \tag{5.85}$$

where A = pore water pressure parameter. For this case,

$$S_{c(p)} = \int m_v \Delta u \, dz = \int (m_v)\{\Delta\sigma_{(3)} + A[\Delta\sigma_{(1)} - \Delta\sigma_{(3)}]\} \, dz \tag{5.86}$$

Thus, we can write

$$K_{cir} = \frac{S_{c(p)}}{S_{c(p)-oed}} = \frac{\displaystyle\int_0^{H_c} m_v \Delta u \, dz}{\displaystyle\int_0^{H_c} m_v \Delta\sigma'_{(1)} dz} = A + (1 - A)\left[\frac{\displaystyle\int_0^{H_c} \Delta\sigma'_{(3)} dz}{\displaystyle\int_0^{H_c} \Delta\sigma'_{(1)} dz}\right] \tag{5.87}$$

where K_{cir} = settlement ratio for circular foundations.

The settlement ratio for a continuous foundation, K_{str}, can be determined in a manner similar to that for a circular foundation. The variation of K_{cir} and K_{str} with A and H_c/B is given in Figure 5.31. (*Note: B* = diameter of a circular foundation, and *B* = width of a continuous foundation.)

The preceding technique is generally referred to as the *Skempton–Bjerrum modification* (1957) for a consolidation settlement calculation.

Leonards (1976) examined the correction factor K_{cr} for a three-dimensional consolidation effect in the field for a circular foundation located over *overconsolidated clay*. Referring to Figure 5.30, we have

$$S_{c(p)} = K_{cr(OC)} \, S_{c(p)-oed} \tag{5.88}$$

where

$$K_{cr(OC)} = f\left(OCR, \frac{B}{H_c}\right) \tag{5.89}$$

Figure 5.31 Settlement ratios for circular (K_{cir}) and continuous (K_{str}) foundations

in which

$$OCR = \text{overconsolidation ratio} = \frac{\sigma'_c}{\sigma'_o} \qquad (5.90)$$

where

σ'_c = preconsolidation pressure
σ'_o = present average effective pressure

The interpolated values of $K_{cr(OC)}$ from Leonard's 1976 work are given in Table 5.14.

Table 5.14 Variation of $K_{cr(OC)}$ with OCR and B/H_c

	$K_{cr(OC)}$		
OCR	$B/H_c = 4.0$	$B/H_c = 1.0$	$B/H_c = 0.2$
1	1	1	1
2	0.986	0.957	0.929
3	0.972	0.914	0.842
4	0.964	0.871	0.771
5	0.950	0.829	0.707
6	0.943	0.800	0.643
7	0.929	0.757	0.586
8	0.914	0.729	0.529
9	0.900	0.700	0.493
10	0.886	0.671	0.457
11	0.871	0.643	0.429
12	0.864	0.629	0.414
13	0.857	0.614	0.400
14	0.850	0.607	0.386
15	0.843	0.600	0.371
16	0.843	0.600	0.357

Example 5.11

A plan of a foundation 1 m × 2 m is shown in Figure 5.32. Estimate the consolidation settlement of the foundation, taking into account the three-dimensional effect. Given: $A = 0.6$.

Solution

The clay is normally consolidated. Thus,

$$S_{c(p)-\text{oed}} = \frac{C_c H_c}{1 + e_o} \log \frac{\sigma'_o + \Delta\sigma'_{av}}{\sigma'_o}$$

so

$$\sigma'_o = (2.5)(16.5) + (0.5)(17.5 - 9.81) + (1.25)(16 - 9.81)$$

$$= 41.25 + 3.85 + 7.74 = 52.84 \text{ kN/m}^2$$

From Eq. (5.84),

$$\Delta\sigma'_{av} = \tfrac{1}{6}(\Delta\sigma'_t + 4\Delta\sigma'_m + \Delta\sigma'_b)$$

Now the following table can be prepared (*Note:* $L = 2$ m; $B = 1$ m):

$m_1 = L/B$	z(m)	$z/(B/2) = n_1$	I_c[a]	$\Delta\sigma' = q_o I_c$[b]
2	2	4	0.190	$28.5 = \Delta\sigma'_t$
2	$2 + 2.5/2 = 3.25$	6.5	≈ 0.085	$12.75 = \Delta\sigma'_m$
2	$2 + 2.5 = 4.5$	9	0.045	$6.75 = \Delta\sigma'_b$

[a] Table 5.3
[b] Eq. (5.10)

$q_0 = 150$ kN/m² (net stress increase)

$B \times L = 1$ m × 2 m

Sand
$\gamma = 16.5$ kN/m³

Groundwater table

Sand
$\gamma_{\text{sat}} = 17.5$ kN/m³

Normally consolidated clay
$\gamma = 16$ kN/m³ $e_a = 0.8$
$E_s = 6{,}000$ kN/m² $C_e = 0.32$
$\mu_s = 0.5$ $C_s = 0.09$

1 m

1.5 m

0.5 m

2.5 m

Figure 5.32 Calculation of primary consolidation settlement for a foundation

Now,

$$\Delta\sigma'_{av} = \tfrac{1}{6}(28.5 + 4 \times 12.75 + 6.75) = 14.38 \text{ kN/m}^2$$

so

$$S_{c(p)-\text{oed}} = \frac{(0.32)(2.5)}{1 + 0.8}\log\left(\frac{52.84 + 14.38}{52.84}\right) = 0.0465 \text{ m}$$

$$= \mathbf{46.5 \text{ mm}}$$

Now assuming that the 2:1 method of stress increase (see Figure 5.5) holds good, the area of distribution of stress at the top of the clay layer will have dimensions

$$B' = \text{width} = B + z = 1 + (1.5 + 0.5) = 3 \text{ m}$$

and

$$L' = \text{width} = L + z = 2 + (1.5 + 0.5) = 4 \text{ m}$$

The diameter of an equivalent circular area, B_{eq}, can be given as

$$\frac{\pi}{4}B_{eq}^2 = B'L'$$

so that

$$B_{eq} = \sqrt{\frac{4B'L'}{\pi}} = \sqrt{\frac{(4)(3)(4)}{\pi}} = 3.91 \text{ m}$$

Also,

$$\frac{H_c}{B_{eq}} = \frac{2.5}{3.91} = 0.64$$

From Figure 5.31, for $A = 0.6$ and $H_c/B_{eq} = 0.64$, the magnitude of $K_{cr} \approx 0.78$. Hence,

$$S_{e(p)} = K_{cr}S_{e(p)-\text{oed}} = (0.78)(46.5) \approx \mathbf{36.3 \text{ mm}}$$ ■

5.17 *Settlement Due to Secondary Consolidation*

At the end of primary consolidation (i.e., after the complete dissipation of excess pore water pressure) some settlement is observed that is due to the plastic adjustment of soil fabrics. This stage of consolidation is called *secondary consolidation*. A plot of deformation against the logarithm of time during secondary consolidation is practically linear

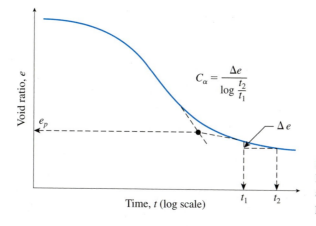

Figure 5.33 Variation of e with log t under a given load increment, and definition of secondary compression index

as shown in Figure 5.33. From the figure, the secondary compression index can be defined as

$$C_\alpha = \frac{\Delta e}{\log t_2 - \log t_1} = \frac{\Delta e}{\log (t_2/t_1)} \tag{5.91}$$

where

C_α = secondary compression index
Δe = change of void ratio
t_1, t_2 = time

The magnitude of the secondary consolidation can be calculated as

$$S_{c(s)} = C'_\alpha H_c \log(t_2/t_1) \tag{5.92}$$

where

$C'_\alpha = C_\alpha/(1 + e_p)$ $\tag{5.93}$
e_p = void ratio at the end of primary consolidation
H_c = thickness of clay layer

Mesri (1973) correlated C'_α with the natural moisture content (w) of several soils, from which it appears that

$$C'_\alpha \approx 0.0001w \tag{5.94}$$

where w = natural moisture content, in percent. For most overconsolidated soils, C'_a varies between 0.0005 to 0.001.

Mesri and Godlewski (1977) compiled the magnitude of C_α/C_c (C_c = compression index) for a number of soils. Based on their compilation, it can be summarized that

• For inorganic clays and silts:

$$C_\alpha/C_c \approx 0.04 \pm 0.01$$

- For organic clays and silts:

$$C_\alpha/C_c \approx 0.05 \pm 0.01$$

- For peats:

$$C_\alpha/C_c \approx 0.075 \pm 0.01$$

Secondary consolidation settlement is more important in the case of all organic and highly compressible inorganic soils. In overconsolidated inorganic clays, the secondary compression index is very small and of less practical significance.

There are several factors that might affect the magnitude of secondary consolidation, some of which are not yet very clearly understood (Mesri, 1973). The ratio of secondary to primary compression for a given thickness of soil layer is dependent on the ratio of the stress increment, $\Delta\sigma'$, to the initial effective overburden stress, σ'_o. For small $\Delta\sigma'/\sigma'_o$ ratios, the secondary-to-primary compression ratio is larger.

5.18   *Field Load Test*

The ultimate load-bearing capacity of a foundation, as well as the allowable bearing capacity based on tolerable settlement considerations, can be effectively determined from the field load test, generally referred to as the *plate load test*. The plates that are used for tests in the field are usually made of steel and are 25 mm thick and 150 mm to 762 mm in diameter. Occasionally, square plates that are 305 mm × 305 mm are also used.

To conduct a plate load test, a hole is excavated with a minimum diameter of $4B$ (B is the diameter of the test plate) to a depth of D_f, the depth of the proposed foundation. The plate is placed at the center of the hole, and a load that is about one-fourth to one-fifth of the estimated ultimate load is applied to the plate in steps by means of a jack. A schematic diagram of the test arrangement is shown in Figure 5.34a. During each step of the application of the load, the settlement of the plate is observed on dial gauges. At least one hour is allowed to elapse between each application. The test should be conducted until failure, or at least until the plate has gone through 25 mm (1 in.) of settlement. Figure 5.34b shows the nature of the load–settlement curve obtained from such tests, from which the ultimate load per unit area can be determined. Figure 5.35 shows a plate load test conducted in the field.

For tests in clay,

$$q_{u(F)} = q_{u(P)} \tag{5.95}$$

where

$q_{u(F)}$ = ultimate bearing capacity of the proposed foundation
$q_{u(P)}$ = ultimate bearing capacity of the test plate

Equation (5.95) implies that the ultimate bearing capacity in clay is virtually independent of the size of the plate.

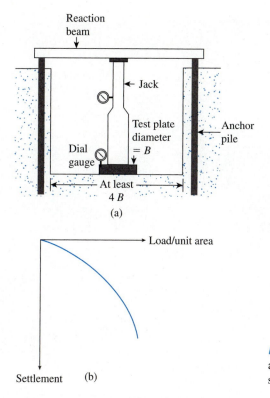

Figure 5.34 Plate load test: (a) test arrangement; (b) nature of load–settlement curve

For tests in sandy soils,

$$q_{u(F)} = q_{u(P)}\frac{B_F}{B_P}$$ (5.96)

where

B_F = width of the foundation
B_P = width of the test plate

The allowable bearing capacity of a foundation, based on settlement considerations and for a given intensity of load, q_o, is

$$S_F = S_P\frac{B_F}{B_P} \quad \text{(for clayey soil)}$$ (5.97)

and

$$S_F = S_P\left(\frac{2B_F}{B_F + B_P}\right)^2 \quad \text{(for sandy soil)}$$ (5.98)

Figure 5.35 Plate load test in the field (*Courtesy of Braja M. Das, Henderson, NV*)

5.19   *Presumptive Bearing Capacity*

Several building codes (e.g., the Uniform Building Code, Chicago Building Code, and New York City Building Code) specify the allowable bearing capacity of foundations on various types of soil. For minor construction, they often provide fairly acceptable guidelines. However, these bearing capacity values are based primarily on the *visual* classification of near-surface soils and generally do not take into consideration factors such as the stress history of the soil, the location of the water table, the depth of the foundation, and the tolerable settlement. So, for large construction projects, the codes' presumptive values should be used only as guides.

5.20   *Tolerable Settlement of Buildings*

In most instances of construction, the subsoil is not homogeneous and the load carried by various shallow foundations of a given structure can vary widely. As a result, it is reasonable to expect varying degrees of settlement in different parts of a given building.

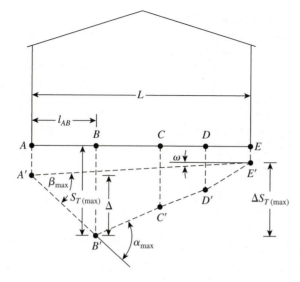

Figure 5.36 Definition of parameters for differential settlement

The *differential settlement* of the parts of a building can lead to damage of the super-structure. Hence, it is important to define certain parameters that quantify differential settlement and to develop limiting values for those parameters in order that the resulting structures be safe. Burland and Worth (1970) summarized the important parameters relating to differential settlement.

Figure 5.36 shows a structure in which various foundations, at A, B, C, D, and E, have gone through some settlement. The settlement at A is AA', at B is BB', etc. Based on this figure, the definitions of the various parameters are as follows:

S_T = total settlement of a given point

ΔS_T = difference in total settlement between any two points

α = gradient between two successive points

β = angular distortion = $\dfrac{\Delta S_{T(ij)}}{l_{ij}}$

(*Note*: l_{ij} = distance between points i and j)

ω = tilt

Δ = relative deflection (i.e. movement from a straight line joining two reference points)

$\dfrac{\Delta}{L}$ = deflection ratio

Since the 1950s, various researchers and building codes have recommended allowable values for the preceding parameters. A summary of several of these recommendations is presented next.

In 1956, Skempton and McDonald proposed the following limiting values for maximum settlement and maximum angular distortion, to be used for building purposes:

Maximum settlement, $S_{T(max)}$

In sand   32 mm

In clay   45 mm

Maximum differential settlement, $\Delta S_{T(\max)}$

Isolated foundations in sand	51 mm
Isolated foundations in clay	76 mm
Raft in sand	51–76 mm
Raft in clay	76–127 mm

Maximum angular distortion, $\beta_{\max}$ 1/300

On the basis of experience, Polshin and Tokar (1957) suggested the following allowable deflection ratios for buildings as a function of L/H, the ratio of the length to the height of a building:

$$\Delta/L = 0.0003 \text{ for } L/H \leq 2$$

$$\Delta/L = 0.001 \text{ for } L/H = 8$$

The 1955 Soviet Code of Practice gives the following allowable values:

Type of building	L/H	Δ/L
Multistory buildings and civil dwellings	≤3	0.0003 (for sand) 0.0004 (for clay)
	≥5	0.0005 (for sand) 0.0007 (for clay)
One-story mills		0.001 (for sand and clay)

Bjerrum (1963) recommended the following limiting angular distortion, $\beta_{\max}$ for various structures:

Category of potential damage	$\beta_{\max}$
Safe limit for flexible brick wall ($L/H > 4$)	1/150
Danger of structural damage to most buildings	1/150
Cracking of panel and brick walls	1/150
Visible tilting of high rigid buildings	1/250
First cracking of panel walls	1/300
Safe limit for no cracking of building	1/500
Danger to frames with diagonals	1/600

If the maximum allowable values of $\beta_{\max}$ are known, the magnitude of the allowable $S_{T(\max)}$ can be calculated with the use of the foregoing correlations.

The European Committee for Standardization has also provided limiting values for serviceability and the maximum accepted foundation movements. (See Table 5.15.)

Table 5.15 Recommendations of European Committee for Standardization on Differential Settlement Parameters

Item	Parameter	Magnitude	Comments
Limiting values for	S_T	25 mm	Isolated shallow foundation
serviceability		50 mm	Raft foundation
(European Committee	ΔS_T	5 mm	Frames with rigid cladding
for Standardization,		10 mm	Frames with flexible cladding
1994a)		20 mm	Open frames
	β	1/500	—
Maximum acceptable	S_T	50	Isolated shallow foundation
foundation movement	ΔS_T	20	Isolated shallow foundation
(European Committee	β	$\approx 1/500$	—
for Standardization, 1994b)			

Problems

5.1 A flexible circular area is subjected to a uniformly distributed load of 150 kN/m². The diameter of the loaded area is 2 m. Determine the stress increase in a soil mass at a point located 3 m below the center of the loaded area.
   a. by using Eq. (5.3)
   b. by using Eq. (5.28). Use $\mu_s = 0$

5.2 Refer to Figure 5.4, which shows a flexible rectangular area. Given: $B_1 = 1.22$ m, $B_2 = 1.83$ m, $L_1 = 2.44$ m, and $L_2 = 3.05$ m. If the area is subjected to a uniform load of 3000 lb/ft², determine the stress increase at a depth of 3.05 m located immediately below point O.

5.3 Repeat Problem 5.2 with the following: $B_1 = 1.22$ m, $B_2 = 3.05$ m, $L_1 = 2.44$ m, $L_2 = 3.66$ m, and the uniform load on the flexible area $= 120$ kN/m². Determine the stress increase below point O at a depth of 6.1 m. Use Eq. (5.29) and $\mu_s = 0$.

5.4 Using Eq. (5.10), determine the stress increase ($\Delta\sigma$) at $z = 3.05$ m below the center of the area described in Problem 5.2.

5.5 Refer to Figure P5.5. Using the procedure outlined in Section 5.5, determine the average stress increase in the clay layer below the center of the foundation due to the net foundation load of 900 kN.

5.6 Solve Problem 5.5 using the 2:1 method [Eqs. (5.14) and (5.88)].

5.7 Figure P5.7 shows an embankment load on a silty clay soil layer. Determine the stress increase at points A, B, and C, which are located at a depth of 5 m below the ground surface.

5.8 A planned flexible load area (see Figure P5.8) is to be 2 m × 3.2 m and carries a uniformly distributed load of 210 kN/m². Estimate the elastic settlement below the center of the loaded area. Assume that $D_f = 1.6$ m and $H = \infty$. Use Eq. (5.33).

5.9 Redo Problem 5.8 assuming that $D_f = 1.2$ m and $H = 4$ m.

5.10 Consider a flexible foundation measuring 1.52 m × 3.05 m in a plan on a soft saturated clay ($\mu_s = 0.5$). The depth of the foundation is 1.22 m below the ground

900 kN (net load)

Sand
$\gamma = 15.7$ kN/m³

1.52 m

1.83 m × 1.83 m

Groundwater table

Sand
1.22 m $\gamma_{sat} = 19.24$ kN/m³

$\gamma_{sat} = 19.24$ kN/m³
$e_o = 0.8$
$C_c = 0.25$
$C_s = 0.06$

3.05 m

Preconsolidation pressure = 100 kN/m²

Figure P5.5

6 m

Center line

1V:2H

1V:2H

10 m

$\gamma = 17$ kN/m³

5 m

C

B A

Figure P5.7

D_f

210 kN/m²

2 m × 3.2 m

Silty sand
$E_s = 8500$ kN/m²
$\mu_s = 0.3$

H

Rock

Figure P5.8

surface. A rigid rock layer is located at 12.2 m below the bottom of the foundation. Given: $q_o = 144$ kN/m² and, for the clay, $E_s = 12{,}938$ kN/m². Determine the average elastic settlement of the foundation. Use Eq. (5.30).

5.11 Figure 5.16 shows a foundation of 3.05 m × 1.91 m resting on a sand deposit. The net load per unit area at the level of the foundation, q_o, is 144 kN/m². For the sand, $\mu_s = 0.3$, $E_s = 22{,}080$ kN/m², $D_f = 0.76$ m and $H = 9.76$ m. Assume that the foundation is rigid, and determine the elastic settlement the foundation would undergo. Use Eqs. (5.33) and (5.41).

5.12 Repeat Problems 5.11 for a foundation of size $= 1.8$ m $\times$ 1.8 m and with $q_o = 190$ kN/m^2, $D_f = 1.0$ m, $H = 15$ m; and soil conditions of $\mu_s = 0.4$, $E_s = 15{,}400$ kN/m^2, and $\gamma = 17$ kN/m^3.

5.13 For a shallow foundation supported by a silty clay, as shown in Figure 5.17, the following are given:

Length, $L = 2$ m
Width, $B = 1$ m
Depth of foundation, $D_f = 1$ m
Thickness of foundation, $t = 0.23$ m
Load per unit area, $q_o = 190$ kN/m^2
$E_f = 15 \times 10^6$ kN/m^2

The silty clay soil has the following properties:

$H = 2$ m
$\mu_s = 0.4$
$E_o = 9000$ kN/m^2
$k = 500$ kN/m^2/m

Using Eq. (5.46), estimate the elastic settlement of the foundation.

5.14 A plan calls for a square foundation measuring 3 m $\times$ 3 m, supported by a layer of sand. (See Figure 5.17.) Led $D_f = 1.5$ m, $t = 0.25$ m, $E_o = 16{,}000$ kN/m^2, $k = 400$ kN/m^2/m, $\mu_s = 0.3$, $H = 20$ m, $E_f = 15 \times 10^6$ kN/m^2, and $q_o = 150$ kN/m^2. Calculate the elastic settlement. Use Eq. (5.46).

5.15 Solve Problem 5.11 with Eq. (5.49). For the correction factor C_2, use a time of 5 years for creep and, for the unit weight of soil, use $\gamma = 18.08$ kN/m^3.

5.16 A continuous foundation on a deposit of sand layer is shown in Figure P5.16 along with the variation of the modulus of elasticity of the soil (E_s). Assuming $\gamma = 18$ kN/m^3 and C_2 for 10 years, calculate the elastic settlement of the foundation using the strain influence factor method.

Figure P5.16

5.17 Following are the results of standard penetration tests in a granular soil deposit.

Depth (m)	Standard penetration number, N_{60}
1.5	11
3.0	10
4.5	12
6.0	9
7.5	14

What will be the net allowable bearing capacity of a foundation planned to be 1.83 m × 1.83 m? Let $D_f = 6.91$ m and the allowable settlement = 25 mm, and use the relationships presented in Eq. (5.60).

5.18 A shallow foundation measuring 1.2 m × 1.2 m in plan is to be constructed over a normally consolidated sand layer. Given: $D_f = 1$ m, N_{60} increases with depth, $\overline{N}_{60}$ (in the depth of stress influence) = 8, and $q_{net} = 210$ kN/m². Estimate the elastic settlement using Burland and Burbidge's method.

5.19 Refer to Figure 5.25. For a foundation on a layer of sand, given: $B = 1.52$ m, $L = 3.05$ m, $d = 1.52$ m, $\beta = 26.6°$, $e = 0.152$ m, and $\delta = 10°$. The Pressuremeter testing at the site produced a mean Pressuremeter curve for which the $p_{p(m)}$ versus $\Delta R/R_o$ points are as follow:

$\Delta R/R_o$ (1)	$p_{p(m)}$ (lb/in²) (2)
0.002	49.68
0.004	166.98
0.008	224.94
0.012	292.56
0.024	475.41
0.05	870.09
0.08	1225.97
0.1	1452.45
0.2	2550.24

What should be the magnitude of Q_o for a settlement (center) of 25 mm?

5.20 Estimate the consolidation settlement of the clay layer shown in Figure P5.5 using the results of Problem 5.5.

5.21 Estimate the consolidation settlement of the clay layer shown in Figure P5.5 using the results of Problem 5.6.

References

AHLVIN, R. G., and ULERY, H. H. (1962). *Tabulated Values of Determining the Composite Pattern of Stresses, Strains, and Deflections beneath a Uniform Load on a Homogeneous Half Space.* Highway Research Board Bulletin 342, pp. 1–13.

BJERRUM, L. (1963). "Allowable Settlement of Structures," *Proceedings, European Conference on Soil Mechanics and Foundation Engineering,* Wiesbaden, Germany, Vol. III, pp. 135–137.

BOUSSINESQ, J. (1883). *Application des Potentials á L'Étude de L'Équilibre et du Mouvement des Solides Élastiques,* Gauthier-Villars, Paris.

BOWLES, J. E. (1987). "Elastic Foundation Settlement on Sand Deposits," *Journal of Geotechnical Engineering,* ASCE, Vol. 113, No. 8, pp. 846–860.

BOWLES, J. E. (1977). *Foundation Analysis and Design,* 2d ed., McGraw-Hill, New York.

BRIAUD, J. L. (2007). "Spread Footings in Sand: Load Settlement Curve Approach," *Journal of Geotechnical and Geoenvironmental Engineering*, American Society of Civil Engineers, Vol. 133, No. 8, pp. 905–920.

BURLAND, J. B., and BURBIDGE, M. C. (1985). "Settlement of Foundations on Sand and Gravel," *Proceedings, Institute of Civil Engineers,* Part I, Vol. 7, pp. 1325–1381.

CHRISTIAN, J. T., and CARRIER, W. D. (1978). "Janbu, Bjerrum, and Kjaernsli's Chart Reinterpreted," *Canadian Geotechnical Journal,* Vol. 15, pp. 124–128.

DAS, B. (2008). *Advanced Soil Mechanics,* 3d ed., Taylor and Francis, London.

DUNCAN, J. M., and BUCHIGNANI, A. N. (1976). *An Engineering Manual for Settlement Studies,* Department of Civil Engineering, University of California, Berkeley.

EUROPEAN COMMITTEE FOR STANDARDIZATION (1994a). *Basis of Design and Actions on Structures,* Eurocode 1, Brussels, Belgium.

EUROPEAN COMMITTEE FOR STANDARDIZATION (1994b). *Geotechnical Design, General Rules—Part 1,* Eurocode 7, Brussels, Belgium.

FOX, E. N. (1948). "The Mean Elastic Settlement of a Uniformly Loaded Area at a Depth below the Ground Surface," *Proceedings, 2nd International Conference on Soil Mechanics and Foundation Engineering,* Rotterdam, Vol. 1, pp. 129–132.

GRIFFITHS, D. V. (1984). "A Chart for Estimating the Average Vertical Stress Increase in an Elastic Foundation below a Uniformly Loaded Rectangular Area," *Canadian Geotechnical Journal,* Vol. 21, No. 4, pp. 710–713.

JANBU, N., BJERRUM, L., and KJAERNSLI, B. (1956). "Veiledning vedlosning av fundamentering—soppgaver," *Publication No. 18,* Norwegian Geotechnical Institute, pp. 30–32.

LEONARDS, G. A. (1976). *Estimating Consolidation Settlement of Shallow Foundations on Overconsolidated Clay,* Special Report No. 163, Transportation Research Board, Washington, D.C., pp. 13–16.

MAYNE, P. W., and POULOS, H. G. (1999). "Approximate Displacement Influence Factors for Elastic Shallow Foundations," *Journal of Geotechnical and Geoenvironmental Engineering,* ASCE, Vol. 125, No. 6, pp. 453–460.

MESRI, G. (1973). "Coefficient of Secondary Compression," *Journal of the Soil Mechanics and Foundations Division,* American Society of Civil Engineers, Vol. 99, No. SM1, pp. 122–137.

MESRI, G., and GODLEWSKI, P. M. (1977). "Time and Stress—Compressibility Interrelationship," *Journal of Geotechnical Engineering Division,* American Society of Civil Engineers, Vol. 103, No. GT5, pp. 417–430.

MEYERHOF, G. G. (1956). "Penetration Tests and Bearing Capacity of Cohesionless Soils," *Journal of the Soil Mechanics and Foundations Division,* American Society of Civil Engineers, Vol. 82, No. SM1, pp. 1–19.

MEYERHOF, G. G. (1965). "Shallow Foundations," *Journal of the Soil Mechanics and Foundations Division,* American Society of Civil Engineers, Vol. 91, No. SM2, pp. 21–31.

NEWMARK, N. M. (1935). *Simplified Computation of Vertical Pressure in Elastic Foundation,* Circular 24, University of Illinois Engineering Experiment Station, Urbana, IL.

OSTERBERG, J. O. (1957). "Influence Values for Vertical Stresses in Semi-Infinite Mass Due to Embankment Loading," *Proceedings, Fourth International Conference on Soil Mechanics and Foundation Engineering,* London, Vol. 1, pp. 393–396.

POLSHIN, D. E., and TOKAR, R. A. (1957). "Maximum Allowable Nonuniform Settlement of Structures," *Proceedings, Fourth International Conference on Soil Mechanics and Foundation Engineering,* London, Vol. 1, pp. 402–405.

SALGADO, R. (2008). *The Engineering of Foundations*, McGraw-Hill, New York.

SCHMERTMANN, J. H., HARTMAN, J. P., and BROWN, P. R. (1978). "Improved Strain Influence Factor Diagrams," *Journal of the Geotechnical Engineering Division,* American Society of Civil Engineers, Vol. 104, No. GT8, pp. 1131–1135.

SKEMPTON, A. W., and BJERRUM, L. (1957). " A Contribution to Settlement Analysis of Foundations in Clay," *Geotechnique,* London, Vol 7, p. 178.

SKEMPTON, A. W., and McDONALD, D. M. (1956). "The Allowable Settlement of Buildings," *Proceedings of Institute of Civil Engineers,* Vol. 5, Part III, p. 727.

STEINBRENNER, W. (1934). "Tafeln zur Setzungsberechnung," *Die Strasse,* Vol. 1, pp. 121–124.

TERZAGHI, K., PECK, R. B., and MESRI, G. (1996). *Soil Mechanics in Engineering Practice, 3rd Edition*, Wiley, New York.

WESTERGAARD, H. M. (1938). "A Problem of Elasticity Suggested by a Problem in Soil Mechanics: Soft Material Reinforced by Numerous Strong Horizontal Sheets," *Contributions to the Mechanics of Solids, Dedicated to Stephen Timoshenko*, pp. 268–277, MacMillan, New York.

6 Mat Foundations

6.1 ## Introduction

Under normal conditions, square and rectangular footings such as those described in Chapters 3 and 4 are economical for supporting columns and walls. However, under certain circumstances, it may be desirable to construct a footing that supports a line of two or more columns. These footings are referred to as *combined footings*. When more than one line of columns is supported by a concrete slab, it is called a *mat foundation*. Combined footings can be classified generally under the following categories:

a. Rectangular combined footing
b. Trapezoidal combined footing
c. Strap footing

Mat foundations are generally used with soil that has a low bearing capacity. A brief overview of the principles of combined footings is given in Section 6.2, followed by a more detailed discussion on mat foundations.

6.2 Combined Footings

Rectangular Combined Footing

In several instances, the load to be carried by a column and the soil bearing capacity are such that the standard spread footing design will require extension of the column foundation beyond the property line. In such a case, two or more columns can be supported on a single rectangular foundation, as shown in Figure 6.1. If the net allowable soil pressure is known, the size of the foundation $(B \times L)$ can be determined in the following manner:

a. Determine the area of the foundation

$$A = \frac{Q_1 + Q_2}{q_{\text{net(all)}}} \tag{6.1}$$

where

Q_1, Q_2 = column loads
$q_{\text{net(all)}}$ = net allowable soil bearing capacity

Figure 6.1 Rectangular combined footing

b. Determine the location of the resultant of the column loads. From Figure 6.1,

$$X = \frac{Q_2 L_3}{Q_1 + Q_2} \qquad (6.2)$$

c. For a uniform distribution of soil pressure under the foundation, the resultant of the column loads should pass through the centroid of the foundation. Thus,

$$L = 2(L_2 + X) \qquad (6.3)$$

where L = length of the foundation.

d. Once the length L is determined, the value of L_1 can be obtained as follows:

$$L_1 = L - L_2 - L_3 \qquad (6.4)$$

Note that the magnitude of L_2 will be known and depends on the location of the property line.

e. The width of the foundation is then

$$B = \frac{A}{L} \qquad (6.5)$$

Trapezoidal Combined Footing

Trapezoidal combined footing (see Figure 6.2) is sometimes used as an isolated spread foundation of columns carrying large loads where space is tight. The size of the foundation that will uniformly distribute pressure on the soil can be obtained in the following manner:

a. If the net allowable soil pressure is known, determine the area of the foundation:

$$A = \frac{Q_1 + Q_2}{q_{\text{net(all)}}}$$

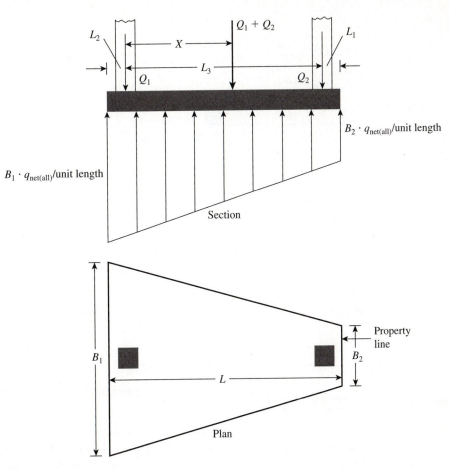

Figure 6.2 Trapezoidal combined footing

From Figure 6.2,

$$A = \frac{B_1 + B_2}{2} L \qquad (6.6)$$

b. Determine the location of the resultant for the column loads:

$$X = \frac{Q_2 L_3}{Q_1 + Q_2}$$

c. From the property of a trapezoid,

$$X + L_2 = \left(\frac{B_1 + 2B_2}{B_1 + B_2} \right) \frac{L}{3} \qquad (6.7)$$

With known values of A, L, X, and L_2, solve Eqs. (6.6) and (6.7) to obtain B_1 and B_2. Note that, for a trapezoid,

$$\frac{L}{3} < X + L_2 < \frac{L}{2}$$

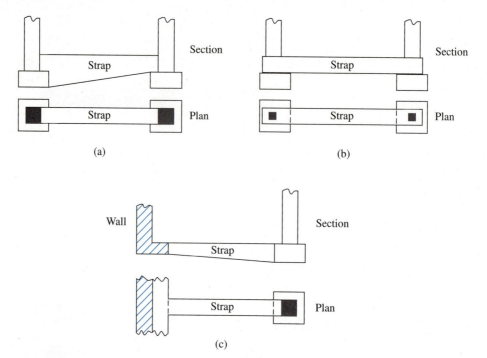

Figure 6.3 Cantilever footing—use of strap beam

Cantilever Footing

Cantilever footing construction uses a *strap beam* to connect an eccentrically loaded column foundation to the foundation of an interior column. (See Figure 6.3). Cantilever footings may be used in place of trapezoidal or rectangular combined footings when the allowable soil bearing capacity is high and the distances between the columns are large.

6.3   Common Types of Mat Foundations

The mat foundation, which is sometimes referred to as a *raft foundation,* is a combined footing that may cover the entire area under a structure supporting several columns and walls. Mat foundations are sometimes preferred for soils that have low load-bearing capacities, but that will have to support high column or wall loads. Under some conditions, spread footings would have to cover more than half the building area, and mat foundations might be more economical. Several types of mat foundations are used currently. Some of the common ones are shown schematically in Figure 6.4 and include the following:

1. Flat plate (Figure 6.4a). The mat is of uniform thickness.
2. Flat plate thickened under columns (Figure 6.4b).
3. Beams and slab (Figure 6.4c). The beams run both ways, and the columns are located at the intersection of the beams.
4. Flat plates with pedestals (Figure 6.4d).
5. Slab with basement walls as a part of the mat (Figure 6.4e). The walls act as stiffeners for the mat.

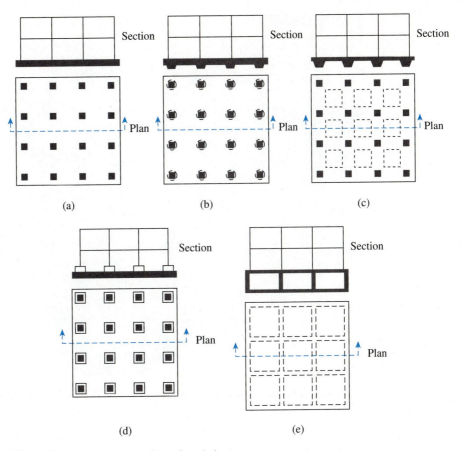

(a)　　　　　　　　(b)　　　　　　　　(c)

(d)　　　　　　　　(e)

Figure 6.4 Common types of mat foundations

Mats may be supported by piles, which help reduce the settlement of a structure built over highly compressible soil. Where the water table is high, mats are often placed over piles to control buoyancy. Figure 6.5 shows the difference between the depth D_f and the width B of isolated foundations and mat foundations. Figure 6.6 shows a flat-plate mat foundation under construction.

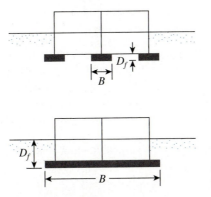

Figure 6.5 Comparison of isolated foundation and mat foundation (B = width, D_f = depth)

Figure 6.6 A flat plate mat foundation under construction (*Courtesy of Dharma Shakya, Geotechnical Solutions, Inc., Irvine, California*)

6.4 Bearing Capacity of Mat Foundations

The *gross ultimate bearing capacity* of a mat foundation can be determined by the same equation used for shallow foundations (see Section 3.6), or

$$q_u = c'N_cF_{cs}F_{cd}F_{ci} + qN_qF_{qs}F_{qd}F_{qi} + \tfrac{1}{2}\gamma B N_\gamma F_{\gamma s}F_{\gamma d}F_{\gamma i} \qquad \text{[Eq. (3.19)]}$$

(Chapter 3 gives the proper values of the bearing capacity factors, as well as the shape depth, and load inclination factors.) The term B in Eq. (3.19) is the smallest dimension of the mat. The *net ultimate capacity* of a mat foundation is

$$q_{\text{net}(u)} = q_u - q \qquad \text{[Eq. (3.14)]}$$

A suitable factor of safety should be used to calculate the net *allowable* bearing capacity. For mats on clay, the factor of safety should not be less than 3 under dead load or maximum live load. However, under the most extreme conditions, the factor of safety should be at least 1.75 to 2. For mats constructed over sand, a factor of safety of 3 should normally be used. Under most working conditions, the factor of safety against bearing capacity failure of mats on sand is very large.

For saturated clays with $\phi = 0$ and a vertical loading condition, Eq. (3.19) gives

$$q_u = c_u N_c F_{cs} F_{cd} + q \qquad (6.8)$$

where c_u = undrained cohesion. (*Note:* N_c = 5.14, N_q = 1, and N_γ = 0.)

From Table 3.4, for ϕ = 0,

$$F_{cs} = 1 + \frac{B}{L}\left(\frac{N_q}{N_c}\right) = 1 + \left(\frac{B}{L}\right)\left(\frac{1}{5.14}\right) = 1 + \frac{0.195B}{L}$$

and

$$F_{cd} = 1 + 0.4\left(\frac{D_f}{B}\right)$$

Substitution of the preceding shape and depth factors into Eq. (6.8) yields

$$q_u = 5.14c_u\left(1 + \frac{0.195B}{L}\right)\left(1 + 0.4\frac{D_f}{B}\right) + q \tag{6.9}$$

Hence, the net ultimate bearing capacity is

$$q_{net(u)} = q_u - q = 5.14c_u\left(1 + \frac{0.195B}{L}\right)\left(1 + 0.4\frac{D_f}{B}\right) \tag{6.10}$$

For FS = 3, the net allowable soil bearing capacity becomes

$$q_{net(all)} = \frac{q_{u(net)}}{FS} = 1.713c_u\left(1 + \frac{0.195B}{L}\right)\left(1 + 0.4\frac{D_f}{B}\right) \tag{6.11}$$

The net allowable bearing capacity for mats constructed over granular soil deposits can be adequately determined from the standard penetration resistance numbers. From Eq. (5.64), for shallow foundations,

$$q_{net}(\text{kN/m}^2) = \frac{N_{60}}{0.08}\left(\frac{B+0.3}{B}\right)^2 F_d\left(\frac{S_e}{25}\right) \tag{Eq. (5.64)}$$

where

N_{60} = standard penetration resistance
B = width (m)
F_d = $1 + 0.33(D_f/B) \leq 1.33$
S_e = settlement, (mm)

When the width B is large, the preceding equation can be approximated as

$$q_{net}(\text{kN/m}^2) = \frac{N_{60}}{0.08}F_d\left(\frac{S_e}{25}\right)$$

$$= \frac{N_{60}}{0.08}\left[1 + 0.33\left(\frac{D_f}{B}\right)\right]\left[\frac{S_e(\text{mm})}{25}\right] \tag{6.12}$$

$$\leq 16.63N_{60}\left[\frac{S_e(\text{mm})}{25}\right]$$

In English units, Eq. (6.12) may be expressed as

$$q_{net(all)}(\text{kip/ft}^2) = 0.25 N_{60}\left[1 + 0.33\left(\frac{D_f}{B}\right)\right][S_e(\text{in.})]$$

$$\leq 0.33 N_{60}[S_e(\text{in.})]$$ (6.13)

Generally, shallow foundations are designed for a maximum settlement of 25 mm and a differential settlement of about 19 mm.

However, the width of the raft foundations are larger than those of the isolated spread footings. As shown in Table 5.3, the depth of significant stress increase in the soil below a foundation depends on the width of the foundation. Hence, for a raft foundation, the depth of the zone of influence is likely to be much larger than that of a spread footing. Thus, the loose soil pockets under a raft may be more evenly distributed, resulting in a smaller differential settlement. Accordingly, the customary assumption is that, for a maximum raft settlement of 50 mm, the differential settlement would be 19 mm. Using this logic and conservatively assuming that $F_d = 1$, we can respectively approximate Eqs. (6.12) and (6.13) as

$$q_{net(all)} = q_{net}(\text{kN/m}^2) \approx 25 N_{60}$$ (6.14a)

The net allowable pressure applied on a foundation (see Figure 6.7) may be expressed as

$$q = \frac{Q}{A} - \gamma D_f$$ (6.15)

where

Q = dead weight of the structure and the live load
A = area of the raft

In all cases, q should be less than or equal to allowable q_{net}.

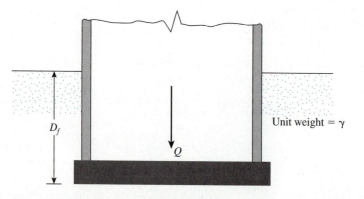

Figure 6.7 Definition of net pressure on soil caused by a mat foundation

Example 6.1

Determine the net ultimate bearing capacity of a mat foundation measuring 15 m $\times$ 10 m on a saturated clay with $c_u = 95$ kN/m^2, $\phi = 0$, and $D_f = 2$ m.

Solution

From Eq. (6.10),

$$q_{\text{net}(u)} = 5.14 c_u \left[1 + \left(\frac{0.195 B}{L} \right) \right] \left[1 + 0.4 \frac{D_f}{B} \right]$$

$$= (5.14)(95) \left[1 + \left(\frac{0.195 \times 10}{15} \right) \right] \left[1 + \left(\frac{0.4 \times 2}{10} \right) \right]$$

$$= \textbf{595.9 kN/m}^2 \qquad \blacksquare$$

Example 6.2

What will be the net allowable bearing capacity of a mat foundation with dimensions of 15 m $\times$ 10 m constructed over a sand deposit? Here, $D_f = 2$ m, the allowable settlement is 25 mm, and the average penetration number $N_{60} = 10$.

Solution

From Eq. (6.12),

$$q_{\text{net}(\text{all})} = \frac{N_{60}}{0.08} \left[1 + 0.33 \left(\frac{D_f}{B} \right) \right] \left(\frac{S_e}{25} \right) \leq 16.63 N_{60} \left(\frac{S_e}{25} \right)$$

or

$$q_{\text{net}(\text{all})} = \frac{10}{0.08} \left[1 + \frac{0.33 \times 2}{10} \right] \left(\frac{25}{25} \right) = \textbf{133.25 kN/m}^2 \qquad \blacksquare$$

6.5 Differential Settlement of Mats

In 1988, the American Concrete Institute Committee 336 suggested a method for calculating the differential settlement of mat foundations. According to this method, the rigidity factor K_r is calculated as

$$K_r = \frac{E' I_b}{E_s B^3} \qquad (6.16)$$

where

$E' =$ modulus of elasticity of the material used in the structure
$E_s =$ modulus of elasticity of the soil
$B =$ width of foundation
$I_b =$ moment of inertia of the structure per unit length at right angles to B

The term $E' I_b$ can be expressed as

$$E' I_b = E' \left(I_F + \sum I_b' + \sum \frac{a h^3}{12} \right) \qquad (6.17)$$

where

$E'I_b$ = flexural rigidity of the superstructure and foundation per unit length at right angles to B

$\Sigma E'I_b'$ = flexural rigidity of the framed members at right angles to B

$\Sigma(E'ah^3/12)$ = flexural rigidity of the shear walls

a = shear wall thickness

h = shear wall height

$E'I_F$ = flexibility of the foundation

Based on the value of K_r, the ratio (δ) of the differential settlement to the total settlement can be estimated in the following manner:

1. If $K_r > 0.5$, it can be treated as a rigid mat, and $\delta = 0$.
2. If $K_r = 0.5$, then $\delta \approx 0.1$.
3. If $K_r = 0$, then $\delta = 0.35$ for square mats $(B/L = 1)$ and $\delta = 0.5$ for long foundations $(B/L = 0)$.

6.6   Field Settlement Observations for Mat Foundations

Several field settlement observations for mat foundations are currently available in the literature. In this section, we compare the observed settlements for some mat foundations constructed over granular soil deposits with those obtained from Eqs. (6.12) and (6.13).

Meyerhof (1965) compiled the observed maximum settlements for mat foundations constructed on sand and gravel, as listed in Table 6.1. In Eq. (6.12), if the depth factor, $1 + 0.33(D_f/B)$, is assumed to be approximately unity, then

$$S_e(\text{mm}) \approx \frac{2q_{\text{net(all)}}}{N_{60}} \tag{6.18}$$

From the values of $q_{\text{net(all)}}$ and N_{60} given in Columns 6 and 5, respectively, of Table 6.1, the magnitudes of S_e were calculated and are given in Column 8.

Column 9 of Table 6.1 gives the ratios of calculated to measured values of S_e. These ratios vary from about 0.79 to 3.39. Thus, calculating the net allowable bearing capacity with the use of Eq. (6.12) or (6.13) will yield safe and conservative values.

6.7   Compensated Foundation

Figure 6.7 and Eq. (6.15) indicate that the net pressure increase in the soil under a mat foundation can be reduced by increasing the depth D_f of the mat. This approach is generally referred to as the *compensated foundation design* and is extremely useful when structures are to be built on very soft clays. In this design, a deeper basement is made below the higher portion of the superstructure, so that the net pressure increase in soil at any depth is relatively uniform. (See Figure 6.8.) From Eq. (6.15) and Figure 6.7, the net average applied pressure on soil is

$$q = \frac{Q}{A} - \gamma D_f$$

Table 6.1 Settlement of Mat Foundations on Sand and Gravel (Based on Meyerhof, 1965) (Based on Meyerhof, G. G., (1965). "Shallow Foundations," *Journal of the Soil Mechanics and Foundation Engineering Division*, American Society of Civil Engineers, Vol. 91, No. 2, pp. 21–31, Table 1. With permission from ASCE.)

Case No. (1)	Structure (2)	Reference (3)	B m (4)	Average N_{60} (5)	$q_{net(all)}$ kN/m² (6)	Observed maximum settlement, S_e mm (7)	Calculated maximum settlement, S_e mm (8)	calculated S_e / observed S_e (9)
1	T. Edison São Paulo, Brazil	Rios and Silva (1948)	18.29	15	229.8	15.24	30.64	2.01
2	Banco do Brazil São Paulo, Brazil	Rios and Silva (1948); Vargas (1961)	22.86	18	239.4	27.94	26.6	0.95
3	Iparanga São Paulo, Brazil	Vargas (1948)	9.14	9	304.4	35.56	67.64	1.9
4	C.B.I., Esplanda São Paulo, Brazil	Vargas (1961)	14.63	22	383.0	27.94	34.82	1.25
5	Riscala São Paulo, Brazil	Vargas (1948)	3.96	20	229.8	12.7	22.98	1.81
6	Thyssen Düsseldorf, Germany	Schultze (1962)	22.55	25	239.4	24.13	19.15	0.79
7	Ministry Düsseldorf, Germany	Schultze (1962)	15.85	20	220.2	20.32	22.02	1.08
8	Chimney Cologne, Germany	Schultze (1962)	20.42	10	172.4	10.16	34.48	3.39

Figure 6.8 Compensated foundation

For no increase in the net pressure on soil below a mat foundation, q should be zero. Thus,

$$D_f = \frac{Q}{A\gamma} \qquad (6.19)$$

This relation for D_f is usually referred to as the depth of a *fully compensated foundation*.

The factor of safety against bearing capacity failure for partially compensated foundations (i.e., $D_f < Q/A\gamma$) may be given as

$$FS = \frac{q_{net(u)}}{q} = \frac{q_{net(u)}}{\dfrac{Q}{A} - \gamma D_f} \qquad (6.20)$$

where $q_{net(u)}$ = net ultimate bearing capacity.

For saturated clays, the factor of safety against bearing capacity failure can thus be obtained by substituting Eq. (6.10) into Eq. (6.20):

$$FS = \frac{5.14 c_u \left(1 + \dfrac{0.195 B}{L}\right)\left(1 + 0.4\dfrac{D_f}{B}\right)}{\dfrac{Q}{A} - \gamma D_f} \qquad (6.21)$$

Example 6.3

The mat shown in Figure 6.7 has dimensions of 18.3 m × 30.5 m. The total dead and live load on the mat is 111 × 10³ kN kip. The mat is placed over a saturated clay having a unit weight of 18.87 kN/m³ and c_u = 134 kN/m². Given that D_f = 1.52 m, determine the factor of safety against bearing capacity failure.

Solution

From Eq. (6.21), the factor of safety

$$FS = \frac{5.14c_u\left(1 + \frac{0.195B}{L}\right)\left(1 + 0.4\frac{D_f}{B}\right)}{\frac{Q}{A} - \gamma D_f}$$

We are given that $c_u = 134 \text{ kN/m}^2$, $D_f = 1.52$ m, $B = 18.3$ m, $L = 30.5$ m, and $\gamma = 18.87 \text{ kN/m}^3$. Hence,

$$FS = \frac{(5.14)(134)\left[1 + \frac{(0.195)(18.3)}{30.5}\right]\left[1 + 0.4\left(\frac{1.52}{18.3}\right)\right]}{\left(\frac{111 \times 10^3 \text{ kN}}{18.3 \times 30.5}\right) - (18.87)(1.52)} = \mathbf{4.66}$$

■

Example 6.4

Consider a mat foundation 30 m $\times$ 40 m in plan, as shown in Figure 6.9. The total dead load and live load on the raft is 200×10^3 kN. Estimate the consolidation settlement at the center of the foundation.

Solution

From Eq. (1.61)

$$S_{c(p)} = \frac{C_c H_c}{1 + e_o} \log\left(\frac{\sigma'_o + \Delta\sigma'_{av}}{\sigma'_o}\right)$$

$$\sigma'_o = (3.67)(15.72) + (13.33)(19.1 - 9.81) + \frac{6}{2}(18.55 - 9.81) \approx 208 \text{ kN/m}^2$$

$$H_c = 6 \text{ m}$$
$$C_c = 0.28$$
$$e_o = 0.9$$

For $Q = 200 \times 10^3$ kN, the net load per unit area is

$$q = \frac{Q}{A} - \gamma D_f = \frac{200 \times 10^3}{30 \times 40} - (15.72)(2) \approx 135.2 \text{ kN/m}^2$$

In order to calculate $\Delta\sigma'_{av}$ we refer to Section 5.5. The loaded area can be divided into four areas, each measuring 15 m $\times$ 20 m. Now using Eq. (5.19), we can calculate the average stress increase in the clay layer below the corner of each rectangular area, or

$$\Delta\sigma'_{av(H_2/H_1)} = q_o\left[\frac{H_2 I_{a(H_2)} - H_1 I_{a(H_1)}}{H_2 - H_1}\right]$$

$$= 135.2\left[\frac{(1.67 + 13.3 + 6)I_{a(H_2)} - (1.67 + 13.33)I_{a(H_1)}}{6}\right]$$

Figure 6.9 Consolidation settlement under a mat foundation

For $I_{a(H_2)}$,

$$m_2 = \frac{B}{H_2} = \frac{15}{1.67 + 13.33 + 6} = 0.71$$

$$n_2 = \frac{L}{H_2} = \frac{20}{21} = 0.95$$

From Fig. 5.7, for $m_2 = 0.71$ and $n_2 = 0.95$, the value of $I_{a(H_2)}$ is 0.21. Again, for $I_{a(H_1)}$,

$$m_2 = \frac{B}{H_1} = \frac{15}{15} = 1$$

$$n_2 = \frac{L}{H_1} = \frac{20}{15} = 1.33$$

From Figure 5.7, $I_{a(H_1)} = 0.225$, so

$$\Delta\sigma'_{av(H_2/H_1)} = 135.2 \left[\frac{(21)(0.21) - (15)(0.225)}{6} \right] = 23.32 \ \text{kN/m}^2$$

So, the stress increase below the center of the 30 m × 40 m area is $(4)(23.32) = 93.28 \ \text{kN/m}^2$. Thus

$$S_{c(p)} = \frac{(0.28)(6)}{1 + 0.9} \log\left(\frac{208 + 93.28}{208} \right) = 0.142 \ \text{m}$$

$$= \mathbf{142 \ mm}$$

$\blacksquare$

6.8 Structural Design of Mat Foundations

The structural design of mat foundations can be carried out by two conventional methods: the conventional rigid method and the approximate flexible method. Finite-difference and finite-element methods can also be used, but this section covers only the basic concepts of the first two design methods.

Conventional Rigid Method

The *conventional rigid method* of mat foundation design can be explained step by step with reference to Figure 6.10:

Step 1. Figure 6.10a shows mat dimensions of $L \times B$ and column loads of $Q_1, Q_2, Q_3, \ldots$ Calculate the total column load as

$$Q = Q_1 + Q_2 + Q_3 + \cdots \tag{6.22}$$

Step 2. Determine the pressure on the soil, q, below the mat at points $A, B, C, D, \ldots$, by using the equation

$$q = \frac{Q}{A} \pm \frac{M_y x}{I_y} \pm \frac{M_x y}{I_x} \tag{6.23}$$

where

$A = BL$
$I_x = (1/12)BL^3$ = moment of inertia about the x-axis
$I_y = (1/12)LB^3$ = moment of inertia about the y-axis
M_x = moment of the column loads about the x-axis = Qe_y
M_y = moment of the column loads about the y-axis = Qe_x

The load eccentricities, e_x and e_y, in the x and y directions can be determined by using (x', y') coordinates:

$$x' = \frac{Q_1 x_1' + Q_2 x_2' + Q_3 x_3' + \cdots}{Q} \tag{6.24}$$

and

$$e_x = x' - \frac{B}{2} \tag{6.25}$$

Similarly,

$$y' = \frac{Q_1 y_1' + Q_2 y_2' + Q_3 y_3' + \cdots}{Q} \tag{6.26}$$

and

$$e_y = y' - \frac{L}{2} \tag{6.27}$$

Step 3. Compare the values of the soil pressures determined in Step 2 with the net allowable soil pressure to determine whether $q \leq q_{\text{all(net)}}$.

Step 4. Divide the mat into several strips in the x and y directions. (See Figure 6.10). Let the width of any strip be B_1.

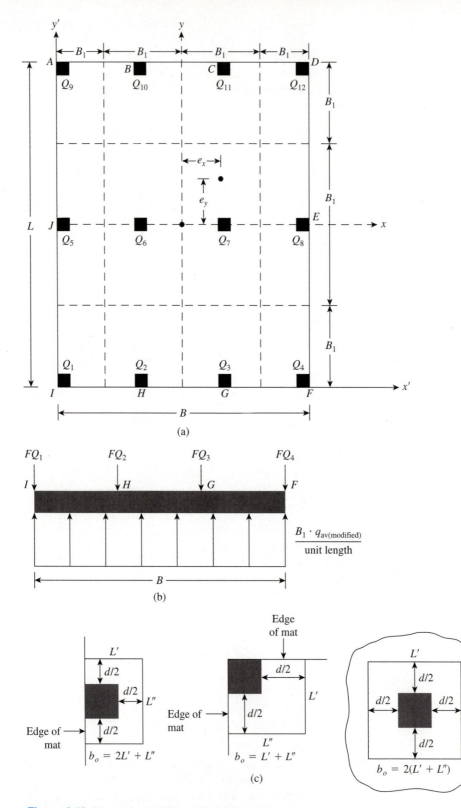

Figure 6.10 Conventional rigid mat foundation design

Step 5. Draw the shear, V, and the moment, M, diagrams for each individual strip (in the x and y directions). For example, the average soil pressure of the bottom strip in the x direction of Figure 6.10a is

$$q_{av} \approx \frac{q_I + q_F}{2} \tag{6.28}$$

where q_I and q_F = soil pressures at points I and F, as determined from Step 2.

The total soil reaction is equal to $q_{av}B_1B$. Now obtain the total column load on the strip as $Q_1 + Q_2 + Q_3 + Q_4$. The sum of the column loads on the strip will not equal $q_{av}B_1B$, because the shear between the adjacent strips has not been taken into account. For this reason, the soil reaction and the column loads need to be adjusted, or

$$\text{Average load} = \frac{q_{av}B_1B + (Q_1 + Q_2 + Q_3 + Q_4)}{2} \tag{6.29}$$

Now, the modified average soil reaction becomes

$$q_{av(modified)} = q_{av}\left(\frac{\text{average load}}{q_{av}B_1B}\right) \tag{6.30}$$

and the column load modification factor is

$$F = \frac{\text{average load}}{Q_1 + Q_2 + Q_3 + Q_4} \tag{6.31}$$

So the modified column loads are FQ_1, FQ_2, FQ_3, and FQ_4. This modified loading on the strip under consideration is shown in Figure 6.10b. The shear and the moment diagram for this strip can now be drawn, and the procedure is repeated in the x and y directions for all strips.

Step 6. Determine the effective depth d of the mat by checking for diagonal tension shear near various columns. According to ACI Code 318-95 (Section 11.12.2.1c, American Concrete Institute, 1995), for the critical section,

$$U = b_o d[\phi(0.34)\sqrt{f_c'}] \tag{6.32}$$

where

U = factored column load (MN), or (column load) $\times$ (load factor)
ϕ = reduction factor = 0.85
f_c' = compressive strength of concrete at 28 days (MN/m^2)

The units of b_o and d in Eq. (6.32a) are in meters.

The expression for b_o in terms of d, which depends on the location of the column with respect to the plan of the mat, can be obtained from Figure 6.10c.

Step 7. From the moment diagrams of all strips *in one direction* (*x* or *y*), obtain the *maximum* positive and negative moments per unit width (i.e., $M' = M/B_1$).

Step 8. Determine the areas of steel per unit width for positive and negative reinforcement in the *x* and *y* directions. We have

$$M_u = (M')(\text{load factor}) = \phi A_s f_y \left(d - \frac{a}{2} \right) \tag{6.33}$$

and

$$a = \frac{A_s f_y}{0.85 f'_c b} \tag{6.34}$$

where

A_s = area of steel per unit width
f_y = yield stress of reinforcement in tension
M_u = factored moment
$\phi = 0.9$ = reduction factor

Examples 6.5 and 6.6 illustrate the use of the conventional rigid method of mat foundation design.

Approximate Flexible Method

In the conventional rigid method of design, the mat is assumed to be infinitely rigid. Also, the soil pressure is distributed in a straight line, and the centroid of the soil pressure is coincident with the line of action of the resultant column loads. (See Figure 6.11a.) In the *approximate flexible method* of design, the soil is assumed to be equivalent to an infinite number of elastic springs, as shown in Figure 6.11b. This assumption is sometimes referred to as the *Winkler foundation*. The elastic constant of these assumed springs is referred to as the *coefficient of subgrade reaction, k.*

To understand the fundamental concepts behind flexible foundation design, consider a beam of width B_1 having infinite length, as shown in Figure 6.11c. The beam is subjected to a single concentrated load Q. From the fundamentals of mechanics of materials,

$$M = E_F I_F \frac{d^2 z}{dx^2} \tag{6.35}$$

where

M = moment at any section
E_F = modulus of elasticity of foundation material
I_F = moment of inertia of the cross section of the beam = $(\frac{1}{12}) B_1 h^3$ (see Figure 6.11c).

However,

$$\frac{dM}{dx} = \text{shear force} = V$$

and

$$\frac{dV}{dx} = q = \text{soil reaction}$$

Figure 6.11 (a) Principles of design by conventional rigid method; (b) principles of approximate flexible method; (c) derivation of Eq. (6.39) for beams on elastic foundation

Hence,

$$\frac{d^2M}{dx^2} = q \tag{6.36}$$

Combining Eqs. (6.35) and (6.36) yields

$$E_F I_F \frac{d^4z}{dx^4} = q \tag{6.37}$$

However, the soil reaction is

$$q = -zk'$$

where

z = deflection
$k' = kB_1$
k = coefficient of subgrade reaction $(kN/m^3 \text{ or } lb/in^3)$

So,

$$E_F I_F \frac{d^4 z}{dx^4} = -zkB_1 \tag{6.38}$$

Solving Eq. (6.38) yields

$$z = e^{-\alpha x}(A' \cos \beta x + A'' \sin \beta x) \tag{6.39}$$

where A' and A'' are constants and

$$\beta = \sqrt[4]{\frac{B_1 k}{4 E_F I_F}} \tag{6.40}$$

The unit of the term β, as defined by the preceding equation, is $(length)^{-1}$. This parameter is very important in determining whether a mat foundation should be designed by the conventional rigid method or the approximate flexible method. According to the American Concrete Institute Committee 336 (1988), mats should be designed by the conventional rigid method if the spacing of columns in a strip is less than $1.75/\beta$. If the spacing of columns is larger than $1.75/\beta$, the approximate flexible method may be used.

To perform the analysis for the structural design of a flexible mat, one must know the principles involved in evaluating the coefficient of subgrade reaction, k. Before proceeding with the discussion of the approximate flexible design method, let us discuss this coefficient in more detail.

If a foundation of width B (see Figure 6.12) is subjected to a load per unit area of q, it will undergo a settlement Δ. The coefficient of subgrade modulus can be defined as

$$k = \frac{q}{\Delta} \tag{6.41}$$

Figure 6.12 Definition of coefficient of subgrade reaction, k

The unit of k is kN/m^3. The value of the coefficient of subgrade reaction is not a constant for a given soil, but rather depends on several factors, such as the length L and width B of the foundation and also the depth of embedment of the foundation. A comprehensive study by Terzaghi (1955) of the parameters affecting the coefficient of subgrade reaction indicated that the value of the coefficient decreases with the width of the foundation. In the field, load tests can be carried out by means of square plates measuring 0.3 m $\times$ 0.3 m, and values of k can be calculated. The value of k can be related to large foundations measuring $B \times B$ in the following ways:

Foundations on Sandy Soils

For foundations on sandy soils,

$$k = k_{0.3}\left(\frac{B + 0.3}{2B}\right)^2 \tag{6.42}$$

where $k_{0.3}$ and k = coefficients of subgrade reaction of foundations measuring 0.3 m $\times$ 0.3 m and $B\,(\text{m}) \times B\,(\text{m})$, respectively (unit is kN/m^3).

Foundations on Clays

For foundations on clays,

$$k\,(\text{kN/m}^3) = k_{0.3}(\text{kN/m}^3)\left[\frac{0.3\,(\text{m})}{B(\text{m})}\right] \tag{6.43}$$

The definitions of k and $k_{0.3}$ in Eq. (6.43) are the same as in Eq. (6.42).

For rectangular foundations having dimensions of $B \times L$ (for similar soil and q),

$$k = \frac{k_{(B \times B)}\left(1 + 0.5\dfrac{B}{L}\right)}{1.5} \tag{6.44}$$

where

k = coefficient of subgrade modulus of the rectangular foundation ($L \times B$)
$k_{(B \times B)}$ = coefficient of subgrade modulus of a square foundation having dimension of $B \times B$

Equation (6.44) indicates that the value of k for a very long foundation with a width B is approximately $0.67k_{(B \times B)}$.

The modulus of elasticity of granular soils increases with depth. Because the settlement of a foundation depends on the modulus of elasticity, the value of k increases with the depth of the foundation.

Table 6.2 provides typical ranges of values for the coefficient of subgrade reaction, $k_{0.3}(k_1)$, for sandy and clayey soils.

For long beams, Vesic (1961) proposed an equation for estimating subgrade reaction, namely,

$$k' = Bk = 0.65 \sqrt[12]{\frac{E_s B^4}{E_F I_F}} \frac{E_s}{1 - \mu_s^2}$$

or

$$k = 0.65 \sqrt[12]{\frac{E_s B^4}{E_F I_F}} \frac{E_s}{B(1 - \mu_s^2)} \tag{6.45}$$

where

E_s = modulus of elasticity of soil
B = foundation width
E_F = modulus of elasticity of foundation material
I_F = moment of inertia of the cross section of the foundation
μ_s = Poisson's ratio of soil

Table 6.2 Typical Subgrade Reaction Values, $k_{0.3}(k_1)$

Soil type	$k_{0.3}(k_1)$ MN/m³
Dry or moist sand:	
Loose	8–25
Medium	25–125
Dense	125–375
Saturated sand:	
Loose	10–15
Medium	35–40
Dense	130–150
Clay:	
Stiff	10–25
Very stiff	25–50
Hard	>50

For most practical purposes, Eq. (6.46) can be approximated as

$$k = \frac{E_s}{B(1 - \mu_s^2)} \tag{6.46}$$

Now that we have discussed the coefficient of subgrade reaction, we will proceed with the discussion of the approximate flexible method of designing mat foundations. This method, as proposed by the American Concrete Institute Committee 336 (1988), is described step by step. The use of the design procedure, which is based primarily on the theory of plates, allows the effects (i.e., moment, shear, and deflection) of a concentrated column load in the area surrounding it to be evaluated. If the zones of influence of two or more columns overlap, superposition can be employed to obtain the net moment, shear, and deflection at any point. The method is as follows:

Step 1. Assume a thickness h for the mat, according to Step 6 of the conventional rigid method. (*Note:* h is the *total* thickness of the mat.)

Step 2. Determine the flexural ridigity R of the mat as given by the formula

$$R = \frac{E_F h^3}{12(1 - \mu_F^2)} \tag{6.47}$$

where

E_F = modulus of elasticity of foundation material
μ_F = Poisson's ratio of foundation material

Step 3. Determine the radius of effective stiffness—that is,

$$L' = \sqrt[4]{\frac{R}{k}} \tag{6.48}$$

where k = coefficient of subgrade reaction. The zone of influence of any column load will be on the order of 3 to 4 L'.

Step 4. Determine the moment (in polar coordinates at a point) caused by a column load (see Figure 6.13a). The formulas to use are

$$M_r = \text{radial moment} = -\frac{Q}{4}\left[A_1 - \frac{(1 - \mu_F)A_2}{\frac{r}{L'}}\right] \tag{6.49}$$

and

$$M_t = \text{tangential moment} = -\frac{Q}{4}\left[\mu_F A_1 + \frac{(1 - \mu_F)A_2}{\frac{r}{L'}}\right] \tag{6.50}$$

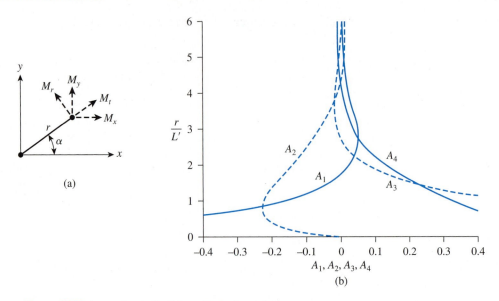

(a)

(b)

Figure 6.13 Approximate flexible method of mat design

where

r = radial distance from the column load
Q = column load
A_1, A_2 = functions of r/L'

The variations of A_1 and A_2 with r/L' are shown in Figure 6.13b. (For details see Hetenyi, 1946.)

In the Cartesian coordinate system (see Figure 6.13a),

$$M_x = M_t \sin^2\alpha + M_r \cos^2\alpha \qquad (6.51)$$

and

$$M_y = M_t \cos^2\alpha + M_r \sin^2\alpha \qquad (6.52)$$

Step 5.   For the unit width of the mat, determine the shear force V caused by a column load:

$$V = \frac{Q}{4L'}A_3 \qquad (6.53)$$

The variation of A_3 with r/L' is shown in Figure 6.13b.

Step 6.   If the edge of the mat is located in the zone of influence of a column, determine the moment and shear along the edge. (Assume that the mat is continuous.) Moment and shear opposite in sign to those determined are applied at the edges to satisfy the known conditions.

Step 7.   The deflection at any point is given by

$$\delta = \frac{QL'^2}{4R}A_4 \qquad (6.54)$$

The variation of A_4 is presented in Figure 6.13b.

Example 6.5

The plan of a mat foundation is shown in Figure 6.14. Calculate the soil pressure at points *A, B, C, D, E,* and *F.* (*Note*: All column sections are planned to be 0.5 m × 0.5 m.)

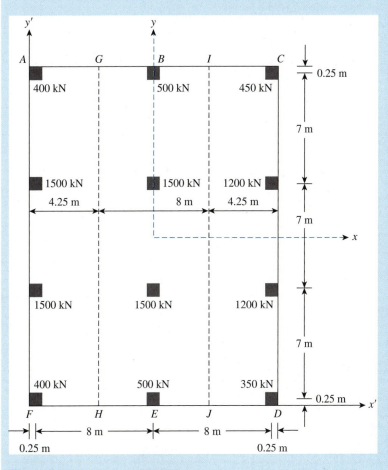

Figure 6.14 Plan of a mat foundation

Solution

Eq. (6.23): $q = \dfrac{Q}{A} \pm \dfrac{M_y\, x}{I_y} \pm \dfrac{M_x\, y}{I_x}$

$A = (16.5)(21.5) = 354.75 \text{ m}^2$

$I_x = \dfrac{1}{12}BL^3 = \dfrac{1}{12}(16.5)(21.5)^3 = 13{,}665 \text{ m}^4$

$I_y = \dfrac{1}{12}LB^3 = \dfrac{1}{12}(21.5)(16.5)^3 = 8{,}050 \text{ m}^4$

$Q = 350 + (2)(400) + 450 + (2)(500) + (2)(1200) + (4)(1500) = 11{,}000 \text{ kN}$

$M_y = Qe_x; \quad e_x = x' - \dfrac{B}{2}$

$$x' = \frac{Q_1 x_1' + Q_2 x_2' + Q_3 x_3' + \cdots}{Q}$$

$$= \frac{1}{11,000} \left[\begin{array}{l} (8.25)(500 + 1500 + 1500 + 500) \\ + (16.25)(350 + 1200 + 1200 + 450) \\ + (0.25)(400 + 1500 + 1500 + 400) \end{array} \right] = 7.814 \text{ m}$$

$$e_x = x' - \frac{B}{2} = 7.814 - 8.25 = -0.435 \text{ m} \approx -0.44 \text{ m}$$

Hence, the resultant line of action is located to the left of the center of the mat. So $M_y = (11,000)(0.44) = 4840$ kN-m. Similarly

$$M_x = Q e_y; \quad e_y = y' - \frac{L}{2}$$

$$y' = \frac{Q_1 y_1' + Q_2 y_2' + Q_3 y_3' + \cdots}{Q}$$

$$= \frac{1}{11,000} \left[\begin{array}{l} (0.25)(400 + 500 + 350) + (7.25)(1500 + 1500 + 1200) \\ + (14.25)(1500 + 1500 + 1200) + (21.25)(400 + 500 + 450) \end{array} \right]$$

$$= 10.85 \text{ m}$$

$$e_y = y' - \frac{L}{2} = 10.85 - 10.75 = 0.1 \text{m}$$

The location of the line of action of the resultant column loads is shown in Figure 6.15.

$M_x = (11,000)(0.1) = 1100$ kN-m. So

$$q = \frac{11,000}{354.75} \pm \frac{4840x}{8050} \pm \frac{1100y}{13,665} = 31.0 \pm 0.6x \pm 0.08y \, (\text{kN/m}^2)$$

Figure 6.15

Therefore,

At A: $q = 31.0 + (0.6)(8.25) + (0.08)(10.75) =$ **36.81 kN/m²**
At B: $q = 31.0 + (0.6)(0) + (0.08)(10.75) =$ **31.86 kN/m²**
At C: $q = 31.0 - (0.6)(8.25) + (0.08)(10.75) =$ **26.91 kN/m²**
At D: $q = 31.0 - (0.6)(8.25) - (0.08)(10.75) =$ **25.19 kN/m²**
At E: $q = 31.0 + (0.6)(0) - (0.08)(10.75) =$ **30.14 kN/m²**
At F: $q = 31.0 + (0.6)(8.25) - (0.08)(10.75) =$ **35.09 kN/m²**   ∎

Example 6.6

Divide the mat shown in Figure 6.14 into three strips, such as $AGHF$ ($B_1 = 4.25$ m), $GIJH$ ($B_1 = 8$ m), and $ICDJ$ ($B_1 = 4.25$ m). Use the result of Example 6.5, and determine the reinforcement requirements in the y direction. Here, $f_c' = 20.7$ MN/m², $f_y = 413.7$ MN/m², and the load factor is 1.7.

Solution
Determination of Shear and Moment Diagrams for Strips:
Strip $AGHF$:

$$\text{Average soil pressure} = q_{av} = q_{(at\ A)} + q_{(at\ F)} = \frac{36.81 + 35.09}{2} = 35.95 \text{ kN/m}^2$$

Total soil reaction $= q_{av}B_1L = (35.95)(4.25)(21.50) = 3285$ kN

$$\text{Average load} = \frac{\text{load due to soil reaction} + \text{column loads}}{2}$$

$$= \frac{3285 + 3800}{2} = 3542.5 \text{ kN}$$

So, modified average soil pressure,

$$q_{av(modified)} = q_{av}\left(\frac{3542.5}{3285}\right) = (35.95)\left(\frac{3542.5}{3285}\right) = 38.768 \text{ kN/m}^2$$

The column loads can be modified in a similar manner by multiplying factor

$$F = \frac{3542.5}{3800} = 0.9322$$

Figure 6.16 shows the loading on the strip and corresponding shear and moment diagrams. Note that the column loads shown in this figure have been multiplied by $F = 0.9322$. Also the load per unit length of the beam is equal to $B_1 q_{av(modified)} = (4.25)(38.768) = 164.76$ kN/m.
Strip $GIJH$: In a similar manner,

$$q_{av} = \frac{q_{(at\ B)} + q_{(at\ E)}}{2} = \frac{31.86 + 30.14}{2} = 31.0 \text{ kN/m}^2$$

$$\text{Total soil reaction} = (31)(8)(21.5) = 5332 \text{ kN}$$
$$\text{Total column load} = 4000 \text{ kN}$$

Figure 6.16 Load, shear, and moment diagrams for strip *AGHF*

$$\text{Average load} = \frac{5332 + 4000}{2} = 4666 \text{ kN}$$

$$q_{\text{av(modified)}} = (31.0)\left(\frac{4666}{5332}\right) = 27.12 \text{ kN/m}^2$$

$$F = \frac{4666}{4000} = 1.1665$$

The load, shear, and moment diagrams are shown in Figure 6.17.

Strip *ICDJ*: Figure 6.18 shows the load, shear, and moment diagrams for this strip.

Determination of the Thickness of the Mat

For this problem, the critical section for diagonal tension shear will be at the column carrying 1500 kN of load at the edge of the mat [Figure 6.19]. So

$$b_o = \left(0.5 + \frac{d}{2}\right) + \left(0.5 + \frac{d}{2}\right) + (0.5 + d) = 1.5 + 2d$$

$$U = (b_o d)\left[(\phi)(0.34)\sqrt{f_c'}\right]$$

$$U = (1.7)(1500) = 2550 \text{ kN} = 2.55 \text{ MN}$$

$$2.55 = (1.5 + 2d)(d)\left[(0.85)(0.34)\sqrt{20.7}\right]$$

or

$$(1.5 + 2d)(d) = 1.94; \quad d = 0.68 \text{ m}$$

Figure 6.17 Load, shear, and moment diagrams for strip *GIJH*

Figure 6.18 Load, shear, and moment diagrams for strip *ICDJ*

Figure 6.19 Critical perimeter column

Assuming a minimum cover of 76 mm over the steel reinforcement and also assuming that the steel bars to be used are 25 mm in diameter, the total thickness of the slab is

$$h = 0.68 + 0.076 + 0.025 = 0.781 \text{ m} \approx \textbf{0.8 m}$$

The thickness of this mat will satisfy the wide beam shear condition across the three strips under consideration.

Determination of Reinforcement

From the moment diagram shown in Figures 6.16, 6.17, and 6.18, it can be seen that the maximum positive moment is located in strip *AGHF*, and its magnitude is

$$M' = \frac{1727.57}{B_1} = \frac{1727.57}{4.25} = 406.5 \text{ kN-m/m}$$

Similarly, the maximum negative moment is located in strip *ICDJ* and its magnitude is

$$M' = \frac{1196.19}{B_1} = \frac{1196.19}{4.25} = 281.5 \text{ kN-m/m}$$

From Eq. (6.33): $M_u = (M')(\text{ load factor }) = \phi A_s f_y \left(d - \dfrac{a}{2} \right)$.

For the positive moment, $M_u = (406.5)(1.7) = (\phi)(A_s)(413.7 \times 1000)\left(0.68 - \dfrac{a}{2}\right)$

$\phi = 0.9$. Also, from Eq. (6.34),

$$a = \frac{A_s f_y}{0.85 f'_c b} = \frac{(A_s)(413.7)}{(0.85)(20.7)(1)} = 23.51 A_s; \text{ or } A_s = 0.0425a$$

$$691.05 = (0.9)(0.0425a)(413,700)\left(0.68 - \frac{a}{2}\right); \text{ or } a \approx 0.0645$$

So, $A_s = (0.0425)(0.0645) = 0.00274 \text{ m}^2/\text{m} = 2740 \text{ mm}^2/\text{m}$.

Use 25-mm diameter bars at 175 mm center-to-center:

$$\left[A_s \text{ provided} = (491)\left(\frac{1000}{175}\right) = 2805.7 \text{ mm}^2/\text{m} \right]$$

Similarly, for negative reinforcement,

$$M_u = (281.5)(1.7) = (\phi)(A_s)(413.7 \times 1000)\left(0.68 - \frac{a}{2}\right)$$

$$\phi = 0.9. \ A_s = 0.0425a$$

So

$$478.55 = (0.9)(0.0425a)(413.7 \times 1000)\left(0.68 - \frac{a}{2}\right); \text{ or } a \approx 0.045$$

So, $A_s = (0.045)(0.0425) = 0.001913 \text{ m}^2/\text{m} = 1913 \text{ mm}^2/\text{m}.$

Use 25-mm diameter bars at 255 mm center-to-center:

$$[A_s \text{ provided} = 1925 \text{ mm}^2]$$

Because negative moment occurs at midbay of strip *ICDJ*, reinforcement should be provided. This moment is

$$M' = \frac{289.95}{4.25} = 68.22 \text{ kN-m/m}$$

Hence,

$$M_u = (68.22)(1.7) = (0.9)(0.0425a)(413.7 \times 1000)\left(0.68 - \frac{a}{2}\right);$$

or $a \approx 0.0108$

$$A_s = (0.0108)(0.0425) = 0.000459 \text{ m}^2/\text{m} = 459 \text{ mm}^2/\text{m}$$

Provide 16-mm diameter bars at 400 mm center-to-center:

$$[A_s \text{ provided} = 502 \text{ mm}^2]$$

For general arrangement of the reinforcement see Figure 6.20.

Top steel   Bottom steel   Top steel

Additional top steel
in strip *ICDJ*

Figure 6.20 General arrangement of reinforcement

Problems

6.1 Determine the net ultimate bearing capacity of mat foundations with the following characteristics:

$$c_u = 120 \text{ kN/m}^2, \phi = 0, B = 8 \text{ m}, L = 18 \text{ m}, D_f = 3 \text{ m}$$

6.2 Following are the results of a standard penetration test in the field (sandy soil):

Depth (m)	Field value of N_{60}
1.5	9
3.0	12
4.5	11
6.0	7
7.5	13
9.0	11
10.5	13

Estimate the net allowable bearing capacity of a mat foundation 6.5 m × 5 m in plan. Here, $D_f = 1.5$ m and allowable settlement = 50 mm. Assume that the unit weight of soil, $\gamma = 16.5 \text{ kN/m}^3$.

6.3 Repeat Problem 6.2 for an allowable settlement of 30 mm.

6.4 A mat foundation on a saturated clay soil has dimensions of 20 m × 20 m. Given: dead and live load = 48 MN, $c_u = 30 \text{ kN/m}^2$, and $\gamma_{clay} = 18.5 \text{ kN/m}^3$.
   a. Find the depth, D_f, of the mat for a fully compensated foundation.
   b. What will be the depth of the mat (D_f) for a factor of safety of 2 against bearing capacity failure?

6.5 Repeat Problem 6.4 part b for $c_u = 20 \text{ kN/m}^2$.

6.6 A mat foundation is shown in Figure P6.6. The design considerations are $L = 12$ m, $B = 10$ m, $D_f = 2.2$ m, $Q = 30$ MN, $x_1 = 2$ m, $x_2 = 2$ m, $x_3 = 5.2$ m, and preconsolidation pressure $\sigma_c' \approx 105 \text{ kN/m}^2$. Calculate the consolidation settlement under the center of the mat.

6.7 For the mat foundation in Problem 6.6, estimate the consolidation settlement under the corner of the mat.

6.8 From the plate load test (plate dimensions 0.3 m × 0.3 m) in the field, the coefficient of subgrade reaction of a sandy soil is determined to be 14,900 kN/m³. What will be the value of the coefficient of subgrade reaction on the same soil for a foundation with dimensions of 7.5 m × 7.5 m?

6.9 Refer to Problem 6.18. If the full-sized foundation had dimensions of 21.3 m × 9.1 m, what will be the value of the coefficient of subgrade reaction?

6.10 The subgrade reaction of a sandy soil obtained from the plate load test (plate dimensions 1 m × 0.7 m) is 18 MN/m³. What will be the value of k on the same soil for a foundation measuring 5 m × 3.5 m?

Size of mat = $B \times L$

D_f

Sand
$\gamma = 16.0$ kN/m³

$\downarrow Q$

x_1

z

Groundwater
table

x_2

Sand
$\gamma_{sat} = 18.0$ kN/m³

x_3

Clay
$\gamma_{sat} = 17.5$ kN/m³
$e_o = 0.88$
$C_c = 0.38$
$C_s = 0.1$

Figure P6.6

References

AMERICAN CONCRETE INSTITUTE (1995). *ACI Standard Building Code Requirements for Reinforced Concrete,* ACI 318–95, Farmington Hills, MI.

AMERICAN CONCRETE INSTITUTE COMMITTEE 336 (1988). "Suggested Design Procedures for Combined Footings and Mats," *Journal of the American Concrete Institute,* Vol. 63, No. 10, pp. 1041–1077.

HETENYI, M. (1946). *Beams of Elastic Foundations,* University of Michigan Press, Ann Arbor, MI.

MEYERHOF, G. G. (1965). "Shallow Foundations," *Journal of the Soil Mechanics and Foundations Division,* American Society of Civil Engineers, Vol. 91, No. SM2, pp. 21–31.

RIOS, L., and SILVA, F. P. (1948). "Foundations in Downtown São Paulo (Brazil)," *Proceedings, Second International Conference on Soil Mechanics and Foundation Engineering,* Rotterdam, Vol. 4, p. 69.

SCHULTZE, E. (1962). "Probleme bei der Auswertung von Setzungsmessungen," *Proceedings, Baugrundtagung,* Essen, Germany, p. 343.

TERZAGHI, K. (1955). "Evaluation of the Coefficient of Subgrade Reactions," *Geotechnique,* Institute of Engineers, London, Vol. 5, No. 4, pp. 197–226.

VARGAS, M. (1948). "Building Settlement Observations in São Paulo," *Proceedings Second International Conference on Soil Mechanics and Foundation Engineering,* Rotterdam, Vol. 4, p. 13.

VARGAS, M. (1961). "Foundations of Tall Buildings on Sand in São Paulo (Brazil)," *Proceedings, Fifth International Conference on Soil Mechanics and Foundation Engineering,* Paris, Vol. 1, p. 841.

VESIC, A. S. (1961). "Bending of Beams Resting on Isotropic Solid," *Journal of the Engineering Mechanics Division,* American Society of Civil Engineers, Vol. 87, No. EM2, pp. 35–53.

7 Lateral Earth Pressure

7.1   *Introduction*

Vertical or near-vertical slopes of soil are supported by retaining walls, cantilever sheet-pile walls, sheet-pile bulkheads, braced cuts, and other, similar structures. The proper design of those structures requires an estimation of lateral earth pressure, which is a function of several factors, such as (a) the type and amount of wall movement, (b) the shear strength parameters of the soil, (c) the unit weight of the soil, and (d) the drainage conditions in the backfill. Figure 7.1 shows a retaining wall of height H. For similar types of backfill,

a. The wall may be restrained from moving (Figure 7.1a). The lateral earth pressure on the wall at any depth is called the *at-rest earth pressure.*
b. The wall may tilt away from the soil that is retained (Figure 7.1b). With sufficient wall tilt, a triangular soil wedge behind the wall will fail. The lateral pressure for this condition is referred to as *active earth pressure.*
c. The wall may be pushed into the soil that is retained (Figure 7.1c). With sufficient wall movement, a soil wedge will fail. The lateral pressure for this condition is referred to as *passive earth pressure.*

Figure 7.2 shows the nature of variation of the lateral pressure, σ'_h, at a certain depth of the wall with the magnitude of wall movement.

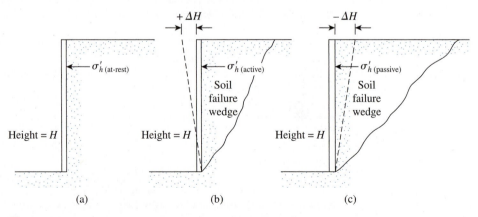

Figure 7.1 Nature of lateral earth pressure on a retaining wall

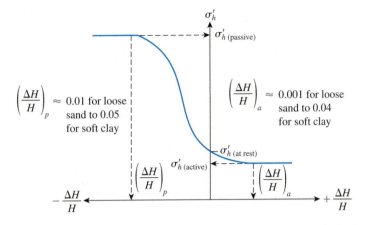

Figure 7.2 Nature of variation of lateral earth pressure at a certain depth

In the sections that follow, we will discuss various relationships to determine the at-rest, active, and passive pressures on a retaining wall. It is assumed that the reader has studied lateral earth pressure in the past, so this chapter will serve as a review.

7.2 *Lateral Earth Pressure at Rest*

Consider a vertical wall of height H, as shown in Figure 7.3, retaining a soil having a unit weight of γ. A uniformly distributed load, q/unit area, is also applied at the ground surface. The shear strength of the soil is

$$s = c' + \sigma' \tan \phi'$$

where

c' = cohesion
ϕ' = effective angle of friction
σ' = effective normal stress

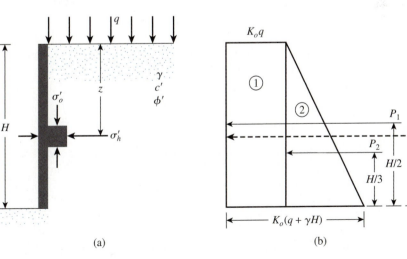

(a)   (b)

Figure 7.3 At-rest earth pressure

At any depth z below the ground surface, the vertical subsurface stress is

$$\sigma'_o = q + \gamma z \tag{7.1}$$

If the *wall is at rest and is not allowed to move at all*, either away from the soil mass or into the soil mass (i.e., there is zero horizontal strain), the lateral pressure at a depth z is

$$\sigma_h = K_o\sigma'_o + u \tag{7.2}$$

where

u = pore water pressure
K_o = coefficient of at-rest earth pressure

For normally consolidated soil, the relation for K_o (Jaky, 1944) is

$$K_o \approx 1 - \sin\phi' \tag{7.3}$$

Equation (7.3) is an empirical approximation.

For overconsolidated soil, the at-rest earth pressure coefficient may be expressed as (Mayne and Kulhawy, 1982)

$$K_o = (1 - \sin\phi')\text{OCR}^{\sin\phi'} \tag{7.4}$$

where OCR = overconsolidation ratio.

With a properly selected value of the at-rest earth pressure coefficient, Eq. (7.2) can be used to determine the variation of lateral earth pressure with depth z. Figure 7.3b shows the variation of σ'_h with depth for the wall depicted in Figure 7.3a. Note that if the surcharge $q = 0$ and the pore water pressure $u = 0$, the pressure diagram will be a triangle. The total force, P_o, *per unit length* of the wall given in Figure 7.3a can now be obtained from the area of the pressure diagram given in Figure 7.3b and is

$$P_o = P_1 + P_2 = qK_oH + \tfrac{1}{2}\gamma H^2 K_o \tag{7.5}$$

where

P_1 = area of rectangle 1
P_2 = area of triangle 2

The location of the line of action of the resultant force, P_o, can be obtained by taking the moment about the bottom of the wall. Thus,

$$\bar{z} = \frac{P_1\left(\dfrac{H}{2}\right) + P_2\left(\dfrac{H}{3}\right)}{P_o} \tag{7.6}$$

If the water table is located at a depth $z < H$, the at-rest pressure diagram shown in Figure 7.3b will have to be somewhat modified, as shown in Figure 7.4. If the effective unit weight of soil below the water table equals γ' (i.e., $\gamma_{sat} - \gamma_w$), then

at $z = 0$, $\quad \sigma'_h = K_o\sigma'_o = K_oq$
at $z = H_1$, $\quad \sigma'_h = K_o\sigma'_o = K_o(q + \gamma H_1)$

and

at $z = H_2$, $\quad \sigma'_h = K_o\sigma'_o = K_o(q + \gamma H_1 + \gamma'H_2)$

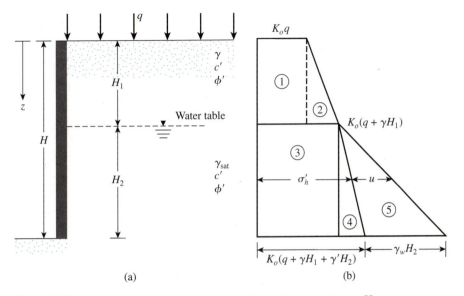

(a)   (b)

Figure 7.4 At-rest earth pressure with water table located at a depth $z < H$

Note that in the preceding equations, σ_o' and σ_h' are effective vertical and horizontal pressures, respectively. Determining the total pressure distribution on the wall requires adding the hydrostatic pressure u, which is zero from $z = 0$ to $z = H_1$ and is $H_2\gamma_w$ at $z = H_2$. The variation of σ_h' and u with depth is shown in Figure 7.4b. Hence, the total force per unit length of the wall can be determined from the area of the pressure diagram. Specifically,

$$P_o = A_1 + A_2 + A_3 + A_4 + A_5$$

where A = area of the pressure diagram.

So,

$$P_o = K_o q H_1 + \tfrac{1}{2}K_o \gamma H_1^2 + K_o(q + \gamma H_1)H_2 + \tfrac{1}{2}K_o \gamma' H_2^2 + \tfrac{1}{2}\gamma_w H_2^2 \qquad (7.7)$$

Example 7.1

For the retaining wall shown in Figure 7.5(a), determine the lateral earth force at rest per unit length of the wall. Also determine the location of the resultant force. Assume OCR = 1.

Solution

$$K_o = 1 - \sin \phi' = 1 - \sin 30° = 0.5$$

At $z = 0$, $\sigma_o' = 0$; $\sigma_h' = 0$

At $z = 2.5$ m, $\sigma_o' = (16.5)(2.5) = 41.25$ kN/m²;

$$\sigma_h' = K_o \sigma_o' = (0.5)(41.25) = 20.63 \text{ kN/m}^2$$

At $z = 5$ m, $\sigma_o' = (16.5)(2.5) + (19.3 - 9.81)2.5 = 64.98$ kN/m²;

$$\sigma_h' = K_o \sigma_o' = (0.5)(64.98) = 32.49 \text{ kN/m}^2$$

Figure 7.5

The hydrostatic pressure distribution is as follows:

From $z = 0$ to $z = 2.5$ m, $u = 0$. At $z = 5$ m, $u = \gamma_w(2.5) = (9.81)(2.5) = 24.53$ kN/m². The pressure distribution for the wall is shown in Figure 7.5b.

The total force per unit length of the wall can be determined from the area of the pressure diagram, or

$$P_o = \text{Area } 1 + \text{Area } 2 + \text{Area } 3 + \text{Area } 4$$

$$= \tfrac{1}{2}(2.5)(20.63) + (2.5)(20.63) + \tfrac{1}{2}(2.5)(32.49 - 20.63)$$

$$+ \tfrac{1}{2}(2.5)(24.53) = \mathbf{122.85 \ kN/m}$$

The location of the center of pressure measured from the bottom of the wall (point O) =

$$\bar{z} = \frac{(\text{Area } 1)\left(2.5 + \dfrac{2.5}{3}\right) + (\text{Area } 2)\left(\dfrac{2.5}{2}\right) + (\text{Area } 3 + \text{Area } 4)\left(\dfrac{2.5}{3}\right)}{P_o}$$

$$= \frac{(25.788)(3.33) + (51.575)(1.25) + (14.825 + 30.663)(0.833)}{122.85}$$

$$= \frac{85.87 + 64.47 + 37.89}{122.85} = \mathbf{1.53 \ m} \qquad \blacksquare$$

Active Pressure

7.3   *Rankine Active Earth Pressure*

The lateral earth pressure described in Section 7.2 involves walls that do not yield at all. However, if a wall tends to move away from the soil a distance Δx, as shown in Figure 7.6a, the soil pressure on the wall at any depth will decrease. For a wall that is *frictionless,* the horizontal stress, σ'_h, at depth z will equal $K_o\sigma'_o(=K_o\gamma z)$ when Δx is zero. However, with $\Delta x > 0$, σ'_h will be less than $K_o\sigma'_o$.

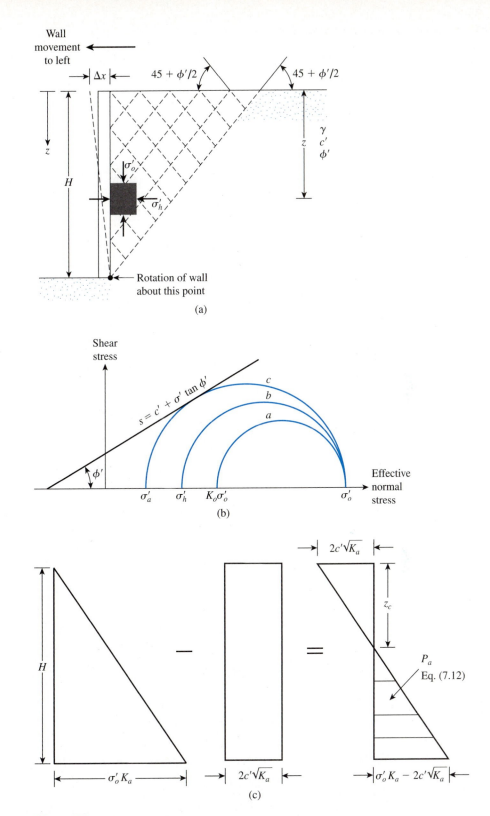

Figure 7.6 Rankine active pressure

The Mohr's circles corresponding to wall displacements of $\Delta x = 0$ and $\Delta x > 0$ are shown as circles a and b, respectively, in Figure 7.6b. If the displacement of the wall, Δx, continues to increase, the corresponding Mohr's circle eventually will just touch the Mohr–Coulomb failure envelope defined by the equation

$$s = c' + \sigma' \tan \phi'$$

This circle, marked c in the figure, represents the failure condition in the soil mass; the horizontal stress then equals σ'_a, referred to as the *Rankine active pressure*. The *slip lines* (failure planes) in the soil mass will then make angles of $\pm(45 + \phi'/2)$ with the horizontal, as shown in Figure 7.6a.

Equation (1.87) relates the principal stresses for a Mohr's circle that touches the Mohr–Coulomb failure envelope:

$$\sigma'_1 = \sigma'_3 \tan^2\left(45 + \frac{\phi'}{2}\right) + 2c' \tan\left(45 + \frac{\phi'}{2}\right)$$

For the Mohr's circle c in Figure 7.6b,

Major principal stress, $\sigma'_1 = \sigma'_o$

and

Minor principal stress, $\sigma'_3 = \sigma'_a$

Thus,

$$\sigma'_o = \sigma'_a \tan^2\left(45 + \frac{\phi'}{2}\right) + 2c' \tan\left(45 + \frac{\phi'}{2}\right)$$

$$\sigma'_a = \frac{\sigma'_o}{\tan^2\left(45 + \frac{\phi'}{2}\right)} - \frac{2c'}{\tan\left(45 + \frac{\phi'}{2}\right)}$$

or

$$\sigma'_a = \sigma'_o \tan^2\left(45 - \frac{\phi'}{2}\right) - 2c' \tan\left(45 - \frac{\phi'}{2}\right)$$

$$= \sigma'_o K_a - 2c'\sqrt{K_a} \tag{7.8}$$

where $K_a = \tan^2(45 - \phi'/2) = $ Rankine   active-pressure coefficient.

The variation of the active pressure with depth for the wall shown in Figure 7.6a is given in Figure 7.6c. Note that $\sigma'_o = 0$ at $z = 0$ and $\sigma'_o = \gamma H$ at $z = H$. The pressure distribution shows that at $z = 0$ the active pressure equals $-2c'\sqrt{K_a}$, indicating a tensile stress that decreases with depth and becomes zero at a depth $z = z_c$, or

$$\gamma z_c K_a - 2c'\sqrt{K_a} = 0$$

and

$$z_c = \frac{2c'}{\gamma \sqrt{K_a}} \tag{7.9}$$

The depth z_c is usually referred to as the *depth of tensile crack,* because the tensile stress in the soil will eventually cause a crack along the soil–wall interface. Thus, the total Rankine active force per unit length of the wall before the tensile crack occurs is

$$\begin{aligned} P_a &= \int_0^H \sigma'_a dz = \int_0^H \gamma z K_a dz - \int_0^H 2c'\sqrt{K_a} dz \\ &= \tfrac{1}{2}\gamma H^2 K_a - 2c'H\sqrt{K_a} \end{aligned} \tag{7.10}$$

After the tensile crack appears, the force per unit length on the wall will be caused only by the pressure distribution between depths $z = z_c$ and $z = H$, as shown by the hatched area in Figure 7.6c. This force may be expressed as

$$P_a = \tfrac{1}{2}(H - z_c)(\gamma H K_a - 2c'\sqrt{K_a}) \tag{7.11}$$

or

$$P_a = \frac{1}{2}\left(H - \frac{2c'}{\gamma\sqrt{K_a}}\right)\left(\gamma H K_a - 2c'\sqrt{K_a}\right) \tag{7.12}$$

However, it is important to realize that the active earth pressure condition will be reached only if the wall is allowed to "yield" sufficiently. The necessary amount of outward displacement of the wall is about $0.001H$ to $0.004H$ for granular soil backfills and about $0.01H$ to $0.04H$ for cohesive soil backfills.

Note further that if the *total stress* shear strength parameters (c, ϕ) were used, an equation similar to Eq. (7.8) could have been derived, namely,

$$\sigma_a = \sigma_o \tan^2\left(45 - \frac{\phi}{2}\right) - 2c \tan\left(45 - \frac{\phi}{2}\right)$$

Example 7.2

A 6-m-high retaining wall is to support a soil with unit weight $\gamma = 17.4$ kN/m^3, soil friction angle $\phi' = 26°$, and cohesion $c' = 14.36$ kN/m^2. Determine the Rankine active force per unit length of the wall both before and after the tensile crack occurs, and determine the line of action of the resultant in both cases.

Solution
For $\phi' = 26°$,

$$K_a = \tan^2\left(45 - \frac{\phi'}{2}\right) = \tan^2(45 - 13) = 0.39$$

$$\sqrt{K_a} = 0.625$$

$$\sigma'_a = \gamma H K_a - 2c'\sqrt{K_a}$$

From Figure 7.6c, at $z = 0$,

$$\sigma'_a = -2c'\sqrt{K_a} = -2(14.36)(0.625) = -17.95 \text{ kN/m}^2$$

and at $z = 6$ m,

$$\sigma'_a = (17.4)(6)(0.39) - 2(14.36)(0.625)$$
$$= 40.72 - 17.95 = 22.77 \text{ kN/m}^2$$

Active Force before the Tensile Crack Appeared: Eq. (7.10)

$$P_a = \tfrac{1}{2}\gamma H^2 K_a - 2c'H\sqrt{K_a}$$
$$= \tfrac{1}{2}(6)(40.72) - (6)(17.95) = 122.16 - 107.7 = \mathbf{14.46 \text{ kN/m}}$$

The line of action of the resultant can be determined by taking the moment of the area of the pressure diagrams about the bottom of the wall, or

$$P_a\bar{z} = (122.16)\left(\frac{6}{3}\right) - (107.7)\left(\frac{6}{2}\right)$$

Thus,

$$\bar{z} = \frac{244.32 - 323.1}{14.46} = \mathbf{-5.45 \text{ m.}}$$

Active Force after the Tensile Crack Appeared: Eq. (7.9)

$$z_c = \frac{2c'}{\gamma\sqrt{K_a}} = \frac{2(14.36)}{(17.4)(0.625)} = 2.64 \text{ m}$$

Using Eq. (7.11) gives

$$P_a = \tfrac{1}{2}(H - z_c)(\gamma H K_a - 2c'\sqrt{K_a}) = \tfrac{1}{2}(6 - 2.64)(22.77) = \mathbf{38.25 \text{ kN/m}}$$

Figure 7.6c indicates that the force $P_a = 38.25$ kN/m is the area of the hatched triangle. Hence, the line of action of the resultant will be located at a height $\bar{z} = (H - z_c)/3$ above the bottom of the wall, or

$$\bar{z} = \frac{6 - 2.64}{3} = \mathbf{1.12 \text{ m}} \qquad\blacksquare$$

Example 7.3

Assume that the retaining wall shown in Figure 7.7a can yield sufficiently to develop an active state. Determine the Rankine active force per unit length of the wall and the location of the resultant line of action.

Solution
If the cohesion, c', is zero, then

$$\sigma'_a = \sigma'_o K_a$$

For the top layer of soil, $\phi_1' = 30°$, so

$$K_{a(1)} = \tan^2\left(45 - \frac{\phi_1'}{2}\right) = \tan^2(45 - 15) = \frac{1}{3}$$

Similarly, for the bottom layer of soil, $\phi_2' = 36°$, and it follows that

$$K_{a(2)} = \tan^2\left(45 - \frac{36}{2}\right) = 0.26$$

The following table shows the calculation of σ_a' and u at various depths below the ground surface.

Depth, z (m)	σ_o' (lb/ft^2)	K_a	$\sigma_a' = K_a\sigma_o'$ (lb/ft^2)	u (lb/ft^2)
0	0	1/3	0	0
3.05$^-$	(16)(3.05) = 48.8	1/3	16.27	0
3.05$^+$	48.8	0.26	12.69	0
6.1	(16)(3.05) + (19 − 9.81)(3.05) = 76.83	0.26	19.98	(9.81)(3.05) = 29.92

The pressure distribution diagram is plotted in Figure 7.7b. The force per unit length is

$$P_a = \text{area 1} + \text{area 2} + \text{area 3} + \text{area 4}$$

$$= \tfrac{1}{2}(3.05)(16.27) + (12.69)(3.05) + \tfrac{1}{2}(19.98 - 12.69)(3.05) + \tfrac{1}{2}(29.92)(3.05)$$

$$= 24.81 + 38.70 + 11.12 + 45.63 = \mathbf{120.26\ kN/m}$$

(a)　　　　　(b)

Figure 7.7 Rankine active force behind a retaining wall

> The distance of the line of action of the resultant force from the bottom of the wall can be determined by taking the moments about the bottom of the wall (point O in Figure 7.7a) and is
>
> $$\bar{z} = \frac{(24.81)\left(3.05 + \dfrac{3.05}{3}\right) + (38.7)\left(\dfrac{3.05}{2}\right) + (11.12 + 45.63)\left(\dfrac{3.05}{3}\right)}{120.26} = \mathbf{1.81\ m} \ \blacksquare$$

7.4 A Generalized Case for Rankine Active Pressure

In Section 7.3, the relationship was developed for Rankine active pressure for a retaining wall with a vertical back and a horizontal backfill. That can be extended to general cases of frictionless walls with inclined backs and inclined backfills. Some of these cases will be discussed in this section.

Granular Backfill

Figure 7.8 shows a retaining wall whose back is inclined at an angle θ with the vertical. The granular backfill is inclined at an angle α with the horizontal.

For a Rankine active case, the lateral earth pressure (σ_a') at a depth z can be given as (Chu, 1991),

$$\sigma_a' = \frac{\gamma z \cos\alpha \sqrt{1 + \sin^2\phi' - 2\sin\phi'\cos\psi_a}}{\cos\alpha + \sqrt{\sin^2\phi' - \sin^2\alpha}} \tag{7.13}$$

where $\psi_a = \sin^{-1}\left(\dfrac{\sin\alpha}{\sin\phi'}\right) - \alpha + 2\theta.$ $\tag{7.14}$

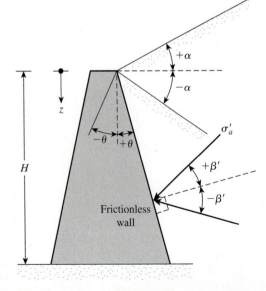

Figure 7.8 General case for a retaining wall with granular backfill

The pressure σ'_a will be inclined at an angle β' with the plane drawn at right angle to the backface of the wall, and

$$\beta' = \tan^{-1}\left(\frac{\sin\phi' \sin\psi_a}{1 - \sin\phi' \cos\psi_a}\right) \tag{7.15}$$

The active force P_a for unit length of the wall then can be calculated as

$$P_a = \frac{1}{2}\gamma H^2 K_a \tag{7.16}$$

where

$$K_a = \frac{\cos(\alpha - \theta)\sqrt{1 + \sin^2\phi' - 2\sin\phi' \cos\psi_a}}{\cos^2\theta\left(\cos\alpha + \sqrt{\sin^2\phi' - \sin^2\alpha}\right)}$$

= Rankine active earth-pressure coefficient for generalized case $\tag{7.17}$

The location and direction of the resultant force P_a is shown in Figure 7.9. Also shown in this figure is the failure wedge, *ABC*. Note that *BC* will be inclined at an angle η. Or

$$\eta = \frac{\pi}{4} + \frac{\phi'}{2} + \frac{\alpha}{2} - \frac{1}{2}\sin^{-1}\left(\frac{\sin\alpha}{\sin\phi'}\right) \tag{7.18}$$

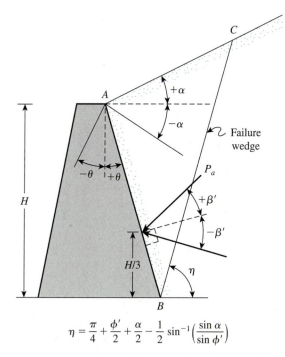

$$\eta = \frac{\pi}{4} + \frac{\phi'}{2} + \frac{\alpha}{2} - \frac{1}{2}\sin^{-1}\left(\frac{\sin\alpha}{\sin\phi'}\right)$$

Figure 7.9 Location and direction of Rankine active force

Granular Backfill with Vertical Back Face

As a special case, for a vertical backface of a wall (that is, $\theta = 0$), as shown in Figure 7.10, Eqs. (7.13), (7.16) and (7.17) simplify to the following.

If the backfill of a frictionless retaining wall is a *granular soil* ($c' = 0$) and rises at an angle α with respect to the horizontal (see Figure 7.10), the *active earth-pressure coefficient* may be expressed in the form

$$K_a = \cos \alpha \frac{\cos \alpha - \sqrt{\cos^2 \alpha - \cos^2 \phi'}}{\cos \alpha + \sqrt{\cos^2 \alpha - \cos^2 \phi'}} \qquad (7.19)$$

where ϕ' = angle of friction of soil.

At any depth z, the *Rankine active pressure* may be expressed as

$$\sigma'_a = \gamma z K_a \qquad (7.20)$$

Also, the total force per unit length of the wall is

$$P_a = \tfrac{1}{2} \gamma H^2 K_a \qquad (7.21)$$

Note that, in this case, the direction of the resultant force P_a is *inclined at an angle α with the horizontal* and intersects the wall at a distance $H/3$ from the base of the wall. Table 7.1 presents the values of K_a (active earth pressure) for various values of α and ϕ'.

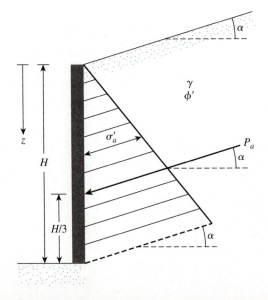

Figure 7.10 Notations for active pressure—Eqs. (7.19), (7.20), (7.21)

Table 7.1 Values of K_a [Eq. (7.19)]

α (deg)	28	29	30	31	32	33	34	35	36	37	38	39	40
0	0.3610	0.3470	0.3333	0.3201	0.3073	0.2948	0.2827	0.2710	0.2596	0.2486	0.2379	0.2275	0.2174
1	0.3612	0.3471	0.3335	0.3202	0.3074	0.2949	0.2828	0.2711	0.2597	0.2487	0.2380	0.2276	0.2175
2	0.3618	0.3476	0.3339	0.3207	0.3078	0.2953	0.2832	0.2714	0.2600	0.2489	0.2382	0.2278	0.2177
3	0.3627	0.3485	0.3347	0.3214	0.3084	0.2959	0.2837	0.2719	0.2605	0.2494	0.2386	0.2282	0.2181
4	0.3639	0.3496	0.3358	0.3224	0.3094	0.2967	0.2845	0.2726	0.2611	0.2500	0.2392	0.2287	0.2186
5	0.3656	0.3512	0.3372	0.3237	0.3105	0.2978	0.2855	0.2736	0.2620	0.2508	0.2399	0.2294	0.2192
6	0.3676	0.3531	0.3389	0.3253	0.3120	0.2992	0.2868	0.2747	0.2631	0.2518	0.2409	0.2303	0.2200
7	0.3701	0.3553	0.3410	0.3272	0.3138	0.3008	0.2883	0.2761	0.2644	0.2530	0.2420	0.2313	0.2209
8	0.3730	0.3580	0.3435	0.3294	0.3159	0.3027	0.2900	0.2778	0.2659	0.2544	0.2432	0.2325	0.2220
9	0.3764	0.3611	0.3463	0.3320	0.3182	0.3049	0.2921	0.2796	0.2676	0.2560	0.2447	0.2338	0.2233
10	0.3802	0.3646	0.3495	0.3350	0.3210	0.3074	0.2944	0.2818	0.2696	0.2578	0.2464	0.2354	0.2247
11	0.3846	0.3686	0.3532	0.3383	0.3241	0.3103	0.2970	0.2841	0.2718	0.2598	0.2482	0.2371	0.2263
12	0.3896	0.3731	0.3573	0.3421	0.3275	0.3134	0.2999	0.2868	0.2742	0.2621	0.2503	0.2390	0.2281
13	0.3952	0.3782	0.3620	0.3464	0.3314	0.3170	0.3031	0.2898	0.2770	0.2646	0.2527	0.2412	0.2301
14	0.4015	0.3839	0.3671	0.3511	0.3357	0.3209	0.3068	0.2931	0.2800	0.2674	0.2552	0.2435	0.2322
15	0.4086	0.3903	0.3729	0.3564	0.3405	0.3253	0.3108	0.2968	0.2834	0.2705	0.2581	0.2461	0.2346
16	0.4165	0.3975	0.3794	0.3622	0.3458	0.3302	0.3152	0.3008	0.2871	0.2739	0.2612	0.2490	0.2373
17	0.4255	0.4056	0.3867	0.3688	0.3518	0.3356	0.3201	0.3053	0.2911	0.2776	0.2646	0.2521	0.2401
18	0.4357	0.4146	0.3948	0.3761	0.3584	0.3415	0.3255	0.3102	0.2956	0.2817	0.2683	0.2555	0.2433
19	0.4473	0.4249	0.4039	0.3842	0.3657	0.3481	0.3315	0.3156	0.3006	0.2862	0.2724	0.2593	0.2467
20	0.4605	0.4365	0.4142	0.3934	0.3739	0.3555	0.3381	0.3216	0.3060	0.2911	0.2769	0.2634	0.2504
21	0.4758	0.4498	0.4259	0.4037	0.3830	0.3637	0.3455	0.3283	0.3120	0.2965	0.2818	0.2678	0.2545
22	0.4936	0.4651	0.4392	0.4154	0.3934	0.3729	0.3537	0.3356	0.3186	0.3025	0.2872	0.2727	0.2590
23	0.5147	0.4829	0.4545	0.4287	0.4050	0.3832	0.3628	0.3438	0.3259	0.3091	0.2932	0.2781	0.2638
24	0.5404	0.5041	0.4724	0.4440	0.4183	0.3948	0.3731	0.3529	0.3341	0.3164	0.2997	0.2840	0.2692
25	0.5727	0.5299	0.4936	0.4619	0.4336	0.4081	0.3847	0.3631	0.3431	0.3245	0.3070	0.2905	0.2750

ϕ' (deg) →

Vertical Backface with c′–ϕ′ Soil Backfill

For a retaining wall with a *vertical back* ($\theta = 0$) and *inclined backfill* of c'–ϕ' soil (Mazindrani and Ganjali, 1997),

$$\sigma_a' = \gamma z K_a = \gamma z K_a' \cos \alpha \qquad (7.22)$$

where

$$K_a' = \frac{1}{\cos^2 \phi'} \left\{ \begin{array}{l} 2\cos^2 \alpha + 2\left(\dfrac{c'}{\gamma z}\right)\cos \phi' \sin \phi' \\[2ex] -\sqrt{\left[4\cos^2\alpha(\cos^2 \alpha - \cos^2 \phi') + 4\left(\dfrac{c'}{\gamma z}\right)^2 \cos^2 \phi' + 8\left(\dfrac{c'}{\gamma z}\right)\cos^2 \alpha \sin \phi' \cos \phi' \right]} \end{array} \right\} - 1$$

$$(7.23)$$

Some values of K_a' are given in Table 7.2. For a problem of this type, the depth of tensile crack is given as

$$z_c = \frac{2c'}{\gamma}\sqrt{\frac{1 + \sin \phi'}{1 - \sin \phi'}} \qquad (7.24)$$

For this case, the active pressure is inclined at an angle α with the horizontal.

Table 7.2 Values of K_a'

ϕ' (deg)	α (deg)	$\dfrac{c'}{\gamma z}$			
		0.025	0.05	0.1	0.5
15	0	0.550	0.512	0.435	−0.179
	5	0.566	0.525	0.445	−0.184
	10	0.621	0.571	0.477	−0.186
	15	0.776	0.683	0.546	−0.196
20	0	0.455	0.420	0.350	−0.210
	5	0.465	0.429	0.357	−0.212
	10	0.497	0.456	0.377	−0.218
	15	0.567	0.514	0.417	−0.229
25	0	0.374	0.342	0.278	−0.231
	5	0.381	0.348	0.283	−0.233
	10	0.402	0.366	0.296	−0.239
	15	0.443	0.401	0.321	−0.250
30	0	0.305	0.276	0.218	−0.244
	5	0.309	0.280	0.221	−0.246
	10	0.323	0.292	0.230	−0.252
	15	0.350	0.315	0.246	−0.263

Example 7.4

Refer to the retaining wall in Figure 7.9. The backfill is granular soil. Given:

$$\text{Wall:} \qquad H = 10\,\text{ft}$$
$$\theta = +10°$$
$$\text{Backfill:} \qquad \alpha = 15°$$
$$\phi' = 35°$$
$$c' = 0$$
$$\gamma = 110\,\text{lb/ft}^3$$

Determine the Rankine active force, P_a, and its location and direction.

Solution

From Eq. (7.14),

$$\psi_a = \sin^{-1}\left(\frac{\sin \alpha}{\sin \phi'}\right) - \alpha + 2\theta = \sin^{-1}\left(\frac{\sin 15}{\sin 35}\right) - 15 + (2)(10) = 31.82°$$

From Eq. (7.17),

$$K_a = \frac{\cos(\alpha - \theta)\sqrt{1 + \sin^2 \phi' - 2\sin\phi' \cos\psi_a}}{\cos^2\theta\left(\cos\alpha + \sqrt{\sin^2 \phi' - \sin^2 \alpha}\right)}$$

$$= \frac{\cos(15 - 10)\sqrt{1 + \sin^2 35 - (2)(\sin 35)(\sin 31.82)}}{\cos^2 10\left(\cos 15 + \sqrt{\sin^2 35 - \sin^2 15}\right)} = 0.59$$

$$P_a = \tfrac{1}{2}\gamma H^2 K_a = (\tfrac{1}{2})(110)(10)^2(0.59) = \mathbf{3245\ lb/ft}$$

From Eq. (7.15),

$$\beta' = \tan^{-1}\left(\frac{\sin\phi' \sin\psi_a}{1 - \sin\phi' \cos\psi_a}\right) = \tan^{-1}\left[\frac{(\sin 35)(\sin 31.82)}{1 - (\sin 35)(\cos 31.82)}\right] = \mathbf{30.5°}$$

The force P_a will act at a distance of $10/3 = 3.33$ ft above the bottom of the wall and will be inclined at an angle of $+30.5°$ to the normal drawn to the back face of the wall. ∎

Example 7.5

For the retaining wall shown in Figure 7.10, $H = 7.5$ m, $\gamma = 18$ kN/m³, $\phi' = 20°$, $c' = 13.5$ kN/m², and $\alpha = 10°$. Calculate the Rankine active force, P_a, per unit length of the wall and the location of the resultant force after the occurrence of the tensile crack.

Solution

From Eq. (7.24).

$$z_r = \frac{2c'}{\gamma}\sqrt{\frac{1 + \sin\phi'}{1 - \sin\phi'}} = \frac{(2)(13.5)}{18}\sqrt{\frac{1 + \sin 20}{1 - \sin 20}} = 2.14\,\text{m}$$

At $z = 7.5$ m,

$$\frac{c'}{\gamma z} = \frac{13.5}{(18)(7.5)} = 0.1$$

From Table 7.2, for $\phi' = 20°$, $c'/\gamma z = 0.1$, and $\alpha = 10°$, the value of K'_a is 0.377, so at $z = 7.5$ m,

$$\sigma'_a = \gamma z K'_a \cos \alpha = (18)(7.5)(0.377)(\cos 10) = 50.1 \text{ kN/m}^2$$

After the occurrence of the tensile crack, the pressure distribution on the wall will be as shown in Figure 7.11, so

$$P_a = \left(\frac{1}{2}\right)(50.1)(7.5 - 2.14) = \textbf{134.3 kN/m}$$

and

$$\bar{z} = \frac{7.5 - 2.14}{3} = \textbf{1.79 m} \qquad \blacksquare$$

Figure 7.11 Calculation of Rankine active force, $c' - \phi'$ soil

7.5 Coulomb's Active Earth Pressure

The Rankine active earth pressure calculations discussed in the preceding sections were based on the assumption that the wall is frictionless. In 1776, Coulomb proposed a theory for calculating the lateral earth pressure on a retaining wall with granular soil backfill. This theory takes wall friction into consideration.

To apply Coulomb's active earth pressure theory, let us consider a retaining wall with its back face inclined at an angle β with the horizontal, as shown in Figure 7.12a. The backfill is a granular soil that slopes at an angle α with the horizontal.

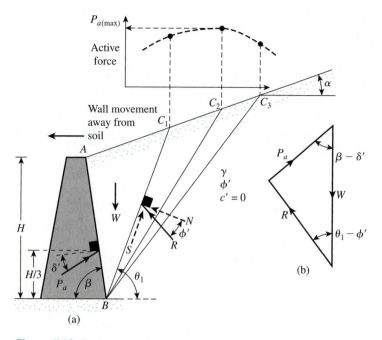

Figure 7.12 Coulomb's active pressure

Also, let δ' be the angle of friction between the soil and the wall (i.e., the angle of wall friction).

Under active pressure, the wall will move away from the soil mass (to the left in the figure). Coulomb assumed that, in such a case, the failure surface in the soil mass would be a plane (e.g., $BC_1, BC_2, \ldots$). So, to find the active force, consider a possible soil failure wedge ABC_1. The forces acting on this wedge (per unit length at right angles to the cross section shown) are as follows:

1. The weight of the wedge, W.
2. The resultant, R, of the normal and resisting shear forces along the surface, BC_1. The force R will be inclined at an angle ϕ' to the normal drawn to BC_1.
3. The active force per unit length of the wall, P_a, which will be inclined at an angle δ' to the normal drawn to the back face of the wall.

For equilibrium purposes, a force triangle can be drawn, as shown in Figure 7.12b. Note that θ_1 is the angle that BC_1 makes with the horizontal. Because the magnitude of W, as well as the directions of all three forces, are known, the value of P_a can now be determined. Similarly, the active forces of other trial wedges, such as $ABC_2, ABC_3, \ldots$, can be determined. The maximum value of P_a thus determined is Coulomb's active force (see top part of Figure 7.12), which may be expressed as

$$P_a = \tfrac{1}{2}K_a\gamma H^2 \qquad (7.25)$$

where

K_a = Coulomb's active earth pressure coefficient

$$= \frac{\sin^2 (\beta + \phi')}{\sin^2 \beta \sin (\beta - \delta') \left[1 + \sqrt{\dfrac{\sin (\phi' + \delta') \sin (\phi' - \alpha)}{\sin (\beta - \delta') \sin (\alpha + \beta)}} \right]^2} \qquad (7.26)$$

and H = height of the wall.

The values of the active earth pressure coefficient, K_a, for a vertical retaining wall ($\beta = 90°$) with horizontal backfill ($\alpha = 0°$) are given in Table 7.3. Note that the line of action of the resultant force (P_a) will act at a distance $H/3$ above the base of the wall and will be inclined at an angle δ' to the normal drawn to the back of the wall.

In the actual design of retaining walls, the value of the wall friction angle δ' is assumed to be between $\phi'/2$ and $\frac{2}{3}\phi'$. The active earth pressure coefficients for various values of ϕ', α, and β with $\delta' = \frac{1}{2}\phi'$ and $\frac{2}{3}\phi'$ are respectively given in Tables 7.4 and 7.5. These coefficients are very useful design considerations.

If a uniform surcharge of intensity q is located above the backfill, as shown in Figure 7.13, the active force, P_a, can be calculated as

$$P_a = \tfrac{1}{2}K_a\gamma_{eq}H^2 \qquad (7.27)$$
$$\uparrow$$
$$\text{Eq. (7.25)}$$

where

$$\gamma_{eq} = \gamma + \left[\frac{\sin\beta}{\sin (\beta + \alpha)} \right] \left(\frac{2q}{H} \right) \qquad (7.28)$$

Table 7.3 Values of K_a [Eq. (7.26)] for $\beta = 90°$ and $\alpha = 0°$

ϕ' (deg)	δ' (deg)					
	0	5	10	15	20	25
28	0.3610	0.3448	0.3330	0.3251	0.3203	0.3186
30	0.3333	0.3189	0.3085	0.3014	0.2973	0.2956
32	0.3073	0.2945	0.2853	0.2791	0.2755	0.2745
34	0.2827	0.2714	0.2633	0.2579	0.2549	0.2542
36	0.2596	0.2497	0.2426	0.2379	0.2354	0.2350
38	0.2379	0.2292	0.2230	0.2190	0.2169	0.2167
40	0.2174	0.2098	0.2045	0.2011	0.1994	0.1995
42	0.1982	0.1916	0.1870	0.1841	0.1828	0.1831

Table 7.4 Values of K_a [from Eq. (7.26)] for $\delta' = \frac{2}{3}\phi'$

α (deg)	φ' (deg)	β (deg)					
		90	85	80	75	70	65
0	28	0.3213	0.3588	0.4007	0.4481	0.5026	0.5662
	29	0.3091	0.3467	0.3886	0.4362	0.4908	0.5547
	30	0.2973	0.3349	0.3769	0.4245	0.4794	0.5435
	31	0.2860	0.3235	0.3655	0.4133	0.4682	0.5326
	32	0.2750	0.3125	0.3545	0.4023	0.4574	0.5220
	33	0.2645	0.3019	0.3439	0.3917	0.4469	0.5117
	34	0.2543	0.2916	0.3335	0.3813	0.4367	0.5017
	35	0.2444	0.2816	0.3235	0.3713	0.4267	0.4919
	36	0.2349	0.2719	0.3137	0.3615	0.4170	0.4824
	37	0.2257	0.2626	0.3042	0.3520	0.4075	0.4732
	38	0.2168	0.2535	0.2950	0.3427	0.3983	0.4641
	39	0.2082	0.2447	0.2861	0.3337	0.3894	0.4553
	40	0.1998	0.2361	0.2774	0.3249	0.3806	0.4468
	41	0.1918	0.2278	0.2689	0.3164	0.3721	0.4384
	42	0.1840	0.2197	0.2606	0.3080	0.3637	0.4302
5	28	0.3431	0.3845	0.4311	0.4843	0.5461	0.6190
	29	0.3295	0.3709	0.4175	0.4707	0.5325	0.6056
	30	0.3165	0.3578	0.4043	0.4575	0.5194	0.5926
	31	0.3039	0.3451	0.3916	0.4447	0.5067	0.5800
	32	0.2919	0.3329	0.3792	0.4324	0.4943	0.5677
	33	0.2803	0.3211	0.3673	0.4204	0.4823	0.5558
	34	0.2691	0.3097	0.3558	0.4088	0.4707	0.5443
	35	0.2583	0.2987	0.3446	0.3975	0.4594	0.5330
	36	0.2479	0.2881	0.3338	0.3866	0.4484	0.5221
	37	0.2379	0.2778	0.3233	0.3759	0.4377	0.5115
	38	0.2282	0.2679	0.3131	0.3656	0.4273	0.5012
	39	0.2188	0.2582	0.3033	0.3556	0.4172	0.4911
	40	0.2098	0.2489	0.2937	0.3458	0.4074	0.4813
	41	0.2011	0.2398	0.2844	0.3363	0.3978	0.4718
	42	0.1927	0.2311	0.2753	0.3271	0.3884	0.4625
10	28	0.3702	0.4164	0.4686	0.5287	0.5992	0.6834
	29	0.3548	0.4007	0.4528	0.5128	0.5831	0.6672
	30	0.3400	0.3857	0.4376	0.4974	0.5676	0.6516
	31	0.3259	0.3713	0.4230	0.4826	0.5526	0.6365
	32	0.3123	0.3575	0.4089	0.4683	0.5382	0.6219
	33	0.2993	0.3442	0.3953	0.4545	0.5242	0.6078
	34	0.2868	0.3314	0.3822	0.4412	0.5107	0.5942
	35	0.2748	0.3190	0.3696	0.4283	0.4976	0.5810
	36	0.2633	0.3072	0.3574	0.4158	0.4849	0.5682
	37	0.2522	0.2957	0.3456	0.4037	0.4726	0.5558
	38	0.2415	0.2846	0.3342	0.3920	0.4607	0.5437
	39	0.2313	0.2740	0.3231	0.3807	0.4491	0.5321
	40	0.2214	0.2636	0.3125	0.3697	0.4379	0.5207
	41	0.2119	0.2537	0.3021	0.3590	0.4270	0.5097
	42	0.2027	0.2441	0.2921	0.3487	0.4164	0.4990
15	28	0.4065	0.4585	0.5179	0.5868	0.6685	0.7670

(*continued*)

Table 7.4 (*Continued*)

α (deg)	φ′ (deg)	β (deg)					
		90	85	80	75	70	65
	29	0.3881	0.4397	0.4987	0.5672	0.6483	0.7463
	30	0.3707	0.4219	0.4804	0.5484	0.6291	0.7265
	31	0.3541	0.4049	0.4629	0.5305	0.6106	0.7076
	32	0.3384	0.3887	0.4462	0.5133	0.5930	0.6895
	33	0.3234	0.3732	0.4303	0.4969	0.5761	0.6721
	34	0.3091	0.3583	0.4150	0.4811	0.5598	0.6554
	35	0.2954	0.3442	0.4003	0.4659	0.5442	0.6393
	36	0.2823	0.3306	0.3862	0.4513	0.5291	0.6238
	37	0.2698	0.3175	0.3726	0.4373	0.5146	0.6089
	38	0.2578	0.3050	0.3595	0.4237	0.5006	0.5945
	39	0.2463	0.2929	0.3470	0.4106	0.4871	0.5805
	40	0.2353	0.2813	0.3348	0.3980	0.4740	0.5671
	41	0.2247	0.2702	0.3231	0.3858	0.4613	0.5541
	42	0.2146	0.2594	0.3118	0.3740	0.4491	0.5415
20	28	0.4602	0.5205	0.5900	0.6714	0.7689	0.8880
	29	0.4364	0.4958	0.5642	0.6445	0.7406	0.8581
	30	0.4142	0.4728	0.5403	0.6195	0.7144	0.8303
	31	0.3935	0.4513	0.5179	0.5961	0.6898	0.8043
	32	0.3742	0.4311	0.4968	0.5741	0.6666	0.7799
	33	0.3559	0.4121	0.4769	0.5532	0.6448	0.7569
	34	0.3388	0.3941	0.4581	0.5335	0.6241	0.7351
	35	0.3225	0.3771	0.4402	0.5148	0.6044	0.7144
	36	0.3071	0.3609	0.4233	0.4969	0.5856	0.6947
	37	0.2925	0.3455	0.4071	0.4799	0.5677	0.6759
	38	0.2787	0.3308	0.3916	0.4636	0.5506	0.6579
	39	0.2654	0.3168	0.3768	0.4480	0.5342	0.6407
	40	0.2529	0.3034	0.3626	0.4331	0.5185	0.6242
	41	0.2408	0.2906	0.3490	0.4187	0.5033	0.6083
	42	0.2294	0.2784	0.3360	0.4049	0.4888	0.5930

Table 7.5 Values of K_a [from Eq. (7.26)] for $\delta' = \phi'/2$

α (deg)	φ′ (deg)	β (deg)					
		90	85	80	75	70	65
0	28	0.3264	0.3629	0.4034	0.4490	0.5011	0.5616
	29	0.3137	0.3502	0.3907	0.4363	0.4886	0.5492
	30	0.3014	0.3379	0.3784	0.4241	0.4764	0.5371
	31	0.2896	0.3260	0.3665	0.4121	0.4645	0.5253
	32	0.2782	0.3145	0.3549	0.4005	0.4529	0.5137
	33	0.2671	0.3033	0.3436	0.3892	0.4415	0.5025
	34	0.2564	0.2925	0.3327	0.3782	0.4305	0.4915
	35	0.2461	0.2820	0.3221	0.3675	0.4197	0.4807
	36	0.2362	0.2718	0.3118	0.3571	0.4092	0.4702

Table 7.5 (*Continued*)

α (deg)	φ' (deg)	β (deg) 90	85	80	75	70	65
	37	0.2265	0.2620	0.3017	0.3469	0.3990	0.4599
	38	0.2172	0.2524	0.2920	0.3370	0.3890	0.4498
	39	0.2081	0.2431	0.2825	0.3273	0.3792	0.4400
	40	0.1994	0.2341	0.2732	0.3179	0.3696	0.4304
	41	0.1909	0.2253	0.2642	0.3087	0.3602	0.4209
	42	0.1828	0.2168	0.2554	0.2997	0.3511	0.4177
5	28	0.3477	0.3879	0.4327	0.4837	0.5425	0.6115
	29	0.3337	0.3737	0.4185	0.4694	0.5282	0.5972
	30	0.3202	0.3601	0.4048	0.4556	0.5144	0.5833
	31	0.3072	0.3470	0.3915	0.4422	0.5009	0.5698
	32	0.2946	0.3342	0.3787	0.4292	0.4878	0.5566
	33	0.2825	0.3219	0.3662	0.4166	0.4750	0.5437
	34	0.2709	0.3101	0.3541	0.4043	0.4626	0.5312
	35	0.2596	0.2986	0.3424	0.3924	0.4505	0.5190
	36	0.2488	0.2874	0.3310	0.3808	0.4387	0.5070
	37	0.2383	0.2767	0.3199	0.3695	0.4272	0.4954
	38	0.2282	0.2662	0.3092	0.3585	0.4160	0.4840
	39	0.2185	0.2561	0.2988	0.3478	0.4050	0.4729
	40	0.2090	0.2463	0.2887	0.3374	0.3944	0.4620
	41	0.1999	0.2368	0.2788	0.3273	0.3840	0.4514
	42	0.1911	0.2276	0.2693	0.3174	0.3738	0.4410
10	28	0.3743	0.4187	0.4688	0.5261	0.5928	0.6719
	29	0.3584	0.4026	0.4525	0.5096	0.5761	0.6549
	30	0.3432	0.3872	0.4368	0.4936	0.5599	0.6385
	31	0.3286	0.3723	0.4217	0.4782	0.5442	0.6225
	32	0.3145	0.3580	0.4071	0.4633	0.5290	0.6071
	33	0.3011	0.3442	0.3930	0.4489	0.5143	0.5920
	34	0.2881	0.3309	0.3793	0.4350	0.5000	0.5775
	35	0.2757	0.3181	0.3662	0.4215	0.4862	0.5633
	36	0.2637	0.3058	0.3534	0.4084	0.4727	0.5495
	37	0.2522	0.2938	0.3411	0.3957	0.4597	0.5361
	38	0.2412	0.2823	0.3292	0.3833	0.4470	0.5230
	39	0.2305	0.2712	0.3176	0.3714	0.4346	0.5103
	40	0.2202	0.2604	0.3064	0.3597	0.4226	0.4979
	41	0.2103	0.2500	0.2956	0.3484	0.4109	0.4858
	42	0.2007	0.2400	0.2850	0.3375	0.3995	0.4740
15	28	0.4095	0.4594	0.5159	0.5812	0.6579	0.7498
	29	0.3908	0.4402	0.4964	0.5611	0.6373	0.7284
	30	0.3730	0.4220	0.4777	0.5419	0.6175	0.7080
	31	0.3560	0.4046	0.4598	0.5235	0.5985	0.6884
	32	0.3398	0.3880	0.4427	0.5059	0.5803	0.6695
	33	0.3244	0.3721	0.4262	0.4889	0.5627	0.6513
	34	0.3097	0.3568	0.4105	0.4726	0.5458	0.6338
	35	0.2956	0.3422	0.3953	0.4569	0.5295	0.6168

(continued)

Table 7.5 (*Continued*)

α (deg)	ϕ' (deg)	β (deg)					
		90	85	80	75	70	65
	36	0.2821	0.3282	0.3807	0.4417	0.5138	0.6004
	37	0.2692	0.3147	0.3667	0.4271	0.4985	0.5846
	38	0.2569	0.3017	0.3531	0.4130	0.4838	0.5692
	39	0.2450	0.2893	0.3401	0.3993	0.4695	0.5543
	40	0.2336	0.2773	0.3275	0.3861	0.4557	0.5399
	41	0.2227	0.2657	0.3153	0.3733	0.4423	0.5258
	42	0.2122	0.2546	0.3035	0.3609	0.4293	0.5122
20	28	0.4614	0.5188	0.5844	0.6608	0.7514	0.8613
	29	0.4374	0.4940	0.5586	0.6339	0.7232	0.8313
	30	0.4150	0.4708	0.5345	0.6087	0.6968	0.8034
	31	0.3941	0.4491	0.5119	0.5851	0.6720	0.7772
	32	0.3744	0.4286	0.4906	0.5628	0.6486	0.7524
	33	0.3559	0.4093	0.4704	0.5417	0.6264	0.7289
	34	0.3384	0.3910	0.4513	0.5216	0.6052	0.7066
	35	0.3218	0.3736	0.4331	0.5025	0.5851	0.6853
	36	0.3061	0.3571	0.4157	0.4842	0.5658	0.6649
	37	0.2911	0.3413	0.3991	0.4668	0.5474	0.6453
	38	0.2769	0.3263	0.3833	0.4500	0.5297	0.6266
	39	0.2633	0.3120	0.3681	0.4340	0.5127	0.6085
	40	0.2504	0.2982	0.3535	0.4185	0.4963	0.5912
	41	0.2381	0.2851	0.3395	0.4037	0.4805	0.5744
	42	0.2263	0.2725	0.3261	0.3894	0.4653	0.5582

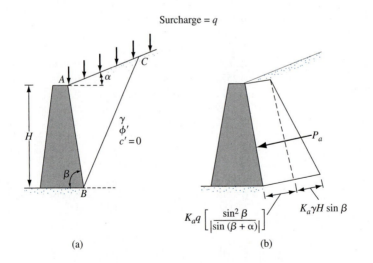

(a)

(b)

Figure 7.13 Coulomb's active pressure with a surcharge on the backfill

Example 7.6

Consider the retaining wall shown in Figure 7.12a. Given: $H = 4.6$ m; unit weight of soil $= 16.5$ kN/m³; angle of friction of soil $= 30°$; wall friction-angle, $\delta' = \frac{2}{3}\phi'$, soil cohesion, $c' = 0$; $\alpha = 0$, and $\beta = 90°$. Calculate the Coulomb's active force per unit length of the wall.

Solution

From Eq. (7.25)

$$P_a = \tfrac{1}{2}\gamma H^2 K_a$$

From Table 7.4, for $\alpha = 0°$, $\beta = 90°$, $\phi' = 30°$, and $\delta' = \frac{2}{3}\phi' = 20°$, $K_a = 0.297$. Hence,

$$P_a = \tfrac{1}{2}(16.5)(4.6)^2(0.297) = \textbf{51.85 kN/m} \qquad \blacksquare$$

Example 7.7

Refer to Figure 7.13a. Given: $H = 6.1$ m, $\phi' = 30°$, $\delta' = 20°$, $\alpha = 5°$, $\beta = 85°$, $q = 96$ kN/m², and $\gamma = 18$ kN/m³. Determine Coulomb's active force and the location of the line of action of the resultant P_a.

Solution

For $\beta = 85°$, $\alpha = 5°$, $\delta' = 20°$, $\phi' = 30°$, and $K_a = 0.3578$ (Table 7.4). From Eqs. (7.27) and (7.28),

$$P_a = \frac{1}{2}K_a\gamma_{eq}H^2 = \frac{1}{2}K_a\left[\gamma + \frac{2q}{H}\frac{\sin\beta}{\sin(\beta + \alpha)}\right]H^2 = \underbrace{\frac{1}{2}K_a\gamma H^2}_{P_{a(1)}}$$

$$+ \underbrace{K_a Hq\left[\frac{\sin\beta}{\sin(\beta + \alpha)}\right]}_{P_{a(2)}}$$

$$= (0.5)(0.3578)(18)(6.1)^2 + (0.3578)(6.1)(96)\left[\frac{\sin 85}{\sin(85 + 5)}\right]$$

$$= 119.8 + 208.7 = \textbf{328.5 kN/m}$$

Location of the line of action of the resultant:

$$P_a\bar{z} = P_{a(1)}\frac{H}{3} + P_{a(2)}\frac{H}{2}$$

or

$$\bar{z} = \frac{(119.8)\left(\dfrac{6.1}{3}\right) + (208.7)\left(\dfrac{6.1}{2}\right)}{328.5}$$

$$= \textbf{2.68 m} \text{ (measured vertically from the bottom of the wall)} \qquad \blacksquare$$

7.6 *Lateral Earth Pressure Due to Surcharge*

In several instances, the theory of elasticity is used to determine the lateral earth pressure on unyielding retaining structures caused by various types of surcharge loading, such as *line loading* (Figure 7.14a) and *strip loading* (Figure 7.14b).

According to the theory of elasticity, the stress at any depth, z, on a retaining structure caused by a line load of intensity q/unit length (Figure 7.14a) may be given as

$$\sigma = \frac{2q}{\pi H} \frac{a^2 b}{(a^2 + b^2)^2} \tag{7.29}$$

where σ = horizontal stress at depth $z = bH$

(See Figure 7.14a for explanations of the terms a and b.)

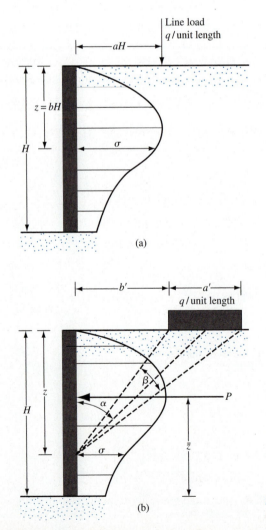

Figure 7.14 Lateral earth pressure caused by (a) line load and (b) strip load

However, because soil is not a perfectly elastic medium, some deviations from Eq. (7.29) may be expected. The modified forms of this equation generally accepted for use with soils are as follows:

$$\sigma = \frac{4a}{\pi H} \frac{a^2 b}{(a^2 + b^2)} \qquad \text{for } a > 0.4 \qquad (7.30)$$

and

$$\sigma = \frac{q}{H} \frac{0.203b}{(0.16 + b^2)^2} \qquad \text{for } a \leq 0.4 \qquad (7.31)$$

Figure 7.14b shows a strip load with an intensity of q/unit area located at a distance b' from a wall of height H. Based on the theory of elasticity, the horizontal stress, σ, at any depth z on a retaining structure is

$$\sigma = \frac{q}{\pi}(\beta - \sin \beta \cos 2\alpha) \qquad (7.32)$$

(The angles α and β are defined in Figure 7.14b.)

However, in the case of soils, the right-hand side of Eq. (7.32) is doubled to account for the yielding soil continuum, or

$$\sigma = \frac{2q}{\pi}(\beta - \sin\beta \cos 2\alpha) \qquad (7.33)$$

The total force per unit length (P) due to the *strip loading only* (Jarquio, 1981) may be expressed as

$$P = \frac{q}{90}[H(\theta_2 - \theta_1)] \qquad (7.34)$$

where

$$\theta_1 = \tan^{-}\left(\frac{b'}{H}\right) \quad (\text{deg}) \qquad (7.35)$$

$$\theta_2 = \tan^{-1}\left(\frac{a' + b'}{H}\right) (\text{deg}) \qquad (7.36)$$

The location $\overline{z}$ (see Figure 7.14b) of the resultant force, P, can be given as

$$\overline{z} = H - \left[\frac{H^2(\theta_2 - \theta_1) + (R - Q) - 57.3a'H}{2H(\theta_2 - \theta_1)}\right] \qquad (7.37)$$

where

$$R = (a' + b')^2(90 - \theta_2) \tag{7.38}$$

$$Q = b'^2(90 - \theta_1) \tag{7.39}$$

Example 7.8

Refer to Figure 7.14b. Here, $a' = 2$ m, $b' = 1$ m, $q = 40$ kN/m^2, and $H = 6$ m. Determine the total force on the wall (kN/m) caused by the strip loading only.

Solution From Eqs. (7.35) and (7.38),

$$\theta_1 = \tan^{-1}\left(\frac{1}{6}\right) = 9.46°$$

$$\theta_2 = \tan^{-1}\left(\frac{2+1}{6}\right) = 26.57°$$

From Eq. (7.34)

$$P = \frac{q}{90}[H(\theta_2 - \theta_1)] = \frac{40}{90}[6(26.57 - 9.46)] = \textbf{45.63 kN/m} \qquad \blacksquare$$

Example 7.9

Refer to Example 7.8. Determine the location of the resultant $\bar{z}$.

Solution
From Eqs. (7.38) and (7.39),

$$R = (a' + b')^2(90 - \theta_2) = (2+1)^2(90 - 26.57) = 570.87$$
$$Q = b'^2(90 - \theta_1) = (1)^2(90 - 9.46) = 80.54$$

From Eq. (7.37),

$$\bar{z} = H - \left[\frac{H^2(\theta_2 - \theta_1) + (R - Q) - 57.3a'H}{2H(\theta_2 - \theta_1)}\right]$$

$$= 6 - \left[\frac{(6)^2(26.57 - 9.46) + (570.87 - 80.54) - (57.3)(2)(6)}{(2)(6)(26.57 - 9.46)}\right] = \textbf{3.96 m} \qquad \blacksquare$$

7.7 *Active Earth Pressure for Earthquake Conditions*

Coulomb's active earth pressure theory (see Section 7.5) can be extended to take into account the forces caused by an earthquake. Figure 7.15 shows a condition of active pressure with a granular backfill ($c' = 0$). Note that the forces acting on the soil failure wedge in Figure 7.15

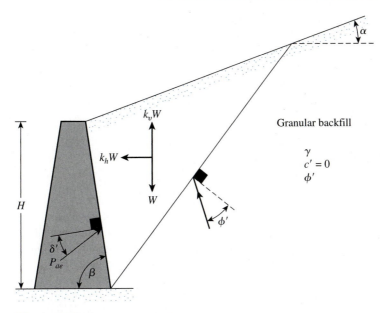

Figure 7.15 Derivation of Eq. (7.42)

are essentially the same as those shown in Figure 7.12a with the addition of k_hW and k_vW in the horizontal and vertical direction respectively; k_h and k_v may be defined as

$$k_h = \frac{\text{horizontal earthquake acceleration component}}{\text{acceleration due to gravity, } g} \tag{7.40}$$

$$k_v = \frac{\text{vertical earthquake acceleration component}}{\text{acceleration due to gravity, } g} \tag{7.41}$$

As in Section 7.5, the relation for the active force per unit length of the wall (P_{ae}) can be determined as

$$P_{ae} = \tfrac{1}{2}\gamma H^2(1 - k_v)K_{ae} \tag{7.42}$$

where

K_{ae} = active earth pressure coefficient

$$= \frac{\sin^2(\phi' + \beta - \theta')}{\cos\theta'\sin^2\beta\sin(\beta - \theta' - \delta')\left[1 + \sqrt{\dfrac{\sin(\phi' + \delta')\sin(\phi' - \theta' - \alpha)}{\sin(\beta - \delta' - \theta')\sin(\alpha + \beta)}}\right]^2}$$

$$\tag{7.43}$$

$$\theta' = \tan^{-1}\left[\frac{k_h}{(1 - k_v)}\right] \tag{7.44}$$

Note that for no earthquake condition

$$k_h = 0, \qquad k_v = 0, \qquad \text{and} \qquad \theta' = 0$$

Hence $K_{ae} = K_a$ [as given by Eq. (7.26)]. Some values of K_{ae} for $\beta = 90°$ and $k_v = 0$ are given in Table 7.6.

The magnitude of P_{ae} as given in Eq. (7.42) also can be determined as (Seed and Whitman, 1970),

$$P_{ae} = \frac{1}{2}\gamma H^2 (1 - k_v)[K_a(\beta',\alpha')]\left(\frac{\sin^2\beta'}{\cos\theta'\sin^2\beta}\right) \tag{7.45}$$

where

$$\beta' = \beta - \theta' \tag{7.46}$$
$$\alpha' = \theta' + \alpha \tag{7.47}$$

$K_a(\beta',\alpha')$ = Coulomb's active earth-pressure coefficient on a wall with a back face inclination of β' with the horizontal and with a back fill inclined at an angle α' with the horizontal (such as Tables 7.4 and 7.5)

Equation (7.42) is usually referred to as the *Mononobe–Okabe* solution. Unlike the case shown in Figure 7.12a, the resultant earth pressure in this situation, as calculated by Eq. (7.42) *does not act* at a distance of $H/3$ from the bottom of the wall. The following procedure may be used to obtain the location of the resultant force P_{ae}:

Step 1. Calculate P_{ae} by using Eq. (7.42)
Step 2. Calculate P_a by using Eq. (7.25)
Step 3. Calculate

$$\Delta P_{ae} = P_{ae} - P_a \tag{7.48}$$

Step 4. Assume that P_a acts at a distance of $H/3$ from the bottom of the wall (Figure 7.16)

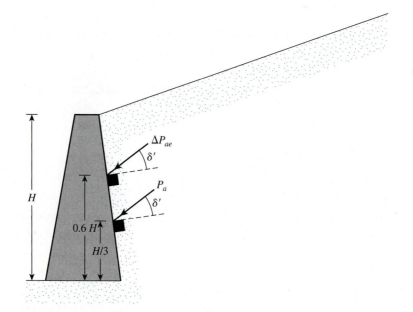

Figure 7.16 Determining the line of action of P_{ae}

Table 7.6 Values of K_{ae} [Eq. (7.43)] for $\beta = 90°$ and $k_v = 0$

k_h	δ' (deg)	α (deg)	ϕ' (deg) 28	30	35	40	45
0.1	0	0	0.427	0.397	0.328	0.268	0.217
0.2			0.508	0.473	0.396	0.382	0.270
0.3			0.611	0.569	0.478	0.400	0.334
0.4			0.753	0.697	0.581	0.488	0.409
0.5			1.005	0.890	0.716	0.596	0.500
0.1	0	5	0.457	0.423	0.347	0.282	0.227
0.2			0.554	0.514	0.424	0.349	0.285
0.3			0.690	0.635	0.522	0.431	0.356
0.4			0.942	0.825	0.653	0.535	0.442
0.5			—	—	0.855	0.673	0.551
0.1	0	10	0.497	0.457	0.371	0.299	0.238
0.2			0.623	0.570	0.461	0.375	0.303
0.3			0.856	0.748	0.585	0.472	0.383
0.4			—	—	0.780	0.604	0.486
0.5			—	—	—	0.809	0.624
0.1	$\phi'/2$	0	0.396	0.368	0.306	0.253	0.207
0.2			0.485	0.452	0.380	0.319	0.267
0.3			0.604	0.563	0.474	0.402	0.340
0.4			0.778	0.718	0.599	0.508	0.433
0.5			1.115	0.972	0.774	0.648	0.522
0.1	$\phi'/2$	5	0.428	0.396	0.326	0.268	0.218
0.2			0.537	0.497	0.412	0.342	0.283
0.3			0.699	0.640	0.526	0.438	0.367
0.4			1.025	0.881	0.690	0.568	0.475
0.5			—	—	0.962	0.752	0.620
0.1	$\phi'/2$	10	0.472	0.433	0.352	0.285	0.230
0.2			0.616	0.562	0.454	0.371	0.303
0.3			0.908	0.780	0.602	0.487	0.400
0.4			—	—	0.857	0.656	0.531
0.5			—	—	—	0.944	0.722
0.1	$\frac{2}{3}\phi'$	0	0.393	0.366	0.306	0.256	0.212
0.2			0.486	0.454	0.384	0.326	0.276
0.3			0.612	0.572	0.486	0.416	0.357
0.4			0.801	0.740	0.622	0.533	0.462
0.5			1.177	1.023	0.819	0.693	0.600
0.1	$\frac{2}{3}\phi'$	5	0.427	0.395	0.327	0.271	0.224
0.2			0.541	0.501	0.418	0.350	0.294
0.3			0.714	0.655	0.541	0.455	0.386
0.4			1.073	0.921	0.722	0.600	0.509
0.5			—	—	1.034	0.812	0.679
0.1	$\frac{2}{3}\phi'$	10	0.472	0.434	0.354	0.290	0.237
0.2			0.625	0.570	0.463	0.381	0.317
0.3			0.942	0.807	0.624	0.509	0.423
0.4			—	—	0.909	0.699	0.573
0.5			—	—	—	1.037	0.800

Step 5. Assume that ΔP_{ae} acts at a distance of $0.6H$ from the bottom of the wall (Figure 7.16)

Step 6. Calculate the location of the resultant as

$$\bar{z} = \frac{(0.6H)(\Delta P_{ae}) + \left(\dfrac{H}{3}\right)(P_a)}{P_{ae}} \tag{7.49}$$

Example 7.10

Refer to Figure 7.17. For $k_v = 0$ and $k_h = 0.3$, determine:

    a. P_{ae} using Eq. (7.45)
    b. The location of the resultant, $\bar{z}$, from the bottom of the wall

$\phi' = 35°$
$\gamma = 16.51 \ \text{kN/m}^3$
$\delta' = 17.5°$

3.05 m

Figure 7.17

Solution
Part a
    From Eq. (7.44),

$$\theta' = \tan^{-1}\left(\frac{k_h}{1 - k_v}\right) = \tan^{-1}\left(\frac{0.3}{1 - 0}\right) = 16.7°$$

From Eqs. (7.46) and (7.47),

$$\beta' = \beta - \theta' = 90 - 16.7 = 73.3°$$

$$\alpha' = \theta' + \alpha = 16.7 + 0 = 16.7°$$

$$\frac{\delta'}{\phi'} = \frac{17.5}{35} = 0.5$$

We will refer to Table 7.5. For $\phi' = 35°$, $\delta'/\phi' = 0.5$, $\beta' = 73.3°$, and $\alpha' = 16.7°$, the value of $K_a(\beta',\alpha') \approx 0.495$. Thus, from Eq. (7.45),

$$P_{ae} = \frac{1}{2}\gamma H^2(1 - k_v)[K_a(\beta',\alpha')]\left(\frac{\sin^2\beta'}{\cos\theta'\sin^2\beta}\right)$$

$$= \frac{1}{2}(16.51)(3.05)^2(1 - 0)(0.495)\left(\frac{\sin^2 73.3}{\cos 16.7\sin^2 90}\right) = \mathbf{36.4\ kN/m}$$

Part b
From Eq. (7.25),

$$P_a = \tfrac{1}{2}\gamma H^2 K_a$$

From Eq. (7.26) with $\delta' = 17.5°$, $\beta = 90°$, and $\alpha = 0°$, $K_a \approx 0.246$ (Table 7.5).

$$P_a = \tfrac{1}{2}(16.51)(3.05)^2(0.246) = 18.89\ kN/m$$

$$\Delta P_{ae} = P_{ae} - P_a = 36.32 - 18.89 = 17.43\ kN/m$$

From Eq. (7.49),

$$\bar{z} = \frac{(0.6H)(\Delta P_{ae}) + (H/3)(P_a)}{P_{ae}}$$

$$= \frac{[(0.6)(3.05)](17.43) + (3.05/3)(18.89)}{36.32} = \mathbf{1.41\ m} \quad \blacksquare$$

7.8 *Active Pressure for Wall Rotation about the Top: Braced Cut*

In the preceding sections, we have seen that a retaining wall rotates about its bottom. (See Figure 7.18a.) With sufficient yielding of the wall, the lateral earth pressure is approximately equal to that obtained by Rankine's theory or Coulomb's theory. In contrast to retaining walls, braced cuts show a different type of wall yielding. (See Figure 7.18b.) In this case, deformation of the wall gradually increases with the depth of excavation. The variation of the amount of deformation depends on several factors, such as the type of soil, the depth of excavation, and the workmanship involved. However, with very little wall yielding at the top of the cut, the lateral earth pressure will be close to the at-rest pressure. At the bottom of the wall, with a much larger degree of yielding, the lateral earth pressure will be substantially lower than the Rankine active earth pressure. As a result, the distribution of lateral earth pressure will vary substantially in comparison to the linear distribution assumed in the case of retaining walls.

The total lateral force per unit length of the wall, P_a, imposed on a wall may be evaluated theoretically by using Terzaghi's (1943) general wedge theory. (See Figure 7.19.) The failure surface is assumed to be the arc of a logarithmic spiral, defined as

$$r = r_o e^{\theta\tan\phi'} \tag{7.50}$$

where ϕ' = effective angle of friction of soil.

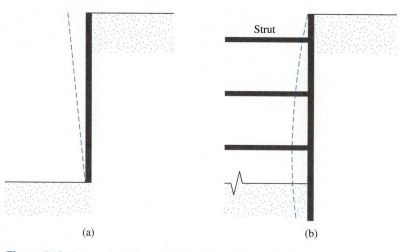

Figure 7.18 Nature of yielding of walls: (a) retaining wall; (b) braced cut

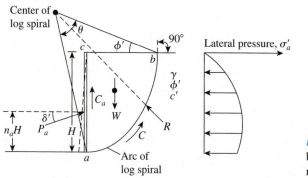

Figure 7.19 Braced cut analysis by general wedge theory: wall rotation about top

In the figure, H is the height of the cut, and the unit weight, angle of friction, and cohesion of the soil are equal to γ, ϕ', and c', respectively. Following are the forces per unit length of the cut acting on the trial failure wedge:

1. Weight of the wedge, W
2. Resultant of the normal and shear forces along ab, R
3. Cohesive force along ab, C
4. Adhesive force along ac, C_a
5. P_a, which is the force acting a distance $n_a H$ from the bottom of the wall and is inclined at an angle δ' to the horizontal

The adhesive force is

$$C_a = c_a' H \tag{7.51}$$

where c_a' = unit adhesion.

A detailed outline for the evaluation of P_a is beyond the scope of this text; those interested should check a soil mechanics text for more information. Kim and Preber (1969) provided tabulated values of $P_a / \frac{1}{2}\gamma H^2$ determined by using the principles of general wedge theory. Table 7.7 gives the variation of $P_a / 0.5\gamma H^2$ for granular soil backfill obtained using the general wedge theory.

Table 7.7 Active Pressure for Wall Rotation—General Wedge Theory (Granular Soil Backfill)

Soil friction angle, ϕ'(deg)	δ'/ϕ'	$P_a/0.5\,\gamma H^2$			
		$n_a = 0.3$	$n_a = 0.4$	$n_a = 0.5$	$n_a = 0.6$
25	0	0.371	0.405	0.447	0.499
	½	0.345	0.376	0.413	0.460
	⅔	0.342	0.373	0.410	0.457
	1	0.344	0.375	0.413	0.461
30	0	0.304	0.330	0.361	0.400
	½	0.282	0.306	0.334	0.386
	⅔	0.281	0.305	0.332	0.367
	1	0.289	0.313	0.341	0.377
35	0	0.247	0.267	0.290	0.318
	½	0.231	0.249	0.269	0.295
	⅔	0.232	0.249	0.270	0.296
	1	0.243	0.262	0.289	0.312
40	0	0.198	0.213	0.230	0.252
	½	0.187	0.200	0.216	0.235
	⅔	0.190	0.204	0.220	0.239
	1	0.197	0.211	0.228	0.248
45	0	0.205	0.220	0.237	0.259
	½	0.149	0.159	0.171	0.185
	⅔	0.153	0.164	0.176	0.196
	1	0.173	0.184	0.198	0.215

7.9 *Active Earth Pressure for Translation of Retaining Wall—Granular Backfill*

Under certain circumstances, retaining walls may undergo lateral translation, as shown in Figure 7.20. A solution to the distribution of active pressure for this case was provided by Dubrova (1963) and was also described by Harr (1966). The solution of Dubrova assumes the validity of Coulomb's solution [Eqs. (7.25) and (7.26)]. In order to understand this procedure, let us consider a vertical wall with a horizontal granular backfill (Figure 7.21). For rotation about the top of the wall, the resultant R of the normal and shear forces along the rupture line AC is inclined at an angle ϕ' to the normal drawn to AC. According to Dubrova there exists infinite number of quasi-rupture lines such as $A'C'$, $A''C''$, ... for which the resultant force R is inclined at an angle ψ, where

$$\psi = \frac{\phi' z}{H} \tag{7.52}$$

Now, refer to Eqs. (7.25) and (7.26) for Coulomb's active pressure. For $\beta = 90°$ and $\alpha = 0$, the relationship for Coulomb's active force can also be rewritten as

$$P_a = \frac{\gamma}{2\cos\delta'} \left[\frac{H}{\dfrac{1}{\cos\phi'} + (\tan^2\phi' + \tan\phi'\tan\delta')^{0.5}} \right]^2 \tag{7.53}$$

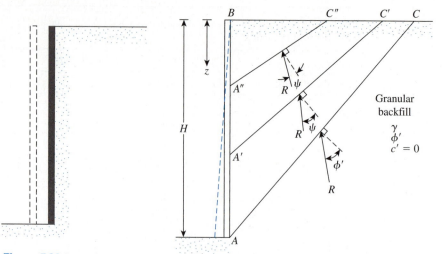

Figure 7.20 Lateral translation of retaining wall

Figure 7.21 Quasi-rupture lines behind a retaining wall

The force against the wall at any z is then given as

$$P_a = \frac{\gamma}{2 \cos \delta'} \left[\frac{z}{\dfrac{1}{\cos \psi} + (\tan^2 \psi + \tan \psi \tan \delta')^{0.5}} \right]^2 \tag{7.54}$$

The active pressure at any depth z for *wall rotation about the top* is

$$\sigma'_a(z) = \frac{dP_a}{dz} \approx \frac{\gamma}{\cos \delta'} \left[\frac{z \cos^2 \psi}{(1 + m \sin \psi)^2} - \frac{z^2 \phi' \cos^2 \psi}{H(1 + m \sin \psi)} (\sin \psi + m) \right] \tag{7.55}$$

where $m = \left(1 + \dfrac{\tan \delta'}{\tan \psi} \right)^{0.5}$. \hfill (7.56)

For frictionless walls, $\delta' = 0$ and Eq. (7.55) simplifies to

$$\sigma'_a(z) = \gamma \tan^2 \left(45 - \frac{\psi}{2} \right) \left(z - \frac{\phi' z^2}{H \cos \psi} \right) \tag{7.57}$$

For wall rotation about the bottom, a similar expression can be found in the form

$$\sigma'_a(z) = \frac{\gamma z}{\cos \delta'} \left(\frac{\cos \phi'}{1 + m \sin \phi'} \right)^2 \tag{7.58}$$

For translation of the wall, the active pressure can then be taken as

$$\sigma'_a(z)_{\text{translation}} = \tfrac{1}{2} [\sigma'_a(z)_{\text{rotation about top}} + \sigma'_a(z)_{\text{rotation about bottom}}] \tag{7.59}$$

Example 7.11

Consider a frictionless wall 5 m high. For the granular backfill, $\gamma = 17.3$ kN/m³ and $\phi' = 36°$. Calculate and plot the variation of $\sigma_a(z)$ for a translation mode of the wall movement.

Solution

For a frictionless wall, $\delta' = 0$. Hence, m is equal to one [Eq. (7.56)]. So for rotation about the top, from Eq. (7.57),

$$\sigma_a'(z) = \sigma_{a(1)}' = \gamma \tan^2\left(45 - \frac{\phi' z}{2H}\right)\left[z - \frac{\phi' z^2}{H \cos\left(\dfrac{\phi' z}{H}\right)} \right]$$

For rotation about the bottom, from Eq. (7.58),

$$\sigma_a'(z) = \sigma_{a(2)}' = \gamma z \left(\frac{\cos \phi'}{1 + \sin \phi'}\right)^2$$

$$\sigma_a'(z)_{\text{translation}} = \frac{\sigma_{a(1)}' + \sigma_{a(2)}'}{2}$$

The following table can now be prepared with $\gamma = 17.3$ kN/m³, $\phi' = 36°$, and $H = 5$ m.

z (m)	$\sigma_{a(1)}'$ (kN/m²)	$\sigma_{a(2)}'$ (kN/m²)	$\sigma_{a(z) \text{ translation}}'$ (kN/m²)
0	0	0	0
1.25	13.26	5.62	9.44
2.5	15.26	11.24	13.25
3.75	11.48	16.86	14.17
5.0	5.02	22.48	13.75

The plot of $\sigma_a(z)$ versus z is shown in Figure 7.22

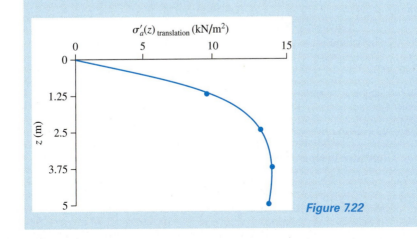

Figure 7.22

Passive Pressure

7.10   Rankine Passive Earth Pressure

Figure 7.23a shows a vertical frictionless retaining wall with a horizontal backfill. At depth z, the effective vertical pressure on a soil element is $\sigma'_o = \gamma z$. Initially, if the wall does not yield at all, the lateral stress at that depth will be $\sigma_h' = K_o \sigma_o'$. This state of stress is illustrated by the Mohr's circle a in Figure 7.23b. Now, if the wall is pushed into the soil mass by an amount Δx, as shown in Figure 7.23a, the vertical stress at depth z will stay the same; however, the horizontal stress will increase. Thus, σ'_h will be greater than $K_o \sigma_o'$. The state of stress can now be represented by the Mohr's circle b in Figure 7.23b. If the wall moves farther inward (i.e., Δx is increased still more), the stresses at depth z will ultimately reach the state represented by Mohr's circle c. Note that this Mohr's circle touches the Mohr–Coulomb failure envelope, which implies that the soil behind the wall will fail by being pushed upward. The horizontal stress, σ'_h, at this point is referred to as the *Rankine passive pressure,* or $\sigma'_h = \sigma'_p$.

For Mohr's circle c in Figure 7.23b, the major principal stress is σ'_p, and the minor principal stress is σ'_o. Substituting these quantities into Eq. (1.87) yields

$$\sigma'_p = \sigma'_o \tan^2\!\left(45 + \frac{\phi'}{2} \right) + 2c'\tan\!\left(45 + \frac{\phi'}{2} \right) \tag{7.60}$$

Now, let

$$K_p = \text{Rankine passive earth pressure coefficient}$$

$$= \tan^2\!\left(45 + \frac{\phi'}{2} \right) \tag{7.61}$$

Then, from Eq. (7.60), we have

$$\sigma'_p = \sigma'_o K_p + 2c'\sqrt{K_p} \tag{7.62}$$

Equation (7.62) produces (Figure 7.23c), the passive pressure diagram for the wall shown in Figure 7.23a. Note that at $z = 0$,

$$\sigma'_o = 0 \quad \text{and} \quad \sigma'_p = 2c'\sqrt{K_p}$$

and at $z = H$,

$$\sigma'_o = \gamma H \quad \text{and} \quad \sigma'_p = \gamma H K_p + 2c'\sqrt{K_p}$$

The passive force per unit length of the wall can be determined from the area of the pressure diagram, or

$$P_p = \tfrac{1}{2}\gamma H^2 K_p + 2c'H\sqrt{K_p} \tag{7.63}$$

The approximate magnitudes of the wall movements, Δx, required to develop failure under passive conditions are as follows:

Soil type	Wall movement for passive condition, Δx
Dense sand	$0.005H$
Loose sand	$0.01H$
Stiff clay	$0.01H$
Soft clay	$0.05H$

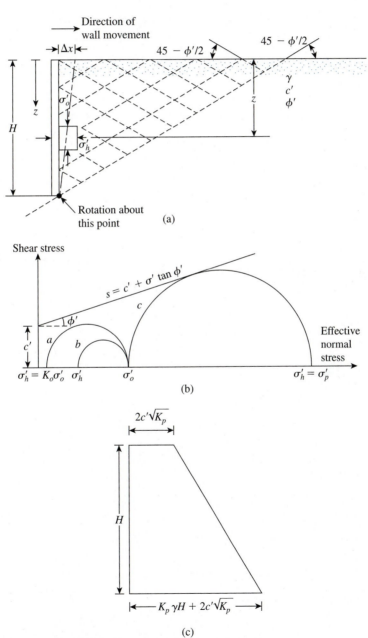

Figure 7.23 Rankine passive pressure

If the backfill behind the wall is a granular soil (i.e., $c' = 0$), then, from Eq. (7.63), the passive force per unit length of the wall will be

$$P_p = \frac{1}{2}\gamma H^2 K_p \qquad (7.64)$$

Example 7.12

A 3-m high wall is shown in Figure 7.24a. Determine the Rankine passive force per unit length of the wall.

Solution

For the top layer

$$K_{p(1)} = \tan^2\left(45 + \frac{\phi_1'}{2}\right) = \tan^2(45 + 15) = 3$$

Figure 7.24

From the bottom soil layer

$$K_{p(2)} = \tan^2\left(45 + \frac{\phi_2'}{2}\right) = \tan^2(45 + 13) = 2.56$$

$$\sigma_p' = \sigma_o' K_p + 2c'\sqrt{K_p}$$

where

σ_o' = effective vertical stress

at $z = 0$, $\sigma_o' = 0$, $c_1' = 0$, $\sigma_p' = 0$

at $z = 2$ m, $\sigma_o' = (15.72)(2) = 31.44$ kN/m^2, $c_1' = 0$

So, for the top soil layer

$$\sigma_p' = 31.44 K_{p(1)} + 2(0)\sqrt{K_{p(1)}} = 31.44(3) = 94.32 \text{ kN/m}^2$$

At this depth, that is $z = 2$ m, for the bottom soil layer

$$\sigma'_p = \sigma'_o K_{p(2)} + 2c'_2\sqrt{K_{p(2)}} = 31.44(2.56) + 2(10)\sqrt{2.56}$$

$$= 80.49 + 32 = 112.49 \text{ kN/m}^2$$

Again, at $z = 3$ m,

$$\sigma'_o = (15.72)(2) + (\gamma_{sat} - \gamma_w)(1)$$

$$= 31.44 + (18.86 - 9.81)(1) = 40.49 \text{ kN/m}^2$$

Hence,

$$\sigma'_p = \sigma'_o K_{p(2)} + 2c'_2\sqrt{K_{p(2)}} = 40.49(2.56) + (2)(10)(1.6)$$

$$= \mathbf{135.65 \text{ kN/m}^2}$$

Note that, because a water table is present, the hydrostatic stress, u, also has to be taken into consideration. For $z = 0$ to 2 m, $u = 0$; $z = 3$ m, $u = (1)(\gamma_w) = 9.81 \text{ kN/m}^2$.

The passive pressure diagram is plotted in Figure 6.24b. The passive force per unit length of the wall can be determined from the area of the pressure diagram as follows:

Area no.	Area	
1	$(\frac{1}{2})(2)(94.32)$	$= 94.32$
2	$(112.49)(1)$	$= 112.49$
3	$(\frac{1}{2})(1)(135.65 - 112.49)$	$= 11.58$
4	$(\frac{1}{2})(9.81)(1)$	$= 4.905$
		$P_P \approx 223.3 \text{ kN/m}$

7.11 Rankine Passive Earth Pressure: Vertical Backface and Inclined Backfill

Granular Soil

For a frictionless vertical retaining wall (Figure 7.10) with a *granular backfill* $(c' = 0)$, the Rankine passive pressure at any depth can be determined in a manner similar to that done in the case of active pressure in Section 7.4. The pressure is

$$\sigma'_p = \gamma z K_p \tag{7.65}$$

and the passive force is

$$P_p = \tfrac{1}{2}\gamma H^2 K_p \tag{7.66}$$

where

$$K_p = \cos \alpha \frac{\cos \alpha + \sqrt{\cos^2 \alpha - \cos^2 \phi'}}{\cos \alpha - \sqrt{\cos^2 \alpha - \cos^2 \phi'}} \tag{7.67}$$

Table 7.8 Passive Earth Pressure Coefficient K_p [from Eq. (7.67)]

		ϕ' (deg)→					
↓α (deg)	28	30	32	34	36	38	40
0	2.770	3.000	3.255	3.537	3.852	4.204	4.599
5	2.715	2.943	3.196	3.476	3.788	4.136	4.527
10	2.551	2.775	3.022	3.295	3.598	3.937	4.316
15	2.284	2.502	2.740	3.003	3.293	3.615	3.977
20	1.918	2.132	2.362	2.612	2.886	3.189	3.526
25	1.434	1.664	1.894	2.135	2.394	2.676	2.987

As in the case of the active force, the resultant force, P_p, is inclined at an angle α with the horizontal and intersects the wall at a distance $H/3$ from the bottom of the wall. The values of K_p (the passive earth pressure coefficient) for various values of α and ϕ' are given in Table 7.8.

$c'-\phi'$ Soil

If the backfill of the frictionless vertical retaining wall is a $c'-\phi'$ soil (see Figure 7.10), then (Mazindrani and Ganjali, 1997)

$$\sigma_a' = \gamma z K_p = \gamma z K_p' \cos \alpha \tag{7.68}$$

where

$$K_p' = \frac{1}{\cos^2 \phi'} \left\{ \frac{2\cos^2 \alpha + 2\left(\dfrac{c'}{\gamma z}\right)\cos \phi' \sin \phi'}{+ \sqrt{4\cos^2 \alpha (\cos^2 \alpha - \cos^2 \phi') + 4\left(\dfrac{c'}{\gamma z}\right)^2 \cos^2 \phi' + 8\left(\dfrac{c'}{\gamma z}\right)\cos^2\alpha \sin \phi' \cos \phi'}} \right\} - 1 \tag{7.69}$$

The variation of K_p' with ϕ', α, and $c'/\gamma z$ is given in Table 7.9 (Mazindrani and Ganjali, 1997).

Table 7.9 Values of K_p'

ϕ' (deg)	α (deg)	$c'/\gamma z$			
		0.025	0.050	0.100	0.500
15	0	1.764	1.829	1.959	3.002
	5	1.716	1.783	1.917	2.971
	10	1.564	1.641	1.788	2.880
	15	1.251	1.370	1.561	2.732
20	0	2.111	2.182	2.325	3.468
	5	2.067	2.140	2.285	3.435
	10	1.932	2.010	2.162	3.339
	15	1.696	1.786	1.956	3.183
25	0	2.542	2.621	2.778	4.034
	5	2.499	2.578	2.737	3.999
	10	2.368	2.450	2.614	3.895
	15	2.147	2.236	2.409	3.726

Table 7.9 (*Continued*)

ϕ' (deg)	α (deg)	$c'/\gamma z$			
		0.025	**0.050**	**0.100**	**0.500**
30	0	3.087	3.173	3.346	4.732
	5	3.042	3.129	3.303	4.674
	10	2.907	2.996	3.174	4.579
	15	2.684	2.777	2.961	4.394

7.12 **Coulomb's Passive Earth Pressure**

Coulomb (1776) also presented an analysis for determining the passive earth pressure (i.e., when the wall moves *into* the soil mass) for walls possessing friction (δ' = angle of wall friction) and retaining a granular backfill material similar to that discussed in Section 7.5.

To understand the determination of Coulomb's passive force, P_p, consider the wall shown in Figure 7.25a. As in the case of active pressure, Coulomb assumed that the potential failure surface in soil is a plane. For a trial failure wedge of soil, such as ABC_1, the forces per unit length of the wall acting on the wedge are

1. The weight of the wedge, W
2. The resultant, R, of the normal and shear forces on the plane BC_1, and
3. The passive force, P_p

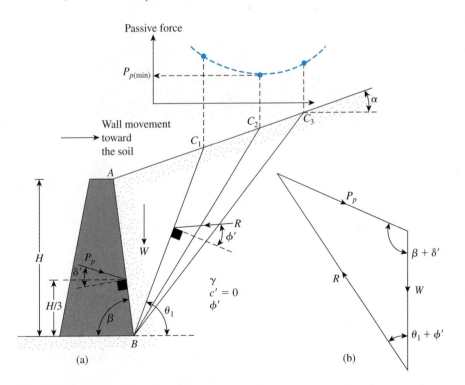

Figure 7.25 Coulomb's passive pressure

Table 7.10 Values of K_p [from Eq. (7.71)] for $\beta = 90°$ and $\alpha = 0°$

ϕ' (deg)	δ' (deg)				
	0	5	10	15	20
15	1.698	1.900	2.130	2.405	2.735
20	2.040	2.313	2.636	3.030	3.525
25	2.464	2.830	3.286	3.855	4.597
30	3.000	3.506	4.143	4.977	6.105
35	3.690	4.390	5.310	6.854	8.324
40	4.600	5.590	6.946	8.870	11.772

Figure 7.25b shows the force triangle at equilibrium for the trial wedge ABC_1. From this force triangle, the value of P_p can be determined, because the direction of all three forces and the magnitude of one force are known.

Similar force triangles for several trial wedges, such as ABC_1, ABC_2, ABC_3, . . . , can be constructed, and the corresponding values of P_p can be determined. The top part of Figure 7.25a shows the nature of variation of the P_p values for different wedges. The *minimum value of P_p* in this diagram is *Coulomb's passive force,* mathematically expressed as

$$P_p = \tfrac{1}{2}\gamma H^2 K_p \tag{7.70}$$

where

$$K_p = \text{Coulomb's passive pressure coefficient}$$

$$= \frac{\sin^2(\beta - \phi')}{\sin^2\beta \sin(\beta + \delta')\left[1 - \sqrt{\dfrac{\sin(\phi' + \delta')\sin(\phi' + \alpha)}{\sin(\beta + \delta')\sin(\beta + \alpha)}}\right]^2} \tag{7.71}$$

The values of the passive pressure coefficient, K_p, for various values of ϕ' and δ' are given in Table 7.10 ($\beta = 90°, \alpha = 0°$).

Note that the resultant passive force, P_p, will act at a distance $H/3$ from the bottom of the wall and will be inclined at an angle δ' to the normal drawn to the back face of the wall.

7.13   *Comments on the Failure Surface Assumption for Coulomb's Pressure Calculations*

Coulomb's pressure calculation methods for active and passive pressure have been discussed in Sections 7.5 and 7.12. The fundamental assumption in these analyses is the acceptance of *plane failure surface.* However, for walls with friction, this assumption does not hold in practice. The nature of *actual* failure surface in the soil mass for active and passive pressure is shown in Figure 7.26a and b, respectively (for a vertical wall with a horizontal backfill). Note that the failure surface BC is curved and that the failure surface CD is a plane.

Although the actual failure surface in soil for the case of active pressure is somewhat different from that assumed in the calculation of the Coulomb pressure, the results are not greatly different. However, in the case of passive pressure, as the value of δ' increases, Coulomb's

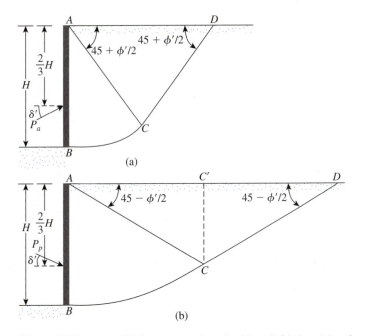

Figure 7.26 Nature of failure surface in soil with wall friction: (a) active pressure; (b) passive pressure

method of calculation gives increasingly erroneous values of P_p. This factor of error could lead to an unsafe condition because the values of P_p would become higher than the soil resistance.

Several studies have been conducted to determine the passive force P_p, assuming that the curved portion BC in Figure 7.26b is an arc of a circle, an ellipse, or a logarithmic spiral.

Shields and Tolunay (1973) analyzed the problem of passive pressure for a *vertical* wall with a *horizontal granular soil* backfill ($c' = 0$). This analysis was done by considering the stability of the wedge $ABCC'$ (see Figure 7.26b), using the *method of slices* and assuming BC as an arc of a logarithmic spiral. From Figure 7.26b, the passive force per unit length of the wall can be expressed as

$$P_p = \frac{1}{2}K_p\gamma H^2 \tag{7.72}$$

The values of the passive earth-pressure coefficient, K_p, obtained by Shields and Tolunay are given in Figure 7.27. These are as good as any for design purposes.

Solution by the Method of Triangular Slices

Zhu and Qian (2000) used the method of triangular slices (such as in the zone of ABC in Fig. 7.28) to obtain the variation of K_p. According to this analysis

$$K_p = K_{p(\delta' = 0)}R \tag{7.73}$$

where

K_p = passive earth-pressure coefficient for a given value of θ, δ', and ϕ'
$K_{p(\delta' = 0)}$ = K_p for a given value of θ, ϕ' with $\delta' = 0$
R = modification factor which is a function of ϕ', θ, δ'/ϕ'

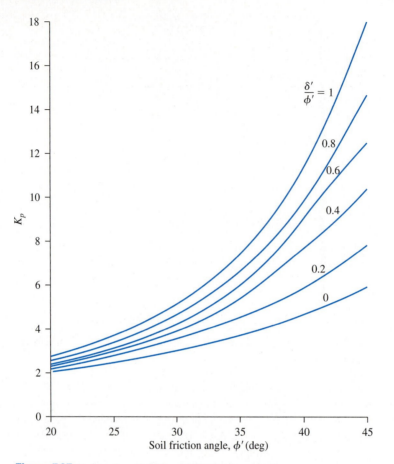

Figure 7.27 K_p based on Shields and Tolunay's analysis

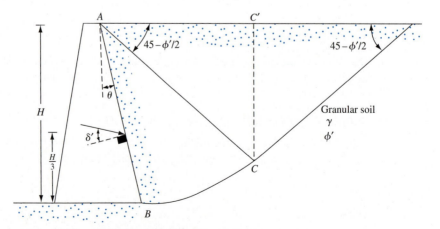

Figure 7.28 Passive pressure solution by the method of triangular slices
(*Note*: *BC* is arc of a logarthimic spiral)

The variations of $K_{p(\delta' = 0)}$ are given in Table 7.11. The interpolated values of R are given in Table 7.12

Table 7.11 Variation of $K_{p(\delta' = 0)}$ [see Eq. (7.73) and Figure 7.28][*]

	θ (deg)						
ϕ' (deg)	30	25	20	15	10	5	0
20	1.70	1.69	1.72	1.77	1.83	1.92	2.04
21	1.74	1.73	1.76	1.81	1.89	1.99	2.12
22	1.77	1.77	1.80	1.87	1.95	2.06	2.20
23	1.81	1.81	1.85	1.92	2.01	2.13	2.28
24	1.84	1.85	1.90	1.97	2.07	2.21	2.37
25	1.88	1.89	1.95	2.03	2.14	2.28	2.46
26	1.91	1.93	1.99	2.09	2.21	2.36	2.56
27	1.95	1.98	2.05	2.15	2.28	2.45	2.66
28	1.99	2.02	2.10	2.21	2.35	2.54	2.77
29	2.03	2.07	2.15	2.27	2.43	2.63	2.88
30	2.07	2.11	2.21	2.34	2.51	2.73	3.00
31	2.11	2.16	2.27	2.41	2.60	2.83	3.12
32	2.15	2.21	2.33	2.48	2.68	2.93	3.25
33	2.20	2.26	2.39	2.56	2.77	3.04	3.39
34	2.24	2.32	2.45	2.64	2.87	3.16	3.53
35	2.29	2.37	2.52	2.72	2.97	3.28	3.68
36	2.33	2.43	2.59	2.80	3.07	3.41	3.84
37	2.38	2.49	2.66	2.89	3.18	3.55	4.01
38	2.43	2.55	2.73	2.98	3.29	3.69	4.19
39	2.48	2.61	2.81	3.07	3.41	3.84	4.38
40	2.53	2.67	2.89	3.17	3.53	4.00	4.59
41	2.59	2.74	2.97	3.27	3.66	4.16	4.80
42	2.64	2.80	3.05	3.38	3.80	4.34	5.03
43	2.70	2.88	3.14	3.49	3.94	4.52	5.27
44	2.76	2.94	3.23	3.61	4.09	4.72	5.53
45	2.82	3.02	3.32	3.73	4.25	4.92	5.80

[*]Based on Zhu and Qian, 2000

Table 7.12 Variation of R [Eq. (7.73)]

θ (deg)	δ'/ϕ'	R for ϕ' (deg)			
		30	35	40	45
0	0.2	1.2	1.28	1.35	1.45
	0.4	1.4	1.6	1.8	2.2
	0.6	1.65	1.95	2.4	3.2
	0.8	1.95	2.4	3.15	4.45
	1.0	2.2	2.85	3.95	6.1
5	0.2	1.2	1.25	1.32	1.4
	0.4	1.4	1.6	1.8	2.1
	0.6	1.6	1.9	2.35	3.0
	0.8	1.9	2.35	3.05	4.3
	1.0	2.15	2.8	3.8	5.7

Table 7.12 (Continued)

θ (deg)	δ'/φ'	R for φ' (deg)			
		30	**35**	**40**	**45**
10	0.2	1.15	1.2	1.3	1.4
	0.4	1.35	1.5	1.7	2.0
	0.6	1.6	1.85	2.25	2.9
	0.8	1.8	2.25	2.9	4.0
	1.0	2.05	2.65	3.6	5.3
15	0.2	1.15	1.2	1.3	1.35
	0.4	1.35	1.5	1.65	1.95
	0.6	1.55	1.8	2.2	2.7
	0.8	1.8	2.2	2.8	3.8
	1.0	2.0	2.6	3.4	4.95
20	0.2	1.15	1.2	1.3	1.35
	0.4	1.35	1.45	1.65	1.9
	0.6	1.5	1.8	2.1	2.6
	0.8	1.8	2.1	2.6	3.55
	1.0	1.9	2.4	3.2	4.8

7.14 Passive Pressure under Earthquake Conditions

The relationship for passive earth pressure on a retaining wall with a granular backfill and under earthquake conditions was evaluated by Subba Rao and Choudhury (2005) by the method of limit equilibrium using the pseudo-static approach. Figure 7.29 shows the nature of failure surface in soil considered in this analysis. The passive pressure, P_{pe}, can be expressed as

$$P_{pe} = [\tfrac{1}{2}\gamma H^2 K_{p\gamma(e)}]\frac{1}{\cos \delta'} \tag{7.74}$$

where $K_{p\gamma(e)}$ = passive earth-pressure coefficient in the normal direction to the wall.

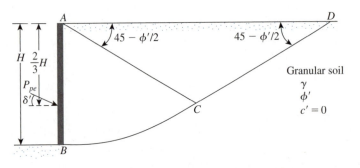

Figure 7.29 Nature of failure surface in soil considered in the analysis to determine P_{pe}

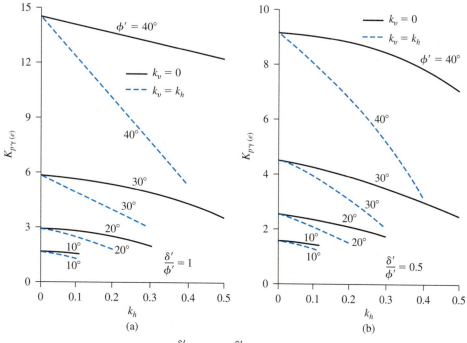

Figure 7.30 Variation of $K_{p\gamma(e)}$: (a) $\dfrac{\delta'}{\phi'} = 1$, (b) $\dfrac{\delta'}{\phi'} = 0.5$

$K_{p\gamma(e)}$ is a function of k_h and k_v that are, respectively, coefficient of horizontal and vertical acceleration due to earthquake. The variations of $K_{p\gamma(e)}$ for $\delta'/\phi' = 0.5$ and 1 are shown in Figures 7.30a and b. The passive pressure P_{pe} will be inclined at an angle δ' to the back face of the wall and will act at a distance of $H/3$ above the bottom of the wall.

Problems

7.1 Refer to Figure 7.3a. Given: $H = 3.5$ m, $q = 20$ kN/m², $\gamma = 18.2$ kN/m³ $c' = 0$, and $\phi' = 35°$. Determine the at-rest lateral earth force per meter length of the wall. Also, find the location of the resultant. Use Eq. (7.4) and OCR = 1.5.

7.2 Use Eq. (7.3), Figure P7.2, and the following values to determine the at-rest lateral earth force per unit length of the wall. Also find the location of the resultant.
$H = 5$ m, $H_1 = 2$ m, $H_2 = 3$ m, $\gamma = 15.5$ kN/m³, $\gamma_{sat} = 18.5$ kN/m³, $\phi' = 34°$, $c' = 0$, $q = 20$ kN/m², and OCR = 1.

7.3 Refer to Figure 7.6a. Given the height of the retaining wall, H is 6.4 m; the backfill is a saturated clay with $\phi = 0°$, $c = 30.2$ kN/m², $\gamma_{sat} = 17.76$ kN/m³,
  a. Determine the Rankine active pressure distribution diagram behind the wall.
  b. Determine the depth of the tensile crack, z_c.
  c. Estimate the Rankine active force per foot length of the wall before and after the occurrence of the tensile crack.

Figure P7.2

7.4 A vertical retaining wall (Figure 7.6a) is 6.3 m high with a horizontal backfill. For the backfill, assume that $\gamma = 17.9$ kN/m^3, $\phi' = 26°$, and $c' = 15$ kN/m^2. Determine the Rankine active force per unit length of the wall after the occurrence of the tensile crack.

7.5 Refer to Problem 7.2. For the retaining wall, determine the Rankine active force per unit length of the wall and the location of the line of action of the resultant.

7.6 Refer to Figure 7.10. For the retaining wall, $H = 6$ m, $\phi' = 34°$, $\alpha = 10°$, $\gamma = 17$ kN/m^3, and $c' = 0$.

   a. Determine the intensity of the Rankine active force at $z = 2$ m, 4 m, and 6 m.

   b. Determine the Rankine active force per meter length of the wall and also the location and direction of the resultant.

7.7 Refer to Figure 7.10. Given: $H = 6.7$ m, $\gamma = 18.08$ kN/m^3, $\phi' = 25°$, $c' = 12$ kN/m^2, and $\alpha = 10°$. Calculate the Rankine active force per unit length of the wall after the occurrence of the tensile crack.

7.8 Refer to Figure 7.12a. Given: $H = 3.66$ m, $\gamma = 16.5$ kN/m^3, $\phi' = 30°$, $c' = 0$, and $\beta = 85°$. Determine the Coulomb's active force per foot length of the wall and the location and direction of the resultant for the following cases:

   a. $\alpha = 10°$ and $\delta' = 20°$

   b. $\alpha = 20°$ and $\delta' = 15°$

7.9 Refer to Figure 7.13a. Given $H = 3.5$ m, $\alpha = 0$, $\beta = 85°$, $\gamma = 18$ kN/m^3, $c' = 0$, $\phi' = 34°$, $\delta'/\phi' = 0.5$, and $q = 30$ kN/m^2. Determine the Coulomb's active force per unit length of the wall.

7.10 Refer to Figure 7.14b. Given $H = 3.3$ m, $a' = 1$ m, $b' = 1.5$ m, and $q = 25$ kN/m^2. Determine the lateral force per unit length of the unyielding wall caused by the surcharge loading only.

7.11 Refer to Figure 7.15. Here, $H = 6$ m, $\gamma = 17$ kN/m^3, $\phi' = 35°$, $\delta' = 17.5°$, $c' = 0$, $\alpha = 10°$, and $\beta = 90°$. Determine the Coulomb's active force for earthquake conditions (P_{ae}) per meter length of the wall and the location and direction of the resultant. Given $k_h = 0.2$ and $k_v = 0$.

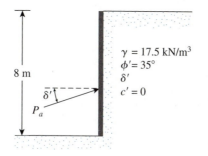

Figure P7.12

7.12 A retaining wall is shown in Figure P7.12. If the wall rotates about its top, determine the magnitude of the active force per unit length of the wall for $n_a = 0.3, 0.4$, and 0.5. Assume $\delta'/\phi' = 0.5$.

7.13 A vertical frictionless retaining wall is 6-m high with a horizontal granular backfill. Given: $\gamma = 16 \text{ kN/m}^3$ and $\phi' = 30°$. For the translation mode of the wall, calculate the active pressure at depths $z = 1.5$ m, 3 m, 4.5 m, and 6 m.

7.14 Refer to Problem 7.3.
    a. Draw the Rankine passive pressure distribution diagram behind the wall.
    b. Estimate the Rankine passive force per foot length of the wall and also the location of the resultant.

7.15 In Figure 7.28, which shows a vertical retaining wall with a horizontal backfill, let $H = 4$ m, $\theta = 10°$, $\gamma = 16.5 \text{ kN/m}^3$, $\phi' = 35°$, and $\delta' = 10°$. Based on Zhu and Qian's work, what would be the passive force per meter length of the wall?

7.16 Consider a 4-m high retaining wall with a vertical back and horizontal granular backfill, as shown in Figure 7.29. Given: $\gamma = 18 \text{ kN/m}^3$, $\phi' = 40°$, $c' = 0$, $\delta' = 20°$, $k_v = 0$ and $k_h = 0.2$. Determine the passive force P_{pe} per unit length of the wall taking the earthquake effect into consideration.

References

CHU, S. C. (1991). "Rankine Analysis of Active and Passive Pressures on Dry Sand," *Soils and Foundations*, Vol. 31, No. 4, pp. 115–120.

COULOMB, C. A. (1776). *Essai sur une Application des Règles de Maximis et Minimum à quelques Problemes de Statique Relatifs à l'Architecture*, Mem. Acad. Roy. des Sciences, Paris, Vol. 3, p. 38.

DUBROVA, G. A. (1963). "Interaction of Soil and Structures," Izd. *Rechnoy Transport*, Moscow.

HARR, M. E. (1966). *Fundamentals of Theoretical Soil Mechanics*, McGraw-Hill, New York.

JAKY, J. (1944). "The Coefficient of Earth Pressure at Rest," *Journal for the Society of Hungarian Architects and Engineers*, October, pp. 355–358.

JARQUIO, R. (1981). "Total Lateral Surcharge Pressure Due to Strip Load," *Journal of the Geotechnical Engineering Division*, American Society of Civil Engineers, Vol. 107, No. GT10, pp. 1424–1428.

KIM, J. S., and Preber, T. (1969). "Earth Pressure against Braced Excavations," *Journal of the Soil Mechanics and Foundations Division*, ASCE, Vol. 96, No. 6, pp. 1581–1584.

MAYNE, P. W., and Kulhawy, F. H. (1982). "K_o–OCR Relationships in Soil," *Journal of the Geotechnical Engineering Division*, ASCE, Vol. 108, No. GT6, pp. 851–872.

MAZINDRANI, Z. H., and Ganjali, M. H. (1997). "Lateral Earth Pressure Problem of Cohesive Backfill with Inclined Surface," *Journal of Geotechnical and Geoenvironmental Engineering*, ASCE, Vol. 123, No. 2, pp. 110–112.

SEED, H. B., and WHITMAN, R. V. (1970). "Design of Earth Retaining Structures for Dynamic Loads," *Proceedings, Specialty Conference on Lateral Stresses in the Ground and Design of Earth Retaining Structures*, American Society of Civil Engineers, pp. 103–147.

SHIELDS, D. H., and Tolunay, A. Z. (1973). "Passive Pressure Coefficients by Method of Slices," *Journal of the Soil Mechanics and Foundations Division*, ASCE, Vol. 99, No. SM12, pp. 1043–1053.

SUBBA RAO, K. S., and CHOUDHURY, D. (2005). "Seismic Passive Earth Pressures in Soil," *Journal of Geotechnical and Geoenvironmental Engineering*, American Society of Civil Engineers, Vo. 131, No. 1, pp. 131–135.

TERZAGHI, K. (1943). *Theoretical Soil Mechanics,* Wiley, New York.

ZHU, D. Y., and Qian, Q. (2000). "Determination of Passive Earth Pressure Coefficient by the Method of Triangular Slices," *Canadian Geotechnical Journal*, Vol. 37, No. 2, pp. 485–491.

8 Retaining Walls

Introduction

In Chapter 7, you were introduced to various theories of lateral earth pressure. Those theories will be used in this chapter to design various types of retaining walls. In general, retaining walls can be divided into two major categories: (a) conventional retaining walls and (b) mechanically stabilized earth walls.

Conventional retaining walls can generally be classified into four varieties:

1. Gravity retaining walls
2. Semigravity retaining walls
3. Cantilever retaining walls
4. Counterfort retaining walls

Gravity retaining walls (Figure 8.1a) are constructed with plain concrete or stone masonry. They depend for stability on their own weight and any soil resting on the masonry. This type of construction is not economical for high walls.

In many cases, a small amount of steel may be used for the construction of gravity walls, thereby minimizing the size of wall sections. Such walls are generally referred to as *semigravity walls* (Figure 8.1b).

Cantilever retaining walls (Figure 8.1c) are made of reinforced concrete that consists of a thin stem and a base slab. This type of wall is economical to a height of about 8 m. Figure 8.2 shows a cantilever retaining wall under construction.

Counterfort retaining walls (Figure 8.1d) are similar to cantilever walls. At regular intervals, however, they have thin vertical concrete slabs known as *counterforts* that tie the wall and the base slab together. The purpose of the counterforts is to reduce the shear and the bending moments.

To design retaining walls properly, an engineer must know the basic parameters—the *unit weight, angle of friction,* and *cohesion*—of the soil retained behind the wall and the soil below the base slab. Knowing the properties of the soil behind the wall enables the engineer to determine the lateral pressure distribution that has to be designed for.

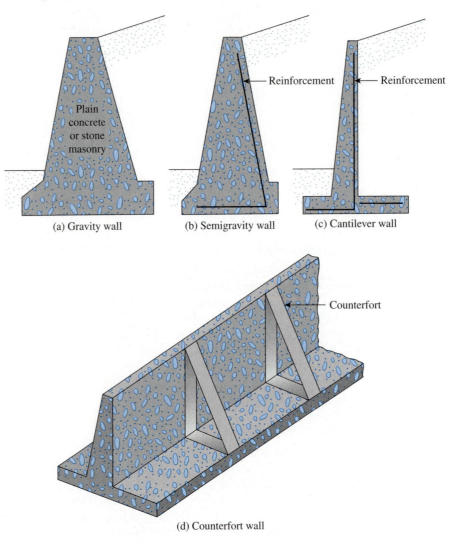

(a) Gravity wall   (b) Semigravity wall   (c) Cantilever wall

(d) Counterfort wall

Figure 8.1 Types of retaining wall

There are two phases in the design of a conventional retaining wall. First, with the lateral earth pressure known, the structure as a whole is checked for *stability*. The structure is examined for possible *overturning, sliding,* and *bearing capacity* failures. Second, each component of the structure is checked for *strength*, and the *steel reinforcement* of each component is determined.

This chapter presents the procedures for determining the stability of the retaining wall. Checks for strength can be found in any textbook on reinforced concrete.

Some retaining walls have their backfills stabilized mechanically by including reinforcing elements such as metal strips, bars, welded wire mats, geotextiles, and

Figure 8.2 A cantilever retaining wall under construction (*Courtesy of Dharma Shakya, Geotechnical Solutions, Inc., Irvine, California*)

geogrids. These walls are relatively flexible and can sustain large horizontal and vertical displacements without much damage.

Gravity and Cantilever Walls

8.2 Proportioning Retaining Walls

In designing retaining walls, an engineer must assume some of their dimensions. Called *proportioning,* such assumptions allow the engineer to check trial sections of the walls for stability. If the stability checks yield undesirable results, the sections can be changed and rechecked. Figure 8.3 shows the general proportions of various retaining-wall components that can be used for initial checks.

Note that the top of the stem of any retaining wall should not be less than about 0.3 m for proper placement of concrete. The depth, *D,* to the bottom of the base slab should be a minimum of 0.6 m. However, the bottom of the base slab should be positioned below the seasonal frost line.

For counterfort retaining walls, the general proportion of the stem and the base slab is the same as for cantilever walls. However, the counterfort slabs may be about 0.3 m thick and spaced at center-to-center distances of 0.3*H* to 0.7*H*.

Figure 8.3 Approximate dimensions for various components of retaining wall for initial stability checks: (a) gravity wall; (b) cantilever wall

8.3   *Application of Lateral Earth Pressure Theories to Design*

The fundamental theories for calculating lateral earth pressure were presented in Chapter 7. To use these theories in design, an engineer must make several simple assumptions. In the case of cantilever walls, the use of the Rankine earth pressure theory for stability checks involves drawing a vertical line *AB* through point *A*, located at the edge of the heel of the base slab in Figure 8.4a. The Rankine active condition is assumed to exist along the vertical plane *AB*. Rankine active earth pressure equations may then be used to calculate the lateral pressure on the face *AB* of the wall. In the analysis of the wall's stability, the force $P_{a(\text{Rankine})}$, the weight of soil above the heel, and the weight W_c of the concrete all should be taken into consideration. The assumption for the development of Rankine active pressure along the soil face *AB* is theoretically correct if the shear zone bounded by the line *AC* is not obstructed by the stem of the wall. The angle, η, that the line *AC* makes with the vertical is

$$\eta = 45 + \frac{\alpha}{2} - \frac{\phi'}{2} - \frac{1}{2}\sin^{-1}\left(\frac{\sin\alpha}{\sin\phi'}\right) \tag{8.1}$$

A similar type of analysis may be used for gravity walls, as shown in Figure 8.4b. However, *Coulomb's active earth pressure theory* also may be used, as shown in Figure 8.4c. If it is used, the only forces to be considered are $P_{a(Coulomb)}$ and the weight of the wall, W_c.

(a)

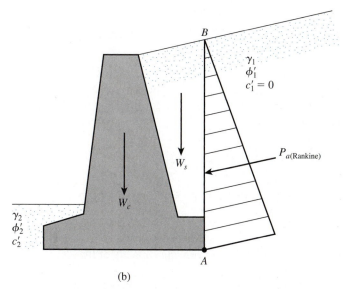

(b)

Figure 8.4 Assumption for the determination of lateral earth pressure: (a) cantilever wall; (b) and (c) gravity wall

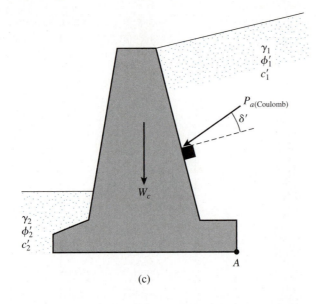

(c)

Figure 8.4 (continued)

If Coulomb's theory is used, it will be necessary to know the range of the wall friction angle δ' with various types of backfill material. Following are some ranges of wall friction angle for masonry or mass concrete walls:

Backfill material	Range of δ' (deg)
Gravel	27–30
Coarse sand	20–28
Fine sand	15–25
Stiff clay	15–20
Silty clay	12–16

In the case of ordinary retaining walls, water table problems and hence hydrostatic pressure are not encountered. Facilities for drainage from the soils that are retained are always provided.

8.4 Stability of Retaining Walls

A retaining wall may fail in any of the following ways:

- It may *overturn* about its toe. (See Figure 8.5a.)
- It may *slide* along its base. (See Figure 8.5b.)
- It may fail due to the loss of *bearing capacity* of the soil supporting the base. (See Figure 8.5c.)
- It may undergo deep-seated shear failure. (See Figure 8.5d.)
- It may go through excessive settlement.

The checks for stability against overturning, sliding, and bearing capacity failure will be described in Sections 8.5, 8.6, and 8.7. The principles used to estimate settlement

(a)

(b)

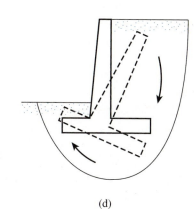

(c)

(d)

Figure 8.5 Failure of retaining wall:
(a) by overturning; (b) by sliding;
(c) by bearing capacity failure;
(d) by deep-seated shear failure

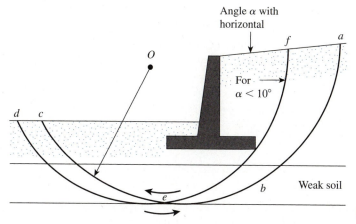

Figure 8.6 Deep-seated shear failure

were covered in Chapter 5 and will not be discussed further. When a weak soil layer is located at a shallow depth—that is, within a depth of 1.5 times the width of the base slab of the retaining wall—the possibility of excessive settlement should be considered. In some cases, the use of lightweight backfill material behind the retaining wall may solve the problem.

Deep shear failure can occur along a cylindrical surface, such as *abc* shown in Figure 8.6, as a result of the existence of a weak layer of soil underneath the wall at a depth of about 1.5 times the width of the base slab of the retaining wall. In such cases, the critical cylindrical failure surface *abc* has to be determined by trial and error, using various centers such as *O*. The failure surface along which the minimum factor of safety is obtained is the *critical surface of sliding*. For the backfill slope with α less than about 10°, the critical failure circle apparently passes through the edge of the heel slab (such as *def* in the figure). In this situation, the minimum factor of safety also has to be determined by trial and error by changing the center of the trial circle.

8.5   *Check for Overturning*

Figure 8.7 shows the forces acting on a cantilever and a gravity retaining wall, based on the assumption that the Rankine active pressure is acting along a vertical plane *AB* drawn through the heel of the structure. P_p is the Rankine passive pressure; recall that its magnitude is [from Eq. (7.63)].

$$P_p = \tfrac{1}{2}K_p\gamma_2 D^2 + 2c_2'\sqrt{K_p}D$$

where

γ_2 = unit weight of soil in front of the heel and under the base slab
K_p = Rankine passive earth pressure coefficient = $\tan^2(45 + \phi_2'/2)$
c_2', ϕ_2' = cohesion and effective soil friction angle, respectively

The factor of safety against overturning about the toe—that is, about point *C* in Figure 8.7—may be expressed as

$$\text{FS}_{(\text{overturning})} = \frac{\Sigma M_R}{\Sigma M_o} \tag{8.2}$$

where

ΣM_o = sum of the moments of forces tending to overturn about point *C*
ΣM_R = sum of the moments of forces tending to resist overturning about point *C*

The overturning moment is

$$\Sigma M_o = P_h\left(\frac{H'}{3}\right) \tag{8.3}$$

where $P_h = P_a\cos\alpha$.

To calculate the resisting moment, ΣM_R (neglecting P_p), a table such as Table 8.1 can be prepared. The weight of the soil above the heel and the weight of the concrete (or masonry) are both forces that contribute to the resisting moment. Note that the force P_v also contributes to the resisting moment. P_v is the vertical component of the active force P_a, or

$$P_v = P_a\sin\alpha$$

The moment of the force P_v about *C* is

$$M_v = P_v B = P_a\sin\alpha B \tag{8.4}$$

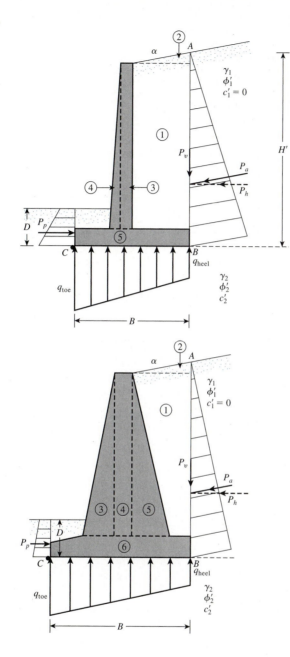

Figure 8.7 Check for overturning, assuming that the Rankine pressure is valid

where B = width of the base slab.

Once ΣM_R is known, the factor of safety can be calculated as

$$\text{FS}_{(\text{overturning})} = \frac{M_1 + M_2 + M_3 + M_4 + M_5 + M_6 + M_v}{P_a \cos\alpha (H'/3)} \qquad (8.5)$$

The usual minimum desirable value of the factor of safety with respect to overturning is 2 to 3.

Table 8.1 Procedure for Calculating ΣM_R

Section (1)	Area (2)	Weight/unit length of wall (3)	Moment arm measured from C (4)	Moment about C (5)
1	A_1	$W_1 = \gamma_1 \times A_1$	X_1	M_1
2	A_2	$W_2 = \gamma_1 \times A_2$	X_2	M_2
3	A_3	$W_3 = \gamma_c \times A_3$	X_3	M_3
4	A_4	$W_4 = \gamma_c \times A_4$	X_4	M_4
5	A_5	$W_5 = \gamma_c \times A_5$	X_5	M_5
6	A_6	$W_6 = \gamma_c \times A_6$	X_6	M_6
		P_v	B	M_v
		ΣV		ΣM_R

(*Note:* γ_l = unit weight of backfill
γ_c = unit weight of concrete)

Some designers prefer to determine the factor of safety against overturning with the formula

$$\text{FS}_{(\text{overturning})} = \frac{M_1 + M_2 + M_3 + M_4 + M_5 + M_6}{P_a \cos\alpha\,(H'/3) - M_v} \tag{8.6}$$

8.6 *Check for Sliding along the Base*

The factor of safety against sliding may be expressed by the equation

$$\text{FS}_{(\text{sliding})} = \frac{\Sigma F_{R'}}{\Sigma F_d} \tag{8.7}$$

where

$\Sigma F_{R'}$ = sum of the horizontal resisting forces
ΣF_d = sum of the horizontal driving forces

Figure 8.8 indicates that the shear strength of the soil immediately below the base slab may be represented as

$$s = \sigma' \tan \delta' + c_a'$$

where

δ' = angle of friction between the soil and the base slab
c_a' = adhesion between the soil and the base slab

Thus, the maximum resisting force that can be derived from the soil per unit length of the wall along the bottom of the base slab is

$$R' = s(\text{area of cross section}) = s(B \times 1) = B\sigma' \tan \delta' + Bc_a'$$

However,

$$B\sigma' = \text{sum of the vertical force} = \Sigma V \text{ (see Table 8.1)}$$

so

$$R' = (\Sigma V)\tan\delta' + Bc_a'$$

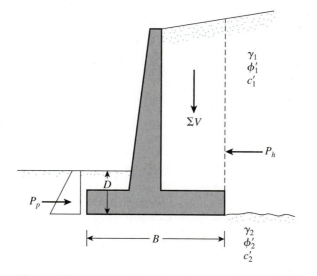

Figure 8.8 Check for sliding along the base

Figure 8.8 shows that the passive force P_p is also a horizontal resisting force. Hence,

$$\Sigma F_{R'} = (\Sigma V) \tan \delta' + Bc_a' + P_p \tag{8.8}$$

The only horizontal force that will tend to cause the wall to slide (a *driving force*) is the horizontal component of the active force P_a, so

$$\Sigma F_d = P_a \cos \alpha \tag{8.9}$$

Combining Eqs. (8.7), (8.8), and (8.9) yields

$$\text{FS}_{(\text{sliding})} = \frac{(\Sigma V) \tan \delta' + Bc_a' + P_p}{P_a \cos \alpha} \tag{8.10}$$

A minimum factor of safety of 1.5 against sliding is generally required.

In many cases, the passive force P_p is ignored in calculating the factor of safety with respect to sliding. In general, we can write $\delta' = k_1 \phi_2'$ and $c_a' = k_2 c_2'$. In most cases, k_1 and k_2 are in the range from $\frac{1}{2}$ to $\frac{2}{3}$. Thus,

$$\text{FS}_{(\text{sliding})} = \frac{(\Sigma V) \tan (k_1 \phi_2') + Bk_2 c_2' + P_p}{P_a \cos \alpha} \tag{8.11}$$

If the desired value of $\text{FS}_{(\text{sliding})}$ is not achieved, several alternatives may be investigated (see Figure 8.9):

- Increase the width of the base slab (i.e., the heel of the footing).
- Use a key to the base slab. If a key is included, the passive force per unit length of the wall becomes

$$P_p = \frac{1}{2} \gamma_2 D_1^2 K_p + 2c_2' D_1 \sqrt{K_p}$$

where $K_p = \tan^2 \left(45 + \dfrac{\phi_2'}{2}\right)$.

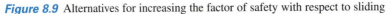

Figure 8.9 Alternatives for increasing the factor of safety with respect to sliding

- Use a *deadman anchor* at the stem of the retaining wall.
- Another possible way to increase the value of $FS_{(sliding)}$ is to consider reducing the value of P_a [see Eq. (8.11)]. One possible way to do so is to use the method developed by Elman and Terry (1988). The discussion here is limited to the case in which the retaining wall has a horizontal granular backfill (Figure 8.10). In Figure 8.10, the active force, P_a, is horizontal ($\alpha = 0$) so that

$$P_a \cos \alpha = P_h = P_a$$

and

$$P_a \sin \alpha = P_v = 0$$

However,

$$P_a = P_{a(1)} + P_{a(2)} \qquad (8.12)$$

The magnitude of $P_{a(2)}$ can be reduced if the heel of the retaining wall is sloped as shown in Figure 8.10. For this case,

$$P_a = P_{a(1)} + AP_{a(2)} \qquad (8.13)$$

The magnitude of A, as shown in Table 8.2, is valid for $\alpha' = 45°$. However note that in Figure 8.10a

$$P_{a(1)} = \frac{1}{2}\gamma_1 K_a (H' - D')^2$$

and

$$P_a = \frac{1}{2}\gamma_1 K_a H'^2$$

Hence,

$$P_{a(2)} = \frac{1}{2}\gamma_1 K_a [H'^2 - (H' - D')^2]$$

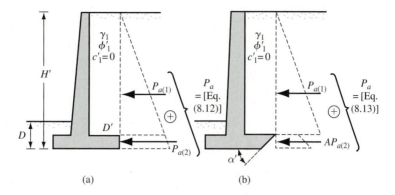

Figure 8.10 Retaining wall with sloped heel

Table 8.2 Variation of A with ϕ_1' (for $\alpha' = 45°$)

Soil friction angle, ϕ_1' (deg)	A
20	0.28
25	0.14
30	0.06
35	0.03
40	0.018

So, for the active pressure diagram shown in Figure 8.10b,

$$P_a = \frac{1}{2}\gamma_1 K_a(H' - D')^2 + \frac{A}{2}\gamma_1 K_a[H'^2 - (H' - D')^2] \tag{8.14}$$

Sloping the heel of a retaining wall can thus be extremely helpful in some cases.

8.7 Check for Bearing Capacity Failure

The vertical pressure transmitted to the soil by the base slab of the retaining wall should be checked against the ultimate bearing capacity of the soil. The nature of variation of the vertical pressure transmitted by the base slab into the soil is shown in Figure 8.11. Note that q_{toe} and q_{heel} are the *maximum* and the *minimum* pressures occurring at the ends of the toe and heel sections, respectively. The magnitudes of q_{toe} and q_{heel} can be determined in the following manner:

The sum of the vertical forces acting on the base slab is ΣV (see column 3 of Table 8.1), and the horizontal force $\mathbf{P}_h$ is $P_a \cos\alpha$. Let

$$\mathbf{R} = \Sigma \mathbf{V} + \mathbf{P}_h \tag{8.15}$$

be the resultant force. The net moment of these forces about point C in Figure 8.11 is

$$M_{net} = \Sigma M_R - \Sigma M_o \tag{8.16}$$

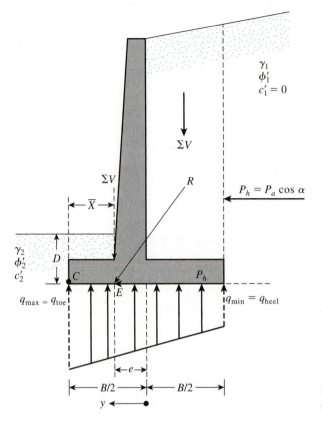

Figure 8.11 Check for bearing capacity failure

Note that the values of ΣM_R and ΣM_o were previously determined. [See Column 5 of Table 8.1 and Eq. (8.3)]. Let the line of action of the resultant R intersect the base slab at E. Then the distance

$$\overline{CE} = \overline{X} = \frac{M_{net}}{\Sigma V} \tag{8.17}$$

Hence, the eccentricity of the resultant R may be expressed as

$$e = \frac{B}{2} - \overline{CE} \tag{8.18}$$

The pressure distribution under the base slab may be determined by using simple principles from the mechanics of materials. First, we have

$$q = \frac{\Sigma V}{A} \pm \frac{M_{net}y}{I} \tag{8.19}$$

where

M_{net} = moment = $(\Sigma V)e$
I = moment of inertia per unit length of the base section
$\quad = \frac{1}{12}(1)(B^3)$

For maximum and minimum pressures, the value of y in Eq. (8.19) equals $B/2$. Substituting into Eq. (8.19) gives

$$q_{max} = q_{toe} = \frac{\Sigma V}{(B)(1)} + \frac{e(\Sigma V)\frac{B}{2}}{\left(\frac{1}{12}\right)(B^3)} = \frac{\Sigma V}{B}\left(1 + \frac{6e}{B}\right) \tag{8.20}$$

Similarly,

$$q_{min} = q_{heel} = \frac{\Sigma V}{B}\left(1 - \frac{6e}{B}\right) \tag{8.21}$$

Note that ΣV includes the weight of the soil, as shown in Table 8.1, and that when the value of the eccentricity e becomes greater than $B/6$, q_{min} [Eq. (8.21)] becomes negative. Thus, there will be some tensile stress at the end of the heel section. This stress is not desirable, because the tensile strength of soil is very small. If the analysis of a design shows that $e > B/6$, the design should be reproportioned and calculations redone.

The relationships pertaining to the ultimate bearing capacity of a shallow foundation were discussed in Chapter 3. Recall that [Eq. (3.40)].

$$q_u = c_2' N_c F_{cd} F_{ci} + q N_q F_{qd} F_{qi} + \tfrac{1}{2} \gamma_2 B' N_\gamma F_{\gamma d} F_{\gamma i} \tag{8.22}$$

where

$q = \gamma_2 D$

$B' = B - 2e$

$F_{cd} = F_{qd} - \dfrac{1 - F_{qd}}{N_c \tan \phi_2'}$

$F_{qd} = 1 + 2 \tan \phi_2'(1 - \sin \phi_2')^2 \dfrac{D}{B'}$

$F_{\gamma d} = 1$

$F_{ci} = F_{qi} = \left(1 - \dfrac{\psi^\circ}{90^\circ}\right)^2$

$F_{\gamma i} = \left(1 - \dfrac{\psi^\circ}{\phi_2'^\circ}\right)^2$

$\psi^\circ = \tan^{-1}\left(\dfrac{P_a \cos\alpha}{\Sigma V}\right)$

Note that the shape factors F_{cs}, F_{qs}, and $F_{\gamma s}$ given in Chapter 3 are all equal to unity, because they can be treated as a continuous foundation. For this reason, the shape factors are not shown in Eq. (8.22).

Once the ultimate bearing capacity of the soil has been calculated by using Eq. (8.22), the factor of safety against bearing capacity failure can be determined:

$$\text{FS}_{(bearing\ capacity)} = \frac{q_u}{q_{max}} \tag{8.23}$$

Generally, a factor of safety of 3 is required. In Chapter 3, we noted that the ultimate bearing capacity of shallow foundations occurs at a settlement of about 10% of the foundation width.

In the case of retaining walls, the width B is large. Hence, the ultimate load q_u will occur at a fairly large foundation settlement. A factor of safety of 3 against bearing capacity failure may not ensure that settlement of the structure will be within the tolerable limit in all cases. Thus, this situation needs further investigation.

An alternate relationship to Eq. (8.22) will be Eq. (3.67), or

$$q_u = c'N_{c(ei)}F_{cd} + qN_{q(ei)}F_{qd} + \tfrac{1}{2}\gamma_2 B N_{\gamma(ei)}F_{\gamma d}$$

Since $F_{\gamma d} = 1$,

$$q_u = c'N_{c(ei)}F_{cd} + qN_{q(ei)}F_{qd} + \tfrac{1}{2}\gamma_2 B N_{\gamma(ei)} \qquad (8.24)$$

The bearing capacity factors, $N_{c(ei)}$, $N_{q(ei)}$, and $N_{\gamma(ei)}$ were given in Figures 3.26 through 3.28.

Example 8.1

The cross section of a cantilever retaining wall is shown in Figure 8.12. Calculate the factors of safety with respect to overturning, sliding, and bearing capacity.

Figure 8.12 Calculation of stability of a retaining wall

Solution
From the figure,

$$H' = H_1 + H_2 + H_3 = 2.6 \tan 10° + 6 + 0.7$$
$$= 0.458 + 6 + 0.7 = 7.158 \text{ m}$$

The Rankine active force per unit length of wall = $P_p = \frac{1}{2}\gamma_1 H'^2 K_a$. For $\phi_1' = 30°$ and $\alpha = 10°$, K_a is equal to 0.3532. (See Table 7.1.) Thus,

$$P_a = \frac{1}{2}(18)(7.158)^2(0.3532) = 162.9 \text{ kN/m}$$

$$P_v = P_a \sin 10° = 162.9 (\sin 10°) = 28.29 \text{ kN/m}$$

and

$$P_h = P_a \cos 10° = 162.9 (\cos 10°) = 160.43 \text{ kN/m}$$

Factor of Safety against Overturning
The following table can now be prepared for determining the resisting moment:

Section no.[a]	Area (m²)	Weight/unit length (kN/m)	Moment arm from point C (m)	Moment (kN-m/m)
1	$6 \times 0.5 = 3$	70.74	1.15	81.35
2	$\frac{1}{2}(0.2)6 = 0.6$	14.15	0.833	11.79
3	$4 \times 0.7 = 2.8$	66.02	2.0	132.04
4	$6 \times 2.6 = 15.6$	280.80	2.7	758.16
5	$\frac{1}{2}(2.6)(0.458) = 0.595$	10.71	3.13	33.52
		$P_v = 28.29$	4.0	113.16
		$\Sigma V = 470.71$		$1130.02 = \Sigma M_R$

[a]For section numbers, refer to Figure 8.12
$\gamma_{concrete} = 23.58 \text{ kN/m}^3$

The overturning moment

$$M_o = P_h\left(\frac{H'}{3}\right) = 160.43\left(\frac{7.158}{3}\right) = 382.79 \text{ kN-m/m}$$

and

$$FS_{(overturning)} = \frac{\Sigma M_R}{M_o} = \frac{1130.02}{382.79} = \textbf{2.95} > \textbf{2, OK}$$

Factor of Safety against Sliding
From Eq. (8.11),

$$FS_{(sliding)} = \frac{(\Sigma V)\tan(k_1\phi_2') + Bk_2c_2' + P_p}{P_a\cos\alpha}$$

Let $k_1 = k_2 = \frac{2}{3}$. Also,

$$P_p = \tfrac{1}{2}K_p\gamma_2 D^2 + 2c_2'\sqrt{K_p}D$$

$$K_p = \tan^2\left(45 + \frac{\phi_2'}{2}\right) = \tan^2(45 + 10) = 2.04$$

and

$$D = 1.5 \text{ m}$$

So

$$P_p = \tfrac{1}{2}(2.04)(19)(1.5)^2 + 2(40)(\sqrt{2.04})(1.5)$$

$$= 43.61 + 171.39 = 215 \text{ kN/m}$$

Hence,

$$\text{FS}_{(\text{sliding})} = \frac{(470.71)\tan\left(\dfrac{2 \times 20}{3}\right) + (4)\left(\dfrac{2}{3}\right)(40) + 215}{160.43}$$

$$= \frac{111.56 + 106.67 + 215}{160.43} = \mathbf{2.7 > 1.5, \; OK}$$

Note: For some designs, the depth D in a passive pressure calculation may be taken to be *equal to the thickness of the base slab*.

Factor of Safety against Bearing Capacity Failure
Combining Eqs. (8.16), (8.17), and (8.18) yields

$$e = \frac{B}{2} - \frac{\Sigma M_R - \Sigma M_o}{\Sigma V} = \frac{4}{2} - \frac{1130.02 - 382.79}{470.71}$$

$$= 0.411 \text{ m} < \frac{B}{6} = \frac{4}{6} = 0.666 \text{ m}$$

Again, from Eqs. (8.20) and (8.21)

$$q_{\text{heel}}^{\text{toe}} = \frac{\Sigma V}{B}\left(1 \pm \frac{6e}{B}\right) = \frac{470.71}{4}\left(1 \pm \frac{6 \times 0.411}{4}\right) = 190.2 \text{ kN/m}^2 \text{ (toe)}$$

$$= 45.13 \text{ kN/m}^2 \text{ (heel)}$$

The ultimate bearing capacity of the soil can be determined from Eq. (8.22)

$$q_u = c_2'N_cF_{cd}F_{ci} + qN_qF_{qd}F_{qi} + \frac{1}{2}\gamma_2 B'N_\gamma F_{\gamma d}F_{\gamma i}$$

For $\phi_2' = 20°$ (see Table 3.3), $N_c = 14.83$, $N_q = 6.4$, and $N_\gamma = 5.39$. Also,

$$q = \gamma_2 D = (19)(1.5) = 28.5 \text{ kN/m}^2$$

$$B' = B - 2e = 4 - 2(0.411) = 3.178 \text{ m}$$

$$F_{cd} = F_{qd} - \frac{1 - F_{qd}}{N_c \tan\phi_2'} = 1.148 - \frac{1 - 1.148}{(14.83)(\tan 20)} = 1.175$$

$$F_{qd} = 1 + 2\tan\phi_2'(1 - \sin\phi_2')^2 \left(\frac{D}{B'}\right) = 1 + 0.315\left(\frac{1.5}{3.178}\right) = 1.148$$

$$F_{\gamma d} = 1$$

$$F_{ci} = F_{qi} = \left(1 - \frac{\psi°}{90°}\right)^2$$

and

$$\psi = \tan^{-1}\left(\frac{P_a \cos\alpha}{\Sigma V}\right) = \tan^{-1}\left(\frac{160.43}{470.71}\right) = 18.82°$$

So

$$F_{ci} = F_{qi} = \left(1 - \frac{18.82}{90}\right)^2 = 0.626$$

and

$$F_{\gamma i} = \left(1 - \frac{\psi}{\phi_2'}\right)^2 = \left(1 - \frac{18.82}{20}\right)^2 \approx 0$$

Hence,

$$q_u = (40)(14.83)(1.175)(0.626) + (28.5)(6.4)(1.148)(0.626)$$
$$+ \tfrac{1}{2}(19)(5.93)(3.178)(1)(0)$$
$$= 436.33 + 131.08 + 0 = 567.41 \text{ kN/m}^2$$

and

$$\text{FS}_{\text{(bearing capacity)}} = \frac{q_u}{q_{\text{toe}}} = \frac{567.41}{190.2} = \textbf{2.98}$$

Note: $\text{FS}_{\text{(bearing capacity)}}$ is less than 3. Some reproportioning will be needed. ∎

Example 8.2

A gravity retaining wall is shown in Figure 8.13. Use $\delta' = 2/3\phi_1'$ and Coulomb's active earth pressure theory. Determine

$\gamma_1 = 18.5 \text{ kN/m}^3$
$\phi_1' = 32°$
$c_1' = 0$

$\gamma_2 = 18 \text{ kN/m}^3$
$\phi_2' = 24°$
$c_2' = 30 \text{ kN/m}^2$

Figure 8.13 Gravity retaining wall (not to scale)

a. The factor of safety against overturning
b. The factor of safety against sliding
c. The pressure on the soil at the toe and heel

Solution
The height

$$H' = 5 + 1.5 = 6.5 \text{ m}$$

Coulomb's active force is

$$P_a = \tfrac{1}{2}\gamma_1 H'^2 K_a$$

With $\alpha = 0°$, $\beta = 75°$, $\delta' = 2/3\phi_1'$, and $\phi_1' = 32°$, $K_a = 0.4023$. (See Table 7.4.) So,

$$P_a = \tfrac{1}{2}(18.5)\,(6.5)^2(0.4023) = 157.22 \text{ kN/m}$$

$$P_h = P_a \cos\left(15 + \tfrac{2}{3}\phi_1'\right) = 157.22 \cos 36.33 = 126.65 \text{ kN/m}$$

and

$$P_v = P_a \sin\left(15 + \tfrac{2}{3}\phi_1'\right) = 157.22 \sin 36.33 = 93.14 \text{ kN/m}$$

Part a: Factor of Safety against Overturning
From Figure 8.13, one can prepare the following table:

Area no.	Area (m²)	Weight* (kN/m)	Moment arm from C (m)	Moment (kN-m/m)
1	$\frac{1}{2}(5.7)(1.53) = 4.36$	102.81	2.18	224.13
2	$(0.6)(5.7) = 3.42$	80.64	1.37	110.48
3	$\frac{1}{2}(0.27)(5.7) = 0.77$	18.16	0.98	17.80
4	$\approx (3.5)(0.8) = 2.8$	66.02	1.75	115.54
		$P_v = 93.14$	2.83	263.59
		$\Sigma V = 360.77$ kN/m		$\Sigma M_R = 731.54$ kN-m/m

*$\gamma_{concrete} = 23.58$ kN/m³

Note that the weight of the soil above the back face of the wall is not taken into account in the preceding table. We have

$$\text{Overturning moment} = M_o = P_h\left(\frac{H'}{3}\right) = 126.65(2.167) = 274.45 \text{ kN-m/m}$$

Hence,

$$\text{FS}_{(overturning)} = \frac{\Sigma M_R}{\Sigma M_o} = \frac{731.54}{274.45} = \mathbf{2.67 > 2, OK}$$

Part b: Factor of Safety against Sliding
We have

$$\text{FS}_{(sliding)} = \frac{(\Sigma V)\tan\left(\frac{2}{3}\phi_2'\right) + \frac{2}{3}c_2'B + P_p}{P_h}$$

$$P_p = \frac{1}{2}K_p\gamma_2 D^2 + 2c_2'\sqrt{K_p}D$$

and

$$K_p = \tan^2\left(45 + \frac{24}{2}\right) = 2.37$$

Hence,

$$P_p = \frac{1}{2}(2.37)(18)(1.5)^2 + 2(30)(1.54)(1.5) = 186.59 \text{ kN/m}$$

So

$$\text{FS}_{(sliding)} = \frac{360.77\tan\left(\frac{2}{3} \times 24\right) + \frac{2}{3}(30)(3.5) + 186.59}{126.65}$$

$$= \frac{103.45 + 70 + 186.59}{126.65} = \textbf{2.84}$$

If P_p is ignored, the factor of safety is **1.37**.

Part c: Pressure on Soil at Toe and Heel
From Eqs. (8.16), (8.17), and (8.18),

$$e = \frac{B}{2} - \frac{\Sigma M_R - \Sigma M_o}{\Sigma V} = \frac{3.5}{2} - \frac{731.54 - 274.45}{360.77} = 0.483 < \frac{B}{6} = 0.583$$

$$q_{toe} = \frac{\Sigma V}{B}\left[1 + \frac{6e}{B}\right] = \frac{360.77}{3.5}\left[1 + \frac{(6)(0.483)}{3.5}\right] = \textbf{188.43 kN/m}^2$$

and

$$q_{heel} = \frac{V}{B}\left[1 - \frac{6e}{B}\right] = \frac{360.77}{3.5}\left[1 - \frac{(6)(0.483)}{3.5}\right] = \textbf{17.73 kN/m}^2 \quad \blacksquare$$

8.8 *Construction Joints and Drainage from Backfill*

Construction Joints

A retaining wall may be constructed with one or more of the following joints:

1. *Construction joints* (see Figure 8.14a) are vertical and horizontal joints that are placed between two successive pours of concrete. To increase the shear at the joints, keys may be used. If keys are not used, the surface of the first pour is cleaned and roughened before the next pour of concrete.
2. *Contraction joints* (Figure 8.14b) are vertical joints (grooves) placed in the face of a wall (from the top of the base slab to the top of the wall) that allow the concrete to shrink without noticeable harm. The grooves may be about 6 to 8 mm wide and 12 to 16 mm deep.
3. *Expansion joints* (Figure 8.14c) allow for the expansion of concrete caused by temperature changes; vertical expansion joints from the base to the top of the wall may also be used. These joints may be filled with flexible joint fillers. In most cases, horizontal reinforcing steel bars running across the stem are continuous through all joints. The steel is greased to allow the concrete to expand.

Drainage from the Backfill

As the result of rainfall or other wet conditions, the backfill material for a retaining wall may become saturated, thereby increasing the pressure on the wall and perhaps creating an unstable condition. For this reason, adequate drainage must be provided by means of *weep holes* or *perforated drainage pipes*. (See Figure 8.15.)

When provided, weep holes should have a minimum diameter of about 0.1 m and be adequately spaced. Note that there is always a possibility that backfill material may

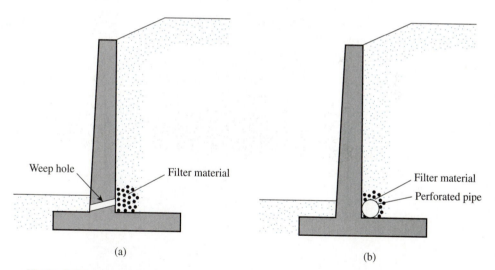

Figure 8.14 (a) Construction joints; (b) contraction joint; (c) expansion joint

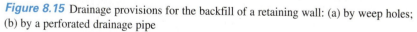

Figure 8.15 Drainage provisions for the backfill of a retaining wall: (a) by weep holes; (b) by a perforated drainage pipe

be washed into weep holes or drainage pipes and ultimately clog them. Thus, a filter material needs to be placed behind the weep holes or around the drainage pipes, as the case may be; geotextiles now serve that purpose.

Two main factors influence the choice of filter material: The grain-size distribution of the materials should be such that (a) the soil to be protected is not washed into the filter and (b) excessive hydrostatic pressure head is not created in the soil with a lower hydraulic conductivity (in this case, the backfill material). The preceding conditions can be satisfied if the following requirements are met (Terzaghi and Peck, 1967):

$$\frac{D_{15(F)}}{D_{85(B)}} < 5 \qquad [\text{to satisfy condition (a)}] \tag{8.25}$$

$$\frac{D_{15(F)}}{D_{15(B)}} > 4 \qquad [\text{to satisfy condition (b)}] \tag{8.26}$$

In these relations, the subscripts F and B refer to the *filter* and the *base* material (i.e., the backfill soil), respectively. Also, D_{15} and D_{85} refer to the diameters through which 15% and 85% of the soil (filter or base, as the case may be) will pass. Example 8.3 gives the procedure for designing a filter.

Example 8.3

Figure 8.16 shows the grain-size distribution of a backfill material. Using the conditions outlined in Section 8.8, determine the range of the grain-size distribution for the filter material.

Figure 8.16 Determination of grain-size distribution of filter material

Solution
From the grain-size distribution curve given in the figure, the following values can be determined:

$$D_{15(B)} = 0.04 \text{ mm}$$

$$D_{85(B)} = 0.25 \text{ mm}$$

$$D_{50(B)} = 0.13 \text{ mm}$$

Conditions of Filter

1. $D_{15(F)}$ should be less than $5D_{85(B)}$; that is, $5 \times 0.25 = 1.25$ mm.
2. $D_{15(F)}$ should be greater than $4D_{15(B)}$; that is, $4 \times 0.04 = 0.16$ mm.
3. $D_{50(F)}$ should be less than $25D_{50(B)}$; that is, $25 \times 0.13 = 3.25$ mm.
4. $D_{15(F)}$ should be less than $20D_{15(B)}$; that is, $20 \times 0.04 = 0.8$ mm.

These limiting points are plotted in Figure 8.16. Through them, two curves can be drawn that are similar in nature to the grain-size distribution curve of the backfill material. These curves define the range of the filter material to be used. ∎

8.9 Gravity Retaining-Wall Design for Earthquake Conditions

Even in mild earthquakes, most retaining walls undergo limited lateral displacement. Richards and Elms (1979) proposed a procedure for designing gravity retaining walls for earthquake conditions that allows limited lateral displacement. This procedure takes into consideration the wall inertia effect. Figure 8.17 shows a retaining wall with various forces acting on it, which are as follows (per unit length of the wall):

a. W_w = weight of the wall
b. P_{ae} = active force with earthquake condition taken into consideration (Section 7.7)

The backfill of the wall and the soil on which the wall is resting are assumed cohesionless. Considering the equilibrium of the wall, it can be shown that

$$W_w = \left[\tfrac{1}{2}\gamma_1 H^2(1 - k_v)K_{ae}\right]C_{IE} \tag{8.27}$$

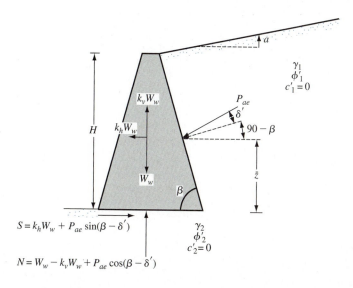

Figure 8.17 Stability of a retaining wall under earthquake forces

where γ_1 = unit weight of the backfill;

$$C_{IE} = \frac{\sin(\beta - \delta') - \cos(\beta - \delta')\tan\phi_2'}{(1 - k_v)(\tan\phi_2' - \tan\theta')}$$ (8.28)

and $\theta' = \tan^{-1}\left(\dfrac{k_k}{1 - k_v}\right)$

For a detailed derivation of Eq. (8.28), see Das (1983).

Based on Eqs. (8.27) and (8.28), the following procedure may be used to determine the weight of the retaining wall, W_w, for tolerable displacement that may take place during an earthquake.

1. Determine the tolerable displacement of the wall, Δ.
2. Obtain a design value of k_k from

$$k_k = A_a\left(\frac{0.2A_v^2}{A_a\Delta}\right)^{0.25}$$ (8.29)

In Eq. (8.29), A and A_a are effective acceleration coefficients and Δ is displacement in inches. The magnitudes of A_a and A_v are given by the Applied Technology Council (1978) for various regions of the United States
3. Assume that $k_v = 0$, and, with the value of k_k obtained, calculate K_{ae} from Eq. (7.43).
4. Use the value of K_{ae} determined in Step 3 to obtain the weight of the wall (W_w)
5. Apply a factor of safety to the value of W_w obtained in Step 4.

Example 8.4

Refer to Figure 8.18. For $k_v = 0$ and $k_k = 0.3$, determine:

a. Weight of the wall for static condition
b. Weight of the wall for zero displacement during an earthquake
c. Weight of the wall for lateral displacement of 38 mm (1.5 in.) during an earthquake

$\phi_1' = 36°$
$\gamma_1 = 16$ kN/m³
$\delta' = 2/3\,\phi_1'$

5 m

$\phi_2' = 36°$
$\gamma_2 = 16$ kN/m³

Figure 8.18

For part c, assume that $A_a = 0.2$ and $A_v = 0.2$. For parts a, b, and c, use a factor of safety of 1.5.

Solution

Part a

For static conditions, $\theta' = 0$ and Eq. (8.28) becomes

$$C_{IE} = \frac{\sin(\beta - \delta') - \cos(\beta - \delta')\tan\phi_2'}{\tan\phi_2'}$$

For $\beta = 90°$, $\delta' = 24°$ and $\phi_2' = 36°$,

$$C_{IE} = \frac{\sin(90 - 24) - \cos(90 - 24)\tan 36}{\tan 36} = 0.85$$

For static conditions, $K_{ae} = K_a$, so

$$W_w = \frac{1}{2}\gamma H^2 K_a C_{IE}$$

For $K_a \approx 0.2349$ [Table 7.4],

$$W_w = \frac{1}{2}(16)(5)^2(0.2349)(0.85) = 39.9 \text{ kN/m}$$

With a factor of safety of 1.5,

$$W_w = (39.9)(1.5) = \textbf{59.9 kN/m}$$

Part b

For zero displacement, $k_v = 0$,

$$C_{IE} = \frac{\sin(\beta - \delta') - \cos(\beta - \delta')\tan\phi_2'}{\tan\phi_2' - \tan\theta'}$$

$$\tan\theta' = \frac{k_h}{1 - k_v} = \frac{0.3}{1 - 0} = 0.3$$

$$C_{IE} = \frac{\sin(90 - 24) - \cos(90 - 24)\tan 36}{\tan 36 - 0.3} = 1.45$$

For $k_h = 0.3$, $\phi_1' = 36°$ and $\delta' = 2\phi_1'/3$, the value of $K_{ae} \approx 0.48$ (Table 7.6).

$$W_w = \tfrac{1}{2}\gamma_1 H^2(1 - k_v)K_{ae}C_{IE} = \tfrac{1}{2}(16)(5)^2(1 - 0)(0.48)(1.45) = 139.2 \text{ kN/m}$$

With a factor of safety of 1.5, $W_w = \textbf{208.8 kN/m}$

Part c

For a lateral displacement of 38 mm,

$$k_h = A_a\left(\frac{0.2A_v^2}{A_a\Delta}\right)^{0.25} = (0.2)\left[\frac{(0.2)(0.2)^2}{(0.2)(38/25.4)}\right]^{0.25} = 0.081$$

$$\tan \theta' = \frac{k_h}{1 - k_v} = \frac{0.081}{1 - 0} = 0.081$$

$$C_{IE} = \frac{\sin(90 - 24) - \cos(90 - 24)\tan 36}{\tan 36 - 0.081} = 0.957$$

$$W_w = \frac{1}{2}\gamma_1 H^2 K_{ae} C_{IE}$$

$$\uparrow$$

$$\approx 0.29 \text{ [Table 7.6]}$$

$$W_w = \frac{1}{2}(16)(5)^2(0.29)(0.957) = 55.5 \text{ kN/m}$$

With a factor of safety of 1.5, $W_w = \mathbf{83.3 \text{ kN/m}}$ ∎

8.10 *Comments on Design of Retaining Walls and a Case Study*

In Section 8.3, it was suggested that the *active earth pressure coefficient* be used to estimate the lateral force on a retaining wall due to the backfill. It is important to recognize the fact that the active state of the backfill can be established only if the wall yields sufficiently, which does not happen in all cases. The degree to which the wall yields depends on its *height* and the *section modulus*. Furthermore, the lateral force of the backfill depends on several factors identified by Casagrande (1973):

1. Effect of temperature
2. Groundwater fluctuation
3. Readjustment of the soil particles due to creep and prolonged rainfall
4. Tidal changes
5. Heavy wave action
6. Traffic vibration
7. Earthquakes

Insufficient wall yielding combined with other unforeseen factors may generate a larger lateral force on the retaining structure, compared with that obtained from the active earth-pressure theory. This is particularly true in the case of gravity retaining walls, bridge abutments, and other heavy structures that have a large section modulus.

Case Study for the Performance of a Cantilever Retaining Wall

Bentler and Labuz (2006) have reported the performance of a cantilever retaining wall built along Interstate 494 in Bloomington, Minnesota. The retaining wall had 83 panels, each having a length of 9.3 m. The panel height ranged from 4.0 m to 7.9 m. One of the 7.9 m high panels was instrumented with earth pressure cells, tiltmeters, strain gauges, and

7.9 m

2.4°

Granular
backfill (SP)
$\gamma_1 = 18.9$ kN/m³
$\phi_1 = 35°$ to 39°
(Av. 37°)

Poorly graded
sand, and sand and
gravel

Figure 8.19 Schematic
diagram of the retaining wall
(drawn to scale)

inclinometer casings. Figure 8.19 shows a schematic diagram (cross section) of the wall panel. Some details on the backfill and the foundation material are:

- Granular Backfill
  Effective size, $D_{10} = 0.13$ mm
  Uniformity coefficient, $C_u = 3.23$
  Coefficient of gradation, $C_c = 1.4$
  Unified soil classification − SP
  Compacted unit weight, $\gamma_1 = 18.9$ kN/m³
  Triaxial friction angle, $\phi_1' − 35°$ to 39° (average 37°)

- Foundation Material
  Poorly graded sand and sand with gravel (medium dense to dense)

The backfill and compaction of the granular material started on October 28, 2001 in stages and reached a height of 7.6 m on November 21, 2001. The final 0.3 m of soil was placed the following spring. During backfilling, the wall was continuously going through translation (see Section 7.9). Table 8.3 is a summary of the backfill height and horizontal translation of the wall.

Table 8.3 Horizontal Translation with Backfill Height

Day	Backfill height (m)	Horizontal translation (mm)
1	0.0	0
2	1.1	0
2	2.8	0
3	5.2	2
4	6.1	4
5	6.4	6
11	6.7	9
24	7.3	12
54	7.6	11

Figure 8.20 Observed lateral pressure distribution after fill height reached 6.1 m (Based on Bentler and Labuz, 2006)

Figure 8.20 shows a typical plot of the variation of lateral earth pressure *after compaction*, σ_a', when the backfill height was 6.1 m (October 31, 2001) along with the plot of Rankine active earth pressure ($\phi_1' = 37°$). Note that the measured lateral (horizontal) pressure is higher at most heights than that predicted by the Rankine active pressure theory, which may be due to residual lateral stresses caused by compaction. The measured lateral stress gradually reduced with time. This is demonstrated in Figure 8.21 which shows a plot of the variation of σ_a' with depth (November 27, 2001) when the height of the backfill was 7.6 m. The lateral pressure was lower at practically all depths compared to the Rankine active earth pressure.

Another point of interest is the nature of variation of q_{max} and q_{min} (see Figure 8.11). As shown in Figure 8.11, if the wall rotates about C, q_{max} will be at the toe and q_{min} will be at the heel. However, for the case of the retaining wall under consideration (undergoing

Figure 8.21 Observed pressure distribution on November 27, 2001 (Based on Bentler and Labuz, 2006)

horizontal translation), q_{max} was at the heel of the wall with q_{min} at the toe. On November 27, 2001, when the height of the fill was 7.6 m, q_{max} at the heel was about 140 kN/m², which was approximately equal to (γ_1)(height of fill) $=(18.9)(7.6) = 143.6$ kN/m². Also, at the toe, q_{min} was about 40 kN/m², which suggests that the moment from lateral force had little effect on the vertical effective stress below the heel.

The lessons learned from this case study are the following:

a. Retaining walls may undergo lateral translation which will affect the variation of q_{max} and q_{min} along the base slab.
b. Initial lateral stress caused by compaction gradually decreases with time and lateral movement of the wall.

Mechanically Stabilized Retaining Walls

More recently, soil reinforcement has been used in the construction and design of foundations, retaining walls, embankment slopes, and other structures. Depending on the type of construction, the reinforcements may be galvanized metal strips, geotextiles, geogrids, or geocomposites. Sections 8.11 and 8.12 provide a general overview of soil reinforcement and various reinforcement materials.

Reinforcement materials such as metallic strips, geotextiles, and geogrids are now being used to reinforce the backfill of retaining walls, which are generally referred to as *mechanically stabilized retaining walls*. The general principles for designing these walls are given in the following sections.

8.11 *Soil Reinforcement*

The use of reinforced earth is a recent development in the design and construction of foundations and earth-retaining structures. *Reinforced earth* is a construction material made from soil that has been strengthened by tensile elements such as metal rods or strips, nonbiodegradable fabrics (geotextiles), geogrids, and the like. The fundamental idea of reinforcing soil is not new; in fact, it goes back several centuries. However, the present concept of systematic analysis and design was developed by a French engineer, H. Vidal (1966). The French Road Research Laboratory has done extensive research on the applicability and the beneficial effects of the use of reinforced earth as a construction material. This research has been documented in detail by Darbin (1970), Schlosser and Long (1974), and Schlosser and Vidal (1969). The tests that were conducted involved the use of metallic strips as reinforcing material.

Retaining walls with reinforced earth have been constructed around the world since Vidal began his work. The first reinforced-earth retaining wall with metal strips as reinforcement in the United States was constructed in 1972 in southern California.

The beneficial effects of soil reinforcement derive from (a) the soil's increased tensile strength and (b) the shear resistance developed from the friction at the soil-reinforcement interfaces. Such reinforcement is comparable to that of concrete structures. Currently, most reinforced-earth design is done with *free-draining granular soil only*. Thus, the effect of pore water development in cohesive soils, which, in turn, reduces the shear strength of the soil, is avoided.

8.12	**Considerations in Soil Reinforcement**

Metal Strips

In most instances, galvanized steel strips are used as reinforcement in soil. However, galvanized steel is subject to corrosion. The rate of corrosion depends on several environmental factors. Binquet and Lee (1975) suggested that the average rate of corrosion of galvanized steel strips varies between 0.025 and 0.050 mm/yr. So, in the actual design of reinforcement, allowance must be made for the rate of corrosion. Thus,

$$t_c = t_{\text{design}} + r \, (\text{life span of structure})$$

where

t_c = actual thickness of reinforcing strips to be used in construction
t_{design} = thickness of strips determined from design calculations
r = rate of corrosion

Further research needs to be done on corrosion-resistant materials such as fiberglass before they can be used as reinforcing strips.

Nonbiodegradable Fabrics

Nonbiodegradable fabrics are generally referred to as *geotextiles*. Since 1970, the use of geotextiles in construction has increased greatly around the world. The fabrics are usually made from petroleum products—polyester, polyethylene, and polypropylene. They may also be made from fiberglass. Geotextiles are not prepared from natural fabrics, because they decay too quickly. Geotextiles may be woven, knitted, or nonwoven.

   Woven geotextiles are made of two sets of parallel filaments or strands of yarn systematically interlaced to form a planar structure. *Knitted geotextiles* are formed by interlocking a series of loops of one or more filaments or strands of yarn to form a planar structure. *Nonwoven geotextiles* are formed from filaments or short fibers arranged in an oriented or random pattern in a planar structure. These filaments or short fibers are arranged into a loose web in the beginning and then are bonded by one or a combination of the following processes:

1. *Chemical bonding*—by glue, rubber, latex, a cellulose derivative, or the like
2. *Thermal bonding*—by heat for partial melting of filaments
3. *Mechanical bonding*—by needle punching

Needle-punched nonwoven geotextiles are thick and have high in-plane permeability.
   Geotextiles have four primary uses in foundation engineering:

1. *Drainage:* The fabrics can rapidly channel water from soil to various outlets, thereby providing a higher soil shear strength and hence stability.
2. *Filtration:* When placed between two soil layers, one coarse grained and the other fine grained, the fabric allows free seepage of water from one layer to the other. However, it protects the fine-grained soil from being washed into the coarse-grained soil.
3. *Separation:* Geotextiles help keep various soil layers separate after construction and during the projected service period of the structure. For example, in the construction of highways, a clayey subgrade can be kept separate from a granular base course.
4. *Reinforcement:* The tensile strength of geofabrics increases the load-bearing capacity of the soil.

Geogrids

Geogrids are high-modulus polymer materials, such as polypropylene and polyethylene, and are prepared by tensile drawing. Netlon, Ltd., of the United Kingdom was the first producer of geogrids. In 1982, the Tensar Corporation, presently Tensar International Corporation, introduced geogrids into the United States.

Commercially available geogrids may be categorized by manufacturing process, principally: extruded, woven, and welded. Extruded geogrids are formed using a thick sheet of polyethylene or polypropylene that is punched and drawn to create apertures and to enhance engineering properties of the resulting ribs and nodes. Woven geogrids are manufactured by grouping polymeric—usually polyester and polypropylene—and weaving them into a mesh pattern that is then coated with a polymeric lacquer. Welded geogrids are manufactured by fusing junctions of polymeric strips. Extruded geogrids have shown good performance when compared to other types for pavement reinforcement applications.

Geogrids generally are of two types: (a) uniaxial and (b) biaxial. Figures 8.22a and b shows these two types of geogrids, which are produced by Tensar International Corporation.

Uniaxial TENSAR grids are manufactured by stretching a punched sheet of extruded high-density polyethylene in one direction under carefully controlled conditions. The process aligns the polymer's long-chain molecules in the direction of draw and results in a product with high one-directional tensile strength and a high modulus. Biaxial TENSAR grids are manufactured by stretching the punched sheet of polypropylene in two orthogonal directions. This process results in a product with high tensile strength and a high modulus in two perpendicular directions. The resulting grid apertures are either square or rectangular.

The commercial geogrids currently available for soil reinforcement have nominal rib thicknesses of about 0.5 to 1.5 mm (0.02 to 0.06 in.) and junctions of about 2.5 to 5 mm (0.1 to 0.2 in.). The grids used for soil reinforcement usually have openings or apertures that are rectangular or elliptical. The dimensions of the apertures vary from about 25 to 150 mm (1 to 6 in.). Geogrids are manufactured so that the open areas of the grids are greater than 50% of the total area. They develop reinforcing strength at low strain levels, such as 2% (Carroll, 1988). Table 8.4 gives some properties of the TENSAR biaxial geogrids that are currently available commercially.

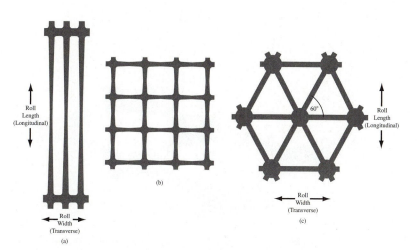

Figure 8.22 Geogrid: (a) uniaxial; (b) biaxial; (c) with triangular apertures (*Courtesy of Tensar International Corporation*)

Table 8.4 Properties of TENSAR Biaxial Geogrids

| Property | Geogrid | | |
	BX1000	BX1100	BX1200
Aperture size			
Machine direction	25 mm (nominal)	25 mm (nominal)	25 mm (nominal)
Cross-machine direction	33 mm (nominal)	33 mm (nominal)	33 mm (nominal)
Open area	70% (minimum)	74% (nominal)	77% (nominal)
Junction			
Thickness	2.3 mm (nominal)	2.8 mm (nominal)	4.1 mm (nominal)
Tensile modulus			
Machine direction	182 kN/m (minimum)	204 kN/m (minimum)	270 kN/m (minimum)
Cross-machine direction	182 kN/m (minimum)	292 kN/m (minimum)	438 kN/m (minimum)
Material			
Polypropylene	97% (minimum)	99% (nominal)	99% (nominal)
Carbon black	2% (minimum)	1% (nominal)	1% (nominal)

The major function of geogrids is *reinforcement*. They are relatively stiff. The apertures are large enough to allow interlocking with surrounding soil or rock (Figure 8.23) to perform the function of reinforcement or segregation (or both). Sarsby (1985) investigated the influence of aperture size on the size of soil particles for maximum frictional efficiency (or efficiency against pullout). According to this study, the highest efficiency occurs when

$$B_{GG} > 3.5D_{50} \tag{8.30}$$

Figure 8.23 Geogrid apertures allowing interlocking with surrounding soil

where

B_{GG} = minimum width of the geogrid aperture

D_{50} = the particle size through which 50% of the backfill soil passes (i.e., the average particle size)

More recently, geogrids with triangular apertures (Figure 8.22c) have been introduced for construction purposes. TENSAR geogrids with triangular apertures are manufactured from a punched polypropylene sheet, which is then oriented in three substantially equilateral directions so that the resulting ribs shall have a high degree of molecular orientation. Table 8.5 gives some properties of TENSAR geogrids with triangular apertures.

Table 8.5 Properties of TENSAR Geogrids with Triangular Apertures

Geogrid	Property	Longitudinal	Diagonal	Transverse	General
TX 160	Rib pitch, (mm)	40	40	—	
	Mid-rib depth, (mm)	—	1.8	1.5	
	Mid-rib width, (mm)	—	1.1	1.3	
	Nodal thickness, (mm)				3.1
	Radial stiffness at low strain, (kN/m @ 0.5% strain)				430
TX 170	Rib pitch, (mm)	40	40	—	
	Mid-rib depth, (mm)	—	2.3	1.8	
	Mid-rib width, (mm)	—	1.2	1.3	
	Nodal thickness, (mm)				4.1
	Radial stiffness at low strain, (kN/m @ 0.5% strain)				475

8.13 *General Design Considerations*

The general design procedure of any mechanically stabilized retaining wall can be divided into two parts:

1. Satisfying *internal stability* requirements
2. Checking the *external stability* of the wall

The internal stability checks involve determining tension and pullout resistance in the reinforcing elements and ascertaining the integrity of facing elements. The external stability checks include checks for overturning, sliding, and bearing capacity failure (Figure 8.24). The sections that follow will discuss the retaining-wall design procedures for use with metallic strips, geotextiles, and geogrids.

(a) Sliding   (b) Overturning

(c) Bearing capacity   (d) Deep-seated stability

Figure 8.24 External stability checks (After Transportation Research Board, 1995) (From Transportation Research Circular 444: Mechanically Stabilized Earth Walls. Transportation Research Board, National Research Council, Washington, D.C., 1995, Figure 3, p. 7. Reproduced with permission of the Transportation Research Board.)

8.14   *Retaining Walls with Metallic Strip Reinforcement*

Reinforced-earth walls are flexible walls. Their main components are

1. *Backfill,* which is granular soil
2. *Reinforcing strips,* which are thin, wide strips placed at regular intervals, and
3. A *cover* or *skin,* on the front face of the wall

Figure 8.25 is a diagram of a reinforced-earth retaining wall. Note that, at any depth, the reinforcing strips or ties are placed with a horizontal spacing of S_H center to center; the vertical spacing of the strips or ties is S_V center to center. The skin can be constructed with sections of relatively flexible thin material. Lee et al. (1973) showed that, with a conservative design, a 5 mm-thick galvanized steel skin would be enough to hold a wall about 14 to 15 m high. In most cases, precast concrete slabs can also be used as skin. The slabs are grooved to fit into each other so that soil cannot flow out between the joints. When metal skins are used, they are bolted together, and reinforcing strips are placed between the skins.

Figures 8.26 and 8.27 show a reinforced-earth retaining wall under construction; its skin (facing) is a precast concrete slab. Figure 8.28 shows a metallic reinforcement tie attached to the concrete slab.

The simplest and most common method for the design of ties is the *Rankine method.* We discuss this procedure next.

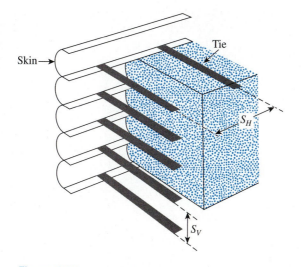

Figure 8.25 Reinforced-earth retaining wall

Figure 8.26 Reinforced-earth retaining wall (with metallic strip) under construction (*Courtesy of Braja M. Das, Henderson, NV*)

Figure 8.27 Another view of the retaining wall shown in Figure 8.26 (*Courtesy of Braja M. Das, Henderson, NV*)

Figure 8.28 Metallic strip attachment to the precast concrete slab used as the skin (*Courtesy of Braja M. Das, Henderson, NV*)

Calculation of Active Horizontal and Vertical Pressure

Figure 8.29 shows a retaining wall with a granular backfill having a unit weight of γ_1 and a friction angle of ϕ_1'. Below the base of the retaining wall, the *in situ* soil has been excavated and recompacted, with granular soil used as backfill. Below the backfill, the *in situ* soil has a unit weight of γ_2, friction angle of ϕ_2', and cohesion of c_2'. A surcharge having an intensity of q per unit area lies atop the retaining wall, which has reinforcement ties at depths $z = 0, S_V, 2S_V, \ldots, NS_V$. The height of the wall is $NS_V = H$.

According to the Rankine active pressure theory (Section 7.3)

$$\sigma_a' = \sigma_o' K_a - 2c' \sqrt{K_a}$$

where σ_a' = Rankine active pressure at any depth z.

For dry granular soils with no surcharge at the top, $c' = 0$, $\sigma_o' = \gamma_1 z$, and $K_a = \tan^2(45 - \phi_1'/2)$. Thus,

$$\sigma_{a(1)}' = \gamma_1 z K_a \tag{8.31}$$

When a surcharge is added at the top, as shown in Figure 8.29,

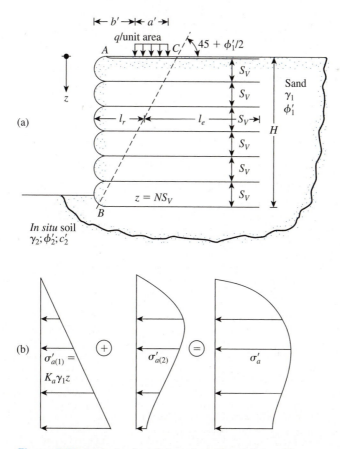

(a)

(b)

Figure 8.29 Analysis of a reinforced-earth retaining wall

$$\sigma'_o = \sigma'_{o(1)} \qquad + \sigma'_{o(2)} \tag{8.32}$$

$$\uparrow \qquad\qquad \uparrow$$

$$= \gamma_1 z \qquad \text{Due to the}$$

Due to   surcharge

soil only

The magnitude of $\sigma'_{o(2)}$ can be calculated by using the 2:1 method of stress distribution described in Eq. (5.14) and Figure 5.5. The 2:1 method of stress distribution is shown in Figure 8.30a. According to Laba and Kennedy (1986),

$$\sigma'_{o(2)} = \frac{qa'}{a' + z} \qquad (\text{for } z \le 2b') \tag{8.33}$$

and

$$\sigma'_{o(2)} = \frac{qa'}{a' + \dfrac{z}{2} + b'} \qquad (\text{for } z > 2b') \tag{8.34}$$

Also, when a surcharge is added at the top, the lateral pressure at any depth is

$$\sigma'_a = \sigma'_{a(1)} \qquad + \sigma'_{a(2)}$$

$$\uparrow \qquad\qquad \uparrow$$

$$= K_a \gamma_1 z \qquad \text{Due to the} \tag{8.35}$$

Due to   surcharge

soil only

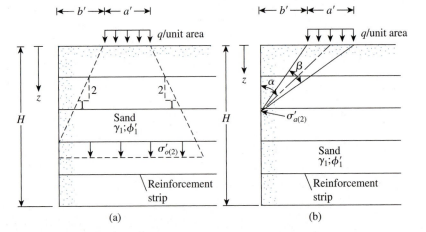

Figure 8.30 (a) Notation for the relationship of $\sigma'_{o(2)}$ in Eqs. (8.33) and (8.34); (b) notation for the relationship of $\sigma'_{a(2)}$ in Eqs. (8.36) and (8.37)

According to Laba and Kennedy (1986), $\sigma'_{a(2)}$ may be expressed (see Figure 8.30b) as

$$\sigma'_{a(2)} = M\left[\frac{2q}{\pi}(\beta - \sin\beta \cos 2\alpha)\right]$$
$$\uparrow$$
$$(\text{in radians})$$

(8.36)

where

$$M = 1.4 - \frac{0.4b'}{0.14H} \geq 1$$

(8.37)

The net active (lateral) pressure distribution on the retaining wall calculated by using Eqs. (8.35), (8.36), and (8.37) is shown in Figure 8.29b.

Tie Force

The tie force *per unit length of the wall* developed at any depth z (see Figure 8.29) is

$$T = \text{active earth pressure at depth } z$$
$$\times \text{ area of the wall to be supported by the tie}$$
$$= (\sigma'_a)(S_V S_H)$$

(8.38)

Factor of Safety against Tie Failure

The reinforcement ties at each level, and thus the walls, could fail by either (a) tie breaking or (b) tie pullout.

The factor of safety against *tie breaking* may be determined as

$$FS_{(B)} = \frac{\text{yield or breaking strength of each tie}}{\text{maximum force in any tie}}$$
$$= \frac{wtf_y}{\sigma'_a S_V S_H}$$

(8.39)

where

w = width of each tie
t = thickness of each tie
f_y = yield or breaking strength of the tie material

A factor of safety of about 2.5 to 3 is generally recommended for ties at all levels.

Reinforcing ties at any depth z will fail by pullout if the frictional resistance developed along the surfaces of the ties is less than the force to which the ties are being subjected. The *effective length* of the ties along which frictional resistance is developed may be conservatively taken as the length that extends *beyond the limits of the Rankine active failure zone,* which is the zone *ABC* in Figure 8.29. Line *BC* makes an angle of $45 + \phi_1'/2$ with the horizontal. Now, the maximum friction force that can be realized for a tie at depth z is

$$F_R = 2l_e w \sigma_o' \tan \phi_\mu' \tag{8.40}$$

where

l_e = effective length
σ_o' = effective vertical pressure at a depth z
ϕ_μ' = soil–tie friction angle

Thus, the factor of safety against *tie pullout* at any depth z is

$$\text{FS}_{(P)} = \frac{F_R}{T} \tag{8.41}$$

Substituting Eqs. (8.38) and (8.40) into Eq. (8.41) yields

$$\text{FS}_{(P)} = \frac{2l_e w \sigma_o' \tan \phi_\mu'}{\sigma_a' S_V S_H} \tag{8.42}$$

Total Length of Tie

The total length of ties at any depth is

$$L = l_r + l_e \tag{8.43}$$

where

l_r = length within the Rankine failure zone
l_e = effective length

For a given $\text{FS}_{(P)}$ from Eq. (8.42),

$$l_e = \frac{\text{FS}_{(P)} \sigma_a' S_V S_H}{2w\sigma_o' \tan \phi_\mu'} \tag{8.44}$$

Again, at any depth z,

$$l_r = \frac{(H - z)}{\tan\left(45 + \dfrac{\phi_1'}{2}\right)} \tag{8.45}$$

So, combining Eqs. (8.43), (8.44), and (8.45) gives

$$L = \frac{(H - z)}{\tan\left(45 + \dfrac{\phi_1'}{2}\right)} + \frac{\text{FS}_{(P)} \sigma_a' S_V S_H}{2w\sigma_o' \tan \phi_\mu'} \tag{8.46}$$

8.15   *Step-by-Step-Design Procedure Using Metallic Strip Reinforcement*

Following is a step-by-step procedure for the design of reinforced-earth retaining walls.

General

Step 1. Determine the height of the wall, *H,* and the properties of the granular backfill material, such as the unit weight (γ_1) and the angle of friction (ϕ_1').

Step 2. Obtain the soil–tie friction angle, ϕ_μ', and the required value of $FS_{(B)}$ and $FS_{(P)}$.

Internal Stability

Step 3. Assume values for horizontal and vertical tie spacing. Also, assume the width of reinforcing strip, *w*, to be used.

Step 4. Calculate σ_a' from Eqs. (8.35), (8.36), and (8.37).

Step 5. Calculate the tie forces at various levels from Eq. (8.38).

Step 6. For the known values of $FS_{(B)}$, calculate the thickness of ties, *t*, required to resist the tie breakout:

$$T = \sigma_a' S_V S_H = \frac{w t f_y}{FS_{(B)}}$$

or

$$t = \frac{(\sigma_a' S_V S_H)[FS_{(B)}]}{w f_y} \tag{8.47}$$

The convention is to keep the magnitude of *t* the same at all levels, so σ_a' in Eq. (8.47) should equal $\sigma_{a(\max)}'$.

Step 7. For the known values of ϕ_μ' and $FS_{(P)}$, determine the length *L* of the ties at various levels from Eq. (8.46).

Step 8. The magnitudes of S_V, S_H, *t*, *w*, and *L* may be changed to obtain the most economical design.

External Stability

Step 9. Check for *overturning,* using Figure 8.31 as a guide. Taking the moment about *B* yields the overturning moment for the unit length of the wall:

$$M_o = P_a z' \tag{8.48}$$

Here,

$$P_a = \text{active force} = \int_0^H \sigma_a' dz$$

The resisting moment per unit length of the wall is

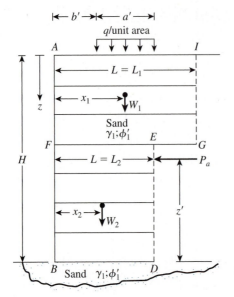

In situ soil
$\gamma_2; \phi_2'; c_2'$

Figure 8.31 Stability check for the retaining wall

$$M_R = W_1 x_1 + W_2 x_2 + \cdots + qa'\left(b' + \frac{a'}{2}\right) \qquad (8.49)$$

where

$W_1 = (\text{area } AFEGI)(1)(\gamma_1)$
$W_2 = (\text{area } FBDE)(1)(\gamma_1)$
$\vdots$

So,

$$FS_{(\text{overturning})} = \frac{M_R}{M_o}$$

$$= \frac{W_1 x_1 + W_2 x_2 + \cdots + qa'\left(b' + \dfrac{a'}{2}\right)}{\left(\displaystyle\int_0^H \sigma_a' \, dz\right) z'} \qquad (8.50)$$

Step 10. The check for *sliding* can be done by using Eq. (8.11), or

$$FS_{(\text{sliding})} = \frac{(W_1 + W_2 + \cdots + qa')[\tan(k\phi_1')]}{P_a} \qquad (8.51)$$

where $k \approx \frac{2}{3}$.

Step 11. Check for ultimate bearing capacity failure, which can be given as

$$q_u = c_2' N_c + \tfrac{1}{2}\gamma_2 L_2 N_\gamma \tag{8.52}$$

The bearing capacity factors N_c and N_γ correspond to the soil friction angle ϕ_2'. (See Table 3.3.)

From Eq. 8.32, the vertical stress at $z = H$ is

$$\sigma_{o(H)}' = \gamma_1 H + \sigma_{o(2)}' \tag{8.53}$$

So the factor of safety against bearing capacity failure is

$$\text{FS}_{(\text{bearing capacity})} = \frac{q_{\text{ult}}}{\sigma_{o(H)}'} \tag{8.54}$$

Generally, minimum values of $\text{FS}_{(\text{overturning})} = 3$, $\text{FS}_{(\text{sliding})} = 3$, and $\text{FS}_{(\text{bearing capacity failure})} = 3$ to 5 are recommended.

Example 8.5

A 10 m high retaining wall with galvanized steel-strip reinforcement in a granular backfill has to be constructed. Referring to Figure 8.29, given:

Granular backfill: $\phi_1' = 36°$
                           $\gamma_1 = 16.5 \text{ kN/m}^3$

Foundation soil: $\phi_2' = 28°$
                           $\gamma_2 = 17.3 \text{ kN/m}^3$
                           $c_2' = 50 \text{ kN/m}^2$

Galvanized steel reinforcement:

Width of strip, $w = 75 \text{ mm}$
                     $S_V = 0.6 \text{ m center-to-center}$
                     $S_H = 1 \text{ m center-to-center}$
                     $f_y = 240,00 \text{ kN/m}^2$
                     $\phi_\mu' = 20°$

Required $\text{FS}_{(B)} = 3$

Required $\text{FS}_{(P)} = 3$

Check for the external and internal stability. Assume the corrosion rate of the galvanized steel to be 0.025 mm/year and the life span of the structure to be 50 years.

Solution
Internal Stability Check

Tie thickness: Maximum tie force, $T_{\text{max}} = \sigma_{a(\text{max})}' S_V S_H$

$$\sigma_{a(max)} = \gamma_1 H K_a = \gamma_1 H \tan^2\left(45 - \frac{\phi_1'}{2}\right)$$

so

$$T_{max} = \gamma_1 H \tan^2\left(45 - \frac{\phi_1'}{2}\right) S_V S_H$$

From Eq. (8.47), for *tie break*,

$$t = \frac{(\sigma_a' S_V S_H)[FS_{(B)}]}{wf_y} = \frac{\left[\gamma_1 H \tan^2\left(45 - \frac{\phi_1'}{2}\right) S_V S_H\right] FS_{(B)}}{wf_y}$$

or

$$t = \frac{\left[(16.5)(10) \tan^2\left(45 - \frac{36}{2}\right)(0.6)(1)\right](3)}{(0.075 \text{ m})(240{,}000 \text{ kN/m}^2)} = 0.00428\text{m} = 4.28 \text{ mm}$$

If the rate of corrosion is 0.025 mm/yr and the life span of the structure is 50 yr, then the actual thickness, *t*, of the ties will be

$$t = 4.28 + (0.025)(50) = 5.53 \text{ mm}$$

So a **tie thickness of 6 mm** would be enough.

Tie length: Refer to Eq. (8.46). For this case, $\sigma_a' = \gamma_1 z K_a$ and $\sigma_o' = \gamma_1 z$, so

$$L = \frac{(H - z)}{\tan\left(45 + \frac{\phi_1'}{2}\right)} + \frac{FS_{(P)} \gamma_1 z K_a S_V S_H}{2w\gamma_1 z \tan\phi_\mu'}$$

Now the following table can be prepared. (Note: $FS_{(P)} = 3$, $H = 10$ m, $w = 75$ mm, and $\phi_\mu' = 20°$.)

z(m)	Tie length L (m) [Eq. (8.46)]
2	12.65
4	11.63
6	10.61
8	9.59
10	8.57

So use a **tie length of $L = 13$ m.**

External Stability Check
Check for overturning: Refer to Figure 8.32. For this case, using Eq. (8.50)

$$FS_{(overturning)} = \frac{W_1 x_1}{\left[\int_0^H \sigma_a' \, dz\right] z'}$$

Figure 8.32 Retaining wall with galvanized steel-strip reinforcement in the backfill

$$W_1 = \gamma_1 HL = (16.5)(10)(13) = 2145 \text{ kN/m}$$

$$x_1 = 6.5 \text{ m}$$

$$P_a = \int_0^H \sigma'_a \, dz = \tfrac{1}{2}\gamma_1 K_a H^2 = \left(\tfrac{1}{2}\right)(16.5)(0.26)(10)^2 = 214.5 \text{ kN/m}$$

$$z' = \frac{10}{3} = 3.33 \text{ m}$$

$$FS_{(\text{overturning})} = \frac{(2145)(6.5)}{(214.5)(3.33)} = 19.52 > 3\text{—OK}$$

Check for sliding: From Eq. (8.51)

$$FS_{(\text{sliding})} = \frac{W_1 \tan(k\phi'_1)}{P_a} = \frac{2145 \tan\left[\left(\frac{2}{3}\right)(36)\right]}{214.5} = 4.45 > 3\text{—OK}$$

Check for bearing capacity: For $\phi'_2 = 28°$, $N_c = 25.8$, $N_\gamma = 16.78$(Table 3.3). From Eq. (8.52),

$$q_{\text{ult}} = c'_2 N_c + \tfrac{1}{2}\gamma_2 L \, N_\gamma$$

$$q_{\text{ult}} = (50)(25.8) + \left(\tfrac{1}{2}\right)(17.3)(13)(16.72) = 3170.16 \text{ kN/m}^2$$

From Eq. (8.53),

$$\sigma'_{o(H)} = \gamma_1 H = (16.5)(10) = 165 \text{ kN/m}^2$$

$$FS_{(\text{bearing capacity})} = \frac{q_{\text{ult}}}{\sigma'_{o(H)}} = \frac{3170.16}{165} = 19.2 > 5\text{—OK}$$

■

Figure 8.33 shows a retaining wall in which layers of geotextile have been used as reinforcement. As in Figure 8.31, the backfill is a granular soil. In this type of retaining wall, the facing of the wall is formed by lapping the sheets as shown with a lap length of l_l. When construction is finished, the exposed face of the wall must be covered; otherwise, the geotextile will deteriorate from exposure to ultraviolet light. *Bitumen emulsion* or *Gunite* is sprayed on the wall face. A wire mesh anchored to the geotextile facing may be necessary to keep the coating on. Figure 8.34 shows the construction of a geotextile-reinforced retaining wall. Figure 8.35 shows a completed geosynthetic-reinforced soil wall. The wall is in DeBeque Canyon, Colorado. Note the versatility of the facing type. In this case, single-tier concrete block facing is integrated with a three-tier facing via rock facing.

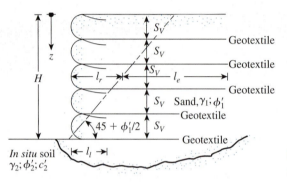

Figure 8.33 Retaining wall with geotextile reinforcement

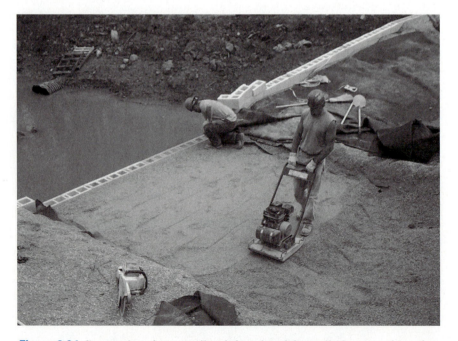

Figure 8.34 Construction of a geotextile-reinforced retaining wall (*Courtesy of Jonathan T. H. Wu, University of Colorado at Denver, Denver, Colorado*)

Figure 8.35 A completed geotextile-reinforced retaining wall in DeBeque Canyon, Colorado (*Courtesy of Jonathan T. H. Wu, University of Colorado at Denver, Denver, Colorado*)

The design of this type of retaining wall is similar to that presented in Section 8.15. Following is a step-by-step procedure for design based on the recommendations of Bell et al. (1975) and Koerner (2005):

Internal Stability

Step 1. Determine the active pressure distribution on the wall from the formula

$$\sigma'_a = K_a \sigma'_o = K_a \gamma_1 z \qquad (8.55)$$

where

K_a = Rankine active pressure coefficient = $\tan^2(45 - \phi'_1/2)$
γ_1 = unit weight of the granular backfill
ϕ'_1 = friction angle of the granular backfill

Step 2. Select a geotextile fabric with an allowable tensile strength, T_{all} (lb/ft or kN/m).

   The allowable tensile strength for retaining wall construction may be expressed as (Koerner, 2005)

$$T_{all} = \frac{T_{ult}}{RF_{id} \times RF_{cr} \times RF_{cbd}} \qquad (8.56)$$

where

T_{ult} = ultimate tensile strength
RF_{id} = reduction factor for installation damage
RF_{cr} = reduction factor for creep
RF_{cbd} = reduction factor for chemical and biological degradation

The recommended values of the reduction factor are as follows (Koerner, 2005)

$$
\begin{array}{ll}
\text{RF}_{id} & 1.1\text{--}2.0 \\
\text{RF}_{cr} & 2\text{--}4 \\
\text{RF}_{cbd} & 1\text{--}1.5
\end{array}
$$

Step 3. Determine the vertical spacing of the layers at any depth z from the formula

$$S_V = \frac{T_{all}}{\sigma'_a \text{FS}_{(B)}} = \frac{T_{all}}{(\gamma_1 z K_a)[\text{FS}_{(B)}]} \tag{8.57}$$

Note that Eq. (8.57) is similar to Eq. (8.39). The magnitude of $\text{FS}_{(B)}$ is generally 1.3 to 1.5.

Step 4. Determine the length of each layer of geotextile from the formula

$$L = l_r + l_e \tag{8.58}$$

where

$$l_r = \frac{H - z}{\tan\left(45 + \dfrac{\phi'_1}{2}\right)} \tag{8.59}$$

and

$$l_e = \frac{S_V \sigma'_a [\text{FS}_{(P)}]}{2\sigma'_o \tan\phi'_F} \tag{8.60}$$

in which

$$\sigma'_a = \gamma_1 z K_a$$

$$\sigma'_o = \gamma_1 z$$

$$\text{FS}_{(P)} = 1.3 \text{ to } 1.5$$

$$\phi'_F = \text{friction angle at geotextile–soil interface}$$

$$\approx \tfrac{2}{3}\phi'_1$$

Note that Eqs. (8.58), (8.59), and (8.60) are similar to Eqs. (8.43), (8.45), and (8.44), respectively.

Based on the published results, the assumption of $\phi'_F/\phi'_1 \approx \tfrac{2}{3}$ is reasonable and appears to be conservative. Martin et al. (1984) presented the following laboratory test results for ϕ'_F/ϕ'_1 between various types of geotextiles and sand.

Type	ϕ'_F/ϕ'_1
Woven—monofilament/concrete sand	0.87
Woven—silt film/concrete sand	0.8
Woven—silt film/rounded sand	0.86
Woven—silt film/silty sand	0.92
Nonwoven—melt-bonded/concrete sand	0.87
Nonwoven—needle-punched/concrete sand	1.0
Nonwoven—needle-punched/rounded sand	0.93
Nonwoven—needle-punched/silty sand	0.91

Step 5. Determine the lap length, l_l, from

$$l_l = \frac{S_V \sigma_a' \mathrm{FS}_{(P)}}{4\sigma_o' \tan \phi_F'} \qquad (8.61)$$

The minimum lap length should be 1 m.

External Stability

Step 6. Check the factors of safety against overturning, sliding, and bearing capacity failure as described in Section 8.15 (Steps 9, 10, and 11).

Example 8.6

A geotextile-reinforced retaining wall 5 m high is shown in Figure 8.36. For the granular backfill, $\gamma_1 = 15.7 \ \mathrm{kN/m^3}$ and $\phi_1' = 36°$. For the geotextile, $T_{\mathrm{ult}} = 52.5 \ \mathrm{kN/m}$. For the design of the wall, determine S_V, L, and l_l. Use $\mathrm{RF}_{\mathrm{id}} = 1.2$, $\mathrm{RF}_{\mathrm{cr}} = 2.5$, and $\mathrm{RF}_{\mathrm{cbd}} = 1.25$.

Solution
We have

$$K_a = \tan^2\left(45 - \frac{\phi_1'}{2}\right) = 0.26$$

Determination of S_V
To find S_V, we make a few trials. From Eq. (8.57),

$$S_V = \frac{T_{\mathrm{all}}}{(\gamma_1 z K_a)[\mathrm{FS}_{(B)}]}$$

$S_V = 0.5$ m

5 m

2.5 m

$\gamma_1 = 15.7 \ \mathrm{kN/m^3}$
$\phi_1' = 36°$

$l_l = 1$ m

$\gamma_2 = 18 \ \mathrm{kN/m^3}$
$\phi_2' = 22°$
$c_2' = 28 \ \mathrm{kN/m^2}$

Figure 8.36 Geotextile-reinforced retaining wall

From Eq. (8.56),

$$T_{all} = \frac{T_{uef}}{RF_{id} \times RF_{cr} \times RF_{cbd}} = \frac{52.5}{1.2 \times 2.5 \times 1.25} = 14 \text{ kN/m}$$

With $FS_{(B)} = 1.5$ at $z = 2$ m,

$$S_V = \frac{14}{(15.7)(2)(0.26)(1.5)} = 1.14 \text{ m}$$

At $z = 4$ m,

$$S_V = \frac{14}{(15.7)(4)(0.26)(1.5)} = 0.57 \text{ m}$$

At $z = 5$ m,

$$S_V = \frac{14}{(15.7)(5)(0.26)(1.5)} = 0.46 \text{ m}$$

So, **use $S_V = 0.5$ m for $z = 0$ to $z = 5$ m** (See Figure 8.36.)

Determination of L
From Eqs. (8.58), (8.59), and (8.60),

$$L = \frac{(H - z)}{\tan\left(45 + \dfrac{\phi_1'}{2}\right)} + \frac{S_V K_a [FS_{(P)}]}{2 \tan \phi_F'}$$

For $FS_{(P)} = 1.5$, $\tan\phi_F' = \tan\left[\left(\frac{2}{3}\right)(36)\right] = 0.445$, and it follows that

$$L = (0.51)(H - z) + 0.438 S_V$$

$H = 5$ m, $S_V = 0.5$ m

At $z = 0.5$ m: $L = (0.51)(5 - 0.5) + (0.438)(0.5) = 2.514$ m
At $z = 2.5$ m: $L = (0.51)(5 - 2.5) + (0.438)(0.5) = 1.494$ m

So, **use $L = 2.5$ m throughout**.

Determination of l_l
From Eq. (8.61),

$$l_l = \frac{S_V \sigma_a' [FS_{(P)}]}{4 \sigma_o' \tan\phi_F'}$$

$\sigma_a' = \gamma_1 z K_a$, $FS_{(P)} = 1.5$; with $\sigma_o' = \gamma_1 z$, $\phi_F' = \frac{2}{3}\phi_1'$. So

$$l_l = \frac{S_V K_a [FS_{(P)}]}{4 \tan\phi_F'} = \frac{S_V (0.26)(1.5)}{4\tan\left[\left(\frac{2}{3}\right)(36)\right]} = 0.219 S_V$$

$$l_l = 0.219 S_V = (0.219)(0.5) = 0.11 \text{ m} \le 1 \text{ m}$$

So, **use $l_l = 1$ m**. ∎

Example 8.7

Consider the results of the internal stability check given in Example 8.6. For the geotextile-reinforced retaining wall, calculate the factor of safety against overturning, sliding, and bearing capacity failure.

Solution
Refer to Figure 8.37.

Factor of Safety Against Overturning

From Eq. (8.50), $\text{FS}_{\text{(overturning)}} = \dfrac{W_1 x_1}{(P_a)\left(\dfrac{H}{3}\right)}$

$$W_1 = (5)(2.5)(15.7) = 196.25 \text{ kN/m}$$

$$x_1 = \frac{2.5}{2} = 1.25 \text{ m}$$

$$P_a = \frac{1}{2}\gamma H^2 K_a = \left(\frac{1}{2}\right)(15.7)(5)^2(0.26) = 51.03 \text{ kN/m}$$

Hence,

$$\text{FS}_{\text{(overturning)}} = \frac{(196.25)(1.25)}{51.03(5/3)} = 2.88 < 3$$

(increase length of geotextile layers to 3 m)

$S_V = 0.5$ m

$\gamma_1 = 15.7$ kN/m^3
$\phi_1' = 36°$

$\gamma_2 = 18$ kN/m^3
$\phi_2' = 22°$
$c_2' = 28$ kN/m^2

Figure 8.37 Stability check

Factor of Safety Against Sliding

From Eq. (8.51),

$$FS_{(sliding)} = \frac{W_1 \tan\left(\frac{2}{3}\phi_1'\right)}{P_a} = \frac{(196.25)\left[\tan\left(\frac{2}{3} \times 36\right)\right]}{51.03} = \mathbf{1.71 > 1.5 - O.K.}$$

Factor of Safety Against Bearing Capacity Failure

From Eq. (8.52), $q_u = c_2' N_c + \frac{1}{2}\gamma_2 L_2 N_\gamma$

Given: $\gamma_2 = 18$ kN/m^3, $L_2 = 2.5$ m, $c_2' = 28$ kN/m^2, and $\phi_2' = 22°$. From Table 3.3, $N_c = 16.88$, and $N_\gamma = 7.13$.

$$q_u = (28)(16.88) + \left(\frac{1}{2}\right)(18)(2.5)(7.13) \approx 633 \text{ kN/m}^2$$

From Eq. (8.54),

$$FS_{(bearing\ capacity)} = \frac{q_u}{\sigma'_{o(H)}} = \frac{633}{\gamma_1 H} = \frac{633}{(15.7)(5)} = \mathbf{8.06 > 3 - O.K.} \quad \blacksquare$$

8.17   *Retaining Walls with Geogrid Reinforcement—General*

Geogrids can also be used as reinforcement in granular backfill for the construction of retaining walls. Figure 8.38 shows typical schematic diagrams of retaining walls with geogrid reinforcement. Figure 8.39 shows some photographs of geogrid-reinforced retaining walls in the field.

Relatively few field measurements are available for lateral earth pressure on retaining walls constructed with geogrid reinforcement. Figure 8.40 shows a comparison of measured and design lateral pressures (Berg et al., 1986) for two retaining walls constructed with precast panel facing. The figure indicates that the measured earth pressures were substantially smaller than those calculated for the Rankine active case.

8.18   *Design Procedure for Geogrid-Reinforced Retaining Wall*

Figure 8.41 shows a schematic diagram of a concrete panel-faced wall with a granular backfill reinforced with layers of geogrid. The design process of the wall is essentially similar to that with geotextile reinforcement of the backfill given in Section 8.16. The following is a brief step-by-step procedure.

Internal Stability

Step 1.   Determine the active pressure at any depth z as [similar to Eq. (8.55)]:

$$\sigma_a' = K_a \gamma_1 z \qquad (8.62)$$

where

$$K_a = \text{Rankine active pressure coefficient} = \tan^2\left(45 - \frac{\phi_1'}{2}\right)$$

Figure 8.38 Typical schematic diagrams of retaining walls with geogrid reinforcement: (a) geogrid wraparound wall; (b) wall with gabion facing; (c) concrete panel-faced wall (After The Tensar Corporation, 1986)

(a)

(b)

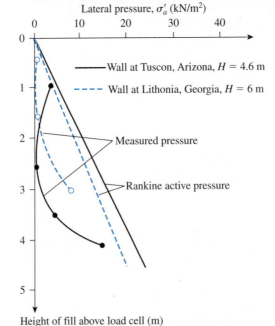

(c)

Figure 8.39 (a) HDPE geogrid-reinforced wall with precast concrete panel facing under construction; (b) Mechanical splice between two pieces of geogrid in the working direction; (c) Segmented concrete-block faced wall reinforced with uniaxial geogrid (*Courtesy of Tensar International Corporation, Atlanta, Georgia*)

Lateral pressure, σ_a' (kN/m²)

Wall at Tuscon, Arizona, $H = 4.6$ m

Wall at Lithonia, Georgia, $H = 6$ m

Measured pressure

Rankine active pressure

Height of fill above load cell (m)

Figure 8.40 Comparison of theoretical and measured lateral pressures in geogrid reinforced retaining walls (Based on Berg et al., 1986)

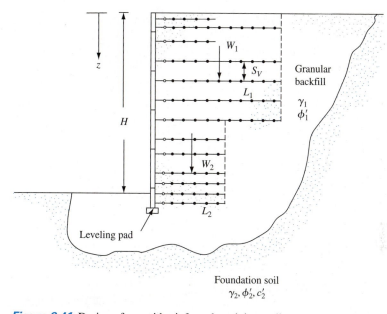

Figure 8.41 Design of geogrid-reinforced retaining wall

Step 2. Select a geogrid with allowable tensile strength, T_{all} [similar to Eq. (8.56)] (Koerner, 2005):

$$T_{all} = \frac{T_{ult}}{RF_{id} \times RF_{cr} \times RF_{cbd}} \tag{8.63}$$

where

RF_{id} = reduction factor for installation damage (1.1 to 1.4)
RF_{cr} = reduction factor for creep (2.0 to 3.0)
RF_{cbd} = reduction factor for chemical and biological degradation (1.1 to 1.5).

Step 3. Obtain the vertical spacing of the geogrid layers, S_V, as

$$S_V = \frac{T_{all}C_r}{\sigma'_a FS_{(B)}} \tag{8.64}$$

where C_r = coverage ratio for geogrid.

The *coverage ratio* is the fractional plan area at any particular elevation that is actually occupied by geogrid. For example, if there is a 0.3 m (1 ft) wide space between each 1.2 m (4 ft) wide piece of geogrid, the coverage ratio is

$$C_r = \frac{1.2 \text{ m}}{1.2 \text{ m} + 0.3 \text{ m}} = 0.8$$

Step 4. Calculate the length of each layer of geogrid at a depth z as [Eq. (8.58)]

$$L = l_r + l_e$$

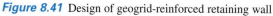

$$l_r = \frac{H - z}{\tan^2\left(45 - \dfrac{\phi'_1}{2}\right)} \tag{8.65}$$

For determination of l_e [similar to Eq. (8.60)],

$$FS_{(P)} = \frac{\text{resistance to pullout at a given normal effective stress}}{\text{pullout force}}$$

$$= \frac{(2)(l_e)(C_i\sigma_0'\tan\phi_1')(C_r)}{S_V\sigma_a'}$$

$$= \frac{(2)(l_e)(C_i\tan\phi_1')(C_r)}{S_V K_a} \tag{8.66}$$

where C_i = interaction coefficient or

$$l_e = \frac{S_V K_a \, FS_{(P)}}{2C_r C_i \tan\phi_1'} \tag{8.67}$$

Thus, at a given depth z, the total length, L, of the geogrid layer is

$$L = l_r + l_e = \frac{H - z}{\tan\left(45 + \dfrac{\phi_1'}{2}\right)} + \frac{S_V K_a \, FS_{(P)}}{2C_r C_i \tan\phi_1'} \tag{8.68}$$

The interaction coefficient, C_i, can be determined experimentally in the laboratory. The following is an approximate range for C_i for various types of backfill.

Gravel, sandy gravel	0.75–0.8
Well graded sand, gravelly sand	0.7–0.75
Fine sand, silty sand	0.55–0.6

External Stability

Check the factors of safety against overturning, sliding, and bearing capacity failure as described in Section 8.15 (Steps 9, 10, and 11).

Example 8.8

Consider a geogrid-reinforced retaining wall. Referring to Figure 8.41, given: $H = 6$ m, $\gamma_1 = 16.5$ kN/m^3, $\phi_1' = 35°$, $T_{all} = 45$ kN/m, $FS_{(B)} = 1.5$, $FS_{(P)} = 1.5$, $C_r = 0.8$, and $C_i = 0.75$. For the design of the wall, determine S_V and L.

Solution

$$K_a = \tan^2\left(45 - \frac{\phi_1'}{2}\right) = \tan^2\left(45 - \frac{35}{2}\right) = 0.27$$

Determination of S_V
From Eq. (8.64),

$$S_v = \frac{T_{all}C_r}{\sigma_a' \, FS_{(B)}} = \frac{T_{all}C_r}{\gamma z K_a \, FS_{(B)}} = \frac{(45)(0.8)}{(16.5)(z)(0.27)(1.5)} = \frac{5.39}{z}$$

At $z = 2$ m: $S_v = \dfrac{5.39}{2} = 2.7$ m

At $z = 4$ m: $S_v = \dfrac{5.39}{4} = 1.35$ m

At $z = 5$ m: $S_v = \dfrac{5.39}{5} = 1.08$ m

Use $S_V \approx$ **1 m**

Determination of L

From Eq. (8.68),

$$L = \dfrac{H - z}{\tan\left(45 + \dfrac{\phi_1'}{2}\right)} + \dfrac{S_V K_a \, FS_{(P)}}{2C_r C_i \tan\phi_1'} = \dfrac{6 - z}{\tan\left(45 + \dfrac{35}{2}\right)} + \dfrac{(1\text{ m})(0.27)(1.5)}{(2)(0.8)(0.75)(\tan 35°)}$$

At $z = 1$ m: $L = 0.52(6 - 1) + 0.482 = 3.08$ m $\approx$ 3.1 m
At $z = 3$ m: $L = 0.52(6 - 3) + 0.482 = 2.04$ m $\approx$ 2.1 m
At $z = 5$ m: $L = 0.52(6 - 5) + 0.482 = 1.0$ m

So, use $L = 3$ m for $z = 0$ to 6 m. ∎

Problems

In Problems 8.1 through 8.4, use $\gamma_{\text{concrete}} = 23.58$ kN/m³. Also, in Eq. (8.11), use $k_1 = k_2 = 2/3$ and $P_p = 0$.

8.1 For the cantilever retaining wall shown in Figure P8.1, let the following data be given:

Wall dimensions: $H = 8$ m, $x_1 = 0.4$ m, $x_2 = 0.6$ m, $x_3 = 1.5$ m, $x_4 = 3.5$ m, $x_5 = 0.96$ m, $D = 1.75$ m, $\alpha = 10°$

Soil properties: $\gamma_1 = 16.5$ kN/m³, $\phi_1' = 32°$, $\gamma_2 = 17.6$ kN/m³, $\phi_2' = 28°$, $c_2' = 30$ kN/m²

Calculate the factor of safety with respect to overturning, sliding, and bearing capacity.

8.2 Repeat Problem 8.1 with the following:

Wall dimensions: $H = 6.5$ m, $x_1 = 0.3$ m, $x_2 = 0.6$ m, $x_3 = 0.8$ m, $x_4 = 2$ m, $x_5 = 0.8$ m, $D = 1.5$ m, $\alpha = 0°$

Soil properties: $\gamma_1 = 18.08$ kN/m³, $\phi_1' = 36°$, $\gamma_2 = 19.65$ kN/m³, $\phi_2' = 15°$, $c_2' = 30$ kN/m²

8.3 A gravity retaining wall is shown in Figure P8.3. Calculate the factor of safety with respect to overturning and sliding, given the following data:

Wall dimensions: $H = 6$ m, $x_1 = 0.6$ m, $x_2 = 2$ m, $x_3 = 2$ m, $x_4 = 0.5$ m, $x_5 = 0.75$ m, $x_6 = 0.8$ m, $D = 1.5$ m

Soil properties: $\gamma_1 = 16.5$ kN/m³, $\phi_1' = 32°$, $\gamma_2 = 18$ kN/m³, $\phi_2' = 22°$, $c_2' = 40$ kN/m²

Use the Rankine active earth pressure in your calculation.

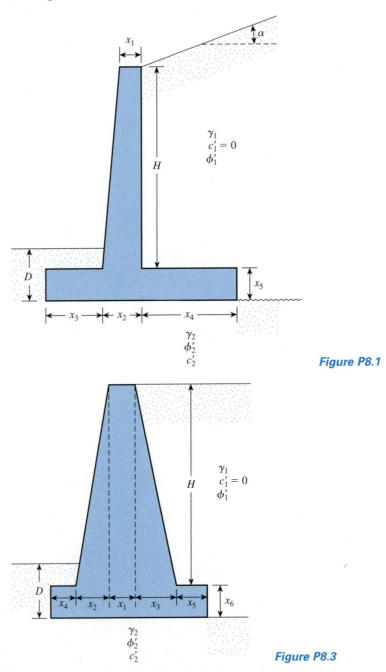

Figure P8.1

Figure P8.3

8.4 Repeat Problem 8.3 using Coulomb's active earth pressure in your calculation and letting $\delta' = 2/3\, \phi'_1$.

8.5 Refer to Figure P8.5 for the design of a gravity retaining wall for earthquake condition Given: $k_v = 0$ and $k_h = 0.3$.

    a. What should be the weight of the wall for a zero displacement condition? Use a factor of safety of 2.

    b. What should be the weight of the wall for an allowable displacement of 50.8 mm?

Sand
$\phi_1' = 30°$
$\gamma_1 = 18$ kN/m³
$\delta_1' = 15°$

7 m

Sand $\phi_2' = 36°$
$\gamma_2 = 18.5$ kN/m³

Figure P8.5

Given: $A_v = 0.15$ and $A_a = 0.25$. Use a factor of safety of 2.

8.6 In Figure 8.29a, use the following parameters:

Wall: $H = 8$ m

Soil: $\gamma_1 = 17$ kN/m³, $\phi_1' = 35°$

Reinforcement: $S_V = 1$ m and $S_H = 1.5$ m

Surcharge: $q = 70$ kN/m², $a' = 1.5$ m, and $b' = 2$ m

Calculate the vertical stress σ_o' [Eqs. (8.32), (8.33) and (8.34)] at $z = 2$ m, 4 m, 6 m and 8 m.

8.7 For the data given in Problem 8.6, calculate the lateral pressure σ_a' at $z = 2$ m, 4 m, 6 m and 8 m. Use Eqs. (8.35), (8.36) and (8.37).

8.8 A reinforced earth retaining wall (Figure 8.29) is to be 10 m. high. Here,

Backfill: unit weight, $\gamma_1 = 16$ kN/m³ and soil friction angle, $\phi_1' = 34°$

Reinforcement: vertical spacing, $S_V = 1$ m; horizontal spacing, $S_H = 1.25$ m; width of reinforcement = 120 mm., $f_y = 260$ MN/m²; $\phi_\mu = 25°$; factor of safety against tie pullout = 3; and factor of safety against tie breaking = 3

Determine:

a. The required thickness of ties

b. The required maximum length of ties

8.9 In Problem 8.8 assume that the ties at all depths are the length determined in Part b. For the *in situ* soil, $\phi_2' = 25°$, $\gamma_2 = 15.5$ kN/m³, $c_2' = 30$ kN/m². Calculate the factor of safety against (a) overturning, (b) sliding, and (c) bearing capacity failure.

8.10 A retaining wall with geotextile reinforcement is 6-m high. For the granular backfill, $\gamma_1 = 15.9$ kN/m³ and $\phi_1' = 30°$. For the geotextile, $T_{all} = 16$ kN/m. For the design of the wall, determine S_V, L, and l_l. Use $FS_{(B)} = FS_{(P)} = 1.5$.

8.11 With the S_V, L, and l_l determined in Problem 8.10, check the overall stability (i.e., factor of safety against overturning, sliding, and bearing capacity failure) of the wall. For the *in situ* soil, $\gamma_2 = 16.8$ kN/m³, $\phi_2' = 20°$, and $c_2' = 55$ kN/m².

References

APPLIED TECHNOLOGY COUNCIL (1978). "Tentative Provisions for the Development of Seismic Regulations for Buildings," *Publication ATC 3-06*, Palo Alto, California.

BELL, J. R., STILLEY, A. N., and VANDRE, B. (1975). "Fabric Retaining Earth Walls," *Proceedings, Thirteenth Engineering Geology and Soils Engineering Symposium,* Moscow, ID.

BENTLER, J. G., and LABUZ, J. F. (2006). "Performance of a Cantilever Retaining Wall," *Journal of Geotechnical and Geoenvironmental Engineering*, American Society of Civil Engineers, Vol. 132, No. 8, pp. 1062–1070.

BERG, R. R., BONAPARTE, R., ANDERSON, R. P., and CHOUERY, V. E. (1986). "Design Construction and Performance of Two Tensar Geogrid Reinforced Walls," *Proceedings, Third International Conference on Geotextiles,* Vienna, pp. 401–406.

BINQUET, J., and LEE, K. L. (1975). "Bearing Capacity Analysis of Reinforced Earth Slabs," *Journal of the Geotechnical Engineering Division,* American Society of Civil Engineers, Vol. 101, No. GT12, pp. 1257–1276.

CARROLL, R., JR. (1988). "Specifying Geogrids," *Geotechnical Fabric Report,* Industrial Fabric Association International, St. Paul, March/April.

CASAGRANDE, L. (1973). "Comments on Conventional Design of Retaining Structure," *Journal of the Soil Mechanics and Foundations Division,* ASCE, Vol. 99, No. SM2, pp. 181–198.

DARBIN, M. (1970). "Reinforced Earth for Construction of Freeways" (in French), *Revue Générale des Routes et Aerodromes,* No. 457, September.

DAS, B. M. (1983). *Fundamentals of Soil Dynamics*, Elsevier, New York.

ELMAN, M. T., and TERRY, C. F. (1988). "Retaining Walls with Sloped Heel," *Journal of Geotechnical Engineering*, American Society of Civil Engineers, Vol. 114, No. GT10, pp. 1194–1199.

KOERNER, R. B. (2005). *Design with Geosynthetics,* 5th. ed., Prentice Hall, Englewood Cliffs, NJ.

LABA, J. T., and KENNEDY, J. B. (1986). "Reinforced Earth Retaining Wall Analysis and Design," *Canadian Geotechnical Journal,* Vol. 23, No. 3, pp. 317–326.

LEE, K. L., ADAMS, B. D., and VAGNERON, J. J. (1973). "Reinforced Earth Retaining Walls," *Journal of the Soil Mechanics and Foundations Division,* American Society of Civil Engineers, Vol. 99, No. SM10, pp. 745–763.

MARTIN, J. P., KOERNER, R. M., and WHITTY, J. E. (1984). "Experimental Friction Evaluation of Slippage Between Geomembranes, Geotextiles, and Soils," *Proceedings,* International Conference on Geomembranes, Denver, pp. 191–196.

RICHARDS, R., and ELMS, D. G. (1979). "Seismic Behavior of Gravity Retaining Walls," *Journal of the Geotechnical Engineering Division*, American Society of Civil Engineers, Vol. 105, No. GT4, pp. 449–464.

SARSBY, R. W. (1985). "The Influence of Aperture Size/Particle Size on the Efficiency of Grid Reinforcement," *Proceedings, 2nd Canadian Symposium on Geotextiles and Geomembranes*, Edmonton, pp. 7–12.

SCHLOSSER, F., and LONG, N. (1974). "Recent Results in French Research on Reinforced Earth," *Journal of the Construction Division,* American Society of Civil Engineers, Vol. 100, No. CO3, pp. 113–237.

SCHLOSSER, F., and VIDAL, H. (1969). "Reinforced Earth" (in French), *Bulletin de Liaison des Laboratoires Routier,* Ponts et Chassées, Paris, France, November, pp. 101–144.

TENSAR CORPORATION (1986). Tensar Technical Note. No. TTN:RW1, August.

TERZAGHI, K., and PECK, R. B. (1967). *Soil Mechanics in Engineering Practice,* Wiley, New York.

TRANSPORTATION RESEARCH BOARD (1995). Transportation Research Circular No. 444, National Research Council, Washington, DC.

VIDAL, H. (1966). "La terre Armée," *Annales de l'Institut Technique du Bâtiment et des Travaux Publiques,* France, July–August, pp. 888–938.

9 Sheet Pile Walls

9.1 Introduction

Connected or semiconnected sheet piles are often used to build continuous walls for waterfront structures that range from small waterfront pleasure boat launching facilities to large dock facilities. (See Figure 9.1.) In contrast to the construction of other types of retaining wall, the building of sheet pile walls does not usually require dewatering of the site. Sheet piles are also used for some temporary structures, such as braced cuts. (See Chapter 10.) The principles of sheet-pile wall design are discussed in the current chapter.

Several types of sheet pile are commonly used in construction: (a) wooden sheet piles, (b) precast concrete sheet piles, and (c) steel sheet piles. Aluminum sheet piles are also marketed.

Wooden sheet piles are used only for temporary, light structures that are above the water table. The most common types are ordinary wooden planks and *Wakefield piles.* The wooden planks are about 50 mm × 300 mm in cross section and are driven edge to edge (Figure 9.2a). Wakefield piles are made by nailing three planks together, with the middle plank offset by 50 to 75 mm (Figure 9.2b). Wooden planks can also be milled to form *tongue-and-groove piles,* as shown in Figure 9.2c. Figure 9.2d shows another type of wooden sheet pile that has precut grooves. Metal *splines* are driven into the grooves of the adjacent sheetings to hold them together after they are sunk into the ground.

Figure 9.1 Example of waterfront sheet-pile wall

Figure 9.2 Various types of wooden and concrete sheet pile

Precast concrete sheet piles are heavy and are designed with reinforcements to withstand the permanent stresses to which the structure will be subjected after construction and also to handle the stresses produced during construction. In cross section, these piles are about 500 to 800 mm wide and 150 to 250 mm thick. Figure 9.2e is a schematic diagram of the elevation and the cross section of a reinforced concrete sheet pile.

Steel sheet piles in the United States are about 10 to 13 mm thick. European sections may be thinner and wider. Sheet-pile sections may be *Z, deep arch, low arch,* or *straight web* sections. The interlocks of the sheet-pile sections are shaped like a *thumb-and-finger* or *ball-and-socket* joint for watertight connections. Figure 9.3a is a schematic diagram of the thumb-and-finger type of interlocking for straight web sections. The ball-and-socket type of interlocking for Z section piles is shown in Figure 9.3b. Figure 9.4 shows a sheet pile wall. Table 9.1 lists the properties of the steel sheet pile sections produced by the Bethlehem Steel Corporation. The allowable design flexural stress for the steel sheet piles is as follows:

Type of steel	Allowable stress
ASTM A-328	170 MN/m^2
ASTM A-572	210 MN/m^2
ASTM A-690	210 MN/m^2

Steel sheet piles are convenient to use because of their resistance to the high driving stress that is developed when they are being driven into hard soils. Steel sheet piles are also light-weight and reusable.

(a)

(b)

Figure 9.3 (a) Thumb-and-finger type sheet pile connection; (b) ball-and-socket type sheet-pile connection

Figure 9.4 A steel sheet pile wall (*Courtesy of N. Sivakugan, James Cook University, Australia*)

Table 9.1 Properties of Some Sheet-Pile Sections Produced by Bethlehem Steel Corporation

Section designation	Sketch of section	Section modulus m^3/m of wall	Moment of inertia m^4/m of wall
PZ-40		326.4×10^{-5}	670.5×10^{-6}

409 mm

12.7 mm

15.2 mm

← Driving distance = 500 mm →

(Continued)

Table 9.1 (*Continued*)

Section designation	Sketch of section	Section modulus m³/m of wall	Moment of inertia m⁴/m of wall
PZ-35		260.5×10^{-5}	493.4×10^{-6}
PZ-27		162.3×10^{-5}	251.5×10^{-6}
PZ-22		97×10^{-5}	115.2×10^{-6}
PSA-31		10.8×10^{-5}	4.41×10^{-6}
PSA-23		12.8×10^{-5}	5.63×10^{-6}

PZ-35: 379 mm, 12.7 mm, 15.2 mm, Driving distance = 575 mm

PZ-27: 304.8 mm, 9.53 mm, 9.53 mm, Driving distance = 457.2 mm

PZ-22: 228.6 mm, 9.53 mm, 9.53 mm, Driving distance = 558.8 mm

PSA-31: 12.7 mm, Driving distance = 500 mm

PSA-23: 9.53 mm, Driving distance = 406.4 mm

9.2   *Construction Methods*

Sheet pile walls may be divided into two basic categories: (a) cantilever and (b) anchored.

In the construction of sheet pile walls, the sheet pile may be driven into the ground and then the backfill placed on the land side, or the sheet pile may first be driven into the ground and the soil in front of the sheet pile dredged. In either case, the soil used for backfill behind the sheet pile wall is usually granular. The soil below the dredge line may be sandy or clayey. The surface of soil on the water side is referred to as the *mud line* or *dredge line*.

Thus, construction methods generally can be divided into two categories (Tsinker, 1983):

1. Backfilled structure
2. Dredged structure

The sequence of construction for a *backfilled structure* is as follows (see Figure 9.5):

Step 1.   Dredge the *in situ* soil in front and back of the proposed structure.
Step 2.   Drive the sheet piles.
Step 3.   Backfill up to the level of the anchor, and place the anchor system.
Step 4.   Backfill up to the top of the wall.

For a cantilever type of wall, only Steps 1, 2, and 4 apply. The sequence of construction for a *dredged structure* is as follows (see Figure 9.6):

Step 1.   Drive the sheet piles.
Step 2.   Backfill up to the anchor level, and place the anchor system.
Step 3.   Backfill up to the top of the wall.
Step 4.   Dredge the front side of the wall.

With cantilever sheet pile walls, Step 2 is not required.

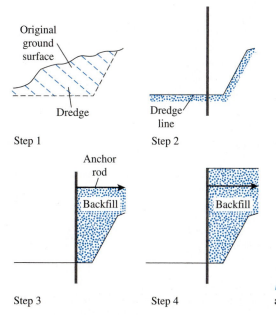

Figure 9.5 Sequence of construction for a backfilled structure

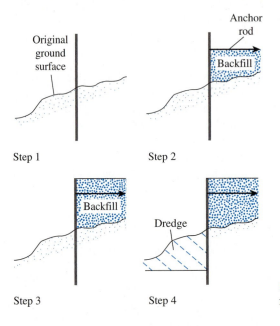

Figure 9.6 Sequence of construction for a dredged structure

9.3 Cantilever Sheet Pile Walls

Cantilever sheet pile walls are usually recommended for walls of moderate height—about 6 m or less, measured above the dredge line. In such walls, the sheet piles act as a wide cantilever beam above the dredge line. The basic principles for estimating net lateral pressure distribution on a cantilever sheet-pile wall can be explained with the aid of Figure 9.7. The figure shows the nature of lateral yielding of a cantilever wall penetrating a sand layer below the dredge line. The wall rotates about point *O* (Figure 9.7a). Because the hydrostatic pressures at any depth from both sides of the wall will cancel each other, we consider only the effective lateral soil pressures. In zone *A,* the lateral pressure is just the active pressure from the land side. In zone *B,* because of the nature of yielding of the wall, there will be active pressure from the land side and passive pressure from the water side. The condition is reversed in zone *C*—that is, below the point of rotation, *O*. The net actual pressure distribution on the wall is like that shown in Figure 9.7b. However, for design purposes, Figure 9.7c shows a simplified version.

Sections 9.4 through 9.7 present the mathematical formulation of the analysis of cantilever sheet pile walls. Note that, in some waterfront structures, the water level may fluctuate as the result of tidal effects. Care should be taken in determining the water level that will affect the net pressure diagram.

9.4 Cantilever Sheet Piling Penetrating Sandy Soils

To develop the relationships for the proper depth of embedment of sheet piles driven into a granular soil, examine Figure 9.8a. The soil retained by the sheet piling above the dredge line also is sand. The water table is at a depth L_1 below the top of the wall.

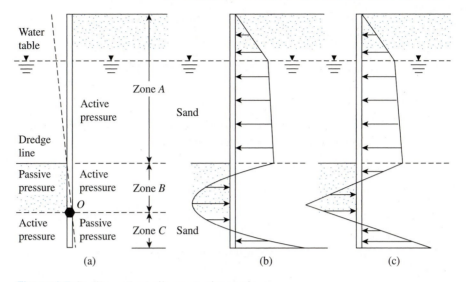

Figure 9.7 Cantilever sheet pile penetrating sand

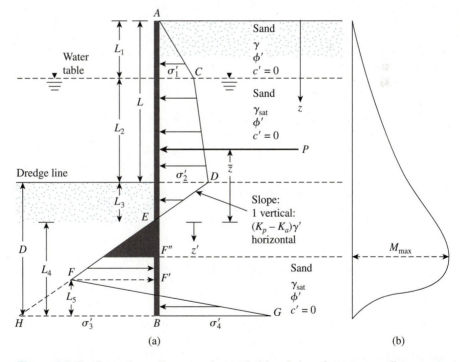

Figure 9.8 Cantilever sheet pile penetrating sand: (a) variation of net pressure diagram; (b) variation of moment

Let the effective angle of friction of the sand be ϕ'. The intensity of the active pressure at a depth $z = L_1$ is

$$\sigma_1' = \gamma L_1 K_a \tag{9.1}$$

where

K_a = Rankine active pressure coefficient = $\tan^2(45 - \phi'/2)$
γ = unit weight of soil above the water table

Similarly, the active pressure at a depth $z = L_1 + L_2$ (i.e., at the level of the dredge line) is

$$\sigma_2' = (\gamma L_1 + \gamma' L_2) K_a \tag{9.2}$$

where γ' = effective unit weight of soil = $\gamma_{sat} - \gamma_w$.

Note that, at the level of the dredge line, the hydrostatic pressures from both sides of the wall are the same magnitude and cancel each other.

To determine the net lateral pressure below the dredge line up to the point of rotation, O, as shown in Figure 9.7a, an engineer has to consider the passive pressure acting from the left side (the water side) toward the right side (the land side) of the wall and also the active pressure acting from the right side toward the left side of the wall. For such cases, ignoring the hydrostatic pressure from both sides of the wall, the active pressure at depth z is

$$\sigma_a' = [\gamma L_1 + \gamma' L_2 + \gamma'(z - L_1 - L_2)]K_a \tag{9.3}$$

Also, the passive pressure at depth z is

$$\sigma_p' = \gamma'(z - L_1 - L_2)K_p \tag{9.4}$$

where K_p = Rankine passive pressure coefficient = $\tan^2(45 + \phi'/2)$.

Combining Eqs. (9.3) and (9.4) yields the net lateral pressure, namely,

$$\sigma' = \sigma_a' - \sigma_p' = (\gamma L_1 + \gamma' L_2) K_a - \gamma'(z - L_1 - L_2)(K_p - K_a)$$

$$= \sigma_2' - \gamma'(z - L)(K_p - K_a) \tag{9.5}$$

where $L = L_1 + L_2$.

The net pressure, σ' equals zero at a depth L_3 below the dredge line, so

$$\sigma_2' - \gamma'(z - L)(K_p - K_a) = 0$$

or

$$(z - L) = L_3 = \frac{\sigma_2'}{\gamma'(K_p - K_a)} \tag{9.6}$$

Equation (9.6) indicates that the slope of the net pressure distribution line DEF is 1 vertical to $(K_p - K_a)\gamma'$ horizontal, so, in the pressure diagram,

$$\overline{HB} = \sigma'_3 = L_4(K_p - K_a)\gamma' \tag{9.7}$$

At the bottom of the sheet pile, passive pressure, σ'_p, acts from the right toward the left side, and active pressure acts from the left toward the right side of the sheet pile, so, at $z = L + D$,

$$\sigma'_p = (\gamma L_1 + \gamma' L_2 + \gamma' D) K_p \tag{9.8}$$

At the same depth,

$$\sigma'_a = \gamma' D K_a \tag{9.9}$$

Hence, the net lateral pressure at the bottom of the sheet pile is

$$
\begin{aligned}
\sigma'_p - \sigma'_a = \sigma'_4 &= (\gamma L_1 + \gamma' L_2) K_p + \gamma' D (K_p - K_a) \\
&= (\gamma L_1 + \gamma' L_2) K_p + \gamma' L_3 (K_p - K_a) + \gamma' L_4 (K_p - K_a) \\
&= \sigma'_5 + \gamma' L_4 (K_p - K_a)
\end{aligned}
\tag{9.10}
$$

where

$$\sigma'_5 = (\gamma L_1 + \gamma' L_2) K_p + \gamma' L_3 (K_p - K_a) \tag{9.11}$$
$$D = L_3 + L_4 \tag{9.12}$$

For the stability of the wall, the principles of statics can now be applied:

$$\Sigma \text{ horizontal forces per unit length of wall} = 0$$

and

$$\Sigma \text{ moment of the forces per unit length of wall about point } B = 0$$

For the summation of the horizontal forces, we have

Area of the pressure diagram $ACDE$ − area of $EFHB$ + area of $FHBG$ = 0

or

$$P - \tfrac{1}{2}\sigma'_3 L_4 + \tfrac{1}{2} L_5 (\sigma'_3 + \sigma'_4) = 0 \tag{9.13}$$

where P = area of the pressure diagram $ACDE$.

Summing the moment of all the forces about point B yields

$$P(L_4 + \bar{z}) - \left(\frac{1}{2} L_4 \sigma'_3\right)\left(\frac{L_4}{3}\right) + \frac{1}{2} L_5 (\sigma'_3 + \sigma'_4)\left(\frac{L_5}{3}\right) = 0 \tag{9.14}$$

From Eq. (9.13),

$$L_5 = \frac{\sigma'_3 L_4 - 2P}{\sigma'_3 + \sigma'_4} \tag{9.15}$$

Combining Eqs. (9.7), (9.10), (9.14), and (9.15) and simplifying them further, we obtain the following fourth-degree equation in terms of L_4:

$$L_4^4 + A_1 L_4^3 - A_2 L_4^2 - A_3 L_4 - A_4 = 0 \tag{9.16}$$

In this equation,

$$A_1 = \frac{\sigma_5'}{\gamma'(K_p - K_a)} \tag{9.17}$$

$$A_2 = \frac{8P}{\gamma'(K_p - K_a)} \tag{9.18}$$

$$A_3 = \frac{6P[2\bar{z}\gamma'(K_p - K_a) + \sigma_5']}{\gamma'^2(K_p - K_a)^2} \tag{9.19}$$

$$A_4 = \frac{P(6\bar{z}\sigma_5' + 4P)}{\gamma'^2(K_p - K_a)^2} \tag{9.20}$$

Step-by-Step Procedure for Obtaining the Pressure Diagram

Based on the preceding theory, a step-by-step procedure for obtaining the pressure diagram for a cantilever sheet pile wall penetrating a granular soil is as follows:

Step 1. Calculate K_a and K_p.
Step 2. Calculate σ_1' [Eq. (9.1)] and σ_2' [Eq. (9.2)]. (*Note:* L_1 and L_2 will be given.)
Step 3. Calculate L_3 [Eq. (9.6)].
Step 4. Calculate P.
Step 5. Calculate $\bar{z}$ (i.e., the center of pressure for the area *ACDE*) by taking the moment about E.
Step 6. Calculate σ_5' [Eq. (9.11)].
Step 7. Calculate A_1, A_2, A_3, and A_4 [Eqs. (9.17) through (9.20)].
Step 8. Solve Eq. (9.16) by trial and error to determine L_4.
Step 9. Calculate σ_4' [Eq. (9.10)].
Step 10. Calculate σ_3' [Eq. (9.7)].
Step 11. Obtain L_5 from Eq. (9.15).
Step 12. Draw a pressure distribution diagram like the one shown in Figure 9.8a.
Step 13. Obtain the theoretical depth [see Eq. (9.12)] of penetration as $L_3 + L_4$. The actual depth of penetration is increased by about 20 to 30%.

Note that some designers prefer to use a factor of safety on the passive earth pressure coefficient at the beginning. In that case, in Step 1,

$$K_{p(\text{design})} = \frac{K_p}{FS}$$

where FS = factor of safety (usually between 1.5 and 2).

For this type of analysis, follow Steps 1 through 12 with the value of $K_a = \tan^2(45 - \phi'/2)$ and $K_{p(\text{design})}$ (instead of K_p). The actual depth of penetration can now be determined by adding L_3, obtained from Step 3, and L_4, obtained from Step 8.

Calculation of Maximum Bending Moment

The nature of the variation of the moment diagram for a cantilever sheet pile wall is shown in Figure 9.8b. The maximum moment will occur between points E and F'. Obtaining the maximum moment ($M_{\max}$) per unit length of the wall requires determining the point of zero shear. For a new axis z' (with origin at point E) for zero shear,

$$P = \tfrac{1}{2}(z')^2(K_p - K_a)\gamma'$$

or

$$z' = \sqrt{\frac{2P}{(K_p - K_a)\gamma'}} \tag{9.21}$$

Once the point of zero shear force is determined (point F'' in Figure 9.8a), the magnitude of the maximum moment can be obtained as

$$M_{\max} = P(\bar{z} + z') - [\tfrac{1}{2}\gamma'z'^2(K_p - K_a)](\tfrac{1}{3})z' \tag{9.22}$$

The necessary profile of the sheet piling is then sized according to the allowable flexural stress of the sheet pile material, or

$$S = \frac{M_{\max}}{\sigma_{\text{all}}} \tag{9.23}$$

where

S = section modulus of the sheet pile required per unit length of the structure
σ_{all} = allowable flexural stress of the sheet pile

Example 9.1

Figure 9.9 shows a cantilever sheet pile wall penetrating a granular soil. Here, $L_1 = 2$ m, $L_2 = 3$ m, $\gamma = 15.9$ kN/m³, $\gamma_{\text{sat}} = 19.33$ kN/m³, and $\phi' = 32°$.

a. What is the theoretical depth of embedment, D?
b. For a 30% increase in D, what should be the total length of the sheet piles?
c. What should be the minimum section modulus of the sheet piles? Use $\sigma_{\text{all}} = 172$ MN/m².

Solution

Part a
Using Figure 9.8a for the pressure distribution diagram, one can now prepare the following table for a step-by-step calculation.

Figure 9.9 Cantilever sheet-pile wall

Quantity required	Eq. no.	Equation and calculation
K_a	—	$\tan^2\left(45 - \dfrac{\phi'}{2}\right) = \tan^2\left(45 - \dfrac{32}{2}\right) = 0.307$
K_p	—	$\tan^2\left(45 + \dfrac{\phi'}{2}\right) = \tan^2\left(45 + \dfrac{32}{2}\right) = 3.25$
σ'_1	9.1	$\gamma L_1 K_a = (15.9)(2)(0.307) = 9.763 \text{ kN/m}^2$
σ'_2	9.2	$(\gamma L_1 + \gamma' L_2)K_a = [(15.9)(2) + (19.33 - 9.81)(3)](0.307) = 18.53 \text{ kN/m}^2$
L_3	9.6	$\dfrac{\sigma'_2}{\gamma'(K_p - K_a)} = \dfrac{18.53}{(19.33 - 9.81)(3.25 - 0.307)} = 0.66 \text{ m}$
P	—	$\frac{1}{2}\sigma'_1 L_1 + \sigma'_1 L_2 + \frac{1}{2}(\sigma'_2 - \sigma'_1)L_2 + \frac{1}{2}\sigma'_2 L_3$
		$= \left(\frac{1}{2}\right)(9.763)(2) + (9.763)(3) + \left(\frac{1}{2}\right)(18.53 - 9.763)(3)$
		$\quad + \left(\frac{1}{2}\right)(18.53)(0.66)$
		$= 9.763 + 29.289 + 13.151 + 6.115 = 58.32 \text{ kN/m}$
$\bar{z}$	—	$\dfrac{\Sigma M_E}{P} = \dfrac{1}{58.32}\left[\begin{array}{l} 9.763\left(0.66 + 3 + \frac{2}{3}\right) + 29.289\left(0.66 + \frac{3}{2}\right) \\ + 13.151\left(0.66 + \frac{3}{3}\right) + 6.115\left(0.66 \times \frac{2}{3}\right)\end{array}\right] = 2.23 \text{ m}$
σ'_5	9.11	$(\gamma L_1 + \gamma' L_2)K_p + \gamma' L_3(K_p - K_a) = [(15.9)(2) + (19.33 - 9.81)(3)](3.25)$
		$\quad\quad + (19.33 - 9.81)(0.66)(3.25$
		$\quad\quad - 0.307) = 214.66 \text{ kN/m}^2$
A_1	9.17	$\dfrac{\sigma'_5}{\gamma'(K_p - K_a)} = \dfrac{214.66}{(19.33 - 9.81)(3.25 - 0.307)} = 7.66$
A_2	9.18	$\dfrac{8P}{\gamma'(K_p - K_a)} = \dfrac{(8)(58.32)}{(19.33 - 9.81)(3.25 - 0.307)} = 16.65$

A_3	9.19	$\dfrac{6P[2\bar{z}\gamma'(K_p - K_a) + \sigma_5']}{\gamma'^2(K_p - K_a)^2}$

$$= \frac{(6)(58.32)[(2)(2.23)(19.33 - 9.81)(3.25 - 0.307) + 214.66]}{(19.33 - 9.81)^2(3.25 - 0.307)^2}$$

$$= 151.93$$

A_4	9.20	$\dfrac{P(6\bar{z}\sigma_5' + 4P)}{\gamma'^2(K_p - K_a)^2} = \dfrac{58.32[(6)(2.23)(214.66) + (4)(58.32)]}{(19.33 - 9.81)^2(3.25 - 0.307)^2}$

$$= 230.72$$

L_4	9.16	$L_4^4 + A_1 L_4^3 - A_2 L_4^2 - A_3 L_4 - A_4 = 0$

$$L_4^4 + 7.66L_4^3 - 16.65L_4^2 - 151.93L_4 - 230.72 = 0; \; L_4 \approx 4.8 \text{ m}$$

Thus,

$$D_{\text{theory}} = L_3 + L_4 = 0.66 + 4.8 = \mathbf{5.46 \text{ m}}$$

Part b

The total length of the sheet piles is

$$L_1 + L_2 + 1.3(L_3 + L_4) = 2 + 3 + 1.3(5.46) = \mathbf{12.1 \text{ m}}$$

Part c

Finally, we have the following table.

Quantity required	Eq. no.	Equation and calculation
z'	9.21	$\sqrt{\dfrac{2P}{(K_p - K_a)\gamma'}} = \sqrt{\dfrac{(2)(58.32)}{(3.25 - 0.307)(19.33 - 9.81)}} = 2.04 \text{ m}$
$M_{\max}$	9.22	$P(\bar{z} + z') - \left[\dfrac{1}{2}\gamma'z'^2(K_p - K_a)\right]\dfrac{z'}{3} = (58.32)(2.23 + 2.04)$ $-\left[\left(\dfrac{1}{2}\right)(19.33 - 9.81)(2.04)^2(3.25 - 0.307)\right]\dfrac{2.04}{3}$ $= 209.39 \text{ kN·m/m}$
S	9.29	$\dfrac{M_{\max}}{\sigma_{\text{all}}} = \dfrac{209.39 \text{ kN·m}}{172 \times 10^3 \text{ kN/m}^2} = \mathbf{1.217 \times 10^{-3} \text{ m}^3\text{/m of wall}}$ ■

9.5 Special Cases for Cantilever Walls Penetrating a Sandy Soil

Sheet Pile Wall with the Absence of Water Table

In the absence of the water table, the net pressure diagram on the cantilever sheet-pile wall will be as shown in Figure 9.10, which is a modified version of Figure 9.8. In this case,

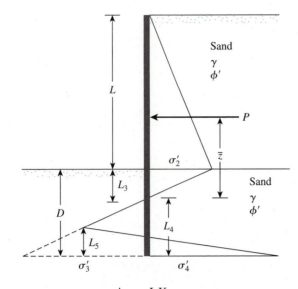

Figure 9.10 Sheet piling penetrating a sandy soil in the absence of the water table

$$\sigma_2' = \gamma L K_a \tag{9.24}$$

$$\sigma_3' = L_4(K_p - K_a)\gamma \tag{9.25}$$

$$\sigma_4' = \sigma_5' + \gamma L_4(K_p - K_a) \tag{9.26}$$

$$\sigma_5' = \gamma L K_p + \gamma L_3(K_p - K_a) \tag{9.27}$$

$$L_3 = \frac{\sigma_2'}{\gamma(K_p - K_a)} = \frac{L K_a}{(K_p - K_a)} \tag{9.28}$$

$$P = \tfrac{1}{2}\sigma_2' L + \tfrac{1}{2}\sigma_2' L_3 \tag{9.29}$$

$$\bar{z} = L_3 + \frac{L}{3} = \frac{L K_a}{K_p - K_a} + \frac{L}{3} = \frac{L(2K_a + K_p)}{3(K_p - K_a)} \tag{9.30}$$

and Eq. (9.16) transforms to

$$L_4^4 + A_1' L_4^3 - A_2' L_4^2 - A_3' L_4 - A_4' = 0 \tag{9.31}$$

where

$$A_1' = \frac{\sigma_5'}{\gamma(K_p - K_a)} \tag{9.32}$$

$$A_2' = \frac{8P}{\gamma(K_p - K_a)} \tag{9.33}$$

$$A_3' = \frac{6P[2\bar{z}\gamma(K_p - K_a) + \sigma_5']}{\gamma^2(K_p - K_a)^2} \tag{9.34}$$

$$A_4' = \frac{P(6\bar{z}\sigma_5' + 4P)}{\gamma^2(K_p - K_a)^2} \tag{9.35}$$

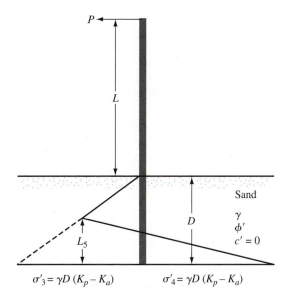

$$\sigma'_3 = \gamma D\,(K_p - K_a) \qquad\qquad \sigma'_4 = \gamma D\,(K_p - K_a)$$

Figure 9.11 Free cantilever sheet piling penetrating a layer of sand

Free Cantilever Sheet Piling

Figure 9.11 shows a free cantilever sheet-pile wall penetrating a sandy soil and subjected to a line load of P per unit length of the wall. For this case,

$$D^4 - \left[\frac{8P}{\gamma(K_p - K_a)}\right]D^2 - \left[\frac{12PL}{\gamma(K_p - K_a)}\right]D - \left[\frac{2P}{\gamma(K_p - K_a)}\right]^2 = 0 \qquad (9.36)$$

$$L_5 = \frac{\gamma(K_p - K_a)D^2 - 2P}{2D(K_p - K_a)\gamma} \qquad (9.37)$$

$$M_{max} = P(L + z') - \frac{\gamma z'^3(K_p - K_a)}{6} \qquad (9.38)$$

and

$$z' = \sqrt{\frac{2P}{\gamma'(K_p - K_a)}} \qquad (9.39)$$

Example 9.2

Redo parts a and b of Example 9.1, assuming the absence of the water table. Use $\gamma = 15.9$ kN/m^3 and $\phi' = 32°$. *Note*: $L = 5$ m.

Solution

Part a

Quantity required	Eq. no.	Equation and calculation
K_a	—	$\tan^2\left(45 - \dfrac{\phi'}{2}\right) = \tan^2\left(45 - \dfrac{32}{2}\right) = 0.307$
K_p	—	$\tan^2\left(45 + \dfrac{\phi'}{2}\right) = \tan^2\left(45 + \dfrac{32}{2}\right) = 3.25$
σ_2'	9.24	$\gamma L K_a = (15.9)(5)(0.307) = 24.41 \text{ kN/m}^2$
L_3	9.28	$\dfrac{L K_a}{K_p - K_a} = \dfrac{(5)(0.307)}{3.25 - 0.307} = 0.521 \text{ m}$
σ_5'	9.27	$\gamma L K_p + \gamma L_3(K_p - K_a) = (15.9)(5)(3.25) + (15.9)(0.521)(3.25 - 0.307)$ $= 282.76 \text{ kN/m}^2$
P	9.29	$\frac{1}{2}\sigma_2' L + \frac{1}{2}\sigma_2' L_3 = \frac{1}{2}\sigma_2'(L + L_3) = \left(\frac{1}{2}\right)(24.41)(5 + 0.521) = 67.38 \text{ kN/m}$
$\bar{z}$	9.30	$\dfrac{L(2K_a - K_p)}{3(K_p - K_a)} = \dfrac{5[(2)(0.307) + 3.25]}{3(3.25 - 0.307)} = 2.188 \text{ m}$
A_1'	9.32	$\dfrac{\sigma_5'}{\gamma(K_p - K_a)} = \dfrac{282.76}{(15.9)(3.25 - 0.307)} = 6.04$
A_2'	9.33	$\dfrac{8P}{\gamma(K_p - K_a)} = \dfrac{(8)(67.38)}{(15.9)(3.25 - 0.307)} = 11.52$
A_3'	9.34	$\dfrac{6P[2\bar{z}\gamma(K_p - K_a) + \sigma_5']}{\gamma^2(K_p - K_a)^2}$ $= \dfrac{(6)(67.38)[(2)(2.188)(15.9)(3.25 - 0.307) + 282.76]}{(15.9)^2(3.25 - 0.307)^2} = 90.01$
A_4'	9.35	$\dfrac{P(6\bar{z}\sigma_5' + 4P)}{\gamma^2(K_p - K_a)^2} = \dfrac{(67.38)[(6)(2.188)(282.76) + (4)(67.38)]}{(15.9)^2(3.25 - 0.307)^2} = 122.52$
L_4	9.31	$L_4^4 + A_1' L_4^3 - A_2' L_4^2 - A_3' L_4 - A_4' = 0$ $L_4^4 + 6.04 L_4^3 - 11.52 L_4^2 - 90.01 L_4 - 122.52 = 0; \; L_4 \approx 4.1 \text{ m}$

$$D_{\text{theory}} = L_3 + L_4 = 0.521 + 4.1 \approx \textbf{4.7 m}$$

Part b
Total length, $L + 1.3(D_{\text{theory}}) = 5 + 1.3(4.7) = \textbf{11.11 m}$ ∎

9.6 Cantilever Sheet Piling Penetrating Clay

At times, cantilever sheet piles must be driven into a clay layer possessing an undrained cohesion $c(\phi = 0)$. The net pressure diagram will be somewhat different from that shown in Figure 9.8a. Figure 9.12 shows a cantilever sheet-pile wall driven into clay with a backfill of granular soil above the level of the dredge line. The water table is at

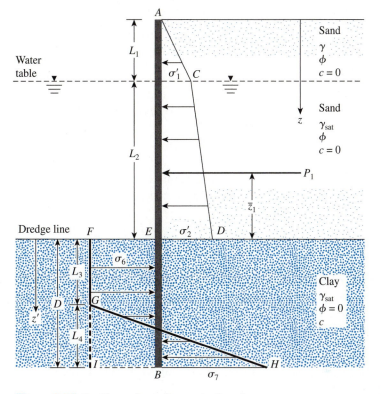

Figure 9.12 Cantilever sheet pile penetrating clay

a depth L_1 below the top of the wall. As before, Eqs. (9.1) and (9.2) give the intensity of the net pressures σ_1' and σ_2', and the diagram for pressure distribution above the level of the dredge line can be drawn. The diagram for net pressure distribution below the dredge line can now be determined as follows.

At any depth greater than $L_1 + L_2$, for $\phi = 0$, the Rankine active earth-pressure coefficient $K_a = 1$. Similarly, for $\phi = 0$, the Rankine passive earth-pressure coefficient $K_p = 1$. Thus, above the point of rotation (point O in Figure 9.7a), the active pressure, from right to left is

$$\sigma_a = [\gamma L_1 + \gamma' L_2 + \gamma_{sat}(z - L_1 - L_2)] - 2c \qquad (9.40)$$

Similarly, the passive pressure from left to right may be expressed as

$$\sigma_p = \gamma_{sat}(z - L_1 - L_2) + 2c \qquad (9.41)$$

Thus, the net pressure is

$$\sigma_6 = \sigma_p - \sigma_a = [\gamma_{sat}(z - L_1 - L_2) + 2c]$$
$$- [\gamma L_1 + \gamma' L_2 + \gamma_{sat}(z - L_1 - L_2)] + 2c$$
$$= 4c - (\gamma L_1 + \gamma' L_2) \qquad (9.42)$$

At the bottom of the sheet pile, the passive pressure from right to left is

$$\sigma_p = (\gamma L_1 + \gamma' L_2 + \gamma_{sat} D) + 2c \qquad (9.43)$$

Similarly, the active pressure from left to right is

$$\sigma_a = \gamma_{sat} D - 2c \tag{9.44}$$

Hence, the net pressure is

$$\sigma_7 = \sigma_p - \sigma_a = 4c + (\gamma L_1 + \gamma' L_2) \tag{9.45}$$

For equilibrium analysis, $\Sigma F_H = 0$; that is, the area of the pressure diagram $ACDE$ minus the area of $EFIB$ plus the area of $GIH = 0$, or

$$P_1 - [4c - (\gamma L_1 + \gamma' L_2)]D + \tfrac{1}{2}L_4[4c - (\gamma L_1 + \gamma' L_2) + 4c + (\gamma L_1 + \gamma' L_2)] = 0$$

where P_1 = area of the pressure diagram $ACDE$.
    Simplifying the preceding equation produces

$$L_4 = \frac{D[4c - (\gamma L_1 + \gamma' L_2)] - P_1}{4c} \tag{9.46}$$

Now, taking the moment about point B ($\Sigma M_B = 0$) yields

$$P_1(D + \bar{z}_1) - [4c - (\gamma L_1 + \gamma' L_2)]\frac{D^2}{2} + \frac{1}{2}L_4(8c)\left(\frac{L_4}{3}\right) = 0 \tag{9.47}$$

where $\bar{z}_1$ = distance of the center of pressure of the pressure diagram $ACDE$, measured from the level of the dredge line.
    Combining Eqs. (9.46) and (9.47) yields

$$D^2[4c - (\gamma L_1 + \gamma' L_2)] - 2DP_1 - \frac{P_1(P_1 + 12c\bar{z}_1)}{(\gamma L_1 + \gamma' L_2) + 2c} = 0 \tag{9.48}$$

Equation (9.48) may be solved to obtain D, the theoretical depth of penetration of the clay layer by the sheet pile.

Step-by-Step Procedure for Obtaining the Pressure Diagram

Step 1.   Calculate $K_a = \tan^2(45 - \phi'/2)$ for the granular soil (backfill).
Step 2.   Obtain σ_1' and σ_2'. [See Eqs. (9.1) and (9.2).]
Step 3.   Calculate P_1 and $\bar{z}_1$.
Step 4.   Use Eq. (9.48) to obtain the theoretical value of D.
Step 5.   Using Eq. (9.46), calculate L_4.
Step 6.   Calculate σ_6 and σ_7. [See Eqs. (9.42) and (9.45).]
Step 7.   Draw the pressure distribution diagram as shown in Figure 9.12.
Step 8.   The actual depth of penetration is

$$D_{actual} = 1.4 \text{ to } 1.6(D_{theoretical})$$

Maximum Bending Moment

According to Figure 9.12, the maximum moment (zero shear) will be between $L_1 + L_2 < z < L_1 + L_2 + L_3$. Using a new coordinate system z' (with $z' = 0$ at the dredge line) for zero shear gives

$$P_1 - \sigma_6 z' = 0$$

or

$$z' = \frac{P_1}{\sigma_6} \tag{9.49}$$

The magnitude of the maximum moment may now be obtained:

$$M_{max} = P_1(z' + \bar{z}_1) - \frac{\sigma_6 z'^2}{2} \tag{9.50}$$

Knowing the maximum bending moment, we determine the section modulus of the sheet pile section from Eq. (9.23).

Example 9.3

In Figure 9.13, for the sheet pile wall, determine

a. The theoretical and actual depth of penetration. Use $D_{actual} = 1.5 D_{theory}$.
b. The minimum size of sheet pile section necessary. Use $\sigma_{all} = 172.5$ MN/m^2.

Figure 9.13 Cantilever sheet pile penetrating into saturated clay

Solution

We will follow the step-by-step procedure given in Section 9.6:

Step 1.

$$K_a = \tan^2\left(45 - \frac{\phi'}{2}\right) = \tan^2\left(45 - \frac{32}{2}\right) = 0.307$$

Step 2.

$$\sigma_1' = \gamma L_1 K_a = (15.9)(2)(0.307) = 9.763 \text{ kN/m}^2$$

$$\sigma_2' = (\gamma L_1 + \gamma' L_2)K_a = [(15.9)(2) + (19.33 - 9.81)3]0.307$$

$$= 18.53 \text{ kN/m}^2$$

Step 3. From the net pressure distribution diagram given in Figure 9.12, we have

$$P_1 = \frac{1}{2}\sigma_1'L_1 + \sigma_1'L_2 + \frac{1}{2}(\sigma_2' - \sigma_1')L_2$$

$$= 9.763 + 29.289 + 13.151 = 52.2 \text{ kN/m}$$

and

$$\bar{z}_1 = \frac{1}{52.2}\left[9.763\left(3 + \frac{2}{3}\right) + 29.289\left(\frac{3}{2}\right) + 13.151\left(\frac{3}{3}\right)\right]$$

$$= 1.78 \text{ m}$$

Step 4. From Eq. (9.48),

$$D^2[4c - (\gamma L_1 + \gamma' L_2)] - 2DP_1 - \frac{P_1(P_1 + 12c\bar{z}_1)}{(\gamma L_1 + \gamma' L_2) + 2c} = 0$$

Substituting proper values yields

$$D^2\{(4)(47) - [(2)(15.9) + (19.33 - 9.81)3]\} - 2D(52.2)$$

$$- \frac{52.2[52.2 + (12)(47)(1.78)]}{[(15.9)(2) + (19.33 - 9.81)3] + (2)(47)} = 0$$

or

$$127.64D^2 - 104.4D - 357.15 = 0$$

Solving the preceding equation, we obtain $D = 2.13$ m.

Step 5. From Eq. (9.46),

$$L_4 = \frac{D[4c - (\gamma L_1 + \gamma' L_2)] - P_1}{4c}$$

and

$$4c - (\gamma L_1 + \gamma' L_2) = (4)(47) - [(15.9)(2) + (19.33 - 9.81)3]$$

$$= 127.64 \text{ kN/m}^2$$

So,

$$L_4 = \frac{2.13(127.64) - 52.2}{(4)(47)} = 1.17 \text{ m}$$

Step 6.

$$\sigma_6 = 4c - (\gamma L_1 + \gamma' L_2) = 127.64 \text{ kN/m}^2$$

$$\sigma_7 = 4c + (\gamma L_1 + \gamma' L_2) = 248.36 \text{ kN/m}^2$$

Step 7. The net pressure distribution diagram can now be drawn, as shown in Figure 9.12.

Step 8. $D_{\text{actual}} \approx 1.5 D_{\text{theoretical}} = 1.5(2.13) \approx \mathbf{3.2 \text{ m}}$

Maximum-Moment Calculation
From Eq. (9.49),

$$z' = \frac{P_1}{\sigma_6} = \frac{52.2}{127.64} \approx 0.41 \text{ m}$$

Again, from Eq. (9.50),

$$M_{\text{max}} = P_1(z' + \bar{z}_1) - \frac{\sigma_6 z'^2}{2}$$

So

$$M_{\text{max}} = 52.2(0.41 + 1.78) - \frac{127.64(0.41)^2}{2}$$

$$= 114.32 - 10.73 = 103.59 \text{ kN-m/m}$$

The minimum required section modulus (assuming that $\sigma_{\text{all}} = 172.5 \text{ MN/m}^2$) is

$$S = \frac{103.59 \text{ kN-m/m}}{172.5 \times 10^3 \text{ kN/m}^2} = \mathbf{0.6 \times 10^{-3} \text{ m}^3/\text{m} \text{ of the wall}} \quad \blacksquare$$

9.7 *Special Cases for Cantilever Walls Penetrating Clay*

Sheet Pile Wall in the Absence of Water Table

As in Section 9.5, relationships for special cases for cantilever walls penetrating clay may also be derived. Referring to Figure 9.14, we can write

$$\sigma_2' = \gamma L K_a \tag{9.51}$$

$$\sigma_6 = 4c - \gamma L \tag{9.52}$$

$$\sigma_7 = 4c + \gamma L \tag{9.53}$$

$$P_1 = \tfrac{1}{2}L\sigma_2' = \tfrac{1}{2}\gamma L^2 K_a \tag{9.54}$$

and

$$L_4 = \frac{D(4c - \gamma L) - \tfrac{1}{2}\gamma L^2 K_a}{4c} \tag{9.55}$$

Figure 9.14 Sheet-pile wall penetrating clay

The theoretical depth of penetration, D, can be calculated [in a manner similar to the calculation of Eq. (9.48)] as

$$D^2(4c - \gamma L) - 2DP_1 - \frac{P_1(P_1 + 12c\bar{z}_1)}{\gamma L + 2c} = 0 \qquad (9.56)$$

where $\bar{z}_1 = \dfrac{L}{3}$. $\qquad (9.57)$

The magnitude of the maximum moment in the wall is

$$M_{max} = P_1(z' + \bar{z}_1) - \frac{\sigma_6 z'^2}{2} \qquad (9.58)$$

where $z' = \dfrac{P_1}{\sigma_6} = \dfrac{\frac{1}{2}\gamma L^2 K_a}{4c - \gamma L}$. $\qquad (9.59)$

Free Cantilever Sheet-Pile Wall Penetrating Clay

Figure 9.15 shows a free cantilever sheet-pile wall penetrating a clay layer. The wall is being subjected to a line load of P per unit length. For this case,

$$\sigma_6 = \sigma_7 = 4c \qquad (9.60)$$

The depth of penetration, D, may be obtained from the relation

$$4D^2c - 2PD - \frac{P(P + 12cL)}{2c} = 0 \qquad (9.61)$$

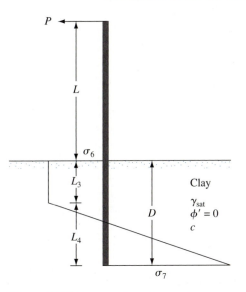

Figure 9.15 Free cantilever sheet piling penetrating clay

Also, note that, for a construction of the pressure diagram,

$$L_4 = \frac{4cD - P}{4c} \tag{9.62}$$

The maximum moment in the wall is

$$M_{max} = P(L + z') - \frac{4cz'^2}{2} \tag{9.63}$$

where $z' = \dfrac{P}{4c}$ (9.64)

Example 9.4

Refer to the free cantilever sheet-pile wall shown in Figure 9.15, for which $P = 32$ kN/m, $L = 3.5$ m, and $c = 12$ kN/m². Calculate the theoretical depth of penetration.

Solution
From Eq. (9.61),

$$4D^2c - 2PD - \frac{P(P + 12cL)}{2c} = 0$$

$$(4)(D^2)(12) - (2)(32)(D) - \frac{32[32 + (12)(12)(3.5)]}{(2)(12)} = 0$$

$$48D^2 - 64D - 714.7 = 0$$

Hence $D \approx$ **4.6 m.** ∎

9.8 *Anchored Sheet-Pile Walls*

When the height of the backfill material behind a cantilever sheet-pile wall exceeds about 6 m, tying the wall near the top to anchor plates, anchor walls, or anchor piles becomes more economical. This type of construction is referred to as *anchored sheet-pile wall* or an *anchored bulkhead*. Anchors minimize the depth of penetration required by the sheet piles and also reduce the cross-sectional area and weight of the sheet piles needed for construction. However, the tie rods and anchors must be carefully designed.

The two basic methods of designing anchored sheet-pile walls are (a) the *free earth support* method and (b) the *fixed earth support* method. Figure 9.16 shows the assumed nature of deflection of the sheet piles for the two methods.

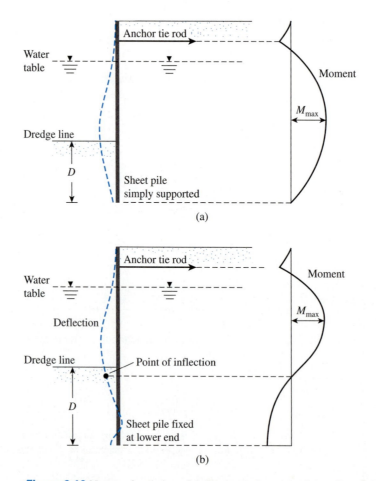

(a)

(b)

Figure 9.16 Nature of variation of deflection and moment for anchored sheet piles: (a) free earth support method; (b) fixed earth support method

The free earth support method involves a minimum penetration depth. Below the dredge line, no pivot point exists for the static system. The nature of the variation of the bending moment with depth for both methods is also shown in Figure 9.16. Note that

$$D_{\text{free earth}} < D_{\text{fixed earth}}$$

9.9 Free Earth Support Method for Penetration of Sandy Soil

Figure 9.17 shows an anchor sheet-pile wall with a granular soil backfill; the wall has been driven into a granular soil. The tie rod connecting the sheet pile and the anchor is located at a depth l_1 below the top of the sheet-pile wall.

The diagram of the net pressure distribution above the dredge line is similar to that shown in Figure 9.8. At depth $z = L_1, \sigma'_1 = \gamma L_1 K_a$, and at $z = L_1 + L_2, \sigma'_2 = (\gamma L_1 + \gamma' L_2) K_a$. Below the dredge line, the net pressure will be zero at $z = L_1 + L_2 + L_3$. The relation for L_3 is given by Eq. (9.6), or

$$L_3 = \frac{\sigma'_2}{\gamma' (K_p - K_a)}$$

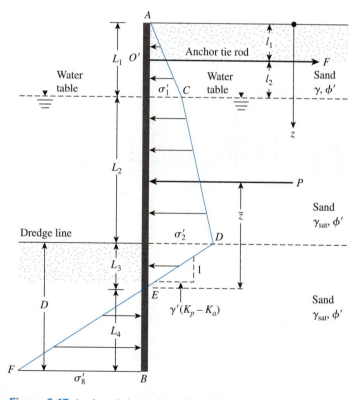

Figure 9.17 Anchored sheet-pile wall penetrating sand

At $z = L_1 + L_2 + L_3 + L_4$, the net pressure is given by

$$\sigma_8' = \gamma'(K_p - K_a)L_4 \tag{9.65}$$

Note that the slope of the line DEF is 1 vertical to $\gamma'(K_p - K_a)$ horizontal.

For equilibrium of the sheet pile, Σ horizontal forces $= 0$, and Σ moment about $O' = 0$. (*Note:* Point O' is located at the level of the tie rod.)

Summing the forces in the horizontal direction (per unit length of the wall) gives

Area of the pressure diagram $ACDE$ − area of EBF − $F = 0$

where F = tension in the tie rod/unit length of the wall, or

$$P - \tfrac{1}{2}\sigma_8'L_4 - F = 0$$

or

$$F = P - \tfrac{1}{2}[\gamma'(K_p - K_a)]L_4^2 \tag{9.66}$$

where P = area of the pressure diagram $ACDE$. Now, taking the moment about point O' gives

$$-P[(L_1 + L_2 + L_3) - (\bar{z} + l_1)] + \tfrac{1}{2}[\gamma'(K_p - K_a)]L_4^2(l_2 + L_2 + L_3 + \tfrac{2}{3}L_4) = 0$$

or

$$L_4^3 + 1.5L_4^2(l_2 + L_2 + L_3) - \frac{3P[(L_1 + L_2 + L_3) - (\bar{z} + l_1)]}{\gamma'(K_p - K_a)} = 0 \tag{9.67}$$

Equation (9.67) may be solved by trial and error to determine the theoretical depth, L_4:

$$D_{\text{theoretical}} = L_3 + L_4$$

The theoretical depth is increased by about 30 to 40% for actual construction, or

$$D_{\text{actual}} = 1.3 \text{ to } 1.4D_{\text{theoretical}} \tag{9.68}$$

The step-by-step procedure in Section 9.4 indicated that a factor of safety can be applied to K_p at the beginning [i.e., $K_{p(\text{design})} = K_p/\text{FS}$]. If that is done, there is no need to increase the theoretical depth by 30 to 40%. This approach is often more conservative.

The maximum theoretical moment to which the sheet pile will be subjected occurs at a depth between $z = L_1$ and $z = L_1 + L_2$. The depth z for zero shear and hence maximum moment may be evaluated from

$$\tfrac{1}{2}\sigma_1' L_1 - F + \sigma_1'(z - L_1) + \tfrac{1}{2}K_a \gamma'(z - L_1)^2 = 0 \tag{9.69}$$

Once the value of z is determined, the magnitude of the maximum moment is easily obtained.

Example 9.5

Let $L_1 = 3.05$ m, $L_2 = 6.1$ m, $l_1 = 1.53$ m, $l_2 = 1.52$ m, $c' = 0$, $\phi' = 30°$, $\gamma = 16$ kN/m³, $\gamma_{sat} = 19.5$ kN/m³, and $E = 207 \times 10^3$ MN/m² in Figure 9.17.

    a. Determine the theoretical and actual depths of penetration. (*Note:* $D_{actual} = 1.3 D_{theory}$.)
    b. Find the anchor force per unit length of the wall.
    c. Determine the maximum moment, M_{max}.

Solution

Part a
We use the following table.

Quantity required	Eq. no.	Equation and calculation
K_a	—	$\tan^2\left(45 - \dfrac{\phi'}{2}\right) = \tan^2\left(45 - \dfrac{30}{2}\right) = \dfrac{1}{3}$
K_P	—	$\tan^2\left(45 + \dfrac{\phi'}{2}\right) = \tan^2\left(45 + \dfrac{30}{2}\right) = 3$
$K_p - K_a$	—	$3 - 0.333 = 2.667$
γ'	—	$\gamma_{sat} - \gamma_w = 19.5 - 9.81 = 9.69$ kN/m³
σ_1'	9.1	$\gamma L_1 K_a = (16)(3.05)\left(\tfrac{1}{3}\right) = 16.27$ kN/m²
σ_2'	9.2	$(\gamma L_1 + \gamma' L_2) K_a = [(16)(3.05) + (9.69)(6.1)]\tfrac{1}{3} = 35.97$ kN/m²
L_3	9.6	$\dfrac{\sigma_2'}{\gamma'(K_p - K_a)} = \dfrac{35.97}{(9.69)(2.667)} = 1.39$ m
P	—	$\tfrac{1}{2}\sigma_1' L_1 + \sigma_2' L_2 + \tfrac{1}{2}(\sigma_2' - \sigma_1') L_2 + \tfrac{1}{2}\sigma_2' L_3 = \left(\tfrac{1}{2}\right)(16.27)(3.05)$ $+ (16.27)(6.1) + \left(\tfrac{1}{2}\right)(35.97 - 16.27)(6.1) + \left(\tfrac{1}{2}\right)(35.97)(1.39)$ $= 24.81 + 99.25 + 60.01 + 25.0 = 209.07$ kN/m
$\bar{z}$	—	$\dfrac{\Sigma M_E}{P} = \left[(24.81)\left(1.39 + 6.1 + \dfrac{3.05}{3}\right) + (99.25)\left(1.39 + \dfrac{6.1}{2}\right) \right. $ $\left. + (60.01)\left(1.39 + \dfrac{6.1}{3}\right) + (25.0)\left(\dfrac{2 \times 1.39}{3}\right) \right] \dfrac{1}{209.07}$ $= 4.21$ m

L_4 9.67 $L_4^3 + 1.5L_4^2(l_2 + L_2 + L_3) - \dfrac{3P[(L_1 + L_2 + L_3) - (\bar{z} + l_1)]}{\gamma'(K_p - K_a)} = 0$

$$L_4^3 + 1.5L_4^2(1.52 + 6.1 + 1.39)$$
$$-\dfrac{(3)(209.07)[(3.05 + 6.1 + 1.39) - (4.21 + 1.53)]}{(9.69)(2.667)} = 0$$
$$L_4 = 2.7 \text{ m}$$

D_{theory} — $L_3 + L_4 = 1.39 + 2.7 = 4.09 \approx \textbf{4.1 m}$

D_{actual} — $1.3D_{\text{theory}} = (1.3)(4.1) = \textbf{5.33 m}$

Part b

The anchor force per unit length of the wall is

$$F = P - \tfrac{1}{2}\gamma'(K_p - K_a)L_4^2$$
$$= 209.07 - \left(\tfrac{1}{2}\right)(9.69)(2.667)(2.7)^2 = 114.87 \text{ kN/m} \approx \textbf{115 kN/m}$$

Part c

From Eq. (9.69), for zero shear,

$$\tfrac{1}{2}\sigma_1'L_1 - F + \sigma_1'(z - L_1) + \tfrac{1}{2}K_a\gamma'(z - L_1)^2 = 0$$

Let $z - L_1 = x$, so that

$$\tfrac{1}{2}\sigma_1'l_1 - F + \sigma_1'x + \tfrac{1}{2}K_a\gamma'x^2 = 0$$

or

$$\left(\tfrac{1}{2}\right)(16.27)(3.05) - 115 + (16.27)(x) + \left(\tfrac{1}{2}\right)\left(\tfrac{1}{3}\right)(9.69)x^2 = 0$$

giving $x^2 + 10.07x - 55.84 = 0$

Now, $x = 4$ m and $z = x + L_1 = 4 + 3.05 = 7.05$ m. Taking the moment about the point of zero shear, we obtain

$$M_{\text{max}} = -\dfrac{1}{2}\sigma_1'L_1\left(x + \dfrac{3.05}{3}\right) + F(x + 1.52) - \sigma_1'\dfrac{x^2}{2} - \dfrac{1}{2}K_a\gamma'x^2\left(\dfrac{x}{3}\right)$$

or

$$M_{\text{max}} = -\left(\dfrac{1}{2}\right)(16.27)(3.05)\left(4 + \dfrac{3.05}{3}\right) + (115)(4 + 1.52) - (16.27)\left(\dfrac{4^2}{2}\right)$$
$$-\left(\dfrac{1}{2}\right)\left(\dfrac{1}{3}\right)(9.69)(4)^2\left(\dfrac{4}{3}\right) = \textbf{344.9 kN·m/m}$$ ∎

Design Charts for Free Earth Support Method (Penetration into Sandy Soil)

Using the free earth support method, Hagerty and Nofal (1992) provided simplified design charts for quick estimation of the depth of penetration, D, anchor force, F, and maximum moment, M_{max}, for anchored sheet-pile walls penetrating into sandy soil, as shown in Figure 9.17. They made the following assumptions for their analysis.

a. The soil friction angle, ϕ', above and below the dredge line is the same.
b. The angle of friction between the sheet-pile wall and the soil is $\phi'/2$.
c. The passive earth pressure below the dredge line has a logarithmic spiral failure surface.
d. For active earth-pressure calculation, Coulomb's theory is valid.

The magnitudes of D, F, and M_{max} may be calculated from the following relationships:

$$\frac{D}{L_1 + L_2} = (GD)(CDL_1) \tag{9.70}$$

$$\frac{F}{\gamma_a(L_1 + L_2)^2} = (GF)(CFL_1) \tag{9.71}$$

$$\frac{M_{max}}{\gamma_a(L_1 + L_2)^3} = (GM)(CML_1) \tag{9.72}$$

where

γ_a = average unit weight of soil

$$= \frac{\gamma L_1^2 + (\gamma_{sat} - \gamma_w) L_2^2 + 2\gamma L_1 L_2}{(L_1 + L_2)^2} \tag{9.73}$$

GD = generalized nondimensional embedment

$$= \frac{D}{L_1 + L_2} \quad \text{(for } L_1 = 0 \text{ and } L_2 = L_1 + L_2\text{)}$$

GF = generalized nondimensional anchor force

$$= \frac{F}{\gamma_a(L_1 + L_2)^2} \quad \text{(for } L_1 = 0 \text{ and } L_2 = L_1 + L_2\text{)}$$

$$GM = \text{generalized nondimensional moment}$$

$$= \frac{M_{max}}{\gamma_a (L_1 + L_2)^3} \quad (\text{for } L_1 = 0 \text{ and } L_2 = L_1 + L_2)$$

$CDL_1, CFL_1, CML_1 = $ correction factors for $L_1 \neq 0$

The variations of GD, GF, GM, CDL_1, CFL_1, and CML_1 are shown in Figures 9.18, 9.19, 9.20, 9.21, 9.22, and 9.23, respectively.

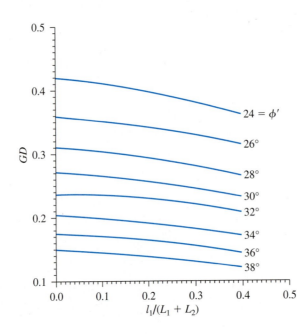

Figure 9.18 Variation of GD with $l_1/(L_1 + L_2)$ and ϕ' (Hagerty, D. J., and Nofal, M. M. (1992). "Design Aids: Anchored Bulkheads in Sand," *Canadian Geotechnical Journal*, Vol. 29, No. 5, pp. 789–795. © 2008 NRC Canada or its licensors. Reproduced with permission.)

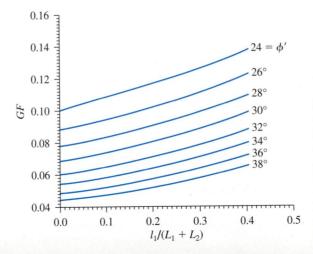

Figure 9.19 Variation of GF with $l_1/(L_1 + L_2)$ and ϕ' (After Hagerty and Nofal, 1992) (Hagerty, D. J., and Nofal, M. M. (1992). "Design Aids: Anchored Bulkheads in Sand," *Canadian Geotechnical Journal*, Vol. 29, No. 5, pp. 789-795. © 2008 NRC Canada or its licensors. Reproduced with permission.

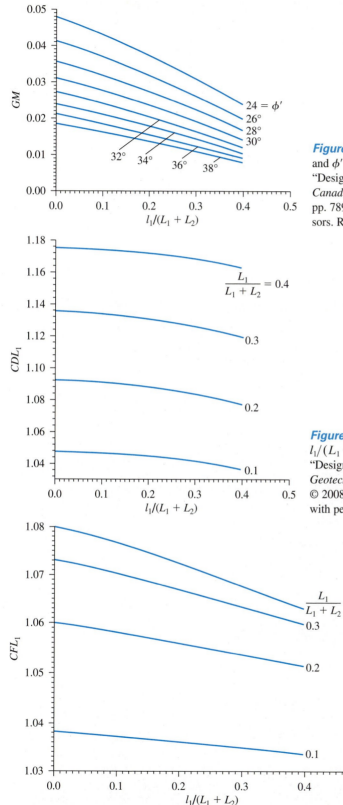

Figure 9.20 Variation of GM with $l_1/(L_1 + L_2)$ and ϕ' (Hagerty, D. J., and Nofal, M. M. (1992). "Design Aids: Anchored Bulkheads in Sand," *Canadian Geotechnical Journal*, Vol. 29, No. 5, pp. 789–795. © 2008 NRC Canada or its licensors. Reproduced with permission.)

Figure 9.21 Variation of CDL_1 with $L_1/(L_1 + L_2)$ and $l_1/(L_1 + L_2)$ (Hagerty, D. J., and Nofal, M. M. (1992). "Design Aids: Anchored Bulkheads in Sand," *Canadian Geotechnical Journal*, Vol. 29, No. 5, pp. 789–795. © 2008 NRC Canada or its licensors. Reproduced with permission.)

Figure 9.22 Variation of CFL_1 with $L_1/(L_1 + L_2)$ and $l_1/(L_1 + L_2)$ (Hagerty, D. J., and Nofal, M. M. (1992). "Design Aids: Anchored Bulkheads in Sand," *Canadian Geotechnical Journal*, Vol. 29, No. 5, pp. 789–795. © 2008 NRC Canada or its licensors. Reproduced with permission.)

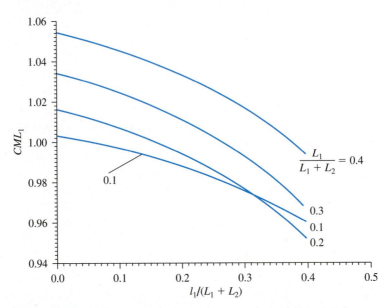

Figure 9.23 Variation of CML_1 with $L_1/(L_1 + L_2)$ and $l_1/(L_1 + L_2)$ (Hagerty, D. J., and Nofal, M. M. (1992). "Design Aids: Anchored Bulkheads in Sand," *Canadian Geotechnical Journal*, Vol. 29, No. 5, pp. 789–795. © 2008 NRC Canada or its licensors. Reproduced with permission.)

Example 9.6

Refer to Figure 9.17. Given: $L_1 = 2$ m, $L_2 = 3$ m, $l_1 = l_2 = 1$ m, $c = 0$, $\phi' = 32°$ $\gamma = 15.9$ kN/m^3, and $\gamma_{sat} = 19.33$ kN/m^3. Determine:

a. Theoretical and actual depth of penetration. *Note:* $D_{actual} = 1.4D_{theory}$.
b. Anchor force per unit length of wall.
c. Maximum moment, M_{max}.

Use the charts presented in Section 9.10.

Solution

Part a
From Eq. (9.70),

$$\frac{D}{L_1 + L_2} = (GD)(CDL_1)$$

$$\frac{l_1}{L_1 + L_2} = \frac{1}{2 + 3} = 0.2$$

From Figure 9.18 for $l_1/(L_1 + L_2) = 0.2$ and $\phi' = 32°$, $GD = 0.22$. From Figure 9.21, for

$$\frac{L_1}{L_1 + L_2} = \frac{2}{2 + 3} = 0.4 \qquad \text{and} \qquad \frac{l_1}{L_1 + L_2} = 0.2$$

$CDL_1 \approx 1.172$. So

$$D_{\text{theory}} = (L_1 + L_2)(GD)(CDL_1) = (5)(0.22)(1.172) \approx 1.3$$

$$D_{\text{actual}} \approx (1.4)(1.3) = 1.82 \approx \mathbf{2\ m}$$

Part b
From Figure 9.19 for $l_1/(L_1 + L_2) = 0.2$ and $\phi' = 32°$, $GF \approx 0.074$. Also, from Figure 9.22, for

$$\frac{L_1}{L_1 + L_2} = \frac{2}{2 + 3} = 0.4, \quad \frac{l_1}{L_1 + L_2} = 0.2, \quad \text{and} \quad \phi' = 32°$$

$CFL_1 = 1.073$. From Eq. (9.73),

$$\gamma_a = \frac{\gamma L_1^2 + \gamma' L_2^2 + 2\gamma L_1 L_2}{(L_1 + L_2)^2}$$

$$= \frac{(15.9)(2)^2 + (19.33 - 9.81)(3)^2 + (2)(15.9)(2)(3)}{(2 + 3)^2} = 13.6\ \text{kN/m}^3$$

Using Eq. (9.71) yields

$$F = \gamma_a(L_1 + L_2)^2(GF)(CFL_1) = (13.6)(5)^2(0.074)(1.073) \approx \mathbf{27\ kN/m}$$

Part c
From Figure 9.20, for $l_1/(L_1 + L_2) = 0.2$ and $\phi = 32°$, $GM = 0.021$. Also, from Figure 9.23, for

$$\frac{L_1}{L_1 + L_2} = \frac{2}{2 + 3} = 0.4, \quad \frac{l_1}{L_1 + L_2} = 0.2, \quad \text{and} \quad \phi' = 32°$$

$CML_1 = 1.036$. Hence from Eq. (9.72),

$$M_{\text{max}} = \gamma_a(L_1 + L_2)^3(GM)(CML_1) = (13.6)(5)^3(0.021)(1.036) = \mathbf{36.99\ kN \cdot m/m} \quad \blacksquare$$

9.11 *Moment Reduction for Anchored Sheet-Pile Walls*

Sheet piles are flexible, and hence sheet-pile walls yield (i.e., become displaced laterally), which redistributes the lateral earth pressure. This change tends to reduce the maximum bending moment, M_{max}, as calculated by the procedure outlined in

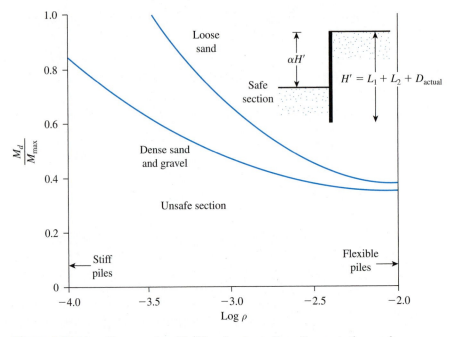

Figure 9.24 Plot of log ρ against M_d/M_{max} for sheet-pile walls penetrating sand (From Rowe, P. W. (1952). "Anchored Sheet Pile Walls," Proceedings, Institute of Civil Engineers, Vol. 1, Part 1, pp. 27–70.)

Section 9.9. For that reason, Rowe (1952, 1957) suggested a procedure for reducing the maximum design moment on the sheet pile walls *obtained from the free earth support method.* This section discusses the procedure of moment reduction for sheet piles *penetrating into sand.*

In Figure 9.24, which is valid for the case of a sheet pile penetrating sand, the following notation is used:

1. H' = total height of pile driven (i.e., $L_1 + L_2 + D_{actual}$)

2. Relative flexibility of pile = $\rho = 10.91 \times 10^{-7}\left(\dfrac{H'^4}{EI}\right)$ (9.74)

where

H' is in meters
E = modulus of elasticity of the pile material (MN/m^2)
I = moment of inertia of the pile section per meter of the wall $(m^4/m$ of wall)

3. M_d = design moment
4. M_{max} = maximum theoretical moment

The procedure for the use of the moment reduction diagram (see Figure 9.24) is as follows:

Step 1. Choose a sheet pile section (e.g., from among those given in Table 9.1).
Step 2. Find the modulus S of the selected section (Step 1) per unit length of the wall.
Step 3. Determine the moment of inertia of the section (Step 1) per unit length of the wall.
Step 4. Obtain H' and calculate ρ [see Eq. (9.74)].
Step 5. Find $\log \rho$.
Step 6. Find the moment capacity of the pile section chosen in Step 1 as $M_d = \sigma_{\text{all}}S$.
Step 7. Determine M_d/M_{max}. Note that M_{max} is the maximum theoretical moment determined before.
Step 8. Plot $\log \rho$ (Step 5) and M_d/M_{max} in Figure 9.24.
Step 9. Repeat Steps 1 through 8 for several sections. The points that fall above the curve (in loose sand or dense sand, as the case may be) are *safe sections.*

The points that fall below the curve are *unsafe sections.* The cheapest section may now be chosen from those points which fall above the proper curve. Note that the section chosen will have an $M_d < M_{\text{max}}$.

Example 9.7

Refer to Example 9.5. Use Rowe's moment reduction diagram (Figure 9.24) to find an appropriate sheet pile section. For the sheet pile, use $E = 207 \times 10^3 \text{ MN/m}^2$ and $\sigma_{\text{all}} = 172,500 \text{ kN/m}^2$.

Solution

$$H' = L_1 + L_2 + D_{\text{actual}} = 3.05 + 6.1 + 5.33 = 14.48 \text{ m}$$

$M_{\text{max}} = 344.9 \text{ kN} \cdot \text{m/m}$. Now the following table can be prepared.

Section	$I(\text{m}^4/\text{m})$	$H'(\text{m})$	$\rho = 10.91 \times 10^{-7}\left(\dfrac{H'^4}{EI}\right)$	$\log \rho$	$S(\text{m}^3/\text{m})$	$M_d = S\sigma_{\text{all}}$ $(\text{kN} \cdot \text{m/m})$	$\dfrac{M_d}{M_{\text{max}}}$
PZ-22	115.2×10^{-6}	14.48	20.11×10^{-4}	-2.7	97×10^{-5}	167.33	0.485
PZ-27	251.5×10^{-6}	14.48	9.21×10^{-4}	-3.04	162.3×10^{-5}	284.84	0.826

Figure 9.25 gives a plot of M_d/M_{max} versus ρ. It can be seen that **PZ-27** will be sufficient.

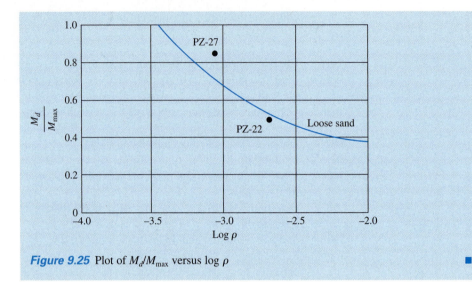

Figure 9.25 Plot of M_d/M_{max} versus log ρ ■

Computational Pressure Diagram Method for Penetration into Sandy Soil

The computational pressure diagram (CPD) method for sheet pile penetrating a sandy soil is a simplified method of design and an alternative to the free earth method described in Sections 9.9 and 9.11 (Nataraj and Hoadley, 1984). In this method, the net pressure diagram shown in Figure 9.17 is replaced by rectangular pressure diagrams, as in Figure 9.26. Note that $\overline{\sigma}'_a$ is the width of the net active pressure diagram above the dredge line and $\overline{\sigma}'_p$ is the width of the net passive pressure diagram below the dredge line. The magnitudes of $\overline{\sigma}'_a$ and $\overline{\sigma}'_p$ may respectively be expressed as

$$\overline{\sigma}'_a = CK_a\gamma'_{av}L \tag{9.75}$$

and

$$\overline{\sigma}'_p = RCK_a\gamma'_{av}L = R\overline{\sigma}'_a \tag{9.76}$$

where

γ'_{av} = average effective unit weight of sand

$$\approx \frac{\gamma L_1 + \gamma' L_2}{L_1 + L_2} \tag{9.77}$$

C = coefficient

$$R = \text{coefficient} = \frac{L(L - 2l_1)}{D(2L + D - 2l_1)} \tag{9.78}$$

The range of values for C and R is given in Table 9.2.

Figure 9.26 Computational pressure diagram method (*Note: $L_1 + L_2 = L$*)

Table 9.2 Range of Values for C and R [from Eqs. (9.75) and (9.76)]

Soil type	C^a	R
Loose sand	0.8–0.85	0.3–0.5
Medium sand	0.7–0.75	0.55–0.65
Dense sand	0.55–0.65	0.60–0.75

[a] Valid for the case in which there is no surcharge above the granular backfill (i.e., on the right side of the wall, as shown in Figure 9.26)

The depth of penetration, D, anchor force per unit length of the wall, F, and maximum moment in the wall, M_{max}, are obtained from the following relationships.

Depth of Penetration

For the depth of penetration, we have

$$D^2 + 2DL\left[1 - \left(\frac{l_1}{L}\right)\right] - \left(\frac{L^2}{R}\right)\left[1 - 2\left(\frac{l_1}{L}\right)\right] = 0 \qquad (9.79)$$

Anchor Force

The anchor force is

$$F = \overline{\sigma}_a'(L - RD) \qquad (9.80)$$

Maximum Moment

The maximum moment is calculated from

$$M_{\max} = 0.5\overline{\sigma}_a' L^2 \left[\left(1 - \frac{RD}{L} \right)^2 - \left(\frac{2l_1}{L} \right)\left(1 - \frac{RD}{L} \right) \right]$$ (9.81)

Note the following qualifications:

1. The magnitude of D obtained from Eq. (9.79) is about 1.25 to 1.5 times the value of D_{theory} obtained by the conventional free earth support method (see Section 9.9), so

$$D \approx D_{\text{actual}}$$
$$\uparrow \qquad \uparrow$$
Eq. (9.79)   Eq. (9.68)

2. The magnitude of F obtained by using Eq. (9.80) is about 1.2 to 1.6 times the value obtained by using Eq. (9.66). Thus, an additional factor of safety for the actual design of anchors need not be used.

3. The magnitude of $M_{\max}$ obtained from Eq. (9.81) is about 0.6 to 0.75 times the value of $M_{\max}$ obtained by the conventional free earth support method. Hence, the former value of $M_{\max}$ can be used as the actual design value, and Rowe's moment reduction need not be applied.

Example 9.8

For the anchored sheet pile wall shown in Figure 9.27, determine (a) D, (b) F, and (c) $M_{\max}$. Use the CPD method; assume that $C = 0.68$ and $R = 0.6$.

Solution

Part a

$$\gamma' = \gamma_{\text{sat}} - \gamma_w = 19.24 - 9.81 = 9.43 \text{ kN/m}^3$$

From Eq. (9.77)

$$\gamma_{\text{av}}' = \frac{\gamma L_1 + \gamma' L_2}{L_1 + L_2} = \frac{(17.3)(3) + (9.43)(6)}{3 + 6} = 12.05 \text{ kN/m}^3$$

$$K_a = \tan^2\left(45 - \frac{\phi'}{2} \right) = \tan^2\left(45 - \frac{35}{2} \right) = 0.271$$

$$\overline{\sigma}_a' = C K_a \gamma_{\text{av}}' L = (0.68)(0.271)(12.05)(9) = 19.99 \text{ kN/m}^2$$

$$\overline{\sigma}_p' = R\overline{\sigma}_a' = (0.6)(19.99) = 11.99 \text{ kN/m}^2$$

From Eq. (9.80)

$$D^2 + 2DL\left[1 - \left(\frac{l_1}{L} \right) \right] - \frac{L^2}{R}\left[1 - 2\left(\frac{l_1}{L} \right) \right] = 0$$

Figure 9.27

or

$$D^2 + 2(D)(9)\left[1 - \left(\frac{1.5}{9}\right)\right] - \frac{(9)^2}{0.6}\left[1 - 2\left(\frac{1.5}{9}\right)\right] = D^2 + 50D - 1000 = 0$$

Hence $D \approx$ **4.6 m**.
Check for the assumption of R:

$$R = \frac{L(L - 2l_1)}{D(2L + D - 2l_1)} = \frac{9[9 - (2)(1.5)]}{4.6[(2)(9) + 4.6 - (2)(1.5)]} \approx \textbf{0.6—OK}$$

Part b
From Eq. (9.80)

$$F = \overline{\sigma}_a'(L - RD) = 19.99[9 - (0.6)(4.6)] = \textbf{124.74 kN/m}$$

Part c
From Eq. (9.81)

$$M_{max} = 0.5\overline{\sigma}_a'L^2\left[\left(1 - \frac{RD}{L}\right)^2 - \left(\frac{2l_1}{L}\right)\left(1 - \frac{RD}{L}\right)\right]$$

$$1 - \frac{RD}{L} = 1 - \frac{(0.6)(4.6)}{9} = 0.693$$

So,

$$M_{max} = (0.5)(19.99)(9)^2\left[(0.693)^2 - \frac{(2)(1.5)(0.693)}{9}\right] = \textbf{201.6 kN-m/m} \quad \blacksquare$$

9.13	*Fixed Earth-Support Method for Penetration into Sandy Soil*

When using the fixed earth support method, we assume that the toe of the pile is restrained from rotating, as shown in Figure 9.28a. In the fixed earth support solution, a simplified method called the *equivalent beam solution* is generally used to calculate L_3 and, thus, D. The development of the equivalent beam method is generally attributed to Blum (1931).

In order to understand this method, compare the sheet pile to a loaded cantilever beam *RSTU*, as shown in Figure 9.29. Note that the support at *T* for the beam is equivalent to the *anchor load reaction* (*F*) on the sheet pile (Figure 9.28). It can be seen that the point *S* of the beam *RSTU* is the inflection point of the elastic line of the beam, which is equivalent to point *I* in Figure 9.28. It the beam is cut at *S* and a free support (reaction P_s) is provided at that point, the bending moment diagram for portion *STU* of the beam will remain unchanged. This beam *STU* will be equivalent to the section *STU* of the beam *RSTU*. The force *P'* shown in Figure 9.28a at *I* will be equivalent to the reaction P_s on the beam (Figure 9.29).

The following is an approximate procedure for the design of an anchored sheet-pile wall (Cornfield, 1975). Refer to Figure 9.28.

Step 1.   Determine L_5, which is a function of the soil friction angle ϕ' below the dredge line, from the following:

ϕ' (deg)	$\dfrac{L_5}{L_1 + L_2}$
30	0.08
35	0.03
40	0

(a) Pressure diagram   (b) Moment diagram

FIGURE 9.28 Fixed earth support method for penetration of sandy soil

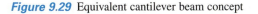

Figure 9.29 Equivalent cantilever beam concept

Step 2. Calculate the span of the equivalent beam as $l_2 + L_2 + L_5 = L'$.

Step 3. Calculate the total load of the span, W. This is the area of the pressure diagram between O' and I.

Step 4. Calculate the maximum moment, M_{max}, as $WL'/8$.

Step 5. Calculate P' by taking the moment about O', or

$$P' = \frac{1}{L'} \, (\text{moment of area } ACDJI \text{ about } O') \tag{9.82}$$

Step 6. Calculate D as

$$D = L_5 + 1.2 \sqrt{\frac{6P'}{(K_p - K_a)\gamma'}} \tag{9.83}$$

Step 7. Calculate the anchor force per unit length, F, by taking the moment about I, or

$$F = \frac{1}{L'} \, (\text{moment of area } ACDJI \text{ about } I)$$

Example 9.9

Consider the anchored sheet-pile structure described in Example 9.5. Using the equivalent beam method described in Section 9.13, determine

   a. Maximum moment
   b. Theoretical depth of penetration
   c. Anchor force per unit length of the structure

Solution

Part a
Determination of L_5: For $\phi' = 30°$,

$$\frac{L_5}{L_1 + L_2} = 0.08$$

$$\frac{L_5}{3.05 + 6.1} = 0.08$$

$$L_5 = 0.73$$

Net Pressure Diagram: From Example 9.5, $K_a = \frac{1}{3}$, $K_p = 3$, $\gamma = 16$ kN/m^3, $\gamma' = 9.69$ kN/m^3, $\sigma_1' = 16.27$ kN/m^2, $\sigma_2' = 35.97$ kN/m^2. The net active pressure at a depth L_5 below the dredge line can be calculated as

$$\sigma_2' - \gamma'(K_p - K_a)L_5 = 35.97 - (9.69)(3 - 0.333)(0.73) = 17.1 \text{ kN/m}^2$$

The net pressure diagram from $z = 0$ to $z = L_1 + L_2 + L_5$ is shown in Figure 9.30.

Maximum Moment:

$$W = \left(\frac{1}{2}\right)(8.16 + 16.27)(1.52) + \left(\frac{1}{2}\right)(6.1)(16.27 + 35.97)$$

$$+ \left(\frac{1}{2}\right)(0.73)(35.97 + 17.1)$$

$$= 197.2 \text{ kN/m}$$

$$L' = l_2 + L_2 + L_5 = 1.52 + 6.1 + 0.73 = 8.35 \text{ m}$$

$$M_{\text{max}} = \frac{WL'}{8} = \frac{(197.2)(8.35)}{8} = \textbf{205.8 kN} \cdot \textbf{m/m}$$

Part b

$$P' = \frac{1}{L'} \text{ (moment of area } ACDJI \text{ about } O')$$

A

8.16 kN/m^2

1.53 m $= l_1$

O' ────────→ F

$l_2 = 1.52$ m

16.27 kN/m^2

C

6.1 m $= L_2$

35.97 kN/m^2

D

$L_5 = 0.73$ m

P'

I 17.1 kN/m^2 J

FIGURE 9.30

$$P' = \frac{1}{8.35} \left[\begin{array}{l} \left(\frac{1}{2}\right)(16.27)(3.05)\left(\frac{2}{3} \times 3.05 - 1.53\right) + (16.27)(6.1)\left(1.52 + \frac{6.1}{2}\right) \\ + \left(\frac{1}{2}\right)(6.1)(35.97 - 16.27)\left(1.52 + \frac{2}{3} \times 6.1\right) + \left(\frac{1}{2}\right)(35.97 + 17.1) \\ \times (0.73)\left(1.52 + 6.1 + \frac{0.73}{2}\right) \end{array} \right]$$

↑
Approximate

$$= 114.48 \text{ kN/m}$$

From Eq. (9.83)

$$D = L_5 + 1.2 \sqrt{\frac{6P'}{(K_p - K_a)\gamma'}} = 0.73 + 1.2 \sqrt{\frac{(6)(114.48)}{(3 - 0.333)(9.69)}} = \textbf{6.92 m}$$

Part c

Taking the moment about I (Figure 9.30)

$$F = \frac{1}{8.35} \left[\begin{array}{l} \left(\frac{1}{2}(16.27)(3.05)\left(0.73 + 6.1 + \frac{3.05}{3}\right) + (16.27)(6.1)\left(0.73 + \frac{6.1}{2}\right) \right. \\ + \left(\frac{1}{2}\right)(6.1)(35.97 - 16.27)\left(0.73 + \frac{6.1}{3}\right) + \left(\frac{1}{2}\right)(35.97 + 17.1)(0.73)\left(\frac{0.73}{2}\right) \end{array} \right]$$

↑
Approximate

$$= \textbf{88.95 kN/m} \qquad \blacksquare$$

9.14 *Field Observations for Anchor Sheet Pile Walls*

In the preceding sections, large factors of safety were used for the depth of penetration, D. In most cases, designers use smaller magnitudes of soil friction angle, ϕ', thereby ensuring a built-in factor of safety for the active earth pressure. This procedure is followed primarily because of the uncertainties involved in predicting the actual earth pressure to which a sheet-pile wall in the field will be subjected. In addition, Casagrande (1973) observed that, if the soil behind the sheet-pile wall has grain sizes that are predominantly smaller than those of coarse sand, the active earth pressure after construction sometimes increases to an at-rest earth-pressure condition. Such an increase causes a large increase in the anchor force, F. The following two case histories are given by Casagrande (1973).

Bulkhead of Pier C—Long Beach Harbor, California (1949)

A typical cross section of the Pier C bulkhead of the Long Beach harbor is shown in Figure 9.31. Except for a rockfill dike constructed with 76 mm (3 in.) maximum-size quarry wastes, the backfill of the sheet-pile wall consisted of fine sand. Figure 9.32 shows the

Figure 9.31 Pier C bulkhead—Long Beach harbor (Adapted after Casagrande, 1973)

variation of the lateral earth pressure between May 24, 1949 (the day construction was completed) and August 6, 1949. On May 24, the lateral earth pressure reached an active state, as shown in Figure 9.32a, due to the wall yielding. Between May 24 and June 3, the anchor resisted further yielding and the lateral earth pressure increased to the at-rest state (Figure 9.32b). However, the flexibility of the sheet piles ultimately resulted in a gradual decrease in the lateral earth-pressure distribution on the sheet piles (see Figure 9.32c).

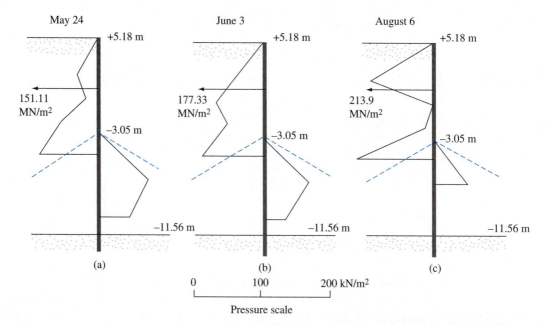

Figure 9.32 Measured stresses at Station 27 + 30—Pier C bulkhead, Long Beach (Adapted after Casagrande, 1973)

With time, the stress on the tie rods for the anchor increased as shown in the following table.

Date	Stress on anchor tie rod (MN/m^2)
May 24, 1949	151.11
June 3, 1949	177.33
June 11, 1949	193.2
July 12, 1949	203.55
August 6, 1949	213.9

These observations show that the magnitude of the active earth pressure may vary with time and depend greatly on the flexibility of the sheet piles. Also, the actual variations in the lateral earth-pressure diagram may not be identical to those used for design.

Bulkhead—Toledo, Ohio (1961)

A typical cross section of a Toledo bulkhead completed in 1961 is shown in Figure 9.33. The foundation soil was primarily fine to medium sand, but the dredge line did cut into highly overconsolidated clay. Figure 9.33 also shows the actual measured values of bending moment along the sheet-pile wall. Casagrande (1973) used the Rankine active earth-pressure distribution to calculate the maximum bending moment according to the free earth support method with and without Rowe's moment reduction.

Design method	Maximum predicted bending moment, M_{max}
Free earth support method	146.5 kN-m
Free earth support method with Rowe's moment reduction	78.6 kN-m

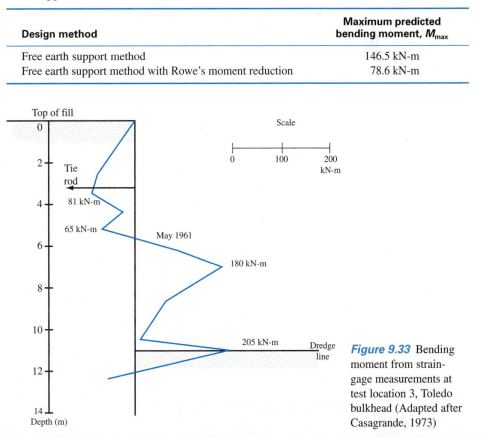

Figure 9.33 Bending moment from strain-gage measurements at test location 3, Toledo bulkhead (Adapted after Casagrande, 1973)

Comparisons of these magnitudes of M_{max} with those actually observed show that the field values are substantially larger. The reason probably is that the backfill was primarily fine sand and the measured active earth-pressure distribution was larger than that predicted theoretically.

9.15 *Free Earth Support Method for Penetration of Clay*

Figure 9.34 shows an anchored sheet-pile wall penetrating a clay soil and with a granular soil backfill. The diagram of pressure distribution above the dredge line is similar to that shown in Figure 9.12. From Eq. (9.42), the net pressure distribution below the dredge line (from $z = L_1 + L_2$ to $z = L_1 + L_2 + D$) is

$$\sigma_6 = 4c - (\gamma L_1 + \gamma' L_2)$$

For static equilibrium, the sum of the forces in the horizontal direction is

$$P_1 - \sigma_6 D = F \tag{9.84}$$

where

P_1 = area of the pressure diagram ACD
F = anchor force per unit length of the sheet pile wall

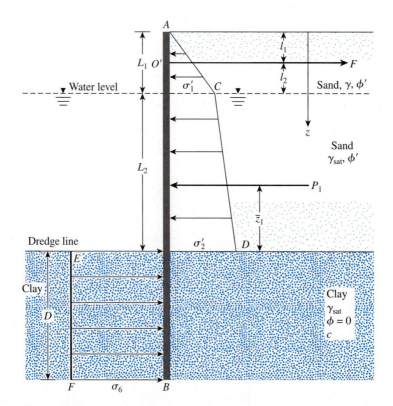

Figure 9.34 Anchored sheet-pile wall penetrating clay

Again, taking the moment about O' produces

$$P_1(L_1 + L_2 - l_1 - \bar{z}_1) - \sigma_6 D\left(l_2 + L_2 + \frac{D}{2}\right) = 0$$

Simplification yields

$$\sigma_6 D^2 + 2\sigma_6 D(L_1 + L_2 - l_1) - 2P_1(L_1 + L_2 - l_1 - \bar{z}_1) = 0 \qquad (9.85)$$

Equation (9.85) gives the theoretical depth of penetration, D.

As in Section 9.9, the maximum moment in this case occurs at a depth $L_1 < z < L_1 + L_2$. The depth of zero shear (and thus the maximum moment) may be determined from Eq. (9.69).

A moment reduction technique similar to that in Section 9.11 for anchored sheet piles penetrating into clay has also been developed by Rowe (1952, 1957). This technique is presented in Figure 9.35, in which the following notation is used:

1. The stability number is

$$S_n = 1.25 \frac{c}{(\gamma L_1 + \gamma' L_2)} \qquad (9.86)$$

where c = undrained cohesion ($\phi = 0$).

For the definition of γ, γ', L_1, and L_2, see Figure 9.34.

2. The nondimensional wall height is

$$\alpha = \frac{L_1 + L_2}{L_1 + L_2 + D_{\text{actual}}} \qquad (9.87)$$

3. The flexibility number is ρ [see Eq. (9.74)]
4. M_d = design moment
   M_{max} = maximum theoretical moment

The procedure for moment reduction, using Figure 9.35, is as follows:

Step 1. Obtain $H' = L_1 + L_2 + D_{\text{actual}}$.
Step 2. Determine $\alpha = (L_1 + L_2)/H'$.
Step 3. Determine S_n [from Eq. (9.86)].
Step 4. For the magnitudes of α and S_n obtained in Steps 2 and 3, determine M_d/M_{max} for various values of log ρ from Figure 9.35, and plot M_d/M_{max} against log ρ.
Step 5. Follow Steps 1 through 9 as outlined for the case of moment reduction of sheet-pile walls penetrating granular soil. (See Section 9.11.)

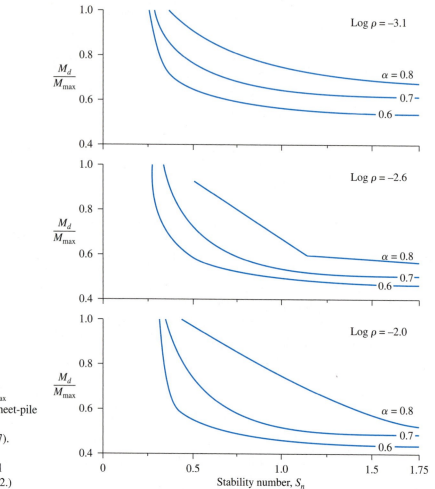

Figure 9.35 Plot of $M_d/M_{\max}$ against stability number for sheet-pile wall penetrating clay (From Rowe, P. W. (1957). "Sheet Pile Walls in Clay," *Proceedings*, Institute of Civil Engineers, Vol. 7, pp. 654–692.)

Example 9.10

In Figure 9.34, let $L_1 = 3$ m, $L_2 = 6$ m, and $l_1 = 1.5$ m. Also, let $\gamma = 17$ kN/m³, $\gamma_{\text{sat}} = 20$ kN/m³, $\phi' = 35°$, and $c = 41$ kN/m².

    a. Determine the theoretical depth of embedment of the sheet-pile wall.
    b. Calculate the anchor force per unit length of the wall.

Solution

Part a
We have

$$K_a = \tan^2\left(45 - \frac{\phi'}{2}\right) = \tan^2\left(45 - \frac{35}{2}\right) = 0.271$$

and

$$K_p = \tan^2\left(45 + \frac{\phi'}{2}\right) = \tan^2\left(45 + \frac{35}{2}\right) = 3.69$$

From the pressure diagram in Figure 9.36,

$$\sigma_1' = \gamma L_1 K_a = (17)(3)(0.271) = 13.82 \text{ kN/m}^2$$

$$\sigma_2' = (\gamma L_1 + \gamma' L_2)K_a = [(17)(3) + (20 - 9.81)(6)](0.271) = 30.39 \text{ kN/m}^2$$

$$P_1 = \text{areas } 1 + 2 + 3 = 1/2(3)(13.82) + (13.82)(6) + 1/2(30.39 - 13.82)(6)$$

$$= 20.73 + 82.92 + 49.71 = 153.36 \text{ kN/m}$$

and

$$\bar{z}_1 = \frac{(20.73)\left(6 + \dfrac{3}{3}\right) + (82.92)\left(\dfrac{6}{2}\right) + (49.71)\left(\dfrac{6}{3}\right)}{153.36} = 3.2 \text{ m}$$

From Eq. (9.85),

$$\sigma_6 D^2 + 2\sigma_6 D(L_1 + L_2 - l_1) - 2P_1(L_1 + L_2 - l_1 - \bar{z}_1) = 0$$

$$\sigma_6 = 4c - (\gamma L_1 + \gamma' L_2) = (4)(41) - [(17)(3)$$

$$+ (20 - 9.81)(6)] = 51.86 \text{ kN/m}^2$$

So,

$$(51.86)D^2 + (2)(51.86)(D)(3 + 6 - 1.5)$$
$$- (2)(153.36)(3 + 6 - 1.5 - 3.2) = 0$$

$l_1 = 1.5$ m

$L_1 = 3$ m

① $l_2 = 1.5$ m

$\sigma_1' = 13.82$ kN/m^2

$L_2 = 6$ m   ②

③

$\sigma_2' = 30.39$ kN/m^2

1.6 m $= D$

$\sigma_6 = 51.86$ kN/m^2

Figure 9.36 Free earth support method, sheet pile penetrating into clay

or

$$D^2 + 15D - 25.43 = 0$$

Hence,

$$D \approx \textbf{1.6 m}$$

Part b
From Eq. (9.84),

$$F = P_1 - \sigma_6 D = 153.36 - (51.86)(1.6) = \textbf{70.38 kN/m} \quad \blacksquare$$

9.16   *Anchors*

Sections 9.9 through 9.15 gave an analysis of anchored sheet-pile walls and discussed how to obtain the force *F* per unit length of the sheet-pile wall that has to be sustained by the anchors. The current section covers in more detail the various types of anchor generally used and the procedures for evaluating their ultimate holding capacities.

The general types of anchor used in sheet-pile walls are as follows:

1. Anchor plates and beams (deadman)
2. Tie backs
3. Vertical anchor piles
4. Anchor beams supported by batter (compression and tension) piles

Anchor plates and beams are generally made of cast concrete blocks. (See Figure 9.37a.) The anchors are attached to the sheet pile by *tie-rods*. A *wale* is placed at the front or back face of a sheet pile for the purpose of conveniently attaching the tie-rod to the wall. To protect the tie rod from corrosion, it is generally coated with paint or asphaltic materials.

In the construction of *tiebacks,* bars or cables are placed in predrilled holes (see Figure 9.37b) with concrete grout (cables are commonly high-strength, prestressed steel tendons). Figures 9.37c and 9.37d show a vertical anchor pile and an anchor beam with batter piles.

Placement of Anchors

The resistance offered by anchor plates and beams is derived primarily from the passive force of the soil located in front of them. Figure 9.37a, in which *AB* is the sheet-pile wall, shows the best location for maximum efficiency of an anchor plate. If the anchor is placed inside wedge *ABC*, which is the Rankine active zone, it would not provide any resistance to failure. Alternatively, the anchor could be placed in zone *CFEH*. Note that line *DFG* is the slip line for the Rankine passive pressure. If part of the passive wedge is located inside the active wedge *ABC*, full passive resistance of the anchor cannot be realized upon

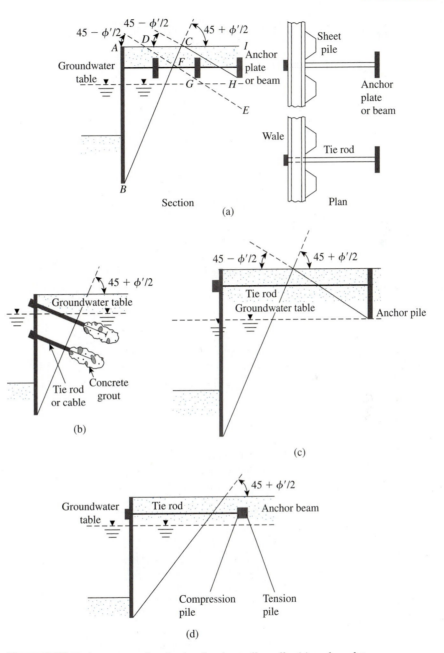

Figure 9.37 Various types of anchoring for sheet-pile walls: (a) anchor plate or beam; (b) tieback; (c) vertical anchor pile; (d) anchor beam with batter piles

failure of the sheet-pile wall. However, if the anchor is placed in zone *ICH,* the Rankine passive zone in front of the anchor slab or plate is located completely outside the Rankine active zone *ABC.* In this case, full passive resistance from the anchor can be realized.

Figures 9.37b, 9.37c, and 9.37d also show the proper locations for the placement of tiebacks, vertical anchor piles, and anchor beams supported by batter piles.

9.17 *Holding Capacity of Anchor Plates in Sand*

Semi-Empirical Method

Ovesen and Stromann (1972) proposed a semi-empirical method for determining the ultimate resistance of anchors in sand. Their calculations, made in three steps, are carried out as follows:

Step 1. **Basic Case.** Determine the depth of embedment, H. Assume that the anchor slab has height H and is continuous (i.e., $B = $ length of anchor slab perpendicular to the cross section $= \infty$), as shown in Figure 9.38, in which the following notation is used:

$P_p = $ passive force per unit length of anchor
$P_a = $ active force per unit length of anchor
$\phi' = $ effective soil friction angle
$\delta' = $ friction angle between anchor slab and soil
$P'_{ult} = $ ultimate resistance per unit length of anchor
$W = $ effective weight per unit length of anchor slab

Also,

$$P'_{ult} = \tfrac{1}{2}\gamma H^2 K_p \cos \delta' - P_a \cos \phi' = \tfrac{1}{2}\gamma H^2 K_p \cos \delta' - \tfrac{1}{2}\gamma H^2 K_a \cos \phi'$$
$$= \tfrac{1}{2}\gamma H^2 (K_p \cos \delta' - K_a \cos \phi') \tag{9.88}$$

where

$K_a = $ active pressure coefficient with $\delta' = \phi'$
    (see Figure 9.39a)
$K_p = $ passive pressure coefficient

To obtain $K_p \cos \delta'$, first calculate

$$K_p \sin \delta' = \frac{W + P_a \sin \phi'}{\tfrac{1}{2}\gamma H^2} = \frac{W + \tfrac{1}{2}\gamma H^2 K_a \sin \phi'}{\tfrac{1}{2}\gamma H^2} \tag{9.89}$$

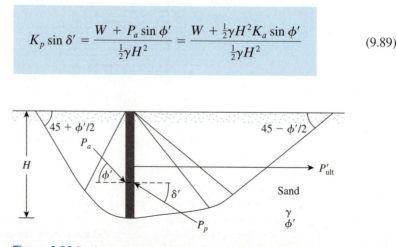

Figure 9.38 Basic case: continuous vertical anchor in granular soil

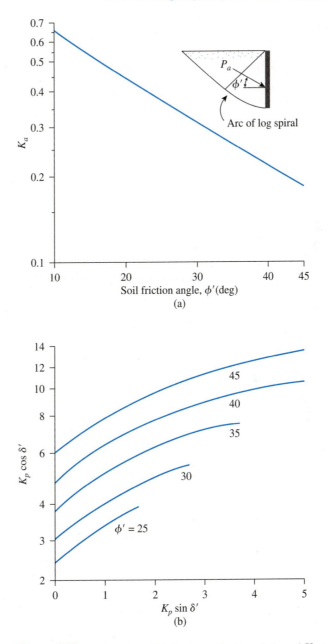

Figure 9.39 (a) Variation of K_a for $\delta' = \phi'$, (b) variation of $K_p \cos \delta'$ with $K_p \sin \delta'$ (Based on Ovesen and Stromann, 1972)

Then use the magnitude of $K_p \sin \delta'$ obtained from Eq. (9.89) to estimate the magnitude of $K_p \cos \delta'$ from the plots given in Figure 9.39b.

Step 2. **Strip Case.** Determine the actual height h of the anchor to be constructed. If a continuous anchor (i.e., an anchor for which $B = \infty$) of height h is placed in the soil so that its depth of embedment is H, as shown in Figure 9.40, the ultimate resistance per unit length is

Figure 9.40 Strip case: vertical anchor

$$P'_{us} = \left[\frac{C_{ov} + 1}{C_{ov} + \left(\dfrac{H}{h}\right)} \right] \underset{\substack{\uparrow \\ \text{Eq. 9.88}}}{P'_{ult}} \tag{9.90}$$

where

P'_{us} = ultimate resistance for the *strip case*
C_{ov} = 19 for dense sand and 14 for loose sand

Step 3. **Actual Case.** In practice, the anchor plates are placed in a row with center-to-center spacing S', as shown in Figure 9.41a. The ultimate resistance of each anchor is

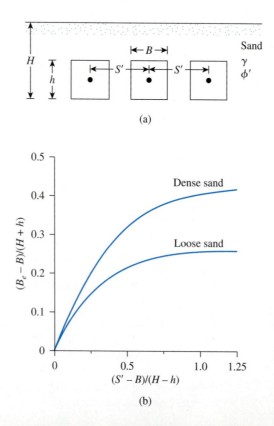

(a)

(b)

Figure 9.41 (a) Actual case for row of anchors; (b) variation of $(B_e - B)/(H + h)$ with $(S' - B)/(H + h)$ (Based on Ovesen and Stromann, 1972)

$$P_{ult} = P'_{us}B_e \qquad (9.91)$$

where B_e = equivalent length.

The equivalent length is a function of S', B, H, and h. Figure 9.41b shows a plot of $(B_e - B)/(H + h)$ against $(S' - B)/(H + h)$ for the cases of loose and dense sand. With known values of S', B, H, and h, the value of B_e can be calculated and used in Eq. (9.91) to obtain P_{ult}.

Stress Characteristic Solution

Neely, Stuart, and Graham (1973) proposed a stress characteristic solution for anchor pull-out resistance using the *equivalent free surface* concept. Figure 9.42 shows the assumed failure surface for a strip anchor. In this figure, OX is the equivalent free surface. The shear stress (s_o) mobilized along OX can be given as

$$m = \frac{s_o}{\sigma'_o \tan \phi'} \qquad (9.92)$$

where

m = shear stress mobilization factor
σ_o' = effective normal stress along OX

Using this analysis, the ultimate resistance (P_{ult}) of an anchor (length = B and height = h) can be given as

$$P_{ult} = M_{\gamma q} (\gamma h^2)BF_s \qquad (9.93)$$

where

$M_{\gamma q}$ = force coefficient
F_s = shape factor
γ = effective unit weight of soil

The variations of $M_{\gamma q}$ for $m = 0$ and 1 are shown in Figure 9.43. For conservative design, $M_{\gamma q}$ with $m = 0$ may be used. The shape factor (F_s) determined experimentally is shown in Figure 9.44 as a function of B/h and H/h.

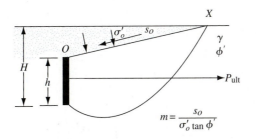

Figure 9.42 Assumed failure surface in soil for stress characteristic solution

Figure 9.43 Variation of $M_{\gamma q}$ with H/h and ϕ' (After Neeley *et al.*, 1973. With permission from ASCE.)

Figure 9.44 Variation of shape factor with H/h and B/h (After Neeley *et al.*, 1973. With permission from ASCE.)

Empirical Correlation Based on Model Tests

Ghaly (1997) used the results of 104 laboratory tests, 15 centrifugal model tests, and 9 field tests to propose an empirical correlation for the ultimate resistance of single anchors. The correlation can be written as

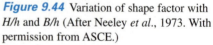

$$P_{\text{ult}} = \frac{5.4}{\tan \phi'} \left(\frac{H^2}{A} \right)^{0.28} \gamma A H \qquad (9.94)$$

where A = area of the anchor = Bh.

Ghaly also used the model test results of Das and Seeley (1975) to develop a load–displacement relationship for single anchors. The relationship can be given as

$$\frac{P}{P_{\text{ult}}} = 2.2\left(\frac{u}{H}\right)^{0.3} \tag{9.95}$$

where u = horizontal displacement of the anchor at a load level P.

Equations (9.94) and (9.95) apply to single anchors (i.e., anchors for which $S'/B = \infty$). For all practical purposes, when $S'/B \approx 2$ the anchors behave as single anchors.

Factor of Safety for Anchor Plates

The allowable resistance per anchor plate may be given as

$$P_{\text{all}} = \frac{P_{\text{ult}}}{\text{FS}}$$

where FS = factor of safety.

Generally, a factor of safety of 2 is suggested when the method of Ovesen and Stromann is used. A factor of safety of 3 is suggested for P_{ult} calculated by Eq. (9.94).

Spacing of Anchor Plates

The center-to-center spacing of anchors, S', may be obtained from

$$S' = \frac{P_{\text{all}}}{F}$$

where F = force per unit length of the sheet pile.

Example 9.11

Refer to Figure 9.41a. Given: $B = h = 0.4$ m, $S' = 1.2$ m, $H = 1$ m, $\gamma = 16.51$ kN/m³, and $\phi' = 35°$. Determine the ultimate resistance for each anchor plate. The anchor plates are made of concrete and have thicknesses of 0.15 m.

Solution
From Figure 9.39a for $\phi' = 35°$, the magnitude of K_a is about 0.26.

$$W = Ht\gamma_{\text{concrete}} = (1 \text{ m})(0.15 \text{ m})(23.5 \text{ kN/m}^3)$$

$$= 3.525 \text{ kN/m}$$

From Eq. (9.89),

$$K_p \sin \delta' = \frac{W + \frac{1}{2}\gamma H^2 K_a \sin \phi'}{\frac{1}{2}\gamma H^2}$$

$$= \frac{3.525 + (0.5)(16.51)(1)^2(0.26)(\sin 35)}{(0.5)(16.51)(1)^2} = 0.576$$

From Figure 9.39b with $\phi' = 35°$ and $K_p \sin \delta' = 0.576$, the value of $K_p \cos \delta'$ is about 4.5. Now, using Eq. (9.88),

$$P'_{ult} = \tfrac{1}{2}\gamma H^2 (K_p \cos \delta' - K_a \cos \phi')$$

$$= (\tfrac{1}{2})(16.51)(1)^2[4.5 - (0.26)(\cos 35)] = 35.39 \text{ kN/m}$$

In order to calculate P'_{us}, let us assume the sand to be loose. So, C_{ov} in Eq. (9.90) is equal to 14. Hence,

$$P'_{us} = \left[\frac{C_{ov} + 1}{C_{ov} + \left(\dfrac{H}{h}\right)} \right] P'_{ult} = \left[\frac{14 + 1}{14 + \left(\dfrac{1}{0.4}\right)} \right] = 32.17 \text{ kN/m}$$

$$\frac{S' - B}{H + h} = \frac{1.2 - 0.4}{1 + 0.4} = \frac{0.8}{1.4} = 0.571$$

For $(S' - B)/(H + h) = 0.571$ and loose sand, Figure 9.41b yields

$$\frac{B_e - B}{H - h} = 0.229$$

So

$$B_e = (0.229)(H + h) + B = (0.229)(1 + 0.4) + 0.4$$

$$= 0.72$$

Hence, from Eq. (9.91)

$$P_{ult} = P'_{us} B_e = (32.17)(0.72) = \textbf{23.16 kN} \qquad \blacksquare$$

Example 9.12

Refer to a *single anchor* given in Example 9.11 using the stress characteristic solution. Estimate the ultimate anchor resistance. Use $m = 0$ in Figure 9.43.

Solution
Given: $B = h = 0.4$ m and $H = 1$ m.
Thus,

$$\frac{H}{h} = \frac{1 \text{ m}}{0.4 \text{ m}} = 2.5$$

$$\frac{B}{h} = \frac{0.4 \text{ m}}{0.4 \text{ m}} = 1$$

From Eq. (9.93),

$$P_{ult} = M_{\gamma q} \gamma h^2 B F_s$$

From Figure 9.43, with $\phi' = 35°$ and $H/h = 2.5$, $M_{\gamma q} \approx 18.2$. Also, from Figure 9.44, with $H/h = 2.5$ and $B/h = 1$, $F_s \approx 1.8$. Hence,

$$P_{ult} = (18.2)(16.51)(0.4)^2(0.4)(1.8) \approx \mathbf{34.62 \ kN}$$

■

Example 9.13

Solve Example Problem 9.12 using Eq. (9.94).

Solution

From Eq. (9.94),

$$P_{ult} = \frac{5.4}{\tan \phi'} \left(\frac{H^2}{A} \right)^{0.28} \gamma A H$$

$$H = 1 \text{ m}$$

$$A = Bh = (0.4 \times 0.4) = 0.16 \text{ m}^2$$

$$P_{ult} = \frac{5.4}{\tan 35} \left[\frac{(1)^2}{0.16} \right]^{0.28} (16.51)(0.16)(1) \approx \mathbf{34.03 \ kN}$$

■

9.18 *Holding Capacity of Anchor Plates in Clay ($\phi = 0$ Condition)*

Relatively few studies have been conducted on the ultimate resistance of anchor plates in clayey soils ($\phi = 0$). Mackenzie (1955) and Tschebotarioff (1973) identified the nature of variation of the ultimate resistance of strip anchors and beams as a function of H, h, and c (undrained cohesion based on $\phi = 0$) in a nondimensional form based on laboratory model test results. This is shown in the form of a nondimensional plot in Figure 9.45 (P_{ult}/hBc versus H/h) and can be used to estimate the ultimate resistance of anchor plates in saturated clay ($\phi = 0$).

9.19 *Ultimate Resistance of Tiebacks*

According to Figure 9.46, the ultimate resistance offered by a tieback in sand is

$$P_{ult} = \pi d l \overline{\sigma}'_o K \tan \phi' \tag{9.96}$$

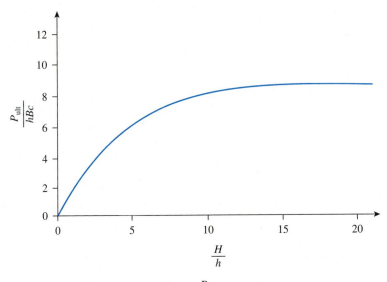

Figure 9.45 Experimental variation of $\dfrac{P_{\text{ult}}}{hBc}$ with H/h for plate anchors in clay
(Based on Mackenzie (1955) and Tschebotarioff (1973))

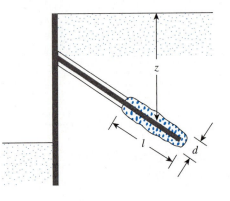

Figure 9.46 Parameters for defining the ultimate resistance of tiebacks

where

ϕ' = effective angle of friction of soil
$\overline{\sigma}'_o$ = average effective vertical stress ($=\gamma z$ in dry sand)
K = earth pressure coefficient

The magnitude of K can be taken to be equal to the earth pressure coefficient at rest (K_o) if the concrete grout is placed under pressure (Littlejohn, 1970). The lower limit of K can be taken to be equal to the Rankine active earth pressure coefficient.

In clays, the ultimate resistance of tiebacks may be approximated as

$$P_{\text{ult}} = \pi dl c_a \qquad (9.97)$$

where c_a = adhesion.

The value of c_a may be approximated as $\frac{2}{3}c_u$ (where c_u = undrained cohesion). A factor of safety of 1.5 to 2 may be used over the ultimate resistance to obtain the allowable resistance offered by each tieback.

Problems

9.1 Figure P9.1 shows a cantilever sheet pile wall penetrating a granular soil. Here, $L_1 = 4$ m, $L_2 = 8$ m, $\gamma = 16.1$ kN/m³, $\gamma_{sat} = 18.2$ kN/m³, and $\phi' = 32°$.
  a. What is the theoretical depth of embedment, D?
  b. For a 30% increase in D, what should be the total length of the sheet piles?
  c. Determine the theoretical maximum moment of the sheet pile.

9.2 Redo Problem 9.1 with the following: $L_1 = 3$ m, $L_2 = 6$ m, $\gamma = 17.3$ kN/m³, $\gamma_{sat} = 19.4$ kN/m³, and $\phi' = 30°$.

9.3 Refer to Figure 9.10. Given: $L = 3$ m, $\gamma = 16.7$ kN/m³, and $\phi' = 30°$. Calculate the theoretical depth of penetration, D, and the maximum moment.

9.4 Refer to Figure P9.4, for which $L_1 = 2.4$ m, $L_2 = 4.6$ m, $\gamma = 15.7$ kN/m³, $\gamma_{sat} = 17.3$ kN/m³, and $\phi' = 30°$, and $c = 29$ kN/m².
  a. What is the theoretical depth of embedment, D?
  b. Increase D by 40%. What length of sheet piles is needed?
  c. Determine the theoretical maximum moment in the sheet pile.

9.5 Refer to Figure 9.14. Given: $L = 4$ m; for sand, $\gamma = 16$ kN/m³; $\phi' = 35°$; and, for clay, $\gamma_{sat} = 19.2$ kN/m³ and $c = 45$ kN/m². Determine the theoretical value of D and the maximum moment.

9.6 An anchored sheet pile bulkhead is shown in Figure P9.6. Let $L_1 = 4$ m, $L_2 = 9$ m, $l_1 = 2$ m, $\gamma = 17$ kN/m³, $\gamma_{sat} = 19$ kN/m³, and $\phi' = 34°$.
  a. Calculate the theoretical value of the depth of embedment, D.
  b. Draw the pressure distribution diagram.
  c. Determine the anchor force per unit length of the wall.
     Use the free earth-support method.

Figure P9.1

Figure P9.4

Figure P9.6

9.7 In Problem 9.6, assume that $D_{actual} = 1.3 D_{theory}$.
   a. Determining the theoretical maximum moment.
   b. Using Rowe's moment reduction technique, choose a sheet pile section. Take
   $E = 210 \times 10^3$ MN/m^2 and $\sigma_{all} = 210,000$ kN/m^2.

9.8 Refer to Figure P9.6. Given: $L_1 = 4$ m, $L_2 = 8$ m, $l_1 = l_2 = 2$ m, $\gamma = 16$ kN/m^3,
   $\gamma_{sat} = 18.5$ kN/m^3, and $\phi' = 35°$. Use the charts presented in Section 9.10 and determine:
   a. Theoretical depth of penetration
   b. Anchor force per unit length
   c. Maximum moment in the sheet pile.

9.9 Refer to Figure P9.6, for which $L_1 = 4$ m, $L_2 = 7$ m, $l_1 = 1.5$ m, $\gamma = 18$ kN/m^3,
   $\gamma_{sat} = 19.5$ kN/m^3, and $\phi' = 30°$. Use the computational diagram method
   (Section 9.12) to determine D, F, and M_{max}. Assume that $C = 0.68$ and $R = 0.6$.

9.10 An anchored sheet-pile bulkhead is shown in Figure P9.10. Let
$L_1 = 2$ m, $L_2 = 6$ m, $l_1 = 1$ m, $\gamma = 16$ kN/m³, $\gamma_{sat} = 18.86$ kN/m³, $\phi' = 32°$,
and $c = 27$ kN/m².

a. Determine the theoretical depth of embedment, D.

b. Calculate the anchor force per unit length of the sheet-pile wall.
Use the free earth support method.

9.11 In Figure 9.41a, for the anchor slab in sand, $H = 1.52$ m, $h = 0.91$ m,
$B = 1.22$ m, $S' = 2.13$ m, $\phi' = 30°$, and $\gamma = 17.3$ kN/m³. The anchor
plates are made of concrete and have a thickness of 76 mm. Using Ovesen
and Stromann's method, calculate the ultimate holding capacity of each anchor.
Take $\gamma_{concrete} = 23.58$ kN/m³.

9.12 A single anchor slab is shown in Figure P9.12. Here, $H = 0.9$ m, $h = 0.3$ m,
$\gamma = 17$ kN/m³, and $\phi' = 32°$. Calculate the ultimate holding capacity of the anchor slab if the width B is (a) 0.3 m, (b) 0.6 m, and (c) $= 0.9$ m.
(*Note*: center-to-center spacing, $S' = \infty$.) Use the empirical correlation given in
Section 9.17 [Eq. (9.94)].

9.13 Repeat Problem 9.12 using Eq. (9.93). Use $m = 0$ in Figure 9.43.

Figure P9.10

Figure P9.12

References

BLUM, H. (1931) *Einspannungsverhältnisse bei Bohlwerken*, W. Ernst und Sohn, Berlin, Germany.

CASAGRANDE, L. (1973). "Comments on Conventional Design of Retaining Structures," *Journal of the Soil Mechanics and Foundations Division,* ASCE, Vol. 99, No. SM2, pp. 181–198.

CORNFIELD, G. M. (1975). "Sheet Pile Structures," in *Foundation Engineering Handbook,* ed. H. F. Wintercorn and H. Y. Fang, Van Nostrand Reinhold, New York, pp. 418–444.

DAS, B. M., and SEELEY, G. R. (1975). "Load–Displacement Relationships for Vertical Anchor Plates," *Journal of the Geotechnical Engineering Division,* American Society of Civil Engineers, Vol. 101, No, GT7, pp. 711–715.

GHALY, A. M. (1997). "Load–Displacement Prediction for Horizontally Loaded Vertical Plates." *Journal of Geotechnical and Geoenvironmental Engineering,* ASCE, Vol. 123, No. 1, pp. 74–76.

HAGERTY, D. J., and NOFAL, M. M. (1992). "Design Aids: Anchored Bulkheads in Sand," *Canadian Geotechnical Journal,* Vol. 29, No. 5, pp. 789–795.

LITTLEJOHN, G. S. (1970). "Soil Anchors," *Proceedings, Conference on Ground Engineering,* Institute of Civil Engineers, London, pp. 33–44.

MACKENZIE, T. R. (1955). *Strength of Deadman Anchors in Clay*, M.S. Thesis, Princeton University, Princeton, N. J.

NATARAJ, M. S., and HOADLEY, P. G. (1984). "Design of Anchored Bulkheads in Sand," *Journal of Geotechnical Engineering,* American Society of Civil Engineers, Vol. 110, No. GT4, pp. 505–515.

NEELEY, W. J., STUART, J. G., and GRAHAM, J. (1973). "Failure Loads of Vertical Anchor Plates is Sand," *Journal of the Soil Mechanics and Foundations Division,* American Society of Civil Engineers, Vol. 99, No. SM9, pp. 669–685.

OVESEN, N. K., and STROMANN, H. (1972). "Design Methods for Vertical Anchor Slabs in Sand," *Proceedings, Specialty Conference on Performance of Earth and Earth-Supported Structures.* American Society of Civil Engineers, Vol. 2.1, pp. 1481–1500.

ROWE, P. W. (1952). "Anchored Sheet Pile Walls," *Proceedings, Institute of Civil Engineers,* Vol. 1, Part 1, pp. 27–70.

ROWE, P. W. (1957). "Sheet Pile Walls in Clay," *Proceedings, Institute of Civil Engineers,* Vol. 7, pp. 654–692.

TSCHEBOTARIOFF, G. P. (1973). *Foundations, Retaining and Earth Structures*, 2nd ed., McGraw-Hill, New York.

TSINKER, G. P. (1983). "Anchored Street Pile Bulkheads: Design Practice," *Journal of Geotechnical Engineering,* American Society of Civil Engineers, Vol. 109, No. GT8, pp. 1021–1038.

10 Braced Cuts

Introduction

Sometimes construction work requires ground excavations with vertical or near-vertical faces—for example, basements of buildings in developed areas or underground transportation facilities at shallow depths below the ground surface (a cut-and-cover type of construction). The vertical faces of the cuts need to be protected by temporary bracing systems to avoid failure that may be accompanied by considerable settlement or by bearing capacity failure of nearby foundations.

Figure 10.1 shows two types of braced cut commonly used in construction work. One type uses the *soldier beam* (Figure 10.1a), which is driven into the ground before excavation and is a vertical steel or timber beam. *Laggings,* which are horizontal timber planks, are placed between soldier beams as the excavation proceeds. When the excavation reaches the desired depth, *wales* and *struts* (horizontal steel beams) are installed. The struts are compression members. Figure 10.1b shows another type of braced excavation. In this case, interlocking *sheet piles* are driven into the soil before excavation. Wales and struts are inserted immediately after excavation reaches the appropriate depth.

Figure 10.2 shows the braced-cut construction used for the Chicago subway in 1940. Timber lagging, timber struts, and steel wales were used. Figure 10.3 shows a braced cut made during the construction of the Washington, DC, metro in 1974. In this cut, timber lagging, steel H-soldier piles, steel wales, and pipe struts were used.

To design braced excavations (i.e., to select wales, struts, sheet piles, and soldier beams), an engineer must estimate the lateral earth pressure to which the braced cuts will be subjected. The theoretical aspects of the lateral earth pressure on a braced cut were discussed in Section 7.8. The total active force per unit length of the wall (P_a) was calculated using the general wedge theory. However, that analysis does not provide the relationships required for estimating the variation of lateral pressure with depth, which is a function of several factors, such as the type of soil, the experience of the construction crew, the type of construction equipment used, and so forth. For that reason, empirical pressure envelopes developed from field observations are used for the design of braced cuts. This procedure is discussed in the next section.

Figure 10.1 Types of braced cut: (a) use of soldier beams; (b) use of sheet piles

10.2   *Pressure Envelope for Braced-Cut Design*

As mentioned in Section 10.1, the lateral earth pressure in a braced cut is dependent on the type of soil, construction method, and type of equipment used. The lateral earth pressure changes from place to place. Each strut should also be designed for the maximum load to

Figure 10.2 Braced cut in Chicago Subway construction, January 1940 (*Courtesy of Ralph B. Peck*)

which it may be subjected. Therefore, the braced cuts should be designed using apparent-pressure diagrams that are envelopes of all the pressure diagrams determined from measured strut loads in the field. Figure 10.4 shows the method for obtaining the apparent-pressure diagram at a section from strut loads. In this figure, let P_1, P_2, P_3, P_4, . . . be the measured strut loads. The apparent horizontal pressure can then be calculated as

$$\sigma_1 = \frac{P_1}{(s)\left(d_1 + \dfrac{d_2}{2}\right)}$$

$$\sigma_2 = \frac{P_2}{(s)\left(\dfrac{d_2}{2} + \dfrac{d_3}{2}\right)}$$

$$\sigma_3 = \frac{P_3}{(s)\left(\dfrac{d_3}{2} + \dfrac{d_4}{2}\right)}$$

$$\sigma_4 = \frac{P_4}{(s)\left(\dfrac{d_4}{2} + \dfrac{d_5}{2}\right)}$$

Figure 10.3 Braced cut in the construction of Washington, DC Metro, May 1974 (*Courtesy of Ralph B. Peck*)

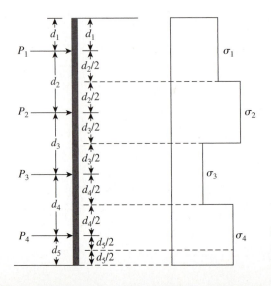

Figure 10.4 Procedure for calculating apparent-pressure diagram from measured strut loads

where

$\sigma_1, \sigma_2, \sigma_3, \sigma_4$ = apparent pressures

s = center-to-center spacing of the struts

Using the procedure just described for strut loads observed from the Berlin subway cut, Munich subway cut, and New York subway cut, Peck (1969) provided the envelope of apparent-lateral-pressure diagrams for design of cuts in *sand*. This envelope is illustrated in Figure 10.5, in which

$$\sigma_a = 0.65\gamma H K_a \tag{10.1}$$

where

γ = unit weight

H = height of the cut

K_a = Rankine active pressure coefficient = $\tan^2(45 - \phi'/2)$

ϕ' = effective friction angle of sand

Cuts in Clay

In a similar manner, Peck (1969) also provided the envelopes of apparent-lateral-pressure diagrams for cuts in *soft to medium clay* and in *stiff clay*. The pressure envelope for soft to medium clay is shown in Figure 10.6 and is applicable to the condition

$$\frac{\gamma H}{c} > 4$$

where c = undrained cohesion ($\phi = 0$).

The pressure, σ_a, is the larger of

$$\sigma_a = \gamma H\left[1 - \left(\frac{4c}{\gamma H}\right)\right]$$

and $\tag{10.2}$

$$\sigma_a = 0.3\gamma H$$

where γ = unit weight of clay.

The pressure envelope for cuts in stiff clay is shown in Figure 10.7, in which

$$\sigma_a = 0.2\gamma H \text{ to } 0.4\gamma H \qquad \text{(with an average of } 0.3\gamma H) \tag{10.3}$$

is applicable to the condition $\gamma H/c \leqslant 4$.

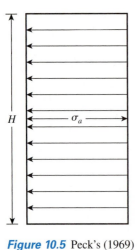

Figure 10.5 Peck's (1969) apparent-pressure envelope for cuts in sand

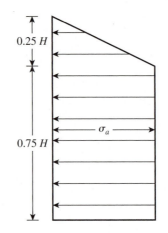

Figure 10.6 Peck's (1969) apparent-pressure envelope for cuts in soft to medium clay

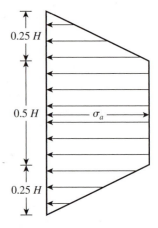

Figure 10.7 Peck's (1969) apparent-pressure envelope for cuts in stiff clay

When using the pressure envelopes just described, keep the following points in mind:

1. They apply to excavations having depths greater than about 6 m.
2. They are based on the assumption that the water table is below the bottom of the cut.
3. Sand is assumed to be drained with zero pore water pressure.
4. Clay is assumed to be undrained and pore water pressure is not considered.

10.3   *Pressure Envelope for Cuts in Layered Soil*

Sometimes, layers of both sand and clay are encountered when a braced cut is being constructed. In this case, Peck (1943) proposed that an equivalent value of cohesion ($\phi = 0$) should be determined according to the formula (see Figure 10.8a).

$$c_{av} = \frac{1}{2H}[\gamma_s K_s H_s^2 \tan \phi_s' + (H - H_s)n'q_u] \tag{10.4}$$

where

H = total height of the cut
γ_s = unit weight of sand
H_s = height of the sand layer
K_s = a lateral earth pressure coefficient for the sand layer (≈ 1)
ϕ_s' = effective angle of friction of sand
q_u = unconfined compression strength of clay
n' = a coefficient of progressive failure (ranging from 0.5 to 1.0; average value 0.75)

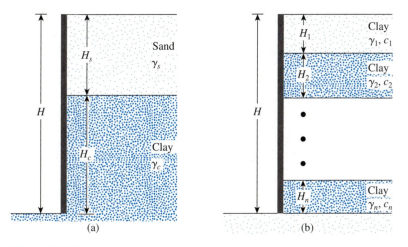

Figure 10.8 Layered soils in braced cuts

The average unit weight of the layers may be expressed as

$$\gamma_a = \frac{1}{H} [\gamma_s H_s + (H - H_s)\gamma_c] \qquad (10.5)$$

where γ_c = saturated unit weight of clay layer.

Once the average values of cohesion and unit weight are determined, the pressure envelopes in clay can be used to design the cuts.

Similarly, when several clay layers are encountered in the cut (Figure 10.8b), the average undrained cohesion becomes

$$c_{av} = \frac{1}{H}(c_1 H_1 + c_2 H_2 + \cdots + c_n H_n) \qquad (10.6)$$

where

$c_1, c_2, \ldots, c_n$ = undrained cohesion in layers 1, 2, . . . , n
$H_1, H_2, \ldots, H_n$ = thickness of layers 1, 2, . . . , n

The average unit weight is now

$$\gamma_a = \frac{1}{H}(\gamma_1 H_1 + \gamma_2 H_2 + \gamma_3 H_3 + \cdots + \gamma_n H_n) \qquad (10.7)$$

10.4 *Design of Various Components of a Braced Cut*

Struts

In construction work, struts should have a minimum vertical spacing of about 2.75 m or more. Struts are horizontal columns subject to bending. The load-carrying capacity of columns depends on their *slenderness ratio,* which can be reduced by providing vertical and horizontal supports at intermediate points. For wide cuts, splicing the struts may be necessary. For braced cuts in clayey soils, the depth of the first strut below the ground surface should be less than the depth of tensile crack, z_c. From Eq. (7.8),

$$\sigma_a' = \gamma z K_a - 2c'\sqrt{K_a}$$

where K_a = coefficient of Rankine active pressure.

For determining the depth of tensile crack,

$$\sigma_a' = 0 = \gamma z_c K_a - 2c'\sqrt{K_a}$$

or

$$z_c = \frac{2c'}{\sqrt{K_a}\gamma}$$

With $\phi = 0$, $K_a = \tan^2(45 - \phi/2) = 1$, so

$$z_c = \frac{2c}{\gamma}$$

A simplified conservative procedure may be used to determine the strut loads. Although this procedure will vary, depending on the engineers involved in the project, the following is a step-by-step outline of the general methodology (see Figure 10.9):

Step 1. Draw the pressure envelope for the braced cut. (See Figures 10.5, 10.6, and 10.7.) Also, show the proposed strut levels. Figure 10.9a shows a pressure envelope for a sandy soil; however, it could also be for a clay. The strut levels are marked *A, B, C,* and *D.* The sheet piles (or soldier beams) are assumed to be hinged at the strut levels, except for the top and bottom ones. In Figure 10.9a, the hinges are at the level of struts *B* and *C.* (Many designers also assume the sheet piles or soldier beams to be hinged at all strut levels except for the top.)

Step 2. Determine the reactions for the two simple cantilever beams (top and bottom) and all the simple beams between. In Figure 10.9b, these reactions are $A, B_1, B_2, C_1, C_2,$ and D.

Step 3. The strut loads in the figure may be calculated via the formulas

$$P_A = (A)(s)$$

$$P_B = (B_1 + B_2)(s) \tag{10.8}$$

$$P_C = (C_1 + C_2)(s)$$

and

$$P_D = (D)(s)$$

Figure 10.9 Determination of strut loads: (a) section and plan of the cut; (b) method for determining strut loads

where

$$P_A, P_B, P_C, P_D = \text{loads to be taken by the individual struts at levels}$$
$$A, B, C, \text{ and } D, \text{ respectively}$$
$$A, B_1, B_2, C_1, C_2, D = \text{reactions calculated in Step 2 (note the unit:}$$
$$\text{force/unit length of the braced cut)}$$
$$s = \text{horizontal spacing of the struts (see plan in}$$
$$\text{Figure 10.9a)}$$

Step 4. Knowing the strut loads at each level and the intermediate bracing conditions allows selection of the proper sections from the steel construction manual.

Sheet Piles

The following steps are involved in designing the sheet piles:

> *Step 1.* For each of the sections shown in Figure 10.9b, determine the maximum bending moment.
>
> *Step 2.* Determine the maximum value of the maximum bending moments (M_{max}) obtained in Step 1. Note that the unit of this moment will be, for example, kN-m/m length of the wall.
>
> *Step 3.* Obtain the required section modulus of the sheet piles, namely,

$$S = \frac{M_{max}}{\sigma_{all}} \tag{10.9}$$

> where σ_{all} = allowable flexural stress of the sheet pile material.
>
> *Step 4.* Choose a sheet pile having a section modulus greater than or equal to the required section modulus from a table such as Table 9.1.

Wales

Wales may be treated as continuous horizontal members if they are spliced properly. Conservatively, they may also be treated as though they are pinned at the struts. For the section shown in Figure 10.9a, the maximum moments for the wales (assuming that they are pinned at the struts) are,

$$\text{At level } A, \quad M_{max} = \frac{(A)(s^2)}{8}$$

$$\text{At level } B, \quad M_{max} = \frac{(B_1 + B_2)s^2}{8}$$

$$\text{At level } C, \quad M_{max} = \frac{(C_1 + C_2)s^2}{8}$$

and

$$\text{At level } D, \quad M_{max} = \frac{(D)(s^2)}{8}$$

where A, B_1, B_2, C_1, C_2, and D are the reactions under the struts per unit length of the wall (see Step 2 of strut design).

Now determine the section modulus of the wales:

$$S = \frac{M_{max}}{\sigma_{all}}$$

The wales are sometimes fastened to the sheet piles at points that satisfy the lateral support requirements.

Example 10.1

The cross section of a long braced cut is shown in Figure 10.10a.

  a. Draw the earth-pressure envelope.
  b. Determine the strut loads at levels A, B, and C.
  c. Determine the section modulus of the sheet pile section required.
  d. Determine a design section modulus for the wales at level B.

(*Note:* The struts are placed at 3 m, center to center, in the plan.) Use

$$\sigma_{all} = 170 \times 10^3 \text{ kN/m}^2$$

Solution

Part a

We are given that $\gamma = 18 \text{ kN/m}^2$, $c = 35 \text{ kN/m}^2$, and $H = 7$ m. So,

$$\frac{\gamma H}{c} = \frac{(18)(7)}{35} = 3.6 < 4$$

Thus, the pressure envelope will be like the one in Figure 10.7. The envelope is plotted in Figure 10.10a with maximum pressure intensity, σ_a, equal to $0.3\gamma H = 0.3(18)(7) = \textbf{37.8 kN/m}^2$.

Part b

To calculate the strut loads, examine Figure 10.10b. Taking the moment about B_1, we have $\Sigma M_{B_1} = 0$, and

$$A(2.5) - \left(\frac{1}{2}\right)(37.8)(1.75)\left(1.75 + \frac{1.75}{3}\right) - (1.75)(37.8)\left(\frac{1.75}{2}\right) = 0$$

or

$$A = 54.02 \text{ kN/m}$$

Also, Σ vertical forces $= 0$. Thus,

$$\tfrac{1}{2}(1.75)(37.8) + (37.8)(1.75) = A + B_1$$

or

$$33.08 + 66.15 - A = B_1$$

So,

$$B_1 = 45.2 \text{ kN/m}$$

Due to symmetry,

$$B_2 = 45.2 \text{ kN/m}$$

and

$$C = 54.02 \text{ kN/m}$$

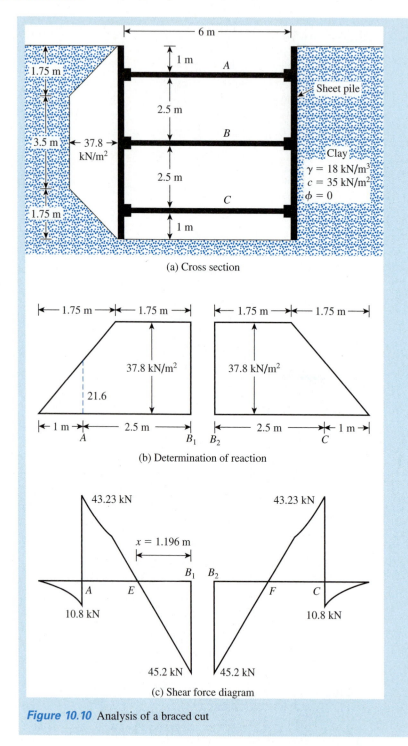

(a) Cross section

(b) Determination of reaction

(c) Shear force diagram

Figure 10.10 Analysis of a braced cut

Hence, the strut loads at the levels indicated by the subscripts are

$$P_A = 54.02 \times \text{horizontal spacing}, s = 54.02 \times 3 = \mathbf{162.06 \ kN}$$

$$P_B = (B_1 + B_2)3 = (45.2 + 45.2)3 = \mathbf{271.2 \ kN}$$

and

$$P_C = 54.02 \times 3 = \mathbf{162.06 \ kN}$$

Part c

At the left side of Figure 10.10b, for the maximum moment, the shear force should be zero. The nature of the variation of the shear force is shown in Figure 10.10c. The location of point E can be given as

$$x = \frac{\text{reaction at } B_1}{37.8} = \frac{45.2}{37.8} = 1.196 \ m$$

Also,

$$\text{Magnitude of moment at } A = \frac{1}{2}(1)\left(\frac{37.8}{1.75} \times 1\right)\left(\frac{1}{3}\right)$$

$$= 3.6 \ \text{kN-m/meter of wall}$$

and

$$\text{Magnitude of moment at } E = (45.2 \times 1.196) - (37.8 \times 1.196)\left(\frac{1.196}{2}\right)$$

$$= 54.06 - 27.03 = 27.03 \ \text{kN-m/meter of wall}$$

Because the loading on the left and right sections of Figure 10.10b are the same, the magnitudes of the moments at F and C (see Figure 10.10c) will be the same as those at E and A, respectively. Hence, the maximum moment is 27.03 kN-m/meter of wall.

The section modulus of the sheet piles is thus

$$S = \frac{M_{max}}{\sigma_{all}} = \frac{27.03 \ \text{kN-m}}{170 \times 10^3 \ \text{kN/m}^2} = \mathbf{15.9 \times 10^{-5} m^3/m \ of \ the \ wall}$$

Part d

The reaction at level B has been calculated in part b. Hence,

$$M_{max} = \frac{(B_1 + B_2)s^2}{8} = \frac{(45.2 + 45.2)3^2}{8} = \mathbf{101.7 \ kN\text{-}m}$$

and

$$\text{Section modulus, } S = \frac{101.7}{\sigma_{all}} = \frac{101.7}{(170 \times 1000)}$$

$$= \mathbf{0.598 \times 10^{-3} \ m^3}$$

Example 10.2

Refer to the braced cut shown in Figure 10.11, for which $\gamma = 17$ kN/m³, $\phi' = 35°$, and $c' = 0$. The struts are located 4 m on center in the plan. Draw the earth-pressure envelope and determine the strut loads at levels *A*, *B*, and *C*.

Solution

For this case, the earth-pressure envelope shown in Figure 10.5 is applicable. Hence,

$$K_a = \tan^2\left(45 - \frac{\phi'}{2}\right) = \tan^2\left(45 - \frac{35}{2}\right) = 0.271$$

From Equation (10.1)

$$\sigma_a = 0.65\,\gamma H K_a = (0.65)(17)(9)(0.271) = 26.95 \text{ kN/m}^2$$

Figure 10.12a shows the pressure envelope. Refer to Figure 10.12b and calculate B_1:

$$\sum M_{B_1} = 0$$

$$A = \frac{(26.95)(5)\left(\dfrac{5}{2}\right)}{3} = 112.29 \text{ kN/m}$$

$$B_1 = (26.95)(5) - 112.29 = 22.46 \text{ kN/m}$$

Now, refer to Figure 10.12c and calculate B_2:

$$\sum M_{B_2} = 0$$

Figure 10.11

Figure 10.12 Load diagrams

$$C = \frac{(26.95)(4)\left(\frac{4}{2}\right)}{3} = 71.87 \text{ kN/m}$$

$$B_2 = (26.95)(4) - 71.87 = 35.93 \text{ kN/m}$$

The strut loads are

At A, $(112.29)(\text{spacing}) = (112.29)(4) = \mathbf{449.16 \text{ kN}}$

At B, $(B_1 + B_2)(\text{spacing}) = (22.46 + 35.93)(4) = \mathbf{233.56 \text{ kN}}$

At C, $(71.87)(\text{spacing}) = (71.87)(4) = \mathbf{287.48 \text{ kN}}$ ∎

10.5 Case Studies of Braced Cuts

The procedure for determining strut loads and the design of sheet piles and wales presented in the preceding sections appears to be fairly straightforward. It is, however, only possible if a proper pressure envelope is chosen for the design, which is difficult. This section describes some case studies of braced cuts and highlights the difficulties and degree of judgment needed for successful completion of various projects.

Subway Extension of the Massachusetts Bay Transportation Authority (MBTA)

Lambe (1970) provided data on the performance of three excavations for the subway extension of the MBTA in Boston (test sections A, B, and D), all of which were well instrumented. Figure 10.13 gives the details of test section B, where the cut was 17.68 m, including subsoil conditions. The subsoil consisted of gravel, sand, silt, and clay (fill) to a depth of about 7.93 m, followed by a light gray, slightly organic silt to a depth of 14.02 m. A layer of coarse sand and gravel with some clay was present from 14.02 m to 16.46 m below the ground surface. Rock was encountered below 16.46 m. The horizontal spacing of the struts was 3.66 m center-to-center.

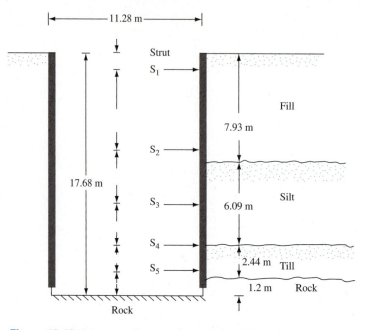

Figure 10.13 Schematic diagram of test section B for subway extension, MTBA

Because the apparent pressure envelopes available (Section 10.2) are for *sand* and *clay* only, questions may arise about how to treat the fill, silt, and till. Figure 10.14 shows the apparent pressure envelopes proposed by Peck (1969), considering the soil as *sand* and also as *clay*, to overcome that problem. For the average soil parameters of the profile, the following values of σ_a were used to develop the pressure envelopes shown in Figure 10.14.

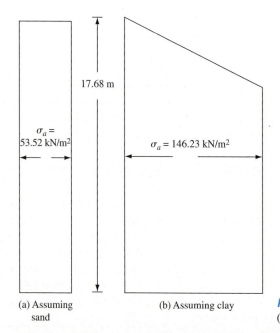

(a) Assuming sand

(b) Assuming clay

Figure 10.14 Pressure envelopes (a) assuming sand; (b) assuming clay

Sand

$$\sigma_a = 0.65\gamma H K_a \qquad (10.10)$$

For $\gamma = 17.92 \text{ kN/m}^3$, $H = 17.68 \text{ m}$, and $K_a = 0.26$,

$$\sigma_a = (0.65)(17.92)(17.68)(0.26) = 53.52 \text{ kN/m}^2$$

Clay

$$\sigma_a = \gamma H\left[1 - \left(\frac{4c}{\gamma H}\right)\right] \qquad (10.11)$$

For $c = 42.65 \text{ kN/m}^2$,

$$\sigma_a = (17.92)(17.68)\left[1 - \frac{(4)(42.65)}{(17.92)(17.68)}\right] = 146.23 \text{ kN/m}^2$$

Table 10.1 shows the variations of the strut load, based on the assumed pressure envelopes shown in Figure 10.14. Also shown in Table 10.1 are the measured strut loads in the field and the design strut loads. This comparison indicates that

1. In most cases the measured strut loads differed widely from those predicted. This result is due primarily to the uncertainties involved in the assumption of the soil parameters.
2. The actual design strut loads were substantially higher than those measured.

B. Construction of National Plaza (South Half) in Chicago

The construction of the south half of the National Plaza in Chicago required a braced cut 21.43 m deep. Swatek et al. (1972) reported the case history for this construction. Figure 10.15 shows a schematic diagram for the braced cut and the subsoil profile. There were six levels of struts. Table 10.2 gives the actual maximum wale and strut loads.

Table 10.1 Computed and Measured Strut Loads at Test Section B

Strut number	Computed load (kip)		Measured strut load (kip)
	Envelope based on sand	Envelope based on clay	
S-1	810	1023	313
S-2	956	2580	956
S-3	685	1868	1352
S-4	480	1299	1023
S-5	334	974	1219

Figure 10.15 Schematic diagram of braced cut—National Plaza of Chicago

Table 10.2 National Plaza Wale and Strut Loads

Strut level	Elevation (m)	Load measured (kN/m)
A	+0.915	233.49
B	−1.83	386.71
C	−4.57	423.20
D	−7.47	423.20
E	−10.37	423.20
F	−13.57	448.0
		Σ2337.8

Figure 10.16 presents a lateral earth-pressure envelope based on the maximum wale loads measured. To compare the theoretical prediction to the actual observation requires making an approximate calculation. To do so, we convert the clayey soil layers from Elevation +0.305 m ft to −17.07 m to a single equivalent layer in Table 10.3 by using Eq. (10.6).

Now, using Eq. (10.4), we can convert the sand layer located between elevations +4.36 m and +0.305 m and the equivalent clay layer of 17.375 m to one equivalent clay layer with a thickness of 21.43 m:

$$c_{av} = \frac{1}{2H}[\gamma_s K_s H_s^2 \tan \phi_s' + (H - H_s)n'q_u]$$

$$= \left[\frac{1}{(2)(21.43)}\right][(17.29)(1)(4.055)^2 \tan 30 + (17.375)(0.75)(2 \times 51.24)]$$

$$\approx 34.99 \text{ kN/m}^2$$

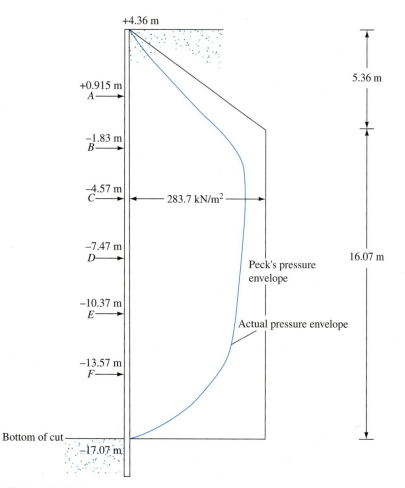

Figure 10.16 Comparison of actual and Peck's pressure envelopes

Table 10.3 Conversion of Soil Layers using Eq. (10.6)

Elevation (m)	Thickness, H (m)	c (kN/m²)	Equivalent c (kN/m²)
+0.305 to −9.67	9.975	19.17	$c_{av} = \dfrac{1}{17.375}[(9.975)(19.17) + (3.44)(33.54)$
			$\quad + (1.83)(95.83)(2.13)(191.67)$
−9.67 to −13.11	3.44	33.54	$= 51.24$ kN/m²
−13.11 to −14.94	1.83	95.83	
−14.94 to −17.07	2.13	191.67	
	Σ17.375		

Equation (10.7) gives

$$\gamma_{av} = \frac{1}{H}[\gamma_1 H_1 + \gamma_2 H_2 + \cdots + \gamma_n H_n)$$

$$= \frac{1}{21.43}[(17.29)(4.055) + (19.97)(10.065) + (20.44)(3.35)$$
$$+ (21.22)(1.83) + (21.22)(2.13)]$$
$$= 19.77 \text{ kN/m}^3$$

For the equivalent clay layer of 21.43 m,

$$\frac{\gamma_{av}H}{c_{av}} = \frac{(19.77)(21.43)}{34.99} = 12.1 > 4$$

Hence the apparent pressure envelope will be of the type shown in Figure 10.6 From Eq. (10.2)

$$\sigma_a = \gamma H \left[1 - \left(\frac{4c_{av}}{\gamma_{av}H} \right) \right] = (19.77)(21.43) \left[1 - \frac{(4)(34.99)}{(19.77)(12.43)} \right] = 283.7 \text{ kN/m}^2$$

The pressure envelope is shown in Figure 10.16. The area of this pressure diagram is 2933 kN/m. Thus Peck's pressure envelope gives a lateral earth pressure of about 1.8 times that actually observed. This result is not surprising because the pressure envelope provided by Figure 10.6 is an envelope developed considering several cuts made at different locations. Under actual field conditions, past experience with the behavior of similar soils can help reduce overdesigning substantially.

10.6 Bottom Heave of a Cut in Clay

Braced cuts in clay may become unstable as a result of heaving of the bottom of the excavation. Terzaghi (1943) analyzed the factor of safety of long braced excavations against bottom heave. The failure surface for such a case in a homogeneous soil is shown in Figure 10.17. In the figure, the following notations are used: B = width of the cut, H = depth of the cut, T = thickness of the clay below the base of excavation, and q = uniform surcharge adjacent to the excavation.

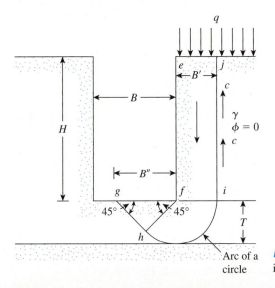

Figure 10.17 Heaving in braced cuts in clay

The ultimate bearing capacity at the base of a soil column with a width of B' can be given as

$$q_{ult} = cN_c \tag{10.12}$$

where $N_c = 5.7$ (for a perfectly rough foundation).

The vertical load per unit area along fi is

$$q = \gamma H + q - \frac{cH}{B'} \tag{10.13}$$

Hence, the factor of safety against bottom heave is

$$FS = \frac{q_{ult}}{q} = \frac{cN_c}{\gamma H + q - \dfrac{cH}{B'}} = \frac{cN_c}{\left(\gamma + \dfrac{q}{H} - \dfrac{c}{B'} \right)H} \tag{10.14}$$

For excavations of limited length L, the factor of safety can be modified to

$$FS = \frac{cN_c\left(1 + 0.2\dfrac{B'}{L}\right)}{\left(\gamma + \dfrac{q}{H} - \dfrac{c}{B'} \right)H} \tag{10.15}$$

where $B' = T$ or $B/\sqrt{2}$ (whichever is smaller).

In 2000, Chang suggested a revision of Eq. (10.15) with the following changes:

1. The shearing resistance along ij may be considered as an increase in resistance rather than a reduction in loading.
2. In Figure 10.17, fg with a width of B'' at the base of the excavation may be treated as a negatively loaded footing.
3. The value of the bearing capacity factor N_c should be 5.14 (not 5.7) for a perfectly smooth footing, because of the restraint-free surface at the base of the excavation.

With the foregoing modifications, Eq. (10.15) takes the form

$$FS = \frac{5.14c\left(1 + \dfrac{0.2B''}{L}\right) + \dfrac{cH}{B'}}{\gamma H + q} \tag{10.16}$$

where

$B' = T$ if $T \le B/\sqrt{2}$
$B' = B/\sqrt{2}$ if $T > B/\sqrt{2}$
$B'' = \sqrt{2}B'$

Bjerrum and Eide (1956) compiled a number of case records for the bottom heave of cuts in clay. Chang (2000) used those records to calculate FS by means of Eq. (10.16); his findings are summarized in Table 10.4. It can be seen from this table that the actual field observations agree well with the calculated factors of safety.

Table 10.4 Calculated Factors of Safety for Selected Case Records Compiled by Bjerrum and Eide (1956) and Calculated by Chang (2000)

Site	B (m)	B/L	H (m)	H/B	γ (kN/m³)	c (kN/m²)	q (kN/m²)	FS [Eq. (10.16)]	Type of failure
Pumping station, Fornebu, Oslo	5.0	1.0	3.0	0.6	17.5	7.5	0	1.05	Total failure
Storehouse, Drammen	4.8	0	2.4	0.5	19.0	12	15	1.05	Total failure
Sewerage tank, Drammen	5.5	0.69	3.5	0.64	18.0	10	10	0.92	Total failure
Excavation, Grey Wedels Plass, Oslo	5.8	0.72	4.5	0.78	18.0	14	10	1.07	Total failure
Pumping station, Jernbanetorget, Oslo	8.5	0.70	6.3	0.74	19.0	22	0	1.26	Partial failure
Storehouse, Freia, Oslo	5.0	0	5.0	1.00	19.0	16	0	1.10	Partial failure
Subway, Chicago	16	0	11.3	0.70	19.0	35	0	1.00	Near failure

Equation (10.16) is recommended for use in this test. In most cases, a factor of safety of about 1.5 is recommended.

In homogeneous clay, if FS becomes less than 1.5, the sheet pile is driven deeper. (See Figure 10.18.) Usually, the depth *d* is kept less than or equal to *B*/2, in which case the force *P* per unit length of the buried sheet pile (*aa'* and *bb'*) may be expressed as (U.S. Department of the Navy, 1971)

$$P = 0.7(\gamma HB - 1.4cH - \pi cB) \qquad \text{for } d > 0.47B \qquad (10.17)$$

Figure 10.18 Force on the buried length of sheet pile

and

$$P = 1.5d\left(\gamma H - \frac{1.4cH}{B} - \pi c\right) \quad \text{for } d < 0.47B \qquad (10.18)$$

Example 10.3

In Figure 10.19. for a braced cut in clay, $B = 3\,$m, $L = 20\,$m, $H = 5.5\,$m, $T = 1.5\,$m, $\gamma = 17\,$kN/m^3, $c = 30\,$kN/m^2, and $q = 0$. Calculate the factor of safety against heave. Use Eq. (10.16).

Solution

From Eq. (10.16),

$$FS = \frac{5.14c\left(1 + \dfrac{0.2B''}{L}\right) + \dfrac{cH}{B'}}{\gamma H + q}$$

with $T = 1.5\,$m,

$$\frac{B}{\sqrt{2}} = \frac{3}{\sqrt{2}} = 2.12\,\text{m}$$

So

$$T \le \frac{B}{\sqrt{2}}$$

Hence, $B' = T = 1.5\,$m, and it follows that

$$B'' = \sqrt{2}B' = (\sqrt{2})(1.5) = 2.12\,\text{m}$$

3 m

5.5 m

Clay
$\gamma = 17\,\text{kN/m}^3$
$c = 30\,\text{kN/m}^3$
$\phi = 0$

1.5 m

Hard stratum

Figure 10.19 Factor of safety against heaving for a braced cut

and

$$FS = \frac{(5.14)(30)\left[1 + \dfrac{(0.2)(2.12)}{20}\right] + \dfrac{(30)(5.5)}{1.5}}{(17)(5.5)} = \textbf{2.86} \quad \blacksquare$$

| 10.7 | *Stability of the Bottom of a Cut in Sand* |

The bottom of a cut in sand is generally stable. When the water table is encountered, the bottom of the cut is stable as long as the water level inside the excavation is higher than the groundwater level. In case dewatering is needed (see Figure 10.20), the factor of safety against piping should be checked. [*Piping* is another term for failure by heave, as defined in Section 1.12; see Eq. (1.45).] Piping may occur when a high hydraulic gradient is created by water flowing into the excavation. To check the factor of safety, draw flow nets and determine the maximum exit gradient $[i_{max(exit)}]$ that will occur at points A and B. Figure 10.21 shows such a flow net, for which the maximum exit gradient is

$$i_{max(exit)} = \frac{\dfrac{h}{N_d}}{a} = \frac{h}{N_d a} \qquad (10.19)$$

where

a = length of the flow element at A (or B)
N_d = number of drops (*Note:* in Figure 10.21, $N_d = 8$; see also Section 1.11)

The factor of safety against piping may be expressed as

$$FS = \frac{i_{cr}}{i_{max(exit)}} \qquad (10.20)$$

where i_{cr} = critical hydraulic gradient.

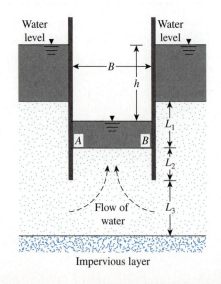

Figure 10.20 Stability of the bottom of a cut in sand

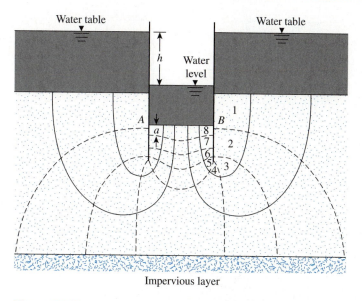

Figure 10.21 Determining the factor of safety against piping by drawing a flow net

The relationship for i_{cr} was given in Chapter 1 as

$$i_{cr} = \frac{G_s - 1}{e + 1}$$

The magnitude of i_{cr} varies between 0.9 and 1.1 in most soils, with an average of about 1. A factor of safety of about 1.5 is desirable.

The maximum exit gradient for sheeted excavations in sands with $L_3 = \infty$ can also be evaluated theoretically (Harr, 1962). (Only the results of these mathematical derivations will be presented here. For further details, see the original work.) To calculate the maximum exit gradient, examine Figures 10.22 and 10.23 and perform the following steps:

1. Determine the modulus, m, from Figure 10.22 by obtaining $2L_2/B$ (or $B/2L_2$) and $2L_1/B$.
2. With the known modulus and $2L_1/B$, examine Figure 10.23 and determine $L_2 i_{exit(max)}/h$. Because L_2 and h will be known, $i_{exit(max)}$ can be calculated.
3. The factor of safety against piping can be evaluated by using Eq. (10.20).

Marsland (1958) presented the results of model tests conducted to study the influence of seepage on the stability of sheeted excavations in sand. The results were summarized by the U.S. Department of the Navy (1971) in NAVFAC DM-7 and are given in Figure 10.24a, b, and c. Note that Figure 10.24b is for the case of determining the sheet pile penetration L_2 needed for the required factor of safety against piping when the sand layer extends to a great depth below the excavation. By contrast, Figure 10.24c represents the case in which an impervious layer lies at a limited depth below the bottom of the excavation.

Figure 10.22 Variation of modulus (From *Groundwater and Seepage*, by M. E. Harr. Copyright 1962 by McGraw-Hill. Used with permission.)

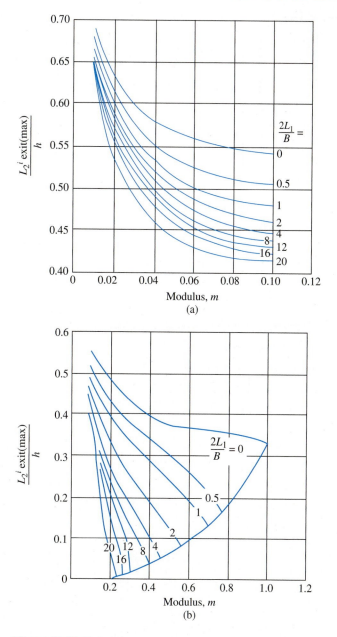

Figure 10.23 Variation of maximum exit gradient with modulus (From *Groundwater and Seepage*, by M. E. Harr. Copyright 1962 by McGraw-Hill. Used with permission.)

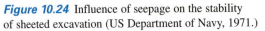

Figure 10.24 Influence of seepage on the stability of sheeted excavation (US Department of Navy, 1971.)

Example 10.4

In Figure 10.20, let $h = 4.5\,\text{m}$, $L_1 = 5\,\text{m}$, $L_2 = 4\,\text{m}$, $B = 5\,\text{m}$, and $L_3 = \infty$. Determine the factor of safety against piping. Use Figures 10.22 and 10.23.

Solution
We have

$$\frac{2L_1}{B} = \frac{2(5)}{5} = 2$$

and

$$\frac{B}{2L_2} = \frac{5}{2(4)} = 0.625$$

According to Figure 10.22b, for $2L_1/B = 2$ and $B/2L_2 = 0.625$, $m \approx 0.033$. From Figure 10.23a, for $m = 0.033$ and $2L_1/B = 2$, $L_2 i_{\text{exit(max)}}/h = 0.54$. Hence,

$$i_{\text{exit(max)}} = \frac{0.54(h)}{L_2} = 0.54(4.5)/4 = 0.608$$

and

$$\text{FS} = \frac{i_{\text{cr}}}{i_{\text{max (exit)}}} = \frac{1}{0.608} = \mathbf{1.645}$$ ∎

| 10.8 | *Lateral Yielding of Sheet Piles and Ground Settlement* |

In braced cuts, some lateral movement of sheet pile walls may be expected. (See Figure 10.25.) The amount of lateral yield (δ_H) depends on several factors, the most important of which is the elapsed time between excavation and the placement of wales and struts. As discussed before, in several instances the sheet piles (or the soldier piles, as the case may be) are driven to a certain depth below the bottom of the excavation. The reason is to reduce the lateral yielding of the walls during the last stages of excavation. Lateral yielding of the walls will cause the ground surface surrounding the cut to settle. The degree of lateral yielding, however, depends mostly on the type of soil below the bottom of the cut. If clay below the cut extends to a great depth and $\gamma H/c$ is less than about 6, extension of the sheet piles or soldier piles below the bottom of the cut will help considerably in reducing the lateral yield of the walls.

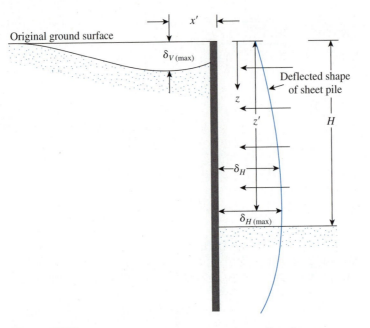

Figure 10.25 Lateral yielding of sheet pile and ground settlement

However, under similar circumstances, if $\gamma H/c$ is about 8, the extension of sheet piles into the clay below the cut does not help greatly. In such circumstances, we may expect a great degree of wall yielding that could result in the total collapse of the bracing systems. If a hard layer of soil lies below a clay layer at the bottom of the cut, the piles should be embedded in the stiffer layer. This action will greatly reduce lateral yield.

The lateral yielding of walls will generally induce ground settlement, δ_V, around a braced cut. Such settlement is generally referred to as *ground loss*. On the basis of several field observations, Peck (1969) provided curves for predicting ground settlement in various types of soil. (See Figure 10.26.) The magnitude of ground loss varies extensively; however, the figure may be used as a general guide.

Moormann (2004) analyzed about 153 case histories dealing mainly with the excavation in soft clay (that is, undrained shear strength, $c \leq 75$ kN/m^2). Following is a summary of his analysis relating to $\delta_{V(max)}$, x', $\delta_{H(max)}$, and z' (see Figure 10.25).

- Maximum Vertical Movement [$\delta_{V(max)}$]

  $\delta_{V(max)}/H \approx 0.1$ to 10.1% with an average of 1.07% (soft clay)
  $\delta_{V(max)}/H \approx 0$ to 0.9% with an average of 0.18% (stiff clay)
  $\delta_{V(max)}/H \approx 0$ to 2.43% with an average of 0.33% (non-cohesive soils)

- Location of $\delta_{V(max)}$, that is x' (Figure 10.25)

  For 70% of all case histories considered, $x' \leq 0.5H$.
  However, in soft clays, x' may be as much as $2H$.

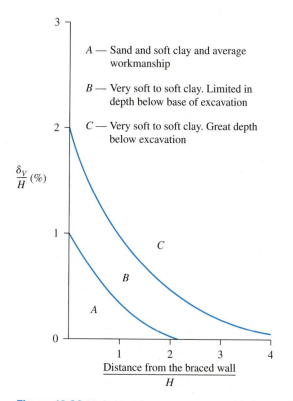

Figure 10.26 Variation of ground settlement with distance (From Peck, R. B. (1969). "Deep Excavation and Tunneling in Soft Ground," Proceedings Seventh International Conference on Soil Mechanics and Foundation Engineering, Mexico City, State-of-the-Art Volume, pp. 225–290. With permission from ASCE.)

- Maximum Horizontal Deflection of Sheet Piles, $\delta_{H(\text{max})}$

   For 40% of excavation in soft clay, $0.5\% \leq \delta_{H(\text{max})}/H \leq 1\%$.
   The average value of $\delta_{H(\text{max})}/H$ is about 0.87%.
   In stiff clays, the average value of $\delta_{H(\text{max})}/H$ is about 0.25%.
   In non-cohesive soils, $\delta_{H(\text{max})}/H$ is about 0.27% of the average.

- Location of $\delta_{H(\text{max})}$, that is z' (Figure 10.25)

   For deep excavation of soft and stiff cohesive soils, z'/H is about 0.5 to 1.0.

Problems

10.1 Refer to the braced cut shown in Figure P10.1. Given: $\gamma = 16$ kN/m³, $\phi' = 38°$, and $c' = 0$. The struts are located at 3.5 m center-to-center in the plan. Draw the earth-pressure envelope and determine the strut loads at levels A, B, and C.

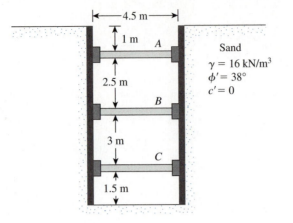

Figure P10.1

10.2 For the braced cut described in Problem 10.1, determine the following:
  a. The sheet-pile section modulus
  b. The section modulus of the wales at level B
  Assume that σ_{all} = 170 MN/m².
10.3 Refer to Fig. P.10.3. Redo Problem 10.1 with γ = 18 kN/m³, ϕ' = 40°, c' = 0, and the center-to-center strut spacing in the plan = 4 m.
10.4 Determine the sheet-pile section modulus for the braced cut described in Problem 10.3. Given: σ_{all} = 170 MN/m².
10.5 Refer to Figure 10.8a. For the braced cut, given H = 6 m; H_s = 2.5 m; γ_s = 16.5 kN/m³; angle of friction of sand, ϕ'_s = 35°; H_c = 3.5 m; γ_c = 17.5 kN/m³; and unconfined compression strength of clay layer, q_u = 62 kN/m².
  a. Estimate the average cohesion (c_{av}) and average unit weight (γ_{av}) for the construction of the earth-pressure envelope.
  b. Plot the earth-pressure envelope.

Figure P10.3

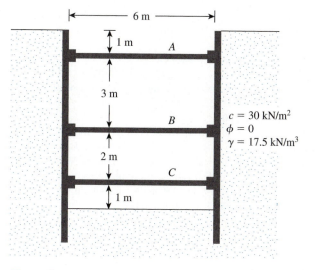

Figure P10.7

10.6 Refer to Figure 10.8b, which shows a braced cut in clay. Given: $H = 7.6$ m, $H_1 = 1.52$ m, $c_1 = 101.8$ kN/m^2, $\gamma_1 = 17.45$ kN/m^3, $H_2 = 3.04$ m, $c_2 = 74.56$ kN/m^2, $\gamma_2 = 16.83$ kN/m^3, $H_3 = 3.04$ m, $c_3 = 80.02$ kN/m^2, and $\gamma_3 = 17.14$ kN/m^3.

   a. Determine the average cohesion (c_{av}) and average unit weight (γ_{av}) for the construction of the earth-pressure envelope.

   b. Plot the earth-pressure envelope.

10.7 Refer to Figure P10.7. Given: $\gamma = 17.5$ kN/m^3, $c = 30$ kN/m^2, and center-to-center spacing of struts in the plan $= 5$ m. Draw the earth-pressure envelope and determine the strut loads at levels A, B, and C.

10.8 Determine the sheet-pile section modulus for the braced cut described in Problem 10.7. Use $\sigma_{all} = 170$ MN/m.^{2}.

10.9 Redo Problem 10.7 assuming that $c = 60$ kN/m^2.

10.10 Determine the factor of safety against bottom heave for the braced cut described in Problem 10.7. Use Eq. (10.16) and assume the length of the cut, $L = 18$ m.

10.11 Determine the factor of safety against bottom heave for the braced cut described in Problem 10.9. Use Eq. (10.15). The length of the cut is 12.5.

References

BJERRUM, L, and EIDE, O. (1956). "Stability of Strutted Excavation in Clay," *Geotechnique,* Vol. 6, No. 1, pp. 32–47.

CHANG, M. F. (2000). "Basal Stability Analysis of Braced Cuts in Clay," *Journal of Geotechnical and Geoenvironmental Engineering,* ASCE, Vol. 126, No. 3, pp. 276–279.

HARR, M. E. (1962). *Groundwater and Seepage,* McGraw-Hill, New York.

LAMBE, T. W. (1970). "Braced Excavations." *Proceedings of the Specialty Conference on Lateral Stresses in the Ground and Design of Earth-Retaining Structures,* American Society of Civil Engineers, pp. 149–218.

MOORMANN, C. (2004). "Analysis of Wall and Ground Movements Due to Deep Excavations in Soft Soil Based on New Worldwide Data Base," *Soils and Foundations*, Vol. 44, No. 1, pp. 87–98.

PECK, R. B. (1943). "Earth Pressure Measurements in Open Cuts, Chicago (ILL.) Subway," *Transactions,* American Society of Civil Engineers, Vol. 108, pp. 1008–1058.

PECK, R. B. (1969). "Deep Excavation and Tunneling in Soft Ground," *Proceedings Seventh International Conference on Soil Mechanics and Foundation Engineering,* Mexico City, State-of-the-Art Volume, pp. 225–290.

SWATEK, E. P., JR., ASROW, S. P., and SEITZ, A. (1972). "Performance of Bracing for Deep Chicago Excavation," *Proceeding of the Specialty Conference on Performance of Earth and Earth Supported Structures*, American Society of Civil Engineers, Vol. 1, Part 2, pp. 1303–1322.

TERZAGHI, K. (1943). *Theoretical Soil Mechanics,* Wiley, New York.

U.S. DEPARTMENT OF THE NAVY (1971). "Design Manual—Soil Mechanics. Foundations, and Earth Structures." NAVFAC DM-7, Washington, D.C.

11 Pile Foundations

11.1 Introduction

Piles are structural members that are made of steel, concrete, or timber. They are used to build pile foundations, which are deep and which cost more than shallow foundations. (See Chapters 3, 4, and 5.) Despite the cost, the use of piles often is necessary to ensure structural safety. The following list identifies some of the conditions that require pile foundations (Vesic, 1977):

1. When one or more upper soil layers are highly compressible and too weak to support the load transmitted by the superstructure, piles are used to transmit the load to underlying bedrock or a stronger soil layer, as shown in Figure 11.1a. When bedrock is not encountered at a reasonable depth below the ground surface, piles are used to transmit the structural load to the soil gradually. The resistance to the applied structural load is derived mainly from the frictional resistance developed at the soil–pile interface. (See Figure 11.1b.)

2. When subjected to horizontal forces (see Figure 11.1c), pile foundations resist by bending, while still supporting the vertical load transmitted by the superstructure. This type of situation is generally encountered in the design and construction of earth-retaining structures and foundations of tall structures that are subjected to high wind or to earthquake forces.

3. In many cases, expansive and collapsible soils may be present at the site of a proposed structure. These soils may extend to a great depth below the ground surface. Expansive soils swell and shrink as their moisture content increases and decreases, and the pressure of the swelling can be considerable. If shallow foundations are used in such circumstances, the structure may suffer considerable damage. However, pile foundations may be considered as an alternative when piles are extended beyond the active zone, which is where swelling and shrinking occur. (See Figure 11.1d)

    Soils such as loess are collapsible in nature. When the moisture content of these soils increases, their structures may break down. A sudden decrease in the void ratio of soil induces large settlements of structures supported by shallow foundations. In such

Figure 11.1 Conditions that require the use of pile foundations

cases, pile foundations may be used in which the piles are extended into stable soil layers beyond the zone where moisture will change.

4. The foundations of some structures, such as transmission towers, offshore platforms, and basement mats below the water table, are subjected to uplifting forces. Piles are sometimes used for these foundations to resist the uplifting force. (See Figure 11.1e.)

5. Bridge abutments and piers are usually constructed over pile foundations to avoid the loss of bearing capacity that a shallow foundation might suffer because of soil erosion at the ground surface. (See Figure 11.1f.)

Although numerous investigations, both theoretical and experimental, have been conducted in the past to predict the behavior and the load-bearing capacity of piles in granular and cohesive soils, the mechanisms are not yet entirely understood and may never be. The design and analysis of pile foundations may thus be considered somewhat of an art as a result of the uncertainties involved in working with some subsoil conditions. This chapter discusses the present state of the art.

Types of Piles and Their Structural Characteristics

Different types of piles are used in construction work, depending on the type of load to be carried, the subsoil conditions, and the location of the water table. Piles can be divided into the following categories: (a) steel piles, (b) concrete piles, (c) wooden (timber) piles, and (d) composite piles.

Steel Piles

Steel piles generally are either *pipe piles* or *rolled steel* H-*section piles*. Pipe piles can be driven into the ground with their ends open or closed. Wide-flange and I-section steel beams can also be used as piles. However, H-section piles are usually preferred because their web and flange thicknesses are equal. (In wide-flange and I-section beams, the web thicknesses are smaller than the thicknesses of the flange.) Table 11.1 gives the dimensions of some standard H-section steel piles used in the United States. Table 11.2 shows selected pipe sections frequency used for piling purposes. In many cases, the pipe piles are filled with concrete after they have been driven.

The allowable structural capacity for steel piles is

$$Q_{all} = A_s f_s \qquad (11.1)$$

where

A_s = cross-sectional area of the steel
f_s = allowable stress of steel (≈ 0.33–$0.5 f_y$)

Once the design load for a pile is fixed, one should determine, on the basis of geotechnical considerations, whether $Q_{(design)}$ is within the allowable range as defined by Eq. 11.1.

When necessary, steel piles are spliced by welding or by riveting. Figure 11.2a shows a typical splice by welding for an H-pile. A typical splice by welding for a pipe pile is shown in Figure 11.2b. Figure 11.2c is a diagram of a splice of an H-pile by rivets or bolts.

When hard driving conditions are expected, such as driving through dense gravel, shale, or soft rock, steel piles can be fitted with driving points or shoes. Figures 11.2d and 11.2e are diagrams of two types of shoe used for pipe piles.

Steel piles may be subject to corrosion. For example, swamps, peats, and other organic soils are corrosive. Soils that have a pH greater than 7 are not so corrosive. To offset the effect of corrosion, an additional thickness of steel (over the actual designed cross-sectional area) is generally recommended. In many circumstances factory-applied epoxy coatings on piles work satisfactorily against corrosion. These coatings are not easily damaged by pile driving. Concrete encasement of steel piles in most corrosive zones also protects against corrosion.

Following are some general facts about steel piles:

- Usual length: 15 m to 60 m
- Usual load: 300 kN to 1200 kN

- Advantages:
  a. Easy to handle with respect to cutoff and extension to the desired length
  b. Can stand high driving stresses
  c. Can penetrate hard layers such as dense gravel and soft rock
  d. High load-carrying capacity
- Disadvantages:
  a. Relatively costly
  b. High level of noise during pile driving
  c. Subject to corrosion
  d. H-piles may be damaged or deflected from the vertical during driving through hard layers or past major obstructions

Table 11.1 Common H-Pile Sections used in the United States

Designation, size (mm) × weight (kg/m)	Depth d_1 (mm)	Section area ($m^2 \times 10^{-3}$)	Flange and web thickness w (mm)	Flange width d_2 (mm)	Moment of inertia ($m^4 \times 10^{-6}$)	
					I_{xx}	I_{yy}
HP 200 × 53	204	6.84	11.3	207	49.4	16.8
HP 250 × 85	254	10.8	14.4	260	123	42
× 62	246	8.0	10.6	256	87.5	24
HP 310 × 125	312	15.9	17.5	312	271	89
× 110	308	14.1	15.49	310	237	77.5
× 93	303	11.9	13.1	308	197	63.7
× 79	299	10.0	11.05	306	164	62.9
HP 330 × 149	334	19.0	19.45	335	370	123
× 129	329	16.5	16.9	333	314	104
× 109	324	13.9	14.5	330	263	86
× 89	319	11.3	11.7	328	210	69
HP 360 × 174	361	22.2	20.45	378	508	184
× 152	356	19.4	17.91	376	437	158
× 132	351	16.8	15.62	373	374	136
× 108	346	13.8	12.82	371	303	109

Table 11.2 Selected Pipe Pile Sections

Outside diameter (mm)	Wall thickness (mm)	Area of steel (cm^2)
219	3.17	21.5
	4.78	32.1
	5.56	37.3
	7.92	52.7
254	4.78	37.5
	5.56	43.6
	6.35	49.4
305	4.78	44.9
	5.56	52.3
	6.35	59.7
406	4.78	60.3
	5.56	70.1
	6.35	79.8
457	5.56	80
	6.35	90
	7.92	112
508	5.56	88
	6.35	100
	7.92	125
610	6.35	121
	7.92	150
	9.53	179
	12.70	238

Figure 11.2 Steel piles: (a) splicing of H-pile by welding; (b) splicing of pipe pile by welding; (c) splicing of H-pile by rivets and bolts; (d) flat driving point of pipe pile; (e) conical driving point of pipe pile

Concrete Piles

Concrete piles may be divided into two basic categories: (a) precast piles and (b) cast-*in-situ* piles. *Precast piles* can be prepared by using ordinary reinforcement, and they can be square or octagonal in cross section. (See Figure 11.3.) Reinforcement is provided to enable the pile to resist the bending moment developed during pickup and transportation, the vertical load, and the bending moment caused by a lateral load. The piles are cast to desired lengths and cured before being transported to the work sites.

Some general facts about concrete piles are as follows:

- Usual length: 10 m to 15 m
- Usual load: 300 kN to 3000 kN
- Advantages:
  a. Can be subjected to hard driving
  b. Corrosion resistant
  c. Can be easily combined with a concrete superstructure

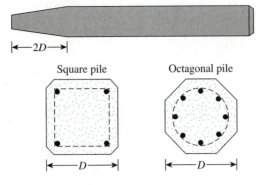

Figure 11.3 Precast piles with ordinary reinforcement

- Disadvantages:
  - **a.** Difficult to achieve proper cutoff
  - **b.** Difficult to transport

Precast piles can also be prestressed by the use of high-strength steel pre-stressing cables. The ultimate strength of these cables is about 1800 MN/m^2. During casting of the piles, the cables are pretensioned to about 900 to 1300 MN/m^2, and concrete is poured around them. After curing, the cables are cut, producing a compressive force on the pile section. Table 11.3 gives additional information about prestressed concrete piles with square and octagonal cross sections.

Some general facts about precast prestressed piles are as follows:

- Usual length: 10 m to 45 m
- Maximum length: 60 m
- Maximum load: 7500 kN to 8500 kN

The advantages and disadvantages are the same as those of precast piles.

Cast-in-situ, or *cast-in-place, piles* are built by making a hole in the ground and then filling it with concrete. Various types of cast-in-place concrete piles are currently used in construction, and most of them have been patented by their manufacturers. These piles may be divided into two broad categories: (a) cased and (b) uncased. Both types may have a pedestal at the bottom.

Cased piles are made by driving a steel casing into the ground with the help of a mandrel placed inside the casing. When the pile reaches the proper depth the mandrel is withdrawn and the casing is filled with concrete. Figures 11.4a, 11.4b, 11.4c, and 11.4d show some examples of cased piles without a pedestal. Figure 11.4e shows a cased pile with a pedestal. The pedestal is an expanded concrete bulb that is formed by dropping a hammer on fresh concrete.

Some general facts about cased cast-in-place piles are as follows:

- Usual length: 5 m to 15 m
- Maximum length: 30 m to 40 m
- Usual load: 200 kN to 500 kN
- Approximate maximum load: 800 kN

Table 11.3 Typical Prestressed Concrete Pile in Use

Pile shape[a]	D (mm)	Area of cross section (cm²)	Perimeter (mm)	Number of strands 12.7-mm diameter	Number of strands 11.1-mm diameter	Minimum effective prestress force (kN)	Section modulus (m³ × 10⁻³)	Design bearing capacity (kN) Strength of concrete (MN/m²) 34.5	Design bearing capacity (kN) Strength of concrete (MN/m²) 41.4
S	254	645	1016	4	4	312	2.737	556	778
O	254	536	838	4	4	258	1.786	462	555
S	305	929	1219	5	6	449	4.719	801	962
O	305	768	1016	4	5	369	3.097	662	795
S	356	1265	1422	6	8	610	7.489	1091	1310
O	356	1045	1168	5	7	503	4.916	901	1082
S	406	1652	1626	8	11	796	11.192	1425	1710
O	406	1368	1346	7	9	658	7.341	1180	1416
S	457	2090	1829	10	13	1010	15.928	1803	2163
O	457	1729	1524	8	11	836	10.455	1491	1790
S	508	2581	2032	12	16	1245	21.844	2226	2672
O	508	2136	1677	10	14	1032	14.355	1842	2239
S	559	3123	2235	15	20	1508	29.087	2694	3232
O	559	2587	1854	12	16	1250	19.107	2231	2678
S	610	3658	2438	18	23	1793	37.756	3155	3786
O	610	3078	2032	15	19	1486	34.794	2655	3186

[a]S = square section; O = octagonal section

- Advantages:
  a. Relatively cheap
  b. Allow for inspection before pouring concrete
  c. Easy to extend
- Disadvantages:
  a. Difficult to splice after concreting
  b. Thin casings may be damaged during driving
- Allowable load:

$$Q_{\text{all}} = A_s f_s + A_c f_c \qquad (11.2)$$

where

A_s = area of cross section of steel
A_c = area of cross section of concrete
f_s = allowable stress of steel
f_c = allowable stress of concrete

Raymond Step-Taper Pile

Corrugated thin cylindrical casing

Maximum usual length: 30 m

(a)

Monotube or Union Metal Pile

Thin, fluted, tapered steel casing driven without mandrel

Maximum usual length: 40 m

(b)

Western Cased Pile

Thin metal casing

Maximum usual length: 30 m–40 m

(c)

Seamless Pile or Armco Pile

Thin metal casing

Maximum usual length: 30 m–40 m

(d)

Franki Cased Pedestal Pile

Straight steel pile casing

Maximum usual length: 30 m–40 m

(e)

Western Uncased Pile without Pedestal

Maximum usual length: 15 m–20 m

(f)

Franki Uncased Pedestal Pile

Maximum usual length: 30 m–40 m

(g)

Figure 11.4 Cast-in-place concrete piles

Figures 11.4f and 11.4g are two types of uncased pile, one with a pedestal and the other without. The uncased piles are made by first driving the casing to the desired depth and then filling it with fresh concrete. The casing is then gradually withdrawn.

Following are some general facts about uncased cast-in-place concrete piles:

- Usual length: 5 m to 15 m
- Maximum length: 30 m to 40 m
- Usual load: 300 kN to 500 kN
- Approximate maximum load: 700 kN
- Advantages:
  a. Initially economical
  b. Can be finished at any elevation
- Disadvantages:
  a. Voids may be created if concrete is placed rapidly
  b. Difficult to splice after concreting
  c. In soft soils, the sides of the hole may cave in, squeezing the concrete
- Allowable load:

$$Q_{all} = A_c f_c \tag{11.3}$$

where

A_c = area of cross section of concrete
f_c = allowable stress of concrete

Timber Piles

Timber piles are tree trunks that have had their branches and bark carefully trimmed off. The maximum length of most timber piles is 10 to 20 m. To qualify for use as a pile, the timber should be straight, sound, and without any defects. The American Society of Civil Engineers' *Manual of Practice,* No. 17 (1959), divided timber piles into three classes:

1. *Class A piles* carry heavy loads. The minimum diameter of the butt should be 356 mm.
2. *Class B piles* are used to carry medium loads. The minimum butt diameter should be 305 to 330 mm.
3. *Class C piles* are used in temporary construction work. They can be used permanently for structures when the entire pile is below the water table. The minimum butt diameter should be 305 mm.

In any case, a pile tip should not have a diameter less than 150 mm.

Timber piles cannot withstand hard driving stress; therefore, the pile capacity is generally limited. Steel shoes may be used to avoid damage at the pile tip (bottom). The tops of timber piles may also be damaged during the driving operation. The crushing of the wooden fibers caused by the impact of the hammer is referred to as *brooming*. To avoid damage to the top of the pile, a metal band or a cap may be used.

Splicing of timber piles should be avoided, particularly when they are expected to carry a tensile load or a lateral load. However, if splicing is necessary, it can be done by using *pipe sleeves* (see Figure 11.5a) or *metal straps and bolts* (see Figure 11.5b). The length of the sleeve should be at least five times the diameter of the pile. The butting ends should be cut square so that full contact can be maintained. The spliced portions should be carefully trimmed so that they fit tightly to the inside of the pipe sleeve. In the case of metal straps and bolts, the butting ends should also be cut square. The sides of the spliced portion should be trimmed plane for putting the straps on.

Figure 11.5 Splicing of timber piles: (a) use of pipe sleeves; (b) use of metal straps and bolts

Timber piles can stay undamaged indefinitely if they are surrounded by saturated soil. However, in a marine environment, timber piles are subject to attack by various organisms and can be damaged extensively in a few months. When located above the water table, the piles are subject to attack by insects. The life of the piles may be increased by treating them with preservatives such as creosote.

The allowable load-carrying capacity of wooden piles is

$$Q_{all} = A_p f_w \tag{11.4}$$

where

A_p = average area of cross section of the pile
f_w = allowable stress on the timber

The following allowable stresses are for pressure-treated round timber piles made from Pacific Coast Douglas fir and Southern pine used in hydraulic structures (ASCE, 1993):

Pacific Coast Douglas Fir

- Compression parallel to grain: 6.04 MN/m²
- Bending: 11.7 MN/m²
- Horizontal shear: 0.66 MN/m²
- Compression perpendicular to grain: 1.31 MN/m²

Southern Pine

- Compression parallel to grain: 5.7 MN/m²
- Bending: 11.4 MN/m²

- Horizontal shear: $0.62 \ MN/m^2$
- Compression perpendicular to grain: $1.41 \ MN/m^2$

The usual length of wooden piles is 5 m to 15 m. The maximum length is about 30 m to 40 m (100 ft to 130 ft). The usual load carried by wooden piles is 300 kN to 500 kN.

Composite Piles

The upper and lower portions of *composite piles* are made of different materials. For example, composite piles may be made of steel and concrete or timber and concrete. Steel-and-concrete piles consist of a lower portion of steel and an upper portion of cast-in-place concrete. This type of pile is used when the length of the pile required for adequate bearing exceeds the capacity of simple cast-in-place concrete piles. Timber-and-concrete piles usually consist of a lower portion of timber pile below the permanent water table and an upper portion of concrete. In any case, forming proper joints between two dissimilar materials is difficult, and for that reason, composite piles are not widely used.

11.3   *Estimating Pile Length*

Selecting the type of pile to be used and estimating its necessary length are fairly difficult tasks that require good judgment. In addition to being broken down into the classification given in Section 11.2, piles can be divided into three major categories, depending on their lengths and the mechanisms of load transfer to the soil: (a) point bearing piles, (b) friction piles, and (c) compaction piles.

Point Bearing Piles

If soil-boring records establish the presence of bedrock or rocklike material at a site within a reasonable depth, piles can be extended to the rock surface. (See Figure 11.6a.) In this case, the ultimate capacity of the piles depends entirely on the load-bearing capacity of the underlying material; thus, the piles are called *point bearing piles*. In most of these cases, the necessary length of the pile can be fairly well established.

If, instead of bedrock, a fairly compact and hard stratum of soil is encountered at a reasonable depth, piles can be extended a few meters into the hard stratum. (See Figure 11.6b.) Piles with pedestals can be constructed on the bed of the hard stratum, and the ultimate pile load may be expressed as

$$Q_u = Q_p + Q_s \tag{11.5}$$

where

Q_p = load carried at the pile point
Q_s = load carried by skin friction developed at the side of the pile (caused by shearing resistance between the soil and the pile)

If Q_s is very small,

$$Q_s \approx Q_p \tag{11.6}$$

In this case, the required pile length may be estimated accurately if proper subsoil exploration records are available.

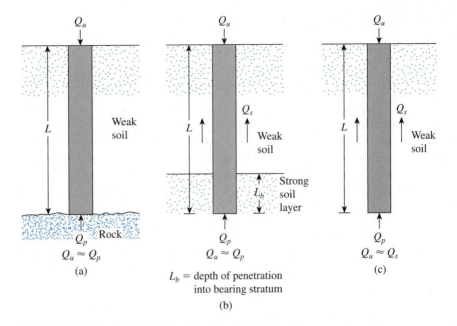

Figure 11.6 (a) and (b) Point bearing piles; (c) friction piles

Friction Piles

When no layer of rock or rocklike material is present at a reasonable depth at a site, point bearing piles become very long and uneconomical. In this type of subsoil, piles are driven through the softer material to specified depths. (See Figure 11.6c.) The ultimate load of the piles may be expressed by Eq. (11.5). However, if the value of Q_p is relatively small, then

$$Q_u \approx Q_s \qquad\qquad (11.7)$$

These piles are called *friction piles*, because most of their resistance is derived from skin friction. However, the term *friction pile*, although used often in the literature, is a misnomer: In clayey soils, the resistance to applied load is also caused by *adhesion*.

The lengths of friction piles depend on the shear strength of the soil, the applied load, and the pile size. To determine the necessary lengths of these piles, an engineer needs a good understanding of soil–pile interaction, good judgment, and experience. Theoretical procedures for calculating the load-bearing capacity of piles are presented later in the chapter.

Compaction Piles

Under certain circumstances, piles are driven in granular soils to achieve proper compaction of soil close to the ground surface. These piles are called *compaction piles*. The lengths of compaction piles depend on factors such as (a) the relative density of the soil before compaction, (b) the desired relative density of the soil after compaction, and (c) the required depth of compaction. These piles are generally short; however, some field tests are necessary to determine a reasonable length.

11.4 *Installation of Piles*

Most piles are driven into the ground by means of *hammers* or *vibratory drivers.* In special circumstances, piles can also be inserted by *jetting* or *partial augering.* The types of hammer used for pile driving include (a) the drop hammer, (b) the single-acting air or steam hammer, (c) the double-acting and differential air or steam hammer, and (d) the diesel hammer. In the driving operation, a cap is attached to the top of the pile. A cushion may be used between the pile and the cap. The cushion has the effect of reducing the impact force and spreading it over a longer time; however, the use of the cushion is optional. A hammer cushion is placed on the pile cap. The hammer drops on the cushion.

Figure 11.7 illustrates various hammers. A drop hammer (see Figure 11.7a) is raised by a winch and allowed to drop from a certain height *H*. It is the oldest type of hammer used for pile driving. The main disadvantage of the drop hammer is its slow rate of blows. The principle of the single-acting air or steam hammer is shown in Figure 11.7b. The striking part, or ram, is raised by air or steam pressure and then drops by gravity. Figure 11.7c shows the operation of the double-acting and differential air or steam hammer. Air or steam is used both to raise the ram and to push it downward, thereby increasing the impact velocity of the ram. The diesel hammer (see Figure 11.7d) consists essentially of a ram, an anvil block, and a fuel-injection system. First the ram is raised and fuel is injected near the anvil. Then the ram is released. When the ram drops, it compresses the air–fuel mixture, which ignites. This action, in effect, pushes the pile downward and raises the ram. Diesel hammers work well under hard driving conditions. In soft soils, the downward movement of the pile is rather large, and the upward movement of the ram is small. This differential may

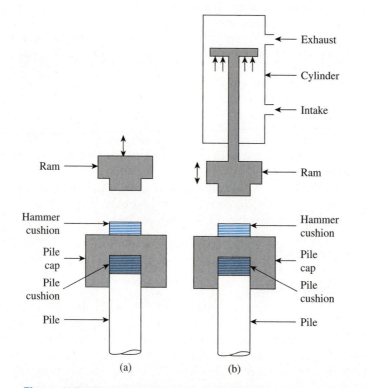

Figure 11.7 Pile-driving equipment: (a) drop hammer; (b) single-acting air or steam hammer

Figure 11.7 (*continued*) Pile-driving equipment: (c) double-acting and differential air or steam hammer; (d) diesel hammer; (e) vibratory pile driver; (f) photograph of a vibratory pile driver (*Courtesy of Michael W. O'Neill, University of Houston*)

Table 11.4 Examples of Commercially Available Pile-Driving Hammers

Maker of hammer[†]	Model No.	Hammer type	Rated energy kN-m	Blows/min	Ram weight kN
V	400C	Single acting	153.9	100	177.9
M	S-20		81.3	60	89.0
M	S-8		35.3	53	35.6
M	S-5		22.0	60	22.2
R	5/O		77.1	44	77.8
R	2/O		44.1	50	44.5
V	200C	Double acting	68.1	98	89.0
V	140C	or	48.8	103	62.3
V	80C	differential	33.1	111	35.6
V	65C		26.0	117	28.9
R	150C		66.1	95–105	66.7
V	4N100	Diesel	58.8	50–60	23.5
V	IN100		33.4	50–60	13.3
M	DE40		43.4	48	17.8
M	DE30		30.4	48	12.5

[†]V—Vulcan Iron Works, Florida
M—McKiernan-Terry, New Jersey
R—Raymond International, Inc., Texas

not be sufficient to ignite the air–fuel system, so the ram may have to be lifted manually. Table 11.4 provides some examples of commercially available pile-driving hammers.

The principles of operation of a vibratory pile driver are shown in Figure 11.7e. This driver consists essentially of two counterrotating weights. The horizontal components of the centrifugal force generated as a result of rotating masses cancel each other. As a result, a sinusoidal dynamic vertical force is produced on the pile and helps drive the pile downward.

Figure 11.7f is a photograph of a vibratory pile driver. Figure 11.8 shows a pile-driving operation in the field.

Jetting is a technique that is sometimes used in pile driving when the pile needs to penetrate a thin layer of hard soil (such as sand and gravel) overlying a layer of softer soil. In this technique, water is discharged at the pile point by means of a pipe 50 to 75 mm in diameter to wash and loosen the sand and gravel.

Piles driven at an angle to the vertical, typically 14 to 20°, are referred to as *batter piles*. Batter piles are used in group piles when higher lateral load-bearing capacity is required. Piles also may be advanced by partial augering, with power augers (see Chapter 2) used to predrill holes part of the way. The piles can then be inserted into the holes and driven to the desired depth.

Piles may be divided into two categories based on the nature of their placement: *displacement piles* and *nondisplacement piles*. Driven piles are displacement piles, because they move some soil laterally; hence, there is a tendency for densification of soil surrounding them. Concrete piles and closed-ended pipe piles are high-displacement piles. However, steel H-piles displace less soil laterally during driving, so they are low-displacement piles. In contrast, bored piles are nondisplacement piles because their placement causes very little change in the state of stress in the soil.

Figure 11.8 A pile-driving operation in the field (*Courtesy of E. C. Shin, University of Incheon, Korea*)

11.5 *Load Transfer Mechanism*

The load transfer mechanism from a pile to the soil is complicated. To understand it, consider a pile of length L, as shown in Figure 11.9a. The load on the pile is gradually increased from zero to $Q_{(z=0)}$ at the ground surface. Part of this load will be resisted by

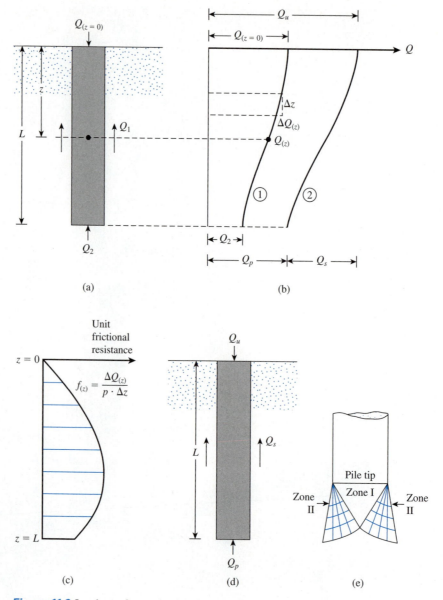

Figure 11.9 Load transfer mechanism for piles

the side friction developed along the shaft, Q_1, and part by the soil below the tip of the pile, Q_2. Now, how are Q_1 and Q_2 related to the total load? If measurements are made to obtain the load carried by the pile shaft, $Q_{(z)}$, at any depth z, the nature of the variation found will be like that shown in curve 1 of Figure 11.9b. The *frictional resistance per unit area* at any depth z may be determined as

$$f_{(z)} = \frac{\Delta Q_{(z)}}{(p)(\Delta z)} \qquad (11.8)$$

where p = perimeter of the cross section of the pile. Figure 11.9c shows the variation of $f_{(z)}$ with depth.

If the load Q at the ground surface is gradually increased, maximum frictional resistance along the pile shaft will be fully mobilized when the relative displacement between the soil and the pile is about 5 to 10 mm, irrespective of the pile size and length L. However, the maximum point resistance $Q_2 = Q_p$ will not be mobilized until the tip of the pile has moved about 10 to 25% of the pile width (or diameter). (The lower limit applies to driven piles and the upper limit to bored piles). At ultimate load (Figure 11.9d and curve 2 in Figure 11.9b), $Q_{(z=0)} = Q_u$. Thus,

$$Q_1 = Q_s$$

and

$$Q_2 = Q_p$$

The preceding explanation indicates that Q_s (or the unit skin friction, f, along the pile shaft) is developed at a *much smaller pile displacement compared with the point resistance, Q_p.* In order to demonstrate this point, let us consider the results of a pile load test conducted in the field by Mansur and Hunter (1970). The details of the pile and subsoil conditions are as follow:

> Type of pile: Steel pile with 406 mm outside diameter with 8.15 mm
> wall thickness
> Type of subsoil: Sand
> Length of pile embedment: 16.8 m

Figure 11.10a shows the load test results, which is a plot of load at the top of the pile $[Q_{(z=0)}]$ versus settlement(s). Figure 11.10b shows the plot of the load carried by the pile shaft $[Q_{(z)}]$ at any depth. It was reported by Mansur and Hunter (1970) that, for this test, at failure

$$Q_u \approx 1601\,\text{kN}$$
$$Q_p \approx 416\,\text{kN}$$

and

$$Q_s \approx 1185\,\text{kN}$$

Now, let us consider the load distribution in Figure 11.10b when the pile settlement(s) is about 2.5 mm. For this condition,

$$Q_{(z=0)} \approx 667\,\text{kN}$$
$$Q_2 \approx 93\,\text{kN}$$
$$Q_1 \approx 574\,\text{kN}$$

Hence, at $s = 2.5\,\text{mm}$,

$$\frac{Q_2}{Q_p} = \frac{93}{416}(100) = 22.4\%$$

and

$$\frac{Q_1}{Q_s} = \frac{574}{1185}(100) = 48.4\%$$

Thus, it is obvious that the skin friction is mobilized faster at low settlement levels as compared to the point load.

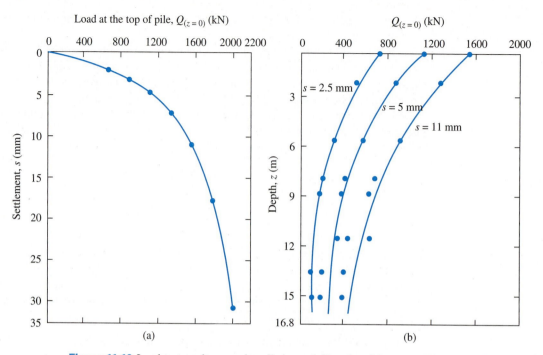

Figure 11.10 Load test results on a pipe pile in sand (Based on Mansur and Hunter, 1970)

At ultimate load, the failure surface in the soil at the pile tip (a bearing capacity failure caused by Q_p) is like that shown in Figure 11.9e. Note that pile foundations are deep foundations and that the soil fails mostly in a *punching mode,* as illustrated previously in Figures 3.1c and 3.3. That is, a *triangular zone,* I, is developed at the pile tip, which is pushed downward without producing any other visible slip surface. In dense sands and stiff clayey soils, a *radial shear zone,* II, may partially develop. Hence, the load displacement curves of piles will resemble those shown in Figure 3.1c.

11.6   *Equations for Estimating Pile Capacity*

The ultimate load-carrying capacity Q_u of a pile is given by the equation

$$Q_u = Q_p + Q_s \tag{11.9}$$

where

Q_p = load-carrying capacity of the pile point
Q_s = frictional resistance (skin friction) derived from the soil–pile interface (see Figure 11.11)

Numerous published studies cover the determination of the values of Q_p and Q_s. Excellent reviews of many of these investigations have been provided by Vesic (1977), Meyerhof (1976), and Coyle and Castello (1981). These studies afford an insight into the problem of determining the ultimate pile capacity.

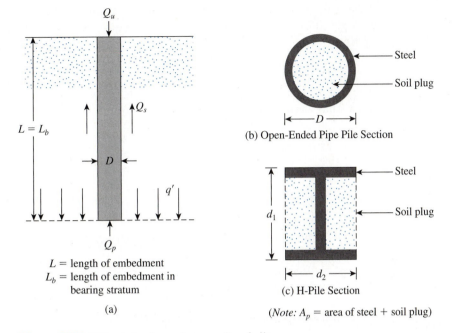

(b) Open-Ended Pipe Pile Section

(c) H-Pile Section

L = length of embedment
L_b = length of embedment in bearing stratum

(a)

(*Note:* A_p = area of steel + soil plug)

Figure 11.11 Ultimate load-carrying capacity of pile

Point Bearing Capacity, Q_p

The ultimate bearing capacity of shallow foundations was discussed in Chapter 3. According to Terzaghi's equations,

$$q_u = 1.3c'N_c + qN_q + 0.4\gamma BN_\gamma \quad \text{(for shallow square foundations)}$$

and

$$q_u = 1.3c'N_c + qN_q + 0.3\gamma BN_\gamma \quad \text{(for shallow circular foundations)}$$

Similarly, the general bearing capacity equation for shallow foundations was given in Chapter 3 (for vertical loading) as

$$q_u = c'N_cF_{cs}F_{cd} + qN_qF_{qs}F_{qd} + \tfrac{1}{2}\gamma BN_\gamma F_{\gamma s}F_{\gamma d}$$

Hence, in general, the ultimate load-bearing capacity may be expressed as

$$q_u = c'N_c^* + qN_q^* + \gamma BN_\gamma^* \tag{11.10}$$

where N_c^*, N_q^*, and N_γ^* are the bearing capacity factors that include the necessary shape and depth factors.

Pile foundations are deep. However, the ultimate resistance per unit area developed at the pile tip, q_p, may be expressed by an equation similar in form to Eq. (11.10), although the values of N_c^*, N_q^*, and N_γ^* will change. The notation used in this chapter for the width of a pile is D. Hence, substituting D for B in Eq. (11.10) gives

$$q_u = q_p = c'N_c^* + qN_q^* + \gamma DN_\gamma^* \tag{11.11}$$

Because the width D of a pile is relatively small, the term $\gamma D N_\gamma^*$ may be dropped from the right side of the preceding equation without introducing a serious error; thus, we have

$$q_p = c' N_c^* + q' N_q^* \tag{11.12}$$

Note that the term q has been replaced by q' in Eq. (11.12), to signify effective vertical stress. Thus, the point bearing of piles is

$$Q_p = A_p q_p = A_p (c' N_c^* + q' N_q^*) \tag{11.13}$$

where

$$A_p = \text{area of pile tip}$$
$$c' = \text{cohesion of the soil supporting the pile tip}$$
$$q_p = \text{unit point resistance}$$
$$q' = \text{effective vertical stress at the level of the pile tip}$$
$$N_c^*, N_q^* = \text{the bearing capacity factors}$$

Frictional Resistance, Q_s

The frictional, or skin, resistance of a pile may be written as

$$Q_s = \Sigma\, p\, \Delta L f \tag{11.14}$$

where

$$p = \text{perimeter of the pile section}$$
$$\Delta L = \text{incremental pile length over which } p \text{ and } f \text{ are taken to be constant}$$
$$f = \text{unit friction resistance at any depth } z$$

The various methods for estimating Q_p and Q_s are discussed in the next several sections. It needs to be reemphasized that, in the field, for full mobilization of the point resistance (Q_p), the pile tip must go through a displacement of 10 to 25% of the pile width (or diameter).

Allowable Load, Q_{all}

After the total ultimate load-carrying capacity of a pile has been determined by summing the point bearing capacity and the frictional (or skin) resistance, a reasonable factor of safety should be used to obtain the total allowable load for each pile, or

$$Q_{all} = \frac{Q_u}{FS}$$

where

$$Q_{all} = \text{allowable load-carrying capacity for each pile}$$
$$FS = \text{factor of safety}$$

The factor of safety generally used ranges from 2.5 to 4, depending on the uncertainties surrounding the calculation of ultimate load.

11.7 ## Meyerhof's Method for Estimating Q$_p$

Sand

The point bearing capacity, q_p, of a pile in sand generally increases with the depth of embedment in the bearing stratum and reaches a maximum value at an embedment ratio of $L_b/D = (L_b/D)_{cr}$. Note that in a homogeneous soil L_b is equal to the actual embedment length of the pile, L. However, where a pile has penetrated into a bearing stratum, $L_b < L$. Beyond the critical embedment ratio, $(L_b/D)_{cr}$, the value of q_p remains constant ($q_p = q_l$). That is, as shown in Figure 11.12 for the case of a homogeneous soil, $L = L_b$.

For piles in sand, $c' = 0$, and Eq. (11.13) simpifies to

$$Q_p = A_p q_p = A_p q' N_q^*$$ (11.15)

The variation of N_q^* with soil friction angle ϕ' is shown in Figure 11.13. The interpolated values of N_q^* for various friction angles are also given in Table 11.5. However, Q_p should not exceed the limiting value $A_p q_l$; that is,

$$Q_p = A_p q' N_q^* \leq A_p q_l$$ (11.16)

Figure 11.13 Variation of the maximum values of N_q^* with soil friction angle ϕ' (From Meyerhof, G. G. (1976). "Bearing Capacity and Settlement of Pile Foundations," Journal of the Geotechnical Engineering Division, American Society of Civil Engineers, Vol. 102, No. GT3, pp. 197–228. With permission from ASCE.)

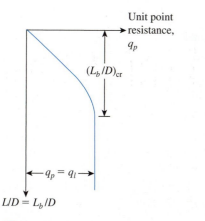

Figure 11.12 Nature of variation of unit point resistance in a homogeneous sand

Table 11.5 Interpolated Values of N_q^* Based on Meyerhof's Theory

Soil friction angle, ϕ (deg)	N_q^*
20	12.4
21	13.8
22	15.5
23	17.9
24	21.4
25	26.0
26	29.5
27	34.0
28	39.7
29	46.5
30	56.7
31	68.2
32	81.0
33	96.0
34	115.0
35	143.0
36	168.0
37	194.0
38	231.0
39	276.0
40	346.0
41	420.0
42	525.0
43	650.0
44	780.0
45	930.0

The limiting point resistance is

$$q_l = 0.5 \, p_a N_q^* \tan \phi' \tag{11.17}$$

where

p_a = atmospheric pressure ($=100 \text{ kN/m}^2$)
ϕ' = effective soil friction angle of the bearing stratum

   A good example of the concept of the critical embedment ratio can be found from the field load tests on a pile in sand at the Ogeechee River site reported by Vesic (1970). The pile tested was a steel pile with a diameter of 457 mm. Table 11.6 shows the ultimate resistance at various depths. Figure 11.14 shows the plot of q_p with depth obtained from the field tests along with the range of standard penetration resistance at the site. From the figure, the following observations can be made.

1. There is a limiting value of q_p. For the tests under consideration, it is about 12,000 kN/m².
2. The $(L/D)_{cr}$ value is about 16 to 18.

Table 11.6 Ultimate Point Resistance, q_p, of Test Pile at the Ogeechee River Site As reported by Vesic (1970)

Pile diameter, D (m)	Depth of embedment, L (m)	L/D	q_p (kN/m²)
0.457	3.02	6.61	3,304
0.457	6.12	13.39	9,365
0.457	8.87	19.4	11,472
0.457	12.0	26.26	11,587
0.457	15.00	32.82	13,971

Figure 11.14 Vesic's pile test (1970) result—variation of q_p and N_{60} with depth

3. The average N_{60} value is about 30 for $L/D \geqslant (L/D)_{cr}$. Using Eq. (11.37), the limiting point resistance is $4p_a N_{60} = (4)(100)(30) = 12,000$ kN/m². This value is generally consistent with the field observation.

Clay ($\phi = 0$)

For piles in *saturated clays* under undrained conditions ($\phi = 0$), the net ultimate load can be given as

$$Q_p \approx N_c^* c_u A_p = 9 c_u A_p \tag{11.18}$$

where c_u = undrained cohesion of the soil below the tip of the pile.

| 11.8 | *Vesic's Method for Estimating Q_p* |

Sand

Vesic (1977) proposed a method for estimating the pile point bearing capacity based on the theory of *expansion of cavities.* According to this theory, on the basis of effective stress parameters, we may write

$$Q_p = A_p q_p = A_p \bar{\sigma}_o' N_\sigma^* \tag{11.19}$$

where

$\bar{\sigma}_o'$ = mean effective normal ground stress at the level of the pile point

$$= \left(\frac{1 + 2K_o}{3} \right) q' \tag{11.20}$$

K_o = earth pressure coefficient at rest = $1 - \sin \phi'$ $\qquad$ (11.21)

and

$$N_\sigma^* = \text{bearing capacity factor}$$

Note that Eq. (11.19) is a modification of Eq. (11.15) with

$$N_\sigma^* = \frac{3N_q^*}{(1 + 2K_o)} \tag{11.22}$$

According to Vesic's theory,

$$N_\sigma^* = f(I_{rr}) \tag{11.23}$$

where I_{rr} = reduced rigidity index for the soil. However,

$$I_{rr} = \frac{I_r}{1 + I_r \Delta} \tag{11.24}$$

where

$$I_r = \text{rigidity index} = \frac{E_s}{2(1 + \mu_s) q' \tan \phi'} = \frac{G_s}{q' \tan \phi'} \tag{11.25}$$

E_s = modulus of elasticity of soil
μ_s = Poisson's ratio of soil
G_s = shear modulus of soil
Δ = average volumatic strain in the plastic zone below the pile point

The general ranges of I_r for various soils are

Sand(relative density = 50% to 80%): 75 to 150
Silt : 50 to 75

In order to estimate I_r [Eq. (11.25)] and hence I_{rr} [Eq. (11.24)], the following approximations may be used (Chen and Kulhawy, 1994)

$$\frac{E_s}{p_a} = m \tag{11.26}$$

where

p_a = atmospheric pressure ($\approx 100 \text{ kN/m}^2$)

$$m = \begin{cases} 100 \text{ to } 200 \,(\text{loose soil}) \\ 200 \text{ to } 500 \,(\text{medium dense soil}) \\ 500 \text{ to } 1000 \,(\text{dense soil}) \end{cases}$$

$$\mu_s = 0.1 + 0.3\left(\frac{\phi' - 25}{20}\right) \text{ (for } 25° \le \phi' \le 45°) \tag{11.27}$$

$$\Delta = 0.005\left(1 - \frac{\phi' - 25}{20}\right)\frac{q'}{p_a} \tag{11.28}$$

On the basis of cone penetration tests in the field, Baldi et al. (1981) gave the following correlations for I_r:

$$I_r = \frac{300}{F_r(\%)} \quad \text{(for mechanical cone penetration)} \tag{11.29}$$

and

$$I_r = \frac{170}{F_r(\%)} \quad \text{(for electric cone penetration)} \tag{11.30}$$

For the definition of F_r, see Eq. (2.41). Table 11.7 gives the values of N_σ^* for various values of I_{rr} and ϕ'.

Clay ($\phi = 0$)

In saturated clay ($\phi = 0$ condition), the net ultimate point bearing capacity of a pile can be approximated as

$$Q_p = A_p q_p = A_p c_u N_c^* \tag{11.31}$$

where c_u = undrained cohesion

According to the *expansion of cavity* theory of Vesic (1977),

$$N_c^* = \frac{4}{3}(\ln I_{rr} + 1) + \frac{\pi}{2} + 1 \tag{11.32}$$

The variations of N_c^* with I_{rr} for $\phi = 0$ condition are given in Table 11.8.

Now, referring to Eq. (11.24) for saturated clay with no volume change, $\Delta = 0$. Hence,

$$I_{rr} = I_r \tag{11.33}$$

Table 11.7 Bearing Capacity Factors N_σ^* Based on the Theory of Expansion of Cavities

ϕ'	I_{rr}									
	10	20	40	60	80	100	200	300	400	500
25	12.12	15.95	20.98	24.64	27.61	30.16	39.70	46.61	52.24	57.06
26	13.18	17.47	23.15	27.30	30.69	33.60	44.53	52.51	59.02	64.62
27	14.33	19.12	25.52	30.21	34.06	37.37	49.88	59.05	66.56	73.04
28	15.57	20.91	28.10	33.40	37.75	41.51	55.77	66.29	74.93	82.40
29	16.90	22.85	30.90	36.87	41.79	46.05	62.27	74.30	84.21	92.80
30	18.24	24.95	33.95	40.66	46.21	51.02	69.43	83.14	94.48	104.33
31	19.88	27.22	37.27	44.79	51.03	56.46	77.31	92.90	105.84	117.11
32	21.55	29.68	40.88	49.30	56.30	62.41	85.96	103.66	118.39	131.24
33	23.34	32.34	44.80	54.20	62.05	68.92	95.46	115.51	132.24	146.87
34	25.28	35.21	49.05	59.54	68.33	76.02	105.90	128.55	147.51	164.12
35	27.36	38.32	53.67	65.36	75.17	83.78	117.33	142.89	164.33	183.16
36	29.60	41.68	58.68	71.69	82.62	92.24	129.87	158.65	182.85	204.14
37	32.02	45.31	64.13	78.57	90.75	101.48	143.61	175.95	203.23	227.26
38	34.63	49.24	70.03	86.05	99.60	111.56	158.65	194.94	225.62	252.71
39	37.44	53.50	76.45	94.20	109.24	122.54	175.11	215.78	250.23	280.71
40	40.47	58.10	83.40	103.05	119.74	134.52	193.13	238.62	277.26	311.50
41	43.74	63.07	90.96	112.68	131.18	147.59	212.84	263.67	306.94	345.34
42	47.27	68.46	99.16	123.16	143.64	161.83	234.40	291.13	339.52	382.53
43	51.08	74.30	108.08	134.56	157.21	177.36	257.99	321.22	375.28	423.39
44	55.20	80.62	117.76	146.97	172.00	194.31	283.80	354.20	414.51	468.28
45	59.66	87.48	128.28	160.48	188.12	212.79	312.03	390.35	457.57	517.58

From "Design of Pile Foundations," by A. S. Vesic. SYNTHESIS OF HIGHWAY PRACTICE by AMERICAN ASSOCIATION OF STATE HIGHWAY AND TRANSPORT. Copyright 1969 by TRANSPORTATION RESEARCH BOARD. Reproduced with permission of TRANSPORTATION RESEARCH BOARD in the format Textbook via Copyright Clearance Center.

Table 11.8 Variation of N_c^* with I_{rr} for $\phi = 0$ Condition based on Vesic's Theory

I_{rr}	N_c^*
10	6.97
20	7.90
40	8.82
60	9.36
80	9.75
100	10.04
200	10.97
300	11.51
400	11.89
500	12.19

For $\phi = 0$,

$$I_r = \frac{E_s}{3c_u} \tag{11.34}$$

O' Neill and Reese (1999) suggested the following approximate relationships for I_r and the undrained cohesion, c_u.

$\dfrac{c_u}{p_a}$	I_r
0.24	50
0.48	150
≥ 0.96	250–300

Note: p_a = atmospheric pressure $\approx 100 \ kN/m^2$.

The preceding values can be approximated as

$$I_r = 347\left(\frac{c_u}{p_a}\right) - 33 \leq 300 \tag{11.35}$$

11.9 *Coyle and Castello's Method for Estimating Q_p in Sand*

Coyle and Castello (1981) analyzed 24 large-scale field load tests of driven piles in sand. On the basis of the test results, they suggested that, in sand,

$$Q_p = q'N_q^* A_p \tag{11.36}$$

where

q' = effective vertical stress at the pile tip
N_q^* = bearing capacity factor

Figure 11.15 shows the variation of N_q^* with L/D and the soil friction angle ϕ'.

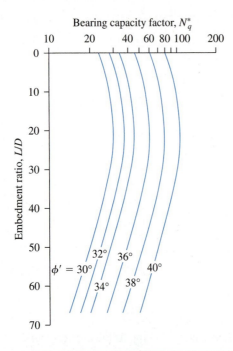

Figure 11.15 Variation of N_q^* with L/D (Redrawn after Coyle and Costello, 1981)

Example 11.1

Consider a 15-m long concrete pile with a cross section of $0.45\,\text{m} \times 0.45\,\text{m}$ fully embedded in sand. For the sand, given: unit weight, $\gamma = 17\text{kN/m}^3$; and soil friction angle, $\phi' = 35°$. Estimate the ultimate point Q_p with each of the following:

    a. Meyerhof's method
    b. Vesic's method
    c. The method of Coyle and Castello
    d. based on the results of parts a, b, and c, adopt a value for Q_p

Solution
Part a
From Eqs. (11.16) and (11.17),

$$Q_p = A_p q' N_q^* \le A_p (0.5 p_a N_q^* \tan \phi')$$

For $\phi' = 35°$, the value of $N_q^* \approx 143$ (Table 11.5). Also, $q' = \gamma L = (17)(15) = 255\text{kN/m}^2$.

Thus,

$$A_p q' N_q^* = (0.45 \times 0.45)(255)(143) \approx 7384\text{kN}$$

Again,

$$A_p (0.5 p_a N_q^* \tan \phi') = (0.45 \times 0.45)[(0.5)(100)(143)(\tan 35)] \approx 1014\text{kN}$$

Hence, $Q_p =$ **1014 kN**.

Part b
From Eq. (11.19),

$$Q_p = A_p \overline{\sigma}'_o N^*_\sigma$$

$$\overline{\sigma}'_o = \left[\frac{1 + 2(1 - \sin \phi')}{3} \right] q' = \left(\frac{1 + 2(1 - \sin 35)}{3} \right) (17 \times 15)$$

$$= 139.96 \text{kN/m}^2$$

From Eq. (11.26),

$$\frac{E_s}{p_a} = m$$

Assume $m \approx 250$ (medium sand). So,

$$E_s = (250)(100) = 25,000 \text{ kN/m}^2$$

From Eq. (11.27),

$$\mu_s = 0.1 + 0.3 \left(\frac{\phi' - 25}{20} \right) = 0.1 + 0.3 \left(\frac{35 - 25}{20} \right) = 0.25$$

From Eq. (11.28),

$$\Delta = 0.005 \left(1 - \frac{\phi' - 25}{20} \right) \left(\frac{q'}{p_a} \right) = 0.005 \left(1 - \frac{35 - 25}{20} \right) \left(\frac{17 \times 15}{100} \right) = 0.0064$$

From Eq. (11.25),

$$I_r = \frac{E_s}{2(1 + \mu_s) q' \tan \phi'} = \frac{25,000}{(2)(1 + 0.25)(17 \times 15)(\tan 35)} = 56$$

From Eq. (11.24),

$$I_{rr} = \frac{I_r}{1 + I_r \Delta} = \frac{56}{1 + (56)(0.0064)} = 41.2$$

From Table 11.7, for $\phi' = 35°$ and $I_{rr} = 41.2$, the value of $N^*_\sigma \approx 55$. Hence,

$$Q_p = A_p \overline{\sigma}'_o N^*_\sigma = (0.45 \times 0.45)(139.96)(55) \approx \textbf{1559 kN}$$

Part c
From Eq. (11.36),

$$Q_p = q' N^*_q A_p$$

$$\frac{L}{D} = \frac{15}{0.45} = 33.3$$

For $\phi' = 35°$ and $L/D = 33.3$, the value of N^*_q is about 48 (Figure 11.15). Thus,

$$Q_p = q' N^*_q A_p = (15 \times 17)(48)(0.45 \times 0.45) \approx \textbf{2479 kN}$$

Part d
It appears that Q_p obtained from the method of Coyle and Castello is too large. Thus, the average of the results from parts a and b is

$$\frac{1014 + 1559}{2} = 1286.5 \text{ kN}$$

$$\text{Use } Q_p = \textbf{1250 kN}.$$ ∎

Example 11.2

Consider a pipe pile (flat driving point—see Figure 11.2d) having an outside diameter of 406 mm. The embedded length of the pile in layered saturated clay is 30 m. The following are the details of the subsoil:

Depth from ground surface (m)	Saturated unit weight, γ (kN/m³)	c_u (kN/m²)
0–5	18	30
5–10	18	30
10–30	19.6	100

The groundwater table is located at a depth of 5 m from the ground surface. Estimate Q_p by using

    a. Meyerhof's method
    b. Vesic's method

Solution
Part a
From Eq. (11.18),

$$Q_p = 9c_u A_p$$

The tip of the pile is resting on a clay with $c_u = 100$ kN/m². So,

$$Q_p = (9)(100)\left[\left(\frac{\pi}{4}\right)\left(\frac{406}{1000}\right)^2\right] = \textbf{116.5 kN}$$

Part b
From Eq. (11.31),

$$Q_p = A_p c_u N_c^*$$

From Eq. (11.35),

$$I_r = I_{rr} = 347\left(\frac{c_u}{p_a}\right) - 33 = 347\left(\frac{100}{100}\right) - 33 = 314$$

So we use $I_{rr} = 300$.
From Table 11.8 for $I_{rr} = 300$, the value of $N_c^* = 11.51$. Thus,

$$Q_p = A_p c_u N_c^* = \left[\left(\frac{\pi}{4}\right)\left(\frac{406}{1000}\right)^2\right](100)(11.51) = \textbf{149.0 kN}$$

Note: The average value of Q_p is

$$\frac{116.5 + 149.0}{2} \approx \mathbf{133\ kN}$$

■

11.10 *Correlations for Calculating Q_p with SPT and CPT Results*

On the basis of field observations, Meyerhof (1976) also suggested that the ultimate point resistance q_p in a homogeneous granular soil $(L = L_b)$ may be obtained from standard penetration numbers as

$$q_p = 0.4 p_a N_{60} \frac{L}{D} \leq 4 p_a N_{60} \tag{11.37}$$

where

N_{60} = the average value of the standard penetration number near the pile point (about $10D$ above and $4D$ below the pile point)
p_a = atmospheric pressure ($\approx 100\ \text{kN/m}^2$ or 2000 lb/ft^2)

Briaud et al. (1985) suggested the following correlation for q_p in granular soil with the standard penetration resistance N_{60}.

$$q_p = 19.7 p_a (N_{60})^{0.36} \tag{11.38}$$

Meyerhof (1956) also suggested that

$$q_p \approx q_c (\text{in granular soil}) \tag{11.39}$$

where q_c = cone penetration resistance.

Example 11.3

Consider a concrete pile that is 0.305 m × 0.305 m in cross section in sand. The pile is 15.2 m long. The following are the variations of N_{60} with depth.

Depth below ground surface (m)	N_{60}
1.5	8
3.0	10
4.5	9
6.0	12
7.5	14
9.0	18
10.5	11
12.0	17
13.5	20
15.0	28
16.5	29
18.0	32
19.5	30
21.0	27

a. Estimate Q_p using Eq. (11.37).
b. Estimate Q_p using Eq. (11.38).

Solution
Part a
The tip of the pile is 15.2 m below the ground surface. For the pile, $D = 0.305$ m. The average of N_{60} 10D above and about 5D below the pile tip is

$$N_{60} = \frac{17 + 20 + 28 + 29}{4} = 23.5 \approx 24$$

From Eq. (11.37)

$$Q_p = A_p(q_p) = A_p\left[0.4 p_a N_{60}\left(\frac{L}{D}\right)\right] \leq A_p(4 p_a N_{60})$$

$$A_p\left[0.4 p_a N_{60}\left(\frac{L}{D}\right)\right] = (0.305 \times 0.305)\left[(0.4)(100)(24)\left(\frac{15.2}{0.305}\right)\right] = 4450.6 \text{ kN}$$

$$A_p(4 p_a N_{60}) = (0.305 \times 0.305)[(4)(100)(24)] = 893 \text{ kN}$$

Thus, $Q_p = 893$ kN

Part b
From Eq. (11.38),

$$Q_p = A_p q_p = A_p[19.7 p_a (N_{60})^{0.36}] = (0.305 \times 0.305)[(19.7)(100)(24)^{0.36}]$$

$$= 575.4 \text{ kN} \qquad \blacksquare$$

11.11   *Frictional Resistance (Q_s) in Sand*

According to Eq. (11.14), the frictional resistance

$$Q_s = \Sigma p \, \Delta L f$$

The unit frictional resistance, f, is hard to estimate. In making an estimation of f, several important factors must be kept in mind:

1. The nature of the pile installation. For driven piles in sand, the vibration caused during pile driving helps densify the soil around the pile. The zone of sand densification may be as much as 2.5 times the pile diameter, in the sand surrounding the pile.
2. It has been observed that the nature of variation of f in the field is approximately as shown in Figure 11.16. The unit skin friction increases with depth more or

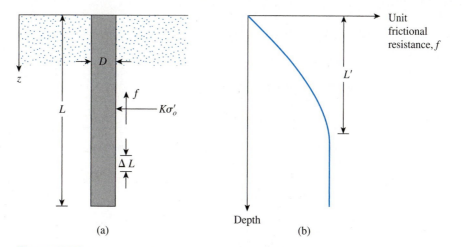

Figure 11.16 Unit frictional resistance for piles in sand

less linearly to a depth of L' and remains constant thereafter. The magnitude of the critical depth L' may be 15 to 20 pile diameters. A conservative estimate would be

$$L' \approx 15D \qquad (11.40)$$

3. At similar depths, the unit skin friction in loose sand is higher for a high-displacement pile, compared with a low-displacement pile.
4. At similar depths, bored, or jetted, piles will have a lower unit skin friction compared with driven piles.

Taking into account the preceding factors, we can give the following approximate relationship for f (see Figure 11.16):

For $z = 0$ to L',

$$f = K\sigma'_o \tan \delta' \qquad (11.41)$$

and for $z = L'$ to L,

$$f = f_{z=L'} \qquad (11.42)$$

In these equations,

K = effective earth pressure coefficient
σ'_o = effective vertical stress at the depth under consideration
δ' = soil-pile friction angle

In reality, the magnitude of K varies with depth; it is approximately equal to the Rankine passive earth pressure coefficient, K_p, at the top of the pile and may be less than the at-rest pressure coefficient, K_o, at a greater depth. Based on presently available results, the following average values of K are recommended for use in Eq. (11.41):

Pile type	K
Bored or jetted	$\approx K_o = 1 - \sin \phi'$
Low-displacement driven	$\approx K_o = 1 - \sin \phi'$ to $1.4K_o = 1.4(1 - \sin \phi')$
High-displacement driven	$\approx K_o = 1 - \sin \phi'$ to $1.8K_o = 1.8(1 - \sin \phi')$

The values of δ' from various investigations appear to be in the range from $0.5\phi'$ to $0.8\phi'$.

Based on load test results in the field, Mansur and Hunter (1970) reported the following average values of K.

$$\text{H-piles}\ldots\ldots K = 1.65$$

$$\text{Steel pipe piles}\ldots\ldots K = 1.26$$

$$\text{Precast concrete piles}\ldots\ldots K = 1.5$$

Coyle and Castello (1981), in conjunction with the material presented in Section 11.9, proposed that

$$Q_s = f_{av}pL = (K\overline{\sigma}_o' \tan \delta')pL \tag{11.43}$$

where

$\overline{\sigma}_o'$ = average effective overburden pressure
δ' = soil–pile friction angle = $0.8\phi'$

The lateral earth pressure coefficient K, which was determined from field observations, is shown in Figure 11.17. Thus, if that figure is used,

$$Q_s = K\overline{\sigma}_o' \tan(0.8\phi')pL \tag{11.44}$$

Correlation with Standard Penetration Test Results

Meyerhof (1976) indicated that the average unit frictional resistance, f_{av}, for high-displacement driven piles may be obtained from average standard penetration resistance values as

$$f_{av} = 0.02p_a(\overline{N}_{60}) \tag{11.45}$$

where

$(\overline{N}_{60})$ = average value of standard penetration resistance
p_a = atmospheric pressure (≈ 100 kN/m^2 or 2000 lb/ft^2)

For low-displacement driven piles

$$f_{av} = 0.01p_a(\overline{N}_{60}) \tag{11.46}$$

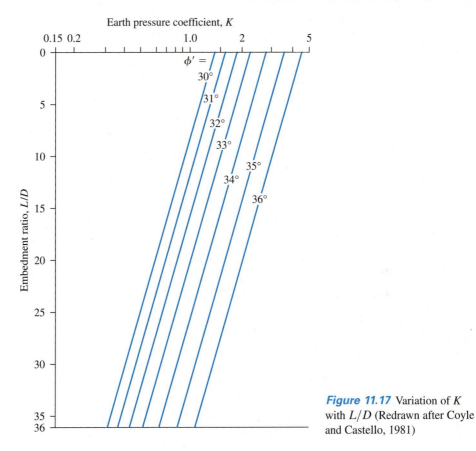

Figure 11.17 Variation of K with L/D (Redrawn after Coyle and Castello, 1981)

Briaud et al. (1985) suggested that

$$f_{av} \approx 0.224 p_a (\overline{N}_{60})^{0.29}$$
(11.47)

Thus,

$$Q_s = pL f_{av}$$
(11.48)

Correlation with Cone Penetration Test Results

Nottingham and Schmertmann (1975) and Schmertmann (1978) provided correlations for estimating Q_s using the frictional resistance (f_c) obtained during cone penetration tests. According to this method

$$f = \alpha' f_c$$
(11.49)

The variations of α' with z/D for electric cone and mechanical cone penetrometers are shown in Figures 11.18 and 11.19, respectively. We have

$$Q_s = \Sigma p(\Delta L) f = \Sigma p(\Delta L) \alpha' f_c$$
(11.50)

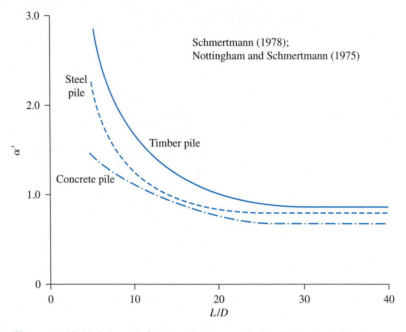

Figure 11.18 Variation of α' with embedment ratio for pile in sand: electric cone penetrometer

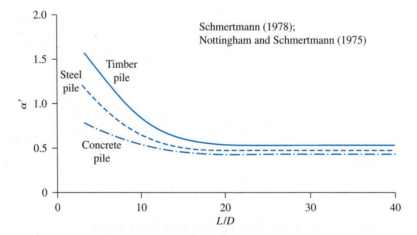

Figure 11.19 Variation of α' with embedment ratio for piles in sand: mechanical cone penetrometer

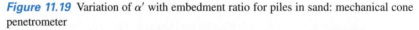

Example 11.4

Refer to the pile described in Example 11.3. Estimate the magnitude of Q_s for the pile.

    a. Use Eq. (11.45).
    b. Use Eq. (11.47).

c. Considering the results in Example 11.3, determine the allowable load-carrying capacity of the pile based on Meyerhof's method and Briaud's method. Use a factor of safety, FS = 3.

Solution

The average N_{60} value for the sand for the top 15.2 m is

$$\overline{N}_{60} = \frac{8 + 10 + 9 + 12 + 14 + 18 + 11 + 17 + 20 + 28}{10} = 14.7 \approx 15$$

Part a
From Eq. (11.45),

$$f_{av} = 0.02 p_a(\overline{N}_{60}) = (0.02)(100)(15) = 30 \text{ kN/m}^2$$
$$Q_s = pLf_{av} = (4 \times 0.305)(15.2)(30) = 556.2 \text{ kN}$$

Part b
From Eq. (11.47),

$$f_{av} = 0.224 p_a(\overline{N}_{60})^{0.29} = (0.224)(100)(15)^{0.29} = 49.13 \text{ kN/m}^2$$
$$Q_s = pLf_{av} = (4 \times 0.305)(15.2)(49.13) = 911.1 \text{ kN}$$

Part c

$$\text{Meyerhof's method: } Q_{all} = \frac{Q_p + Q_s}{FS} = \frac{893 + 556.2}{3} = \textbf{483 kN}$$

$$\text{Briaud's method: } Q_{all} = \frac{Q_p + Q_s}{FS} = \frac{575.4 + 911.1}{3} = \textbf{495.5 kN}$$

So the allowable pile capacity may be taken to be about **490 kN**. ∎

Example 11.5

Refer to Example 11.1. For the pile, estimate the frictional resistance Q_s

a. Based on Eqs. (11.41) and (11.42). Use $K = 1.3$ and $\delta' = 0.8\phi'$.
b. Based on Eq. (11.44).
c. Using the results of Part d of Example 11.1, estimate the allowable bearing capacity of the pile. Use FS = 3.

Solution

Part a
From Eq. (11.40), $L' = 15D = (15)(0.45) = 6.75$m. Refer to Eq. (11.41):
At $z = 0$: $\sigma'_o = 0$
$f = 0$

At $z = 6.75$ m: $\sigma'_o = (6.75)(17) = 114.75 \text{ kN/m}^2$

So

$$f = K\sigma'_o \tan \delta = (1.3)(114.75)[\tan(0.8 \times 35)] = 79.3 \text{ kN/m}^3$$

Thus,

$$Q_s = \frac{(f_{z=0} + f_{z=6.75m})}{2} pL' + f_{z=6.75m} p(L - L')$$

$$= \left(\frac{0 + 79.3}{2}\right)(4 \times 0.45)(6.75) + (79.3)(4 \times 0.45)(15 - 6.75)$$

$$= 481.75 + 1177.61 = 1659.36 \text{ kN} \approx \textbf{1659 kN}$$

Part b

From Eq. (11.44),

$$Q_s = K\overline{\sigma}'_o \tan(0.8\phi') pL$$

$$\overline{\sigma}'_o = \frac{(15)(17)}{2} = 127.5 \text{ kN/m}^2$$

$$\frac{L}{D} = \frac{15}{0.45} = 33.3; \phi' = 35°$$

From Figure 11.17, $K = 0.93$

$$Q_s = (1.3)(127.5)\tan[(0.8 \times 35)](4 \times 0.45)(15) = \textbf{2380 kN}$$

Part c

The average value of Q_s from parts a and b is

$$Q_{s(\text{average})} = \frac{1659 + 2380}{2} = 2019.5 \approx 2020 \text{ kN} - \text{USE}$$

From part d of Example 11.1, $Q_p = 1250$ kN. Thus,

$$Q_{\text{all}} = \frac{Q_p + Q_s}{\text{FS}} = \frac{1250 + 2020}{3} = \textbf{1090 kN} \qquad \blacksquare$$

Example 11.6

Consider an 18-m long concrete pile (cross section: 0.305 m × 0.305 m) fully embedded in a sand layer. For the sand layer, the following is an approximation of the cone penetration resistance q_c (mechanical cone) and the frictional resistance f_c with depth. Estimate the allowable load that the pile can carry. Use FS = 3.

Depth from ground surface (m)	q_c (kN/m²)	f_c (kN/m²)
0–5	3040	73
5–15	4560	102
15–25	9500	226

Solution

$$Q_u = Q_p + Q_s$$

From Eq. (11.39),

$$q_p \approx q_c$$

At the pile tip (i.e., at a depth of 18 m), $q_c \approx 9500 \text{ kN/m}^2$. Thus,

$$Q_p = A_p q_c = (0.305 \times 0.305)(9500) = 883.7 \text{ kN}$$

To determine Q_s, the following table can be prepared. (*Note*: $L/D = 18/0.305 = 59$.)

Depth from ground surface (m)	ΔL (m)	f_c (kN/m²)	α' (Figure 11.19)	$p\Delta L\alpha' f_c$ (kN)
0–5	5	73	0.44	195.9
5–15	10	102	0.44	547.5
15–18	3	226	0.44	363.95
				$Q_s = 1107.35 \text{ kN}$

Hence,

$$Q_u = Q_p + Q_s = 883.7 + 1107.35 = 1991.05 \text{ kN}$$

$$Q_{\text{all}} = \frac{Q_u}{\text{FS}} = \frac{1991.05}{3} = 663.68 \approx \textbf{664 kN}$$

11.12 *Frictional (Skin) Resistance in Clay*

Estimating the frictional (or skin) resistance of piles in clay is almost as difficult a task as estimating that in sand (see Section 11.11), due to the presence of several variables that cannot easily be quantified. Several methods for obtaining the unit frictional resistance of piles are described in the literature. We examine some of them next.

λ *Method*

This method, proposed by Vijayvergiya and Focht (1972), is based on the assumption that the displacement of soil caused by pile driving results in a passive lateral pressure at any depth and that the average unit skin resistance is

$$f_{\text{av}} = \lambda(\overline{\sigma}'_o + 2c_u) \tag{11.51}$$

where

$\overline{\sigma}'_o$ = mean effective vertical stress for the entire embedment length
c_u = mean undrained shear strength ($\phi = 0$)

Table 11.9 Variation of λ with pile embedment length, L

Embedment length, L (m)	λ
0	0.5
5	0.336
10	0.245
15	0.200
20	0.173
25	0.150
30	0.136
35	0.132
40	0.127
50	0.118
60	0.113
70	0.110
80	0.110
90	0.110

The value of λ changes with the depth of penetration of the pile. (See Table 11.9.) Thus, the total frictional resistance may be calculated as

$$Q_s = pLf_{av}$$

Care should be taken in obtaining the values of $\bar{\sigma}'_o$ and c_u in layered soil. Figure 11.20 helps explain the reason. Figure 11.20a shows a pile penetrating three layers of clay. According to Figure 11.20b, the mean value of c_u is $(c_{u(1)}L_1 + c_{u(2)}L_2 + \cdots)/L$. Similarly, Figure 11.20c shows the plot of the variation of effective stress with depth. The mean effective stress is

$$\bar{\sigma}'_o = \frac{A_1 + A_2 + A_3 + \cdots}{L} \tag{11.52}$$

where $A_1, A_2, A_3, \ldots$ = areas of the vertical effective stress diagrams.

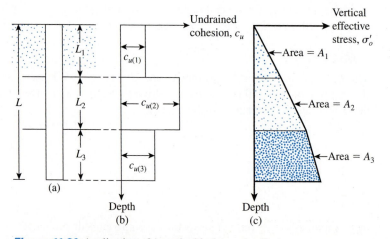

Figure 11.20 Application of λ method in layered soil

α *Method*

According to the α method, the unit skin resistance in clayey soils can be represented by the equation

$$f = \alpha c_u \tag{11.53}$$

where α = empirical adhesion factor. The approximate variation of the value of α is shown in Table 11.10. It is important to realize that the values of α given in Table 11.10 may vary somewhat, since α is actually a function of vertical effective stress and the undrained cohesion. Sladen (1992) has shown that

$$\alpha = C\left(\frac{\overline{\sigma}_o'}{c_u}\right)^{0.45} \tag{11.54}$$

where

$\overline{\sigma}_o'$ = average vertical effective stress
$C \approx 0.4$ to 0.5 for bored piles and ≥ 0.5 for driven piles

The ultimate side resistance can thus be given as

$$Q_s = \Sigma f p \, \Delta L = \Sigma \alpha c_u p \, \Delta L \tag{11.55}$$

Table 11.10 Variation of α (interpolated values based on Terzaghi, Peck and Mesri, 1996)

$\dfrac{c_u}{p_a}$	α
≤ 0.1	1.00
0.2	0.92
0.3	0.82
0.4	0.74
0.6	0.62
0.8	0.54
1.0	0.48
1.2	0.42
1.4	0.40
1.6	0.38
1.8	0.36
2.0	0.35
2.4	0.34
2.8	0.34

Note: p_a = atmospheric pressure $\approx 100 \text{ kN/m}^2$

β Method

When piles are driven into saturated clays, the pore water pressure in the soil around the piles increases. The excess pore water pressure in normally consolidated clays may be four to six times c_u. However, within a month or so, this pressure gradually dissipates. Hence, the unit frictional resistance for the pile can be determined on the basis of the effective stress parameters of the clay in a remolded state $(c' = 0)$. Thus, at any depth,

$$f = \beta \sigma_o' \tag{11.56}$$

where

$$\begin{aligned}
\sigma_o' &= \text{vertical effective stress} \\
\beta &= K \tan \phi_R' \\
\phi_R' &= \text{drained friction angle of remolded clay} \\
K &= \text{earth pressure coefficient}
\end{aligned} \tag{11.57}$$

Conservatively, the magnitude of K is the earth pressure coefficient at rest, or

$$K = 1 - \sin \phi_R' \qquad \text{(for normally consolidated clays)} \tag{11.58}$$

and

$$K = (1 - \sin \phi_R')\sqrt{\text{OCR}} \qquad \text{(for overconsolidated clays)} \tag{11.59}$$

where OCR = overconsolidation ratio.

Combining Eqs. (11.56), (11.57), (11.58), and (11.59), for normally consolidated clays yields

$$f = (1 - \sin \phi_R') \tan \phi_R' \sigma_o' \tag{11.60}$$

and for overconsolidated clays,

$$f = (1 - \sin \phi_R') \tan \phi_R' \sqrt{\text{OCR}}\, \sigma_o' \tag{11.61}$$

With the value of f determined, the total frictional resistance may be evaluated as

$$Q_s = \Sigma f p\, \Delta L$$

Correlation with Cone Penetration Test Results

Nottingham and Schmertmann (1975) and Schmertmann (1978) found the correlation for unit skin friction in clay (with $\phi = 0$) to be

$$f = \alpha' f_c \tag{11.62}$$

The variation of α' with the frictional resistance f_c is shown in Figure 11.21. Thus,

$$Q_s = \Sigma f p(\Delta L) = \Sigma \alpha' f_c p(\Delta L) \tag{11.63}$$

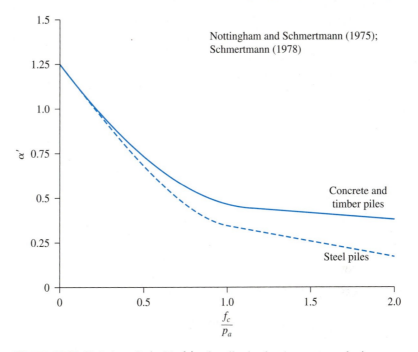

Figure 11.21 Variation of α' with f_c/p_a for piles in clay (p_a = atmospheric pressure $\approx 100 \ kN/m^2$)

11.13 Point Bearing Capacity of Piles Resting on Rock

Sometimes piles are driven to an underlying layer of rock. In such cases, the engineer must evaluate the bearing capacity of the rock. The ultimate unit point resistance in rock (Goodman, 1980) is approximately

$$q_p = q_u(N_\phi + 1) \tag{11.64}$$

where

$N_\phi = \tan^2(45 + \phi'/2)$
q_u = unconfined compression strength of rock
ϕ' = drained angle of friction

The unconfined compression strength of rock can be determined by laboratory tests on rock specimens collected during field investigation. However, extreme caution should be used in obtaining the proper value of q_u, because laboratory specimens usually are small in diameter. As the diameter of the specimen increases, the unconfined compression strength decreases—a phenomenon referred to as the *scale effect*. For specimens larger than about 1 m in diameter, the value of q_u remains approximately constant. There appears to be a fourfold to fivefold reduction of the magnitude of q_u in this process. The scale effect in rock is caused primarily by randomly distributed large and small fractures and also by progressive ruptures along the slip lines. Hence, we always recommend that

$$q_{u(design)} = \frac{q_{u(lab)}}{5} \qquad (11.65)$$

Table 11.11 lists some representative values of (laboratory) unconfined compression strengths of rock. Representative values of the rock friction angle ϕ' are given in Table 11.12.

A factor of safety of at least 3 should be used to determine the allowable point bearing capacity of piles. Thus,

$$Q_{p(all)} = \frac{[q_{u(design)}(N_\phi + 1)]A_p}{FS} \qquad (11.66)$$

Table 11.11 Typical Unconfined Compressive Strength of Rocks

Type of rock	q_u MN/m^2
Sandstone	70–140
Limestone	105–210
Shale	35–70
Granite	140–210
Marble	60–70

Table 11.12 Typical Values of Angle of Friction ϕ' of Rocks

Type of rock	Angle of friction, ϕ' (deg)
Sandstone	27–45
Limestone	30–40
Shale	10–20
Granite	40–50
Marble	25–30

Example 11.7

Refer to the pile in saturated clay shown in Figure 11.22. For the pile,

a. Calculate the skin resistance (Q_s) by (1) the α method, (2) the λ method, and (3) the β method. For the β method, use $\phi'_R = 30°$ for all clay layers. The top 10 m of clay is normally consolidated. The bottom clay layer has an OCR = 2. (Note: diameter of pile = 406 mm)

b. Using the results of Example 11.2, estimate the allowable pile capacity (Q_{all}). Use FS = 4.

Figure 11.22 Estimation of the load bearing capacity of a driven-pipe pile

Solution

Part a

(1) From Eq. (11.55),

$$Q_s = \Sigma \alpha c_u p \Delta L$$

[*Note*: $p = \pi(0.406) = 1.275\text{m}$] Now the following table can be prepared.

Depth (m)	ΔL (m)	c_u (kN/m²)	α (Table 11.10)	$\alpha c_u p \Delta L$ (kN)
0–5	5	30	0.82	156.83
5–10	5	30	0.82	156.83
10–30	20	100	0.48	1224.0

$$Q_s \approx \textbf{1538 kN}$$

(2) From Eq. 11.51, $f_{av} = \lambda(\overline{\sigma}_o' + 2c_u)$. Now, the average value of c_u is

$$\frac{c_{u(1)}(10) + c_{u(2)}(20)}{30} = \frac{(30)(10) + (100)(20)}{30} = 76.7 \text{ kN/m}^2$$

To obtain the average value of $\overline{\sigma}_o'$, the diagram for vertical effective stress variation with depth is plotted in Figure 11.22b. From Eq. (11.52),

$$\overline{\sigma}_o' = \frac{A_1 + A_2 + A_3}{L} = \frac{225 + 552.38 + 4577}{30} = 178.48 \text{ kN/m}^2$$

From Table 11.9, the magnitude of λ is 0.136. So

$$f_{av} = 0.136[178.48 + (2)(76.7)] = 45.14 \text{ kN/m}^2$$

Hence,

$$Q_s = pLf_{av} = \pi(0.406)(30)(45.14) = \textbf{1727 kN}$$

(3) The top layer of clay (10 m) is normally consolidated, and $\phi'_R = 30°$. For $z = 0$–5 m, from Eq. (11.60), we have

$$f_{av(1)} = (1 - \sin \phi'_R) \tan \phi'_R \, \overline{\sigma}'_o$$
$$= (1 - \sin 30°)(\tan 30°)\left(\frac{0 + 90}{2}\right) = 13.0 \text{ kN/m}^2$$

Similarly, for $z = 5$–10 m.

$$f_{av(2)} = (1 - \sin 30°)(\tan 30°)\left(\frac{90 + 130.95}{2}\right) = 31.9 \text{ kN/m}^2$$

For $z = 10$–30 m from Eq. (11.61),

$$f_{av} = (1 - \sin \phi'_R)\tan \phi'_R\sqrt{OCR}\,\overline{\sigma}'_o$$

For OCR = 2,

$$f_{av(3)} = (1 - \sin 30°)(\tan 30°)\sqrt{2}\left(\frac{130.95 + 326.75}{2}\right) = 93.43 \text{ kN/m}^2$$

So,

$$Q_s = p[f_{av(1)}(5) + f_{av(2)}(5) + f_{av(3)}(20)]$$
$$= (\pi)(0.406)[(13)(5) + (31.9)(5) + (93.43)(20)] = \textbf{2670 kN}$$

Part b

$$Q_u = Q_p + Q_s$$

From Example 11.2,

$$Q_p \approx \frac{116.5 + 149}{2} \approx 133 \text{ kN}$$

Again, the values of Q_s from the α method and λ method are close. So,

$$Q_s \approx \frac{1538 + 1727}{2} = 1632.5 \text{ kN}$$
$$Q_{all} = \frac{Q_u}{FS} = \frac{133 + 1632.5}{4} = 441.4 \text{ kN} \approx \textbf{441 kN}$$

■

Example 11.8

A concrete pile 305 mm $\times$ 305 mm in cross section is driven to a depth of 20 m below the ground surface in a saturated clay soil. A summary of the variation of frictional resistance f_c obtained from a cone penetration test is as follows:

Depth (m)	Friction resistance, f_c (kg/cm²)
0–6	0.35
6–12	0.56
12–20	0.72

Estimate the frictional resistance Q_s for the pile.

Solution
We can prepare the following table:

Depth (m)	f_c (kN/m²)	α' (Figure 11.21)	ΔL (m)	$\alpha' f_c p (\Delta L)$ [Eq. (11.63)] (kN)
0–6	34.34	0.84	6	211.5
6–12	54.94	0.71	6	285.5
12–20	70.63	0.63	8	434.2

(*Note:* $p = (4)(0.305) = 1.22$ m)

Thus,

$$Q_s = \Sigma \alpha' f_c p (\Delta L) = \textbf{931 kN}$$ ∎

11.14 *Pile Load Tests*

In most large projects, a specific number of load tests must be conducted on piles. The primary reason is the unreliability of prediction methods. The vertical and lateral load-bearing capacity of a pile can be tested in the field. Figure 11.23a shows a schematic diagram of the pile load arrangement for testing *axial compression* in the field. The load is applied to the pile by a hydraulic jack. Step loads are applied to the pile, and sufficient time is allowed to elapse after each load so that a small amount of settlement occurs. The settlement of the pile is measured by dial gauges. The amount of load to be applied for each step will vary, depending on local building codes. Most building codes require that each step load be about one-fourth of the proposed working load. The load test should be carried out to at least a total load of two times the proposed working load. After the desired pile load is reached, the pile is gradually unloaded.

(a)

(b)

(c)

Figure 11.23 (a) Schematic diagram of pile load test arrangement (b) plot of load against total settlement; (c) plot of load against net settlement

Figure 11.23b shows a load–settlement diagram obtained from field loading and unloading. For any load Q, the net pile settlement can be calculated as follows:

When $Q = Q_1$,

$$\text{Net settlement, } s_{\text{net}(1)} = s_{t(1)} - s_{e(1)}$$

When $Q = Q_2$,

$$\text{Net settlement, } s_{\text{net}(2)} = s_{t(2)} - s_{e(2)}$$

where

s_{net} = net settlement
s_e = elastic settlement of the pile itself
s_t = total settlement

These values of Q can be plotted in a graph against the corresponding net settlement, s_{net}, as shown in Figure 11.23c. The ultimate load of the pile can then be determined from the graph.

Pile settlement may increase with load to a certain point, beyond which the load–settlement curve becomes vertical. The load corresponding to the point where the curve of Q versus s_{net} becomes vertical is the ultimate load, Q_u, for the pile; it is shown by curve 1 in Figure 11.23c. In many cases, the latter stage of the load–settlement curve is almost linear, showing a large degree of settlement for a small increment of load; this is shown by curve 2 in the figure. The ultimate load, Q_u, for such a case is determined from the point of the curve of Q versus s_{net} where this steep linear portion starts.

One of the methods to obtain the ultimate load Q_u from the load-settlement plot is that proposed by Davisson (1973). Davisson's method is used more often in the field and is described here. Referring to Figure 11.24, the ultimate load occurs at a settlement level (s_u) of

$$s_u(\text{mm}) = 0.012D_r + 0.1\left(\frac{D}{D_r}\right) + \frac{Q_uL}{A_pE_p} \tag{11.67}$$

where

Q_u is in kN

D is in mm

D_r = reference pile diameter or width ($= 300$mm)

L = pile length (mm)

A_p = area of pile cross section (mm^2)

E_p = Young's modulus of pile material (kN/mm^2)

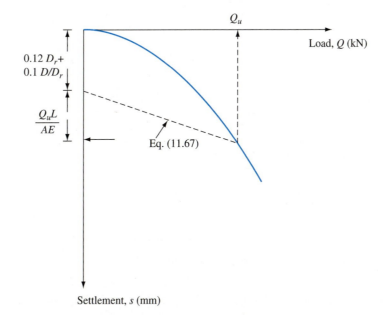

Figure 11.24 Davisson's method for determination of Q_u

The application of this procedure is shown in Example 11.9.

The load test procedure just described requires the application of step loads on the piles and the measurement of settlement and is called a *load-controlled* test. Another technique used for a pile load test is the *constant-rate-of-penetration* test, wherein the load on the pile is continuously increased to maintain a constant rate of penetration, which can vary from 0.25 to 2.5 mm/min (0.01 to 0.1 in./min). This test gives a load–settlement plot similar to that obtained from the load-controlled test. Another type of pile load test is *cyclic loading,* in which an incremental load is repeatedly applied and removed.

In order to conduct a load test on piles, it is important to take into account the time lapse after the end of driving (EOD). When piles are driven into soft clay, a certain zone surrounding the clay becomes remolded or compressed, as shown in Figure 11.25a. This results in a reduction of undrained shear strength, c_u (Figure 11.25b). With time, the loss of undrained shear strength is partially or fully regained. The time lapse may range from 30 to 60 days.

For piles driven in dilative (dense to very dense) saturated fine sands, relaxation is possible. Negative pore water pressure, if developed during pile driving, will dissipate over time, resulting in a reduction in pile capacity with time after the driving operation is completed. At the same time, excess pore water pressure may be generated in contractive fine

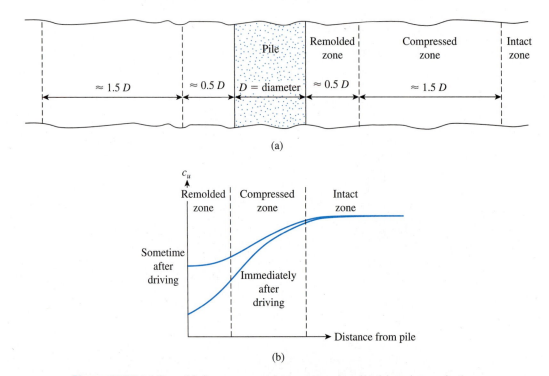

Figure 11.25 (a) Remolded or compacted zone around a pile driven into soft clay; (b) Nature of variation of undrained shear strength (c_u) with time around a pile driven into soft clay

sands during pile driving. The excess pore water pressure will dissipate over time, which will result in greater pile capacity.

Several empirical relationships have been developed to predict changes in pile capacity with time.

Skov and Denver (1988)

Skov and Denver proposed the equation

$$Q_t = Q_{\text{OED}}\left[A \log\left(\frac{t}{t_o}\right) + 1 \right]$$ (11.68)

where Q_t = pile capacity t days after the end of driving
  Q_{OED} = pile capacity at the end of driving
    t = time, in days

For sand, $A = 0.2$ and $t_o = 0.5$ days; for clay, $A = 0.6$ and $t_o = 1.0$ days.

Guang-Yu (1988)

According to Guang-Yu,

$$Q_{14} = (0.375S_t + 1)Q_{\text{OED}} \quad \text{(applicable to clay soil)}$$ (11.69)

where Q_{14} = pile capacity 14 days after pile driving
    S_t = sensitivity of clay

Svinkin (1996)

Svinkin suggests the relationship

$$Q_t = 1.4Q_{\text{OED}}t^{0.1} \quad \text{(upper limit for sand)}$$ (11.70)
$$Q_t = 1.025Q_{\text{OED}}t^{0.1} \quad \text{(lower limit for sand)}$$ (11.71)

where t = time after driving, in days

Example 11.9

Figure 11.26 shows the load test results of a 20-m long concrete pile (406 mm $\times$ 406 mm) embedded in sand. Using Davisson's method, determine the ultimate load Q_u. Given: $E_p = 30 \times 10^6 \text{ kN/m}^2$.

Solution
From Eq. (11.67),

$$s_u = 0.012D_r + 0.1\left(\frac{D}{D_r}\right) + \frac{Q_u L}{A_p E_p}$$

$D_r = 300$ mm, $D = 406$ mm, $L = 20$ m $= 20{,}000$ mm, $A_p = 406$ mm $\times$ 406 mm $= 164{,}836$ mm^2, and $E_p = 30 \times 10^6$ kN/m^2. Hence,

$$s_u = (0.012)(300) + (0.1)\left(\frac{406}{300}\right) + \frac{(Q_u)(20{,}000)}{(30)(164{,}836)}$$

$$= 3.6 + 0.135 + 0.004Q_u = 3.735 + 0.004Q_u$$

The line $s_u(\text{mm}) = 3.735 + 0.004Q_u$ is drawn in Figure 11.26. The intersection of this line with the load-settlement curve gives the failure load $Q_u = \mathbf{1640}$ **kN**.

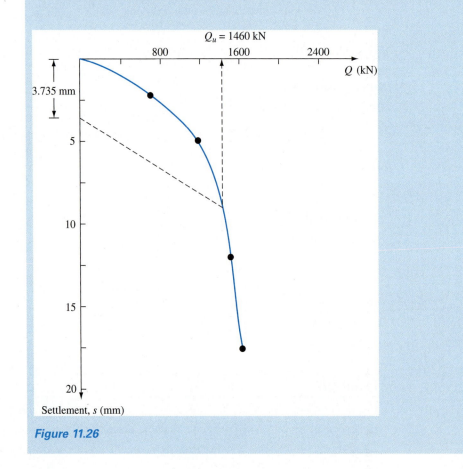

Figure 11.26

11.15 *Elastic Settlement of Piles*

The total settlement of a pile under a vertical working load Q_w is given by

$$s_e = s_{e(1)} + s_{e(2)} + s_{e(3)} \tag{11.72}$$

where

$s_{e(1)}$ = elastic settlement of pile
$s_{e(2)}$ = settlement of pile caused by the load at the pile tip
$s_{e(3)}$ = settlement of pile caused by the load transmitted along the pile shaft

If the pile material is assumed to be elastic, the deformation of the pile shaft can be evaluated, in accordance with the fundamental principles of mechanics of materials, as

$$s_{e(1)} = \frac{(Q_{wp} + \xi Q_{ws})L}{A_p E_p}$$

(11.73)

where

Q_{wp} = load carried at the pile point under working load condition
Q_{ws} = load carried by frictional (skin) resistance under working load condition
A_p = area of cross section of pile
L = length of pile
E_p = modulus of elasticity of the pile material

The magnitude of ξ varies between 0.5 and 0.67 and will depend on the nature of the distribution of the unit friction (skin) resistance f along the pile shaft.

The settlement of a pile caused by the load carried at the pile point may be expressed in the form:

$$s_{e(2)} = \frac{q_{wp}D}{E_s}(1 - \mu_s^2)I_{wp}$$

(11.74)

where

D = width or diameter of pile
q_{wp} = point load per unit area at the pile point = Q_{wp}/A_p
E_s = modulus of elasticity of soil at or below the pile point
μ_s = Poisson's ratio of soil
I_{wp} = influence factor ≈ 0.85

Vesic (1977) also proposed a semi-empirical method for obtaining the magnitude of the settlement of $s_{e(2)}$. His equation is

$$s_{e(2)} = \frac{Q_{wp}C_p}{Dq_p}$$

(11.75)

where

q_p = ultimate point resistance of the pile
C_p = an empirical coefficient

Representative values of C_p for various soils are given in Table 11.13.

Table 11.13 Typical Values of C_p [from Eq. (11.75)]

Type of soil	Driven pile	Bored pile
Sand (dense to loose)	0.02–0.04	0.09–0.18
Clay (stiff to soft)	0.02–0.03	0.03–0.06
Silt (dense to loose)	0.03–0.05	0.09–0.12

From "Design of Pile Foundations," by A. S. Vesic. SYNTHESIS OF HIGHWAY PRACTICE by AMERICAN ASSOCIATION OF STATE HIGHWAY AND TRANSPORT. Copyright 1969 by TRANSPORTATION RESEARCH BOARD. Reproduced with permission of TRANSPORTATION RESEARCH BOARD in the format Textbook via Copyright Clearance Center.

The settlement of a pile caused by the load carried by the pile shaft is given by a relation similar to Eq. (11.74), namely,

$$s_{e(3)} = \left(\frac{Q_{ws}}{pL}\right)\frac{D}{E_s}(1 - \mu_s^2)I_{ws} \tag{11.76}$$

where

p = perimeter of the pile
L = embedded length of pile
I_{ws} = influence factor

Note that the term Q_{ws}/pL in Eq. (11.76) is the average value of f along the pile shaft. The influence factor, I_{ws}, has a simple empirical relation (Vesic, 1977):

$$I_{ws} = 2 + 0.35\sqrt{\frac{L}{D}} \tag{11.77}$$

Vesic (1977) also proposed a simple empirical relation similar to Eq. (11.75) for obtaining $s_{e(3)}$:

$$s_{e(3)} = \frac{Q_{ws}C_s}{Lq_p} \tag{11.78}$$

In this equation, C_s = an empirical constant = $(0.93 + 0.16\sqrt{L/D})C_p$ (11.79)

The values of C_p for use in Eq. (11.75) may be estimated from Table 11.13.

Example 11.10

The allowable working load on a prestressed concrete pile 21-m long that has been driven into sand is 502 kN. The pile is octagonal in shape with D = 356 mm (see Table 11.3a). Skin resistance carries 350 kN of the allowable load, and point bearing carries the rest. Use $E_p = 21 \times 10^6$ kN/m², $E_s = 25 \times 10^3$ kN/m², μ_s 0.35, and ξ = 0.62. Determine the settlement of the pile.

Solution

From Eq. (11.73),

$$S_{e(1)} = \frac{(Q_{wp} + \xi Q_{ws})L}{A_p E_p}$$

From Table 11.3a for $D = 356$ mm, the area of pile cross section. $A_p = 1045$ cm^2, Also, perimeter $p = 1.168$ m. Given: $Q_{ws} = 350$ kN, so

$$Q_{wp} = 502 - 350 = 152 \text{ kN}$$

$$S_{e(1)} = \frac{[152 + 0.62(350)](21)}{(0.1045 \text{ m}^2)(21 \times 10^6)} = 0.00353 \text{ m} = 3.35 \text{ mm}$$

From Eq. (11.74),

$$S_{e(2)} = \frac{q_{wp}D}{E_s}(1 - \mu_s^2)I_{wp} = \left(\frac{152}{0.1045}\right)\left(\frac{0.356}{25 \times 10^3}\right)(1 - 0.35^2)(0.85)$$

$$= 0.0155 \text{ m} = 15.5 \text{ mm}$$

Again, from Eq. (11.76),

$$S_{e(3)} = \left(\frac{Q_{ws}}{pL}\right)\left(\frac{D}{E_s}\right)(1 - \mu_s^2)I_{ws}$$

$$I_{ws} = 2 + 0.35\sqrt{\frac{L}{D}} = 2 + 0.35\sqrt{\frac{21}{0.356}} = 4.69$$

$$S_{e(3)} = \left[\frac{350}{(1.168)(21)}\right]\left(\frac{0.356}{25 \times 10^3}\right)(1 - 0.35^2)(4.69)$$

$$= 0.00084 \text{ m} = 0.84 \text{ mm}$$

Hence, total settlement is

$$S_e = S_{e(1)} + S_{e(2)} + S_{e(3)} = 3.35 + 15.5 + 0.84 = \textbf{19.69 mm} \quad \blacksquare$$

11.16 *Laterally Loaded Piles*

A vertical pile resists a lateral load by mobilizing passive pressure in the soil surrounding it. (See Figure 11.1c.) The degree of distribution of the soil's reaction depends on (a) the stiffness of the pile, (b) the stiffness of the soil, and (c) the fixity of the ends of the pile. In general, laterally loaded piles can be divided into two major categories: (1) short or rigid piles and (2) long or elastic piles. Figures 11.27a and 11.27b show the nature of the variation of the pile deflection and the distribution of the moment and shear force along the pile length when the pile is subjected to lateral loading. We next summarize the current solutions for laterally loaded piles.

Elastic Solution

A general method for determining moments and displacements of a vertical pile embedded in a *granular soil* and subjected to lateral load and moment at the ground surface was given by Matlock and Reese (1960). Consider a pile of length L subjected to a lateral force Q_g and

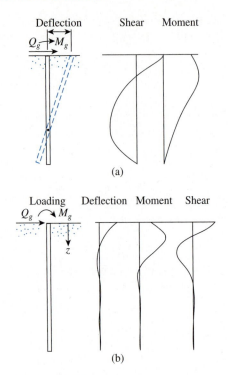

Figure 11.27 Nature of variation of pile deflection, moment, and shear force for (a) a rigid pile and (b) and elastic pile

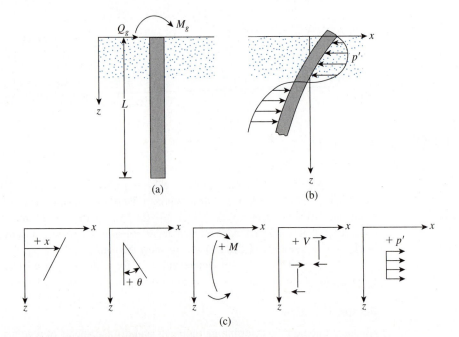

Figure 11.28 (a) Laterally loaded pile; (b) soil resistance on pile caused by lateral load; (c) sign conventions for displacement, slope, moment, shear, and soil reaction

a moment M_g at the ground surface $(z = 0)$, as shown in Figure 11.28a. Figure 11.28b shows the general deflected shape of the pile and the soil resistance caused by the applied load and the moment.

According to a simpler Winkler's model, an elastic medium (soil in this case) can be replaced by a series of infinitely close independent elastic springs. Based on this assumption,

$$k = \frac{p'(kN/m)}{x(m)} \tag{11.80}$$

where

  k = modulus of subgrade reaction
  p' = pressure on soil
  x = deflection

The subgrade modulus for *granular soils* at a depth z is defined as

$$k_z = n_h z \tag{11.81}$$

where n_h = constant of modulus of horizontal subgrade reaction.

Referring to Figure 11.28b and using the theory of beams on an elastic foundation, we can write

$$E_p I_p \frac{d^4x}{dz^4} = p' \tag{11.82}$$

where

  E_p = modulus of elasticity in the pile material
  I_p = moment of inertia of the pile section

Based on Winkler's model

$$p' = -kx \tag{11.83}$$

The sign in Eq. (11.83) is negative because the soil reaction is in the direction opposite that of the pile deflection.

Combining Eqs. (11.82) and (11.83) gives

$$E_p I_p \frac{d^4x}{dz^4} + kx = 0 \tag{11.84}$$

The solution of Eq. (11.84) results in the following expressions:

Pile Deflection at Any Depth $[x_z(z)]$

$$x_z(z) = A_x \frac{Q_g T^3}{E_p I_p} + B_x \frac{M_g T^2}{E_p I_p} \tag{11.85}$$

Slope of Pile at Any Depth $[\theta_z(z)]$

$$\theta_z(z) = A_\theta \frac{Q_g T^2}{E_p I_p} + B_\theta \frac{M_g T}{E_p I_p} \tag{11.86}$$

Moment of Pile at Any Depth [$M_z(z)$]

$$M_z(z) = A_m Q_g T + B_m M_g \tag{11.87}$$

Shear Force on Pile at Any Depth [$V_z(z)$]

$$V_z(z) = A_v Q_g + B_v \frac{M_g}{T} \tag{11.88}$$

Soil Reaction at Any Depth [$p'_z(z)$]

$$p'_z(z) = A_{p'} \frac{Q_g}{T} + B_{p'} \frac{M_g}{T^2} \tag{11.89}$$

where

A_x, B_x, A_θ, B_θ, A_m, B_m, A_v, B_v, $A_{p'}$, and $B_{p'}$ are coefficients
T = characteristic length of the soil–pile system

$$= \sqrt[5]{\frac{E_p I_p}{n_h}} \tag{11.90}$$

n_h has been defined in Eq. (11.81)

When $L \geq 5T$, the pile is considered to be a *long pile*. For $L \leq 2T$, the pile is considered to be a *rigid pile*. Table 11.14 gives the values of the coefficients for long piles ($L/T \geq 5$) in Eqs. (11.85) through (11.89). Note that, in the first column of the table,

$$Z = \frac{z}{T} \tag{11.91}$$

is the nondimensional depth.

The positive sign conventions for $x_z(z)$, $\theta_z(z)$, $M_z(z)$, $V_z(z)$, and $p'_z(z)$ assumed in the derivations in Table 11.14 are shown in Figure 11.28c. Figure 11.29 shows the variation of A_x, B_x, A_m, and B_m for various values of $L/T = Z_{\max}$. It indicates that, when L/T is greater than about 5, the coefficients do not change, which is true of long piles only.

Calculating the characteristic length T for the pile requires assuming a proper value of n_h. Table 11.15 gives some representative values.

Elastic solutions similar to those given in Eqs. 11.85 through 11.89 for piles embedded in *cohesive soil* were developed by Davisson and Gill (1963). Their equations are

$$x_z(z) = A'_x \frac{Q_g R^3}{E_p I_p} + B'_x \frac{M_g R^2}{E_p I_p} \tag{11.92}$$

Table 11.14 Coefficients for Long Piles, $k_z = n_h z$

z	A_x	A_θ	A_m	A_v	A_p'	B_x	B_θ	B_m	B_v	B_p'
0.0	2.435	−1.623	0.000	1.000	0.000	1.623	−1.750	1.000	0.000	0.000
0.1	2.273	−1.618	0.100	0.989	−0.227	1.453	−1.650	1.000	−0.007	−0.145
0.2	2.112	−1.603	0.198	0.956	−0.422	1.293	−1.550	0.999	−0.028	−0.259
0.3	1.952	−1.578	0.291	0.906	−0.586	1.143	−1.450	0.994	−0.058	−0.343
0.4	1.796	−1.545	0.379	0.840	−0.718	1.003	−1.351	0.987	−0.095	−0.401
0.5	1.644	−1.503	0.459	0.764	−0.822	0.873	−1.253	0.976	−0.137	−0.436
0.6	1.496	−1.454	0.532	0.677	−0.897	0.752	−1.156	0.960	−0.181	−0.451
0.7	1.353	−1.397	0.595	0.585	−0.947	0.642	−1.061	0.939	−0.226	−0.449
0.8	1.216	−1.335	0.649	0.489	−0.973	0.540	−0.968	0.914	−0.270	−0.432
0.9	1.086	−1.268	0.693	0.392	−0.977	0.448	−0.878	0.885	−0.312	−0.403
1.0	0.962	−1.197	0.727	0.295	−0.962	0.364	−0.792	0.852	−0.350	−0.364
1.2	0.738	−1.047	0.767	0.109	−0.885	0.223	−0.629	0.775	−0.414	−0.268
1.4	0.544	−0.893	0.772	−0.056	−0.761	0.112	−0.482	0.688	−0.456	−0.157
1.6	0.381	−0.741	0.746	−0.193	−0.609	0.029	−0.354	0.594	−0.477	−0.047
1.8	0.247	−0.596	0.696	−0.298	−0.445	−0.030	−0.245	0.498	−0.476	0.054
2.0	0.142	−0.464	0.628	−0.371	−0.283	−0.070	−0.155	0.404	−0.456	0.140
3.0	−0.075	−0.040	0.225	−0.349	0.226	−0.089	0.057	0.059	−0.213	0.268
4.0	−0.050	0.052	0.000	−0.106	0.201	−0.028	0.049	−0.042	0.017	0.112
5.0	−0.009	0.025	−0.033	0.015	0.046	0.000	−0.011	−0.026	0.029	−0.002

From *Drilled Pier Foundations*, by R. J. Woodward, W. S. Gardner, and D. M. Greer. Copyright 1972 McGraw-Hill. Used with permission of the McGraw-Hill Book Company.

Table 11.15 Representative Values of n_h

Soil	n_h kN/m³
Dry or moist sand	
Loose	1800–2200
Medium	5500–7000
Dense	15,000–18,000
Submerged sand	
Loose	1000–1400
Medium	3500–4500
Dense	9000–12,000

and

$$M_z(z) = A_m' Q_g R + B_m' M_g \tag{11.93}$$

where A_x', B_x, A_m', and B_m' are coefficients.
and

$$R = \sqrt[4]{\frac{E_p I_p}{k}} \tag{11.94}$$

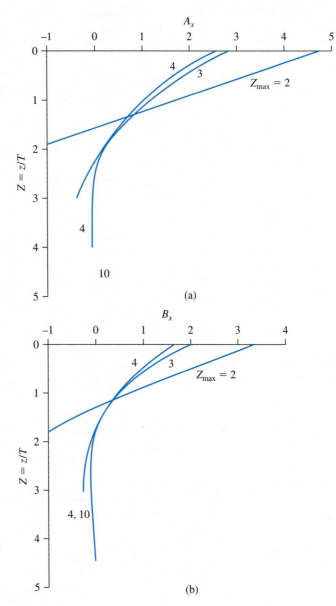

Figure 11.29 Variation of A_x, B_x, A_m, and B_m with Z (From Matlock, H. and Reese, L. C. (1960). "Generalized Solution for Laterally Loaded Piles," *Journal of the Soil Mechanics and Foundations Division*, American Society of Civil Engineers, Vol. 86, No. SM5, Part I, pp. 63–91. With permission from ASCE.)

The values of the coefficients A' and B' are given in Figure 11.30 Note that

$$Z = \frac{z}{R} \tag{11.95}$$

and

$$Z_{\max} = \frac{L}{R} \tag{11.96}$$

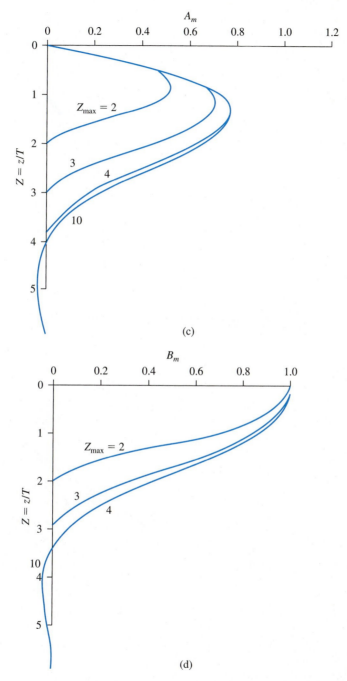

(c)

(d)

Figure 11.29 *(continued)*

The use of Eqs. (11.92) and (11.93) requires knowing the magnitude of the characteristic length, R. This can be calculated from Eq. (11.94), provided that the coefficient of the subgrade reaction is known. For sands, the coefficient of the subgrade reaction was given by Eq. (11.81), which showed a linear variation with depth. However, in cohesive

(a)

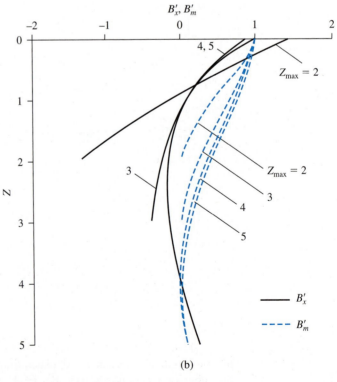

Figure 11.30 Variation of A_x, B'_x, A'_m, and B'_m with Z (From Davisson, M. T. and Gill, H. L. (1963). "Laterally Loaded Piles in a Layered Soil System," *Journal of the Soil Mechanics and Foundations Division*, American Society of Civil Engineers, Vol. 89, No. SM3, pp. 63–94. With permission from ASCE.)

(b)

soils, the subgrade reaction may be assumed to be approximately constant with depth. Vesic (1961) proposed the following equation to estimate the value of k:

$$k = 0.65 \sqrt[12]{\frac{E_s D^4}{E_p I_p}} \frac{E_s}{1 - \mu_s^2} \tag{11.97}$$

Here,

E_s = modulus of elasticity of soil
D = pile width (or diameter)
μ_s = Poisson's ratio for the soil

For all practical puroses, Eq. (11.97) can be written as

$$k \approx \frac{E_s}{1 - \mu_s^2} \tag{11.98}$$

Ultimate Load Analysis: Broms's Method

For laterally loaded piles, Broms (1965) developed a simplified solution based on the assumptions of (a) shear failure in soil, which is the case for short piles, and (b) bending of the pile, governed by the plastic yield resistance of the pile section, which is applicable to long piles. Broms's solution for calculating the ultimate load resistance, $Q_{u(g)}$, for *short piles* is given in Figure 11.31a. A similar solution for piles embedded in cohesive soil is shown in Figure 11.31b. In Figure 11.31a, note that

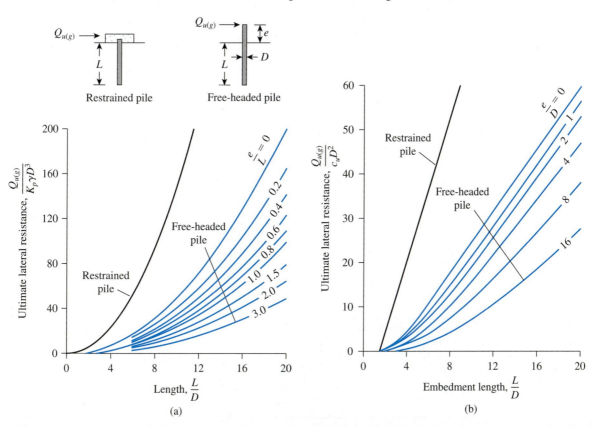

Figure 11.31 Broms's solution for ultimate lateral resistance of short piles (a) in sand and (b) in clay

$$K_p = \text{Rankine passive earth pressure coefficient} = \tan^2\left(45 + \frac{\phi'}{2}\right) \quad (11.99)$$

Similarly, in Figure 11.31b,

$$c_u = \text{undrained cohesion} \approx \frac{0.75q_u}{FS} = \frac{0.75q_u}{2} = 0.375q_u \quad (11.100)$$

where

FS = factor of safety (=2)
q_u = unconfined compression strength

Figure 11.32 shows Broms's analysis of long piles. In the figure, the yield moment for the pile is

$$M_y = SF_Y \quad (11.101)$$

where

S = section modulus of the pile section
F_Y = yield stress of the pile material

In solving a given problem, both cases (i.e., Figure 11.31 and Figure 11.32) should be checked.

The deflection of the pile head, $x_z(z = 0)$, under working load conditions can be estimated from Figure 11.33. In Figure 11.33a, the term η can be expressed as

$$\eta = \sqrt[5]{\frac{n_h}{E_p I_p}} \quad (11.102)$$

The range of n_h for granular soil is given in Table 11.15. Similarly, in Figure 11.33b, which is for clay, the term K is the horizontal soil modulus and can be defined as

$$K = \frac{\text{pressure } (kN/m^2)}{\text{displacement } (m)} \quad (11.103)$$

Also, the term β can be defined as

$$\beta = \sqrt[4]{\frac{KD}{4E_p I_p}} \quad (11.104)$$

Note that, in Figure 11.33, Q_g is the working load.

The following is a general range of values of K for clay soils.

Unconfined compression strength, q_u kN/m²	K kN/m³
200	10,000–20,000
200–800	20,000–40,000
> 800	> 40,000

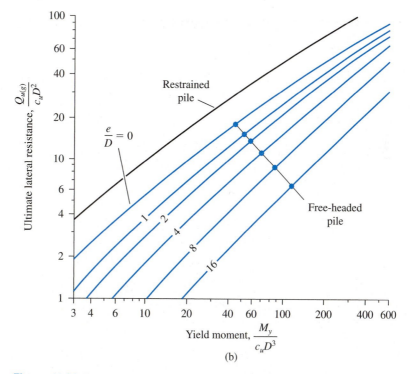

Figure 11.32 Broms's solution for ultimate lateral resistance of long piles (a) in sand (b) in clay

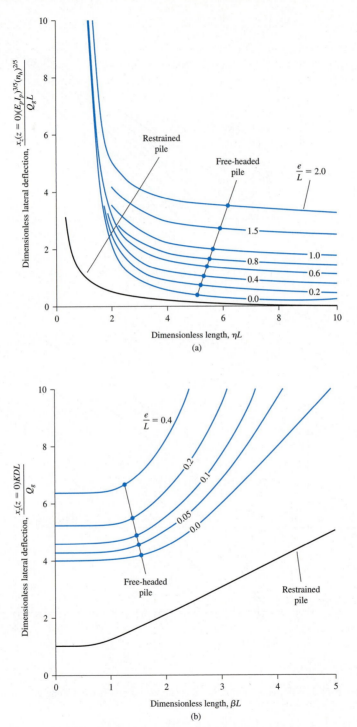

Figure 11.33 Broms's solution for estimating deflection of pile head (a) in sand and (b) in clay

Example 11.11

Consider a steel H-pile (HP 250 $\times$ 85) 25 m long, embedded fully in a granular soil. Assume that $n_h = 12{,}000$ kN/m^3. The allowable displacement at the top of the pile is 8 mm. Determine the allowable lateral load, Q_g. Let $M_g = 0$. Use the elastic solution.

Solution

From Table 11.1a, for an HP 250 $\times$ 85 pile,

$$I_p = 123 \times 10^{-6}\,\text{m}^4 \qquad (\text{about the strong axis})$$

and let

$$E_p = 207 \times 10^6\,\text{kN/m}^2$$

From Eq. (11.90),

$$T = \sqrt[5]{\frac{E_p I_p}{n_h}} = \sqrt[5]{\frac{(207 \times 10^6)(123 \times 10^{-6})}{12{,}000}} = 1.16\,\text{m}$$

Here, $L/T = 25/1.16 = 21.55 > 5$, so the pile is a long one. Because $M_g = 0$, Eq. (11.85) takes the form

$$x_z(z) = A_x \frac{Q_g T^3}{E_p I_p}$$

and it follows that

$$Q_g = \frac{x_z(z) E_p I_p}{A_x T^3}$$

At $z = 0$, $x_z = 8$ mm $= 0.008$ m and $A_x = 2.435$ (see Table 11.14), so

$$Q_g = \frac{(0.008)(207 \times 10^6)(123 \times 10^{-6})}{(2.435)(1.16^3)} = 53.59\,\text{kN}$$

This magnitude of Q_g is based on the *limiting displacement condition only*. However, the magnitude of Q_g based on the *moment capacity* of the pile also needs to be determined. For $M_g = 0$, Eq. (11.87) becomes

$$M_z(z) = A_m Q_g T$$

According to Table 11.14, the maximum value of A_m at any depth is 0.772. The maximum allowable moment that the pile can carry is

$$M_{z(\text{max})} = F_Y \frac{I_p}{\dfrac{d_1}{2}}$$

Let $F_Y = 248,000$ kN/m². From Table 11.1a, $I_p = 123 \times 10^{-6}$ m⁴ and $d_1 = 0.254$ m, so

$$\frac{I_p}{\left(\dfrac{d_1}{2}\right)} = \frac{123 \times 10^{-6}}{\left(\dfrac{0.254}{2}\right)} = 968.5 \times 10^{-6}\,\text{m}^3$$

Now,

$$Q_g = \frac{M_{z(\max)}}{A_m T} = \frac{(968.5 \times 10^{-6})(248,000)}{(0.772)(1.16)} = 268.2 \text{ kN}$$

Because $Q_g = 268.2$ kN > 53.59 kN, the deflection criteria apply. Hence, $Q_g = \mathbf{53.59}$ **kN**. ∎

Example 11.12

Solve Example 11.11 by Broms's method. Assume that the pile is flexible and is free headed. Let the yield stress of the pile material, $F_y = 248$ MN/m²; the unit weight of soil, $\gamma = 18$ kN/m³; and the soil friction angle $\phi' = 35°$.

Solution

We check for bending failure. From Eq. (11.101),

$$M_y = SF_y$$

From Table 11.1a,

$$S = \frac{I_p}{\dfrac{d_1}{2}} = \frac{123 \times 10^{-6}}{\dfrac{0.254}{2}}$$

Also,

$$M_y = \left[\frac{123 \times 10^{-6}}{\dfrac{0.254}{2}}\right](248 \times 10^3) = 240.2 \text{ kN-m}$$

and

$$\frac{M_y}{D^4 \gamma K_p} = \frac{M_y}{D^4 \gamma \tan^2\left(45 + \dfrac{\phi'}{2}\right)} = \frac{240.2}{(0.254)^4 (18) \tan^2\left(45 + \dfrac{35}{2}\right)} = 868.8$$

From Figure 11.32a, for $M_y/D^4 \gamma K_p = 868.8$, the magnitude of $Q_{u(g)}/K_p D^3 \gamma$ (for a free-headed pile with $e/D = 0$) is about 140, so

$$Q_{u(g)} = 140 K_p D^3 \gamma = 140 \tan^2\left(45 + \frac{35}{2}\right)(0.254)^3 (18) = 152.4 \text{ kN}$$

Next, we check for pile head deflection. From Eq. (11.102),

$$\eta = \sqrt[5]{\frac{n_h}{E_p I_p}} = \sqrt[5]{\frac{12{,}000}{(207 \times 10^6)(123 \times 10^{-6})}} = 0.86 \text{ m}^{-1}$$

so

$$\eta L = (0.86)(25) = 21.5$$

From Figure 11.33a, for $\eta L = 21.5$, $e/L = 0$ (free-headed pile): thus,

$$\frac{x_o (E_p I_p)^{3/5} (n_h)^{2/5}}{Q_g L} \approx 0.15 \qquad \text{(by interpolation)}$$

and

$$
\begin{aligned}
Q_g &= \frac{x_o (E_p I_p)^{3/5} (n_h)^{2/5}}{0.15 L} \\[2mm]
&= \frac{(0.008)[(207 \times 10^6)(123 \times 10^{-6})]^{3/5}(12{,}000)^{2/5}}{(0.15)(25)} = 40.2 \text{ kN}
\end{aligned}
$$

Hence, $Q_g =$ **40.2 kN (<152.4 kN)**. ∎

Example 11.13

Assume that the 25-m long pile described in Example 11.11 is a restrained pile and is embedded in clay soil. Given: $c_u = 100 \text{ kN/m}^2$ and $K = 5{,}000 \text{ kN/m}^3$. The allowable lateral displacement at the top of the pile is 10 mm. Determine the allowable lateral load Q_g. Given $M_y \mu_g = 0$. Use Broms's method.

Solution
From Example 11.12, $M_y = 240.2$ kN-m. So

$$\frac{M_y}{c_u D^3} = \frac{240.2}{(100)(0.254)^3} = 146.6$$

For the unrestrained pile, from Figure 11.32b,

$$\frac{Q_{u(g)}}{c_u D^2} \approx 65$$

or

$$Q_{u(g)} = (65)(100)(0.254)^2 = 419.3 \text{ kN}$$

Check Pile-Head Deflection

From Eq. (11.104),

$$\beta = \sqrt[4]{\frac{KD}{4E_pI_p}} = \sqrt[4]{\frac{(5000)(0.254)}{(4)(207 \times 10^6)(123 \times 10^{-6})}} = 0.334$$

$$\beta L = (0.334)(25) = 8.35$$

From Figure 11.33b for $\beta L = 8.35$, by extrapolation the magnitude of

$$\frac{x_z(z=0)KDL}{Q_g} \approx 8$$

$$Q_g = \frac{x_z(z=0)KDL}{8} = \frac{\left(\frac{10}{1000}\right)(5000)(0.254)(25)}{8} = 39.7 \text{ kN}$$

Hence, $Q_g = \mathbf{39.7\ kN\,(< 419.3\ kN)}$. ∎

11.17 *Pile-Driving Formulas*

To develop the desired load-carrying capacity, a point bearing pile must penetrate the dense soil layer sufficiently or have sufficient contact with a layer of rock. This requirement cannot always be satisfied by driving a pile to a predetermined depth, because soil profiles vary. For that reason, several equations have been developed to calculate the ultimate capacity of a pile during driving. These dynamic equations are widely used in the field to determine whether a pile has reached a satisfactory bearing value at the predetermined depth. One of the earliest such equations—commonly referred to as the *Engineering News (EN) Record formula*—is derived from the work—energy theory. That is,

Energy imparted by the hammer per blow =
                    (pile resistance)(penetration per hammer blow)

According to the EN formula, the pile resistance is the ultimate load Q_u, expressed as

$$Q_u = \frac{W_R h}{S + C} \tag{11.105}$$

where

W_R = weight of the ram
h = height of fall of the ram
S = penetration of pile per hammer blow
C = a constant

The pile penetration, S, is usually based on the average value obtained from the last few driving blows. In the equation's original form, the following values of C were recommended:

For drop hammers,

$$C = 25.4 \text{ mm if } S \text{ and } h \text{ are in mm}$$

For steam hammers,

$$C = 2.54 \text{ mm if } S \text{ and } h \text{ are in mm}$$

Also, a factor of safety FS = 6 was recommended for estimating the allowable pile capacity. Note that, for single- and double-acting hammers, the term $W_R h$ can be replaced by EH_E, where E is the efficiency of the hammer and H_E is the rated energy of the hammer. Thus,

$$Q_u = \frac{EH_E}{S + C} \tag{11.106}$$

The EN formula has been revised several times over the years, and other pile-driving formulas also have been suggested. Three of the other relationships generally used are tabulated in Table 11.16.

The maximum stress developed on a pile during the driving operation can be estimated from the pile-driving formulas presented in Table 11.16. To illustrate, we use the modified EN formula:

$$Q_u = \frac{EW_R h}{S + C} \frac{W_R + n^2 W_p}{W_R + W_p}$$

In this equation, S is the average penetration per hammer blow, which can also be expressed as

$$S = \frac{25.4}{N} \tag{11.107}$$

where

S is in mm
N = number of hammer blows per 25.4 mm of penetration

Thus,

$$Q_u = \frac{EW_R h}{(25.4/N) + 2.54} \frac{W_R + n^2 W_p}{W_R + W_p} \tag{11.108}$$

Different values of N may be assumed for a given hammer and pile, and Q_u may be calculated. The driving stress Q_u/A_p can then be calculated for each value of N. This

Table 11.16 Pile-Driving Formulas

Name	Formula

Modified EN formula

$$Q_u = \frac{EW_R h}{S + C} \frac{W_R + n^2 W_p}{W_R + W_p}$$

where E = efficiency of hammer
C = 2.54 mm if the units of S and h are in mm
W_p = weight of the pile
n = coefficient of restitution between the ram and the pile cap

Typical values for E

Single- and double-acting hammers	0.7–0.85
Diesel hammers	0.8–0.9
Drop hammers	0.7–0.9

Typical values for n

Cast-iron hammer and concrete piles (without cap)	0.4–0.5
Wood cushion on steel piles	0.3–0.4
Wooden piles	0.25–0.3

Danish formula (Olson and Flaate, 1967)

$$Q_u = \frac{EH_E}{S + \sqrt{\dfrac{EH_E L}{2A_p E_p}}}$$

where E = efficiency of hammer
H_E = rated hammer energy
E_p = modulus of elasticity of the pile material
L = length of the pile
A_p = cross-sectional area of the pile

Janbu's formula (Janbu, 1953)

$$Q_u = \frac{EH_E}{K'_u S}$$

where $K'_u = C_d \left(1 + \sqrt{1 + \dfrac{\lambda'}{C_d}} \right)$

$$C_d = 0.75 + 0.14 \left(\frac{W_p}{W_R} \right)$$

$$\lambda' = \left(\frac{EH_E L}{A_p E_p S^2} \right)$$

procedure can be demonstrated with a set of numerical values. Suppose that a prestressed concrete pile 24.4 m in length has to be driven by a hammer. The pile sides measure 254 mm. From Table 11.3a, for this pile,

$$A_p = 645 \times 10^{-4}\,\mathrm{m}^2$$

The weight of the pile is

$$A_p L \gamma_c = (645 \times 10^{-4})(24.4\ \mathrm{m})(23.58\ \mathrm{kN/m}^3) = 37.1\ \mathrm{kN}$$

If the weight of the cap is 2.98 kN, then

$$W_p = 37.1 + 2.98 = 40.08\ \mathrm{kN}$$

For the hammer, let

$$\text{Rated energy} = 26.03\ \mathrm{kN\text{-}m} = H_E = W_R h$$
$$\text{Weight of ram} = 22.24\ \mathrm{kN}$$

Assume that the hammer efficiency is 0.85 and that $n = 0.35$. Substituting these values into Eq. (11.108) yields

$$Q_u = \left[\frac{(0.85)(26.03 \times 1000)}{\dfrac{25.4}{N} + 2.54}\right]\left[\frac{22.24 + (0.35)^2(40.08)}{22.24 + 40.08}\right] = \frac{9639.08}{\dfrac{25.4}{N} + 2.54}\ \mathrm{kip}$$

Now the following table can be prepared:

N	Q_u (kN)	A_p (m²)	Q_u/A_p (MN/m²)
0	0	645×10^{-4}	0
2	632	645×10^{-4}	9.79
4	1084	645×10^{-4}	16.8
6	1423	645×10^{-4}	22.06
8	1687	645×10^{-4}	26.16
10	1898	645×10^{-4}	29.43
12	2070	645×10^{-4}	32.12
20	2530	645×10^{-4}	39.22

Both the number of hammer blows per inch and the stress can be plotted in a graph, as shown in Figure 11.34. If such a curve is prepared, the number of blows per inch of pile penetration corresponding to the allowable pile-driving stress can easily be determined.

Actual driving stresses in wooden piles are limited to about $0.7f_u$. Similarly, for concrete and steel piles, driving stresses are limited to about $0.6f'_c$ and $0.85f_y$, respectively.

In most cases, wooden piles are driven with a hammer energy of less than 60 kN-m. Driving resistances are limited mostly to 4 to 5 blows per inch of pile penetration. For concrete and steel piles, the usual values of N are 6 to 8 and 12 to 14, respectively.

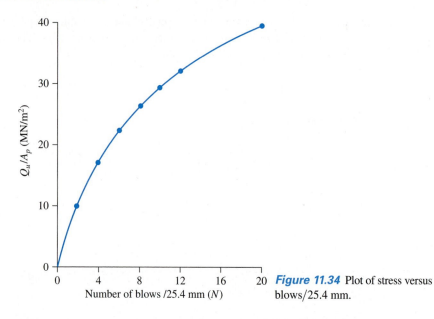

Figure 11.34 Plot of stress versus blows/25.4 mm.

Example 11.14

A precast concrete pile 0.305 m × 0.305 m in cross section is driven by a hammer. Given

Maximum rated hammer energy = 40.67 kN-m
Hammer efficiency = 0.8
Weight of ram = 33.36 kN
Pile length = 24.39 m
Coefficient of restitution = 0.4
Weight of pile cap = 2.45 kN
$E_p = 20.7 \times 10^6$ kN/m²
Number of blows for last 25.4 mm of penetration = 8

Estimate the allowable pile capacity by the

a. Modified EN formula (use FS = 6)
b. Danish formula (use FS = 4)

Solution
Part a

$$Q_u = \frac{E W_R h}{S + C} \frac{W_R + n^2 W_p}{W_R + W_p}$$

Weight of pile + cap = $(0.305 \times 0.305 \times 24.39)(23.58 \text{ kN/m}^3) + 2.45$

$= 55.95$ kN

Given: $W_R h = 40.67$ kN-m

$$Q_u = \frac{(0.8)(40.67 \times 1000)}{\frac{25.4}{8} + 2.54} \times \frac{33.36 + (0.4)^2(55.95)}{33.36 + 55.95} = 2697 \text{ kN}$$

$$Q_{all} = \frac{Q_u}{FS} = \frac{2697}{6} \approx \textbf{449.5 kN}$$

Part b

$$Q_u = \frac{EH_E}{S + \sqrt{\dfrac{EH_E L}{2A_p E_p}}}$$

Use $E_p = 20.7 \times 10^6 \text{ kN/m}^2$

$$\sqrt{\frac{EH_E L}{2A_p E_p}} = \sqrt{\frac{(0.8)(40.67)(24.39)}{2(0.305 \times 0.305)(20.7 \times 10^6 \text{ kN/m}^2)}} = 0.01435 \text{ m} = 14.35 \text{ mm}$$

$$Q_u = \frac{(0.8)(40.67)}{\frac{25.4}{8 \times 1000} + 0.01435} \approx 1857 \text{ kN}$$

$$Q_{all} = \frac{1857}{4} \approx \textbf{464 kN} \qquad \blacksquare$$

11.18 *Pile Capacity For Vibration-Driven Piles*

The principles of vibratory pile drivers (Figure 11.7e) were discussed briefly in Section 11.4. As mentioned there, the driver essentially consists of two counterrotating weights. The amplitude of the centrifugal driving force generated by a vibratory hammer can be given as

$$F_c = me\omega^2 \tag{11.109}$$

where

m = total eccentric rotating mass
e = distance between the center of each rotating mass and the center of rotation
ω = operating circular frequency

Vibratory hammers typically include an isolated *bias weight* that can range from 4 to 40 kN. The bias weight is isolated from oscillation by springs, so it acts as a net downward load helping the driving efficiency by increasing the penetration rate of the pile.

The use of vibratory pile drivers began in the early 1930s. Installing piles with vibratory drivers produces less noise and damage to the pile, compared with impact driving. However, because of a limited understanding of the relationships between the load, the rate of penetration, and the bearing capacity of piles, this method has not gained popularity in the United States.

Vibratory pile drivers are patented. Some examples are the Bodine Resonant Driver (BRD), the Vibro Driver of the McKiernan-Terry Corporation, and the Vibro Driver of the L. B. Foster Company. Davisson (1970) provided a relationship for estimating the ultimate pile capacity in granular soil:

In SI units,

$$Q_u(\text{kN}) = \frac{0.746(H_p) + 98(v_p \, \text{m/s})}{(v_p \, \text{m/s}) + (S_L \, \text{m/cycle})(f \, \text{Hz})} \qquad (11.110)$$

where

H_p = horsepower delivered to the pile
v_p = final rate of pile penetration
S_L = loss factor
f = frequency, in Hz

The loss factor S_L for various types of granular soils is as follows (Bowles, 1996):

Closed-End Pipe Piles

- Loose sand: 0.244×10^{-3} m/cycle
- Medium dense sand: 0.762×10^{-3} m/cycle
- Dense sand: 2.438×10^{-3} m/cycle

H-Piles

- Loose sand: -0.213×10^{-3} m/cycle
- Medium dense sand: 0.762×10^{-3} m/cycle
- Dense sand: 2.134×10^{-3} m/cycle

In 2000, Feng and Deschamps provided the following relationship for the ultimate capacity of vibrodriven piles in granular soil:

$$Q_u = \frac{3.6(F_c + 11W_B)}{1 + 1.8 \times 10^{10}\dfrac{v_p}{c}\sqrt{\text{OCR}}} \frac{L_E}{L} \qquad (11.111)$$

Here,

F_c = centrifugal force
W_B = bias weight
v_p = final rate of pile penetration
c = speed of light $[1.8 \times 10^{10}$ m/min$]$
OCR = overconsolidation ratio
L_E = embedded length of pile
L = pile length

Example 11.15

Consider a 20-m-long steel pile driven by a Bodine Resonant Driver (Section HP 310 × 125) in a medium dense sand. If $H_p = 350$ horsepower, $v_p = 0.0016$ m/s, and $f = 115$ Hz, calculate the ultimate pile capacity, Q_u.

Solution
From Eq. (11.110),

$$Q_u = \frac{0.746 H_p + 98 v_p}{v_p + S_L f}$$

For an HP pile in medium dense sand, $S_L \approx 0.762 \times 10^{-3}$ m/cycle. So

$$Q_u = \frac{(0.746)(350) + (98)(0.0016)}{0.0016 + (0.762 \times 10^{-3})(115)} = \textbf{2928 kN} \quad \blacksquare$$

11.19 *Negative Skin Friction*

Negative skin friction is a downward drag force exerted on a pile by the soil surrounding it. Such a force can exist under the following conditions, among others:

1. If a fill of clay soil is placed over a granular soil layer into which a pile is driven, the fill will gradually consolidate. The consolidation process will exert a downward drag force on the pile (see Figure 11.35a) during the period of consolidation.
2. If a fill of granular soil is placed over a layer of soft clay, as shown in Figure 11.35b, it will induce the process of consolidation in the clay layer and thus exert a downward drag on the pile.
3. Lowering of the water table will increase the vertical effective stress on the soil at any depth, which will induce consolidation settlement in clay. If a pile is located in the clay layer, it will be subjected to a downward drag force.

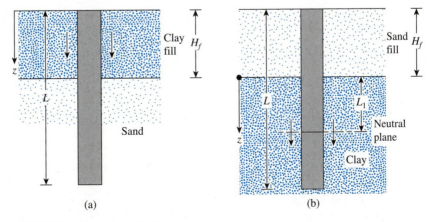

(a)   (b)

Figure 11.35 Negative skin friction

In some cases, the downward drag force may be excessive and cause foundation failure. This section outlines two tentative methods for the calculation of negative skin friction.

Clay Fill over Granular Soil (Figure 11.35a)

Similar to the β method presented in Section 11.12, the negative (downward) skin stress on the pile is

$$f_n = K'\sigma_o' \tan \delta' \tag{11.112}$$

where

K' = earth pressure coefficient = $K_o = 1 - \sin \phi'$
σ_o' = vertical effective stress at any depth $z = \gamma_f' z$
γ_f' = effective unit weight of fill
δ' = soil–pile friction angle ≈ 0.5–$0.7\phi'$

Hence, the total downward drag force on a pile is

$$Q_n = \int_0^{H_f} (pK'\gamma_f' \tan \delta') z \, dz = \frac{pK'\gamma_f' H_f^2 \tan \delta'}{2} \tag{11.113}$$

where H_f = height of the fill. If the fill is above the water table, the effective unit weight, γ_f', should be replaced by the moist unit weight.

Granular Soil Fill over Clay (Figure 11.35b)

In this case, the evidence indicates that the negative skin stress on the pile may exist from $z = 0$ to $z = L_1$, which is referred to as the *neutral depth*. (See Vesic, 1977, pp. 25–26.) The neutral depth may be given as (Bowles, 1982)

$$L_1 = \frac{(L - H_f)}{L_1} \left[\frac{L - H_f}{2} + \frac{\gamma_f' H_f}{\gamma'} \right] - \frac{2\gamma_f' H_f}{\gamma'} \tag{11.114}$$

where γ_f' and γ' = effective unit weights of the fill and the underlying clay layer, respectively.

For end-bearing piles, the neutral depth may be assumed to be located at the pile tip (i.e., $L_1 = L - H_f$).

Once the value of L_1 is determined, the downward drag force is obtained in the following manner: The unit negative skin friction at any depth from $z = 0$ to $z = L_1$ is

$$f_n = K'\sigma_o' \tan \delta' \tag{11.115}$$

where

$K' = K_o = 1 - \sin \phi'$
$\sigma_o' = \gamma_f' H_f + \gamma' z$
$\delta' = 0.5$–$0.7\phi'$

$$Q_n = \int_0^{L_1} pf_n \, dz = \int_0^{L_1} pK'(\gamma_f'H_f + \gamma'z)\tan\delta' \, dz$$

$$= (pK'\gamma_f'H_f \tan\delta')L_1 + \frac{L_1^2 pK'\gamma' \tan\delta'}{2} \qquad (11.116)$$

If the soil and the fill are above the water table, the effective unit weights should be replaced by moist unit weights. In some cases, the piles can be coated with bitumen in the downdrag zone to avoid this problem.

A limited number of case studies of negative skin friction is available in the literature. Bjerrum et al. (1969) reported monitoring the downdrag force on a test pile at Sorenga in the harbor of Oslo, Norway (noted as pile G in the original paper). The study of Bjerrum et al. (1969) was also discussed by Wong and Teh (1995) in terms of the pile being driven to bedrock at 40 m. Figure 11.36a shows the soil profile and the pile. Wong and Teh estimated the following quantities:

- *Fill:* Moist unit weight, $\gamma_f = 16 \text{ kN/m}^3$
  Saturated unit weight, $\gamma_{\text{sat}(f)} = 18.5 \text{ kN/m}^3$
  So
  $$\gamma_f' = 18.5 - 9.81 = 8.69 \text{ kN/m}^3$$
  and
  $$H_f = 13 \text{ m}$$

Figure 11.36 Negative skin friction on a pile in the harbor of Oslo, Norway (Based on Bjerrum et al., (1969) and Wong and The (1995))

- *Clay:* $K' \tan \delta' \approx 0.22$
    Saturated effective unit weight, $\gamma' = 19 - 9.81 = 9.19 \text{ kN/m}^3$
- *Pile:* $L = 40$ m
    Diameter, $D = 500$ m

Thus, the maximum downdrag force on the pile can be estimated from Eq. (11.116). Since in this case the pile is a point bearing pile, the magnitude of $L_1 = 27$ m, and

$$Q_n = (p)(K' \tan \delta')[\gamma_f \times 2 + (13 - 2)\gamma_f'](L_1) + \frac{L_1^2 p\gamma'(K' \tan \delta')}{2}$$

or

$$Q_n = (\pi \times 0.5)(0.22)[(16 \times 2) + (8.69 \times 11)](27) + \frac{(27)^2(\pi \times 0.5)(9.19)(0.22)}{2}$$

$$= 2348 \text{ kN}$$

The measured value of the maximum Q_n was about 2500 kN (Figure 11.36b), which is in good agreement with the calculated value.

Example 11.16

In Figure 11.35a, let $H_f = 2$m. The pile is circular in cross section with a diameter of 0.305 m. For the fill that is above the water table, $\gamma_f = 16 \text{ kN/m}^3$ and $\phi' = 32°$. Determine the total drag force. Use $\delta' = 0.6\phi'$.

Solution
From Eq. (11.113),

$$Q_n = \frac{pK'\gamma_f H_f^2 \tan \delta'}{2}$$

with

$$p = \pi(0.305) = 0.958\text{m}$$

$$K' = 1 - \sin \phi' = 1 - \sin 32 = 0.47$$

and

$$\delta' = (0.6)(32) = 19.2°$$

Thus,

$$Q_n = \frac{(0.958)(0.47)(16)(2)^2 \tan 19.2}{2} = \textbf{5.02 kN} \qquad \blacksquare$$

Example 11.17

In Figure 11.35b, let $H_f = 2$ m, pile diameter $= 0.305$ m, $\gamma_f = 16.5 \text{ kN/m}^3$, $\phi'_{\text{clay}} = 34°$, $\gamma_{\text{sat(clay)}} = 17.2 \text{ kN/m}^3$, and $L = 20$ m. The water table coincides with the top of the clay layer. Determine the downward drag force. Assume that $\delta' = 0.6\phi'_{\text{clay}}$.

Solution

The depth of the neutral plane is given in Eq. (11.114) as

$$L_1 = \frac{L - H_f}{L_1}\left(\frac{L - H_f}{2} + \frac{\gamma_f H_f}{\gamma'}\right) - \frac{2\gamma_f H_f}{\gamma'}$$

Note that γ'_f in Eq. (11.114) has been replaced by γ_f because the fill is above the water table, so

$$L_1 = \frac{(20 - 2)}{L_1}\left[\frac{(20 - 2)}{2} + \frac{(16.5)(2)}{17.2 - 9.81)}\right] - \frac{(2)(16.5)(2)}{(17.2 - 9.81)}$$

or

$$L_1 = \frac{242.4}{L_1} - 8.93; \; L_1 = 11.75 \text{ m}$$

Now, from Eq. (11.116), we have

$$Q_n = (pK'\gamma_f H_f \tan \delta')L_1 + \frac{L_1^2 pK'\gamma' \tan \delta'}{2}$$

with

$$p = \pi(0.305) = 0.958\text{m}$$

and

$$K' = 1 - \sin 34° = 0.44$$

Hence,

$$Q_n = (0.958)(0.44)(16.5)(2)[\tan(0.6 \times 34)](11.75)$$

$$+ \frac{(11.75)^2(0.958)(0.44)(17.2 - 9.81)[\tan(0.6 \times 34)]}{2}$$

$$= 60.78 + 79.97 = \textbf{140.75 kN}$$

Group Piles

11.20 Group Efficiency

In most cases, piles are used in groups, as shown in Figure 11.37, to transmit the structural load to the soil. A *pile cap* is constructed over *group piles*. The cap can be in contact with the ground, as in most cases (see Figure 11.37a), or well above the ground, as in the case of offshore platforms (see Figure 11.37b).

Determining the load-bearing capacity of group piles is extremely complicated and has not yet been fully resolved. When the piles are placed close to each other, a

Section

Pile cap

L

$\leftarrow d \rightarrow|\leftarrow d \rightarrow|$

Plan $\quad|\leftarrow d \rightarrow|\leftarrow d \rightarrow|$

B_g

L_g

d

d

(a)

(b)

Water table

L

Number of piles in group = $n_1 \times n_2$
(*Note:* $L_g \geqslant B_g$)

$L_g = (n_1 - 1)d + 2(D/2)$
$B_g = (n_2 - 1)d + 2(D/2)$

(c)

Figure 11.37 Group piles

reasonable assumption is that the stresses transmitted by the piles to the soil will overlap (see Figure 11.37c), reducing the load-bearing capacity of the piles. Ideally, the piles in a group should be spaced so that the load-bearing capacity of the group is not less than the sum of the bearing capacity of the individual piles. In practice, the minimum center-to-center pile spacing, d, is $2.5D$ and, in ordinary situations, is actually about 3 to $3.5D$.

The efficiency of the load-bearing capacity of a group pile may be defined as

$$\eta = \frac{Q_{g(u)}}{\Sigma\, Q_u} \tag{11.117}$$

where

η = group efficiency
$Q_{g(u)}$ = ultimate load-bearing capacity of the group pile
Q_u = ultimate load-bearing capacity of each pile without the group effect

Many structural engineers use a simplified analysis to obtain the group efficiency for *friction piles,* particularly in sand. This type of analysis can be explained with the aid of Figure 11.37a. Depending on their spacing within the group, the piles may act in one of two ways: (1) as a *block,* with dimensions $L_g \times B_g \times L$, or (2) as *individual piles.* If the piles act as a block, the frictional capacity is $f_{av}\,p_g L \approx Q_{g(u)}$. [*Note:* p_g = perimeter of the cross section of block = $2(n_1 + n_2 - 2)d + 4D$, and f_{av} = average unit frictional resistance.] Similarly, for each pile acting individually, $Q_u \approx pLf_{av}$. (*Note:* p = perimeter of the cross section of each pile.) Thus,

$$\eta = \frac{Q_{g(u)}}{\Sigma\, Q_u} = \frac{f_{av}[2(n_1 + n_2 - 2)d + 4D]L}{n_1 n_2 pLf_{av}}$$
$$= \frac{2(n_1 + n_2 - 2)d + 4D}{pn_1 n_2} \tag{11.118}$$

Hence,

$$Q_{g(u)} = \left[\frac{2(n_1 + n_2 - 2)d + 4D}{pn_1 n_2} \right] \Sigma\, Q_u \tag{11.119}$$

From Eq. (11.119), if the center-to-center spacing d is large enough, $\eta > 1$. In that case, the piles will behave as individual piles. Thus, in practice, if $\eta < 1$, then

$$Q_{g(u)} = \eta \Sigma\, Q_u$$

and if $\eta \geq 1$, then

$$Q_{g(u)} = \Sigma\, Q_u$$

There are several other equations like Eq. (11.119) for calculating the group efficiency of friction piles. Some of these are given in Table 11.17.

It is important, however, to recognize that relationships such as Eq. (11.119) are simplistic and should not be used. In fact, in a group pile, the magnitude of f_{av} depends on the location of the pile in the group (ex., Figure 11.38).

Table 11.17 Equations for Group Efficiency of Friction Piles

Name	Equation
Converse–Labarre equation	$\eta = 1 - \left[\dfrac{(n_1 - 1)n_2 + (n_2 - 1)n_1}{90n_1n_2}\right]\theta$ where $\theta(\deg) = \tan^{-1}(D/d)$
Los Angeles Group Action equation	$\eta = 1 - \dfrac{D}{\pi dn_1n_2}[n_1(n_2 - 1)$ $+ n_2(n_1 - 1) + \sqrt{2}(n_1 - 1)(n_2 - 1)]$
Seiler–Keeney equation (Seiler and Keeney, 1944)	$\eta = \left\{1 - \left[\dfrac{11d}{7(d^2 - 1)}\right]\left[\dfrac{n_1 + n_2 - 2}{n_1 + n_2 - 1}\right]\right\} + \dfrac{0.3}{n_1 + n_2}$ where d is in ft

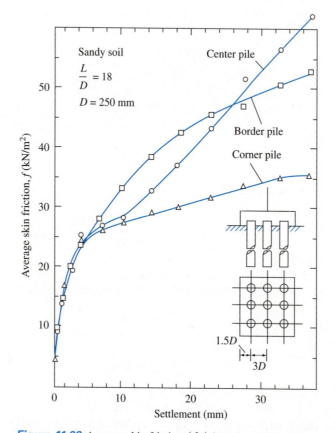

Figure 11.38 Average skin friction (f_{av}) based on pile location (Based on Liu et al., 1985)

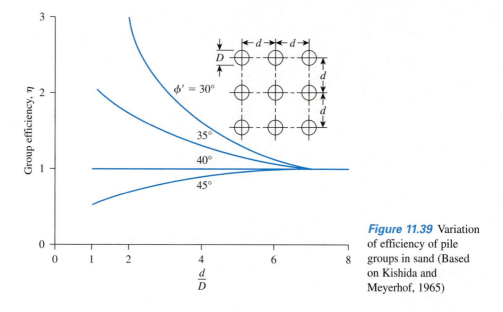

Figure 11.39 Variation of efficiency of pile groups in sand (Based on Kishida and Meyerhof, 1965)

Figure 11.39 shows the variation of the group efficiency η for a 3×3 group pile in sand (Kishida and Meyerhof, 1965). It can be seen that, for loose and medium sands, the magnitude of the group efficiency can be larger than unity. This is due primarily to the densification of sand surrounding the pile.

11.21 *Ultimate Capacity of Group Piles in Saturated Clay*

Figure 11.40 shows a group pile in saturated clay. Using the figure, one can estimate the ultimate load-bearing capacity of group piles in the following manner:

Step 1.　Determine $\Sigma Q_u = n_1 n_2 (Q_p + Q_s)$. From Eq. (11.18),

$$Q_p = A_p [9c_{u(p)}]$$

where $c_{u(p)}$ = undrained cohesion of the clay at the pile tip.
Also, from Eq. (11.55),

$$Q_s = \Sigma \alpha p c_u \Delta L$$

So,

$$\Sigma Q_u = n_1 n_2 [9 A_p c_{u(p)} + \Sigma \alpha p c_u \Delta L] \qquad (11.120)$$

Step 2.　Determine the ultimate capacity by assuming that the piles in the group act as a block with dimensions $L_g \times B_g \times L$. The skin resistance of the block is

$$\Sigma p_g c_u \Delta L = \Sigma 2(L_g + B_g) c_u \Delta L$$

Calculate the point bearing capacity:

$$A_p q_p = A_p c_{u(p)} N_c^* = (L_g B_g) c_{u(p)} N_c^*$$

$$\Sigma 2(L_g + B_g)c_u \Delta L$$

Figure 11.40 Ultimate capacity of group piles in clay

Obtain the value of the bearing capacity factor N_c^* from Figure 11.41. Thus, the ultimate load is

$$\Sigma Q_u = L_g B_g c_{u(p)} N_c^* + \Sigma 2(L_g + B_g)c_u \Delta L \qquad (11.121)$$

Step 3. Compare the values obtained from Eqs. (11.120) and (11.121). The *lower* of the two values is $Q_{g(u)}$.

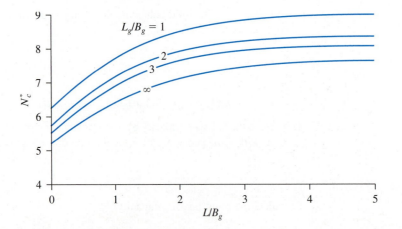

Figure 11.41 Variation of N_c^* with L_g/B_g and L/B_g

Example 11.18

The section of a 3 × 4 group pile in a layered saturated clay is shown in Figure 11.42. The piles are square in cross section (356 mm × 356 mm). The center-to-center spacing, d, of the piles is 889 mm. Determine the allowable load-bearing capacity of the pile group. Use FS = 4. Note that the groundwater table coincides with the ground surface.

Solution

From Eq. (11.120),

$$\Sigma Q_u = n_1 n_2 \left[9 A_p c_{u(p)} + \alpha_1 p c_{u(1)} L_1 + \alpha_2 p c_{u(2)} L_2 \right]$$

From Figure 11.42, $c_{u(1)} = 50.3 \text{ kN/m}^2$ and $c_{u(2)} = 85.1 \text{ kN/m}^2$.

For the top layer with $c_{u(1)} = 50.3 \text{ kN/m}^2$,

$$\frac{c_{u(1)}}{p_a} = \frac{50.3}{100} = 0.503$$

From Table 11.10, $\alpha_1 \approx 0.68$. Similarly,

$$\frac{c_{u(2)}}{p_a} = \frac{85.1}{100} \approx 0.85$$

$$\alpha_2 = 0.51$$

$$\Sigma Q_u = (3)(4) \left[\begin{array}{l} (9)(0.356)^2(85.1) + (0.68)(4 \times 0.356)(50.3)(4.57) \\ + (0.51)(4 \times 0.356)(85.1)(13.72) \end{array} \right]$$

$$= 14011 \text{ kN}$$

For piles acting as a group.

$$L_g = (3)(0.889) + 0.356 = 3.023 \text{ m}$$

$$B_g = (2)(0.889) + 0.356 = 2.134 \text{ m}$$

Figure 11.42 Group pile of layered saturated clay

$$\frac{L_g}{B_g} = \frac{3.023}{2.134} = 1.42$$

$$\frac{L}{B_g} = \frac{18.29}{2.134} = 8.57$$

From Figure 11.41, $N_c^* = 8.75$. From Eq. (11.121),

$$\Sigma Q_u = L_g B_g c_{u(p)} N_c^* + \Sigma 2 (L_g + B_g) c_u \, \Delta L$$

$$= (3.023)(2.134)(85.1)(8.75) + (2)(3.023 + 2.134)[(50.3)(4.57)$$

$$+ \; (85.1)(13.72)]$$

$$= 19217 \text{ kN}$$

Hence, $\Sigma Q_u = 14{,}011$ kN.

$$\Sigma Q_{\text{all}} = \frac{14{,}011}{\text{FS}} = \frac{14{,}011}{4} \approx \textbf{3503 kN} \qquad \blacksquare$$

11.22 *Elastic Settlement of Group Piles*

In general, the settlement of a group pile under a similar working load per pile increases with the width of the group (B_g) and the center-to-center spacing of the piles (d). Several investigations relating to the settlement of group piles have been reported in the literature, with widely varying results. The simplest relation for the settlement of group piles was given by Vesic (1969), namely,

$$s_{g(e)} = \sqrt{\frac{B_g}{D}} \, s_e \qquad (11.122)$$

where

$s_{g(e)}$ = elastic settlement of group piles
B_g = width of group pile section
D = width or diameter of each pile in the group
s_e = elastic settlement of each pile at comparable working load (see Section 11.15)

For group piles in sand and gravel, for elastic settlement, Meyerhof (1976) suggested the empirical relation

$$s_{g(e)}(\text{mm}) = \frac{0.96 q \sqrt{B_g} I}{N_{60}} \qquad (11.123)$$

where

$$q = Q_g/(L_g B_g) \text{ (in kN/m}^2) \qquad (11.124)$$

and

L_g and B_g = length and width of the group pile section, respectively (m)

N_{60} = average standard penetration number within seat of settlement ($\approx B_g$ deep below the tip of the piles)

I = influence factor = $1 - L/8B_g \geqslant 0.5$ $\qquad (11.125)$

L = length of embedment of piles (m)

Similarly, the group pile settlement is related to the cone penetration resistance by the formula

$$S_{g(e)} = \frac{q B_g I}{2 q_c} \qquad (11.126)$$

where q_c = average cone penetration resistance within the seat of settlement. (Note that, in Eq. (11.126), all quantities are expressed in consistent units.)

Example 11.19

Consider a 3 × 4 group of prestressed concrete piles, each 21 m long, in a sand layer. The details of each pile and the sand are similar to that described in Example 11.10. The working load for the pile group is 6024 kN (3 × 4 × Q_{all}—where Q_{all} = 502 kN as in Example 11.10), and $d/D = 3$. Estimate the elastic settlement of the pile group. Use Eq. (11.123).

Solution

$$S_{e(g)} = \sqrt{\frac{B_g}{D}} s_e$$

$$B_g = (3 - 1)d + \frac{2D}{2} = (2)(3D) + D = 7D = (7)(0.356 \text{ m}) = 2.492 \text{ m}$$

From Example 11.10, s_e = 19.69 mm. Hence,

$$S_{e(g)} = \sqrt{\frac{2.492}{0.356}}(19.69) = \textbf{52.09 mm} \qquad \blacksquare$$

11.23   *Consolidation Settlement of Group Piles*

The consolidation settlement of a group pile in clay can be estimated by using the 2:1 stress distribution method. The calculation involves the following steps (see Figure 11.43):

> *Step 1.*   Let the depth of embedment of the piles be L. The group is subjected to a total load of Q_g. If the pile cap is below the original ground surface, Q_g equals the total load of the superstructure on the piles, minus the effective weight of soil above the group piles removed by excavation.
>
> *Step 2.*   Assume that the load Q_g is transmitted to the soil beginning at a depth of $2L/3$ from the top of the pile, as shown in the figure. The load Q_g spreads out along two vertical to one horizontal line from this depth. Lines aa' and bb' are the two 2:1 lines.
>
> *Step 3.*   Calculate the increase in effective stress caused at the middle of each soil layer by the load Q_g. The formula is

$$\Delta\sigma_i' = \frac{Q_g}{(B_g + z_i)(L_g + z_i)} \qquad (11.127)$$

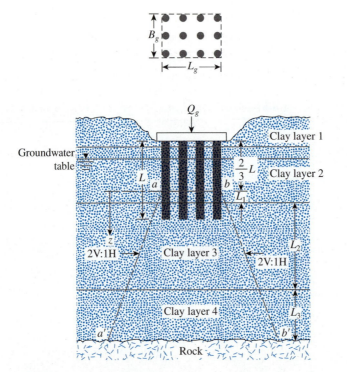

Figure 11.43 Consolidation settlement of group piles

where

$\Delta\sigma_i'$ = increase in effective stress at the middle of layer i
L_g, B_g = length and width, respectively of the planned group piles
z_i = distance from $z = 0$ to the middle of the clay layer i

For example, in Figure 11.43, for layer 2, $z_i = L_1/2$; for layer 3, $z_i = L_1 + L_2/2$; and for layer 4, $z_i = L_1 + L_2 + L_3/2$. Note, however, that there will be no increase in stress in clay layer 1, because it is above the horizontal plane $(z = 0)$ from which the stress distribution to the soil starts.

Step 4. Calculate the consolidation settlement of each layer caused by the increased stress. The formula is

$$\Delta s_{c(i)} = \left[\frac{\Delta e_{(i)}}{1 + e_{o(i)}} \right] H_i \tag{11.128}$$

where

$\Delta s_{c(i)}$ = consolidation settlement of layer i
$\Delta e_{(i)}$ = change of void ratio caused by the increase in stress in layer i
$e_{o(i)}$ = initial void ratio of layer i (before construction)
H_i = thickness of layer i (*Note:* In Figure 11.43, for layer 2, $H_i = L_1$; for layer 3, $H_i = L_2$; and for layer 4, $H_i = L_3$.)

Relationships involving $\Delta e_{(i)}$ are given in Chapter 1.

Step 5. The total consolidation settlement of the group piles is then

$$\Delta s_{c(g)} = \Sigma \Delta s_{c(i)} \tag{11.129}$$

Note that the consolidation settlement of piles may be initiated by fills placed nearby, adjacent floor loads, or the lowering of water tables.

Example 11.20

A group pile in clay is shown in Figure 11.44. Determine the consolidation settlement of the piles. All clays are normally consolidated.

Solution
Because the lengths of the piles are 15 m each, the stress distribution starts at a depth of 10 m below the top of the pile. We are given that $Q_g = 2000$ kN.

Calculation of Settlement of Clay Layer 1
For normally consolidated clays,

$$\Delta s_{c(1)} = \left[\frac{(C_{c(1)}H_1)}{1 + e_{o(1)}} \right] \log \left[\frac{\sigma_{o(1)}' + \Delta\sigma_{(1)}'}{\sigma_{o(1)}'} \right]$$

$$\Delta\sigma_{(1)}' = \frac{Q_g}{(L_g + z_1)(B_g + z_1)} = \frac{2000}{(3.3 + 3.5)(2.2 + 3.5)} = 51.6 \text{ kN/m}^2$$

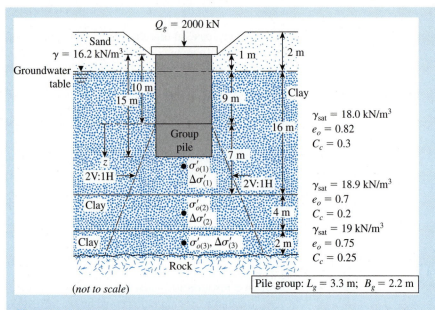

Figure 11.44 Consolidation settlement of a pile group

and

$$\sigma'_{o(1)} = 2(16.2) + 12.5(18.0 - 9.81) = 134.8 \text{ kN/m}^2$$

So

$$\Delta s_{c(1)} = \frac{(0.3)(7)}{1 + 0.82} \log \left[\frac{134.8 + 51.6}{134.8} \right] = 0.1624 \text{ m} = \mathbf{162.4 \text{ mm}}$$

Settlement of Layer 2
As with layer 1,

$$\Delta s_{c(2)} = \frac{C_{c(2)}H_2}{1 + e_{o(2)}} \log \left[\frac{\sigma'_{o(2)} + \Delta\sigma_{(2)}}{\sigma'_{o(2)}} \right]$$

$$\sigma'_{\sigma(2)} = 2(16.2) + 16(18.0 - 9.81) + 2(18.9 - 9.81) = 181.62 \text{ kN/m}^2$$

and

$$\Delta\sigma'_{(2)} = \frac{2000}{(3.3 + 9)(2.2 + 9)} = 14.52 \text{ kN/m}^2$$

Hence,

$$\Delta s_{c(2)} = \frac{(0.2)(4)}{1 + 0.7} \log \left[\frac{181.62 + 14.52}{181.62} \right] = 0.0157 \text{ m} = \mathbf{15.7 \text{ mm}}$$

Settlement of Layer 3
Continuing analogously, we have

$$\sigma'_{o(3)} = 181.62 + 2(18.9 - 9.81) + 1(19 - 9.81) = 208.99 \text{ kN/m}^2$$

$$\Delta\sigma'_{(3)} = \frac{2000}{(3.3 + 12)(2.2 + 12)} = 9.2 \text{ kN/m}^2$$

$$\Delta s_{c(3)} = \frac{(0.25)(2)}{1 + 0.75} \log\left(\frac{208.99 + 9.2}{208.99}\right) = 0.0054 \text{ m} = \textbf{5.4 mm}$$

Hence, the total settlement is

$$\Delta s_{c(g)} = 162.4 + 15.7 + 5.4 = \textbf{183.5 mm}$$ ∎

11.24 *Piles in Rock*

For point bearing piles resting on rock, most building codes specify that $Q_{g(u)} = \Sigma\, Q_u$, provided that the minimum center-to-center spacing of the piles is $D + 300$ mm. For H-piles and piles with square cross sections, the magnitude of D is equal to the diagonal dimension of the cross section of the pile.

Problems

11.1 A 12 m long concrete pile is shown in Figure P11.1. Estimate the ultimate point load Q_p by
  a. Meyerhof's method
  b. Vesic's method
  c. Coyle and Castello's method
  Use $m = 600$ in Eq. (11.26).

11.2 Refer to the pile shown in Figure P11.1. Estimate the side resistance Q_s by
  a. Using Eqs. (11.40) through (11.42). Use $K = 1.3$ and $\delta' = 0.8\phi'$
  b. Coyle and Castello's method [Eq. (11.44)]

11.3 Based on the results of Problems 11.1 and 11.2, recommend an allowable load for the pile. Use FS = 4.

11.4 A driven closed-ended pile, circular in cross section, is shown in Figure P11.4. Calculate the following.
  a. The ultimate point load using Meyerhof's procedure.
  b. The ultimate point load using Vesic's procedure. Take $I_{rr} = 50$.
  c. An approximate ultimate point load on the basis of parts (a) and (b).
  d. The ultimate frictional resistance Q_s. [Use Eqs. (11.40) through (11.42), and take $K = 1.4$ and $\delta' = 0.6\phi'$.]
  e. The allowable load of the pile (use FS = 4).

11.5 Following is the variation of N_{60} with depth in a granular soil deposit. A concrete pile 9 m long (0.305 m × 0.305 m in cross section) is driven into the sand and fully embedded in the sand.

Concrete pile
356 mm × 356 mm

Loose sand
$\phi'_1 = 30°$
$\gamma = 17.5$ kN/m³

12 m

Dense sand
$\phi'_2 = 42°$
$\gamma = 18.5$ kN/m³

Figure P11.1

3.05 m

$\gamma = 15.72$ kN/m³
$\phi' = 32°$
$c' = 0$

▼ Groundwater
table

$\gamma_{sat} = 18.24$ kN/m³
$\phi' = 32°$
$c' = 0$

3.05 m

$\gamma_{sat} = 19.24$ kN/m³
$\phi' = 40°$
$c' = 0$

15.24 m

15 in.

Figure P11.4

Depth (m)	N_{60}
1.5	4
3.0	8
4.5	7
6.0	5
7.5	16
9.0	18
10.5	21
11.0	24
12.5	20
14.0	19

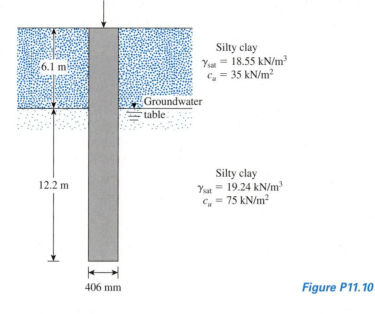

Silty clay
$\gamma_{sat} = 18.55$ kN/m³
$c_u = 35$ kN/m²

6.1 m

Groundwater
table

Silty clay
$\gamma_{sat} = 19.24$ kN/m³
$c_u = 75$ kN/m²

12.2 m

406 mm

Figure P11.10

Estimate the allowable load-carrying capacity of the pile (Q_{all}). Use FS = 4 and Meyerhof's equations [Eqs. (11.37) and (11.45)].

11.6 Solve Problem 11.5 using the equation of Briaud et al. [Eqs. (11.38) and (11.47)].

11.7 A concrete pile 15.24 m long having a cross section of 406 mm × 406 mm is fully embedded in a saturated clay layer for which $\gamma_{sat} = 19.02$ kN/m³, $\phi = 0$, and $c_u = 76.7$ kN/m². Determine the allowable load that the pile can carry. (Let FS = 3.) Use the α method to estimate the skin friction and Veric's method for point load estimation.

11.8 Redo Problem 11.7 using the λ method for estimating the skin friction and Meyerhof's method for the point load estimation.

11.9 A concrete pile 15 m long having a cross section of 0.38 m × 0.38 m is fully embedded in a saturated clay layer. For the clay, given: $\gamma_{sat} = 18$ kN/m³, $\phi = 0$, and $c_u = 80$ kN/m². Determine the allowable load that the pile can carry (FS = 3). Use the λ method to estimate the skin resistance.

11.10 A concrete pile 406 mm × 406 mm in cross section is shown in Figure P11.10. Calculate the ultimate skin friction resistance by using the
  a. α method
  b. λ method
  c. β method
  Use $\phi'_R = 20°$ for all clays, which are normally consolidated.

11.11 A steel pile (H-section; HP 360 × 152; see Table 11.1) is driven into a layer of sandstone. The length of the pile is 18.9 m. Following are the properties of the sandstone: unconfined compression strength = $q_{u(lab)} = 78.7$ MN/m² and angle of friction = 36°. Using a factor of safety of 3, estimate the allowable point load that can be carried by the pile. Use $[q_{u(design)} = q_{u(lab)}/5]$.

11.12 A concrete pile is 18 m long and has a cross section of 0.406 m × 0.406 m. The pile is embedded in a sand having $\gamma = 16$ kN/m³ and $\phi' = 37°$. The allowable working load is 900 kN. If 600 kN are contributed by the frictional resistance and 300 kN are from the point load, determine the elastic settlement of the pile. Given: $E_p = 2.1 \times 10^6$ kN/m.^{2}, $E_s = 30 \times 10^3$ kN/m.^{2}, $\mu_s = 0.38$, and $\xi = 0.57$ [Eq. (11.73)].

11.13 Solve Problem 11.12 with the following: length of pile = 15 m, pile cross section = 0.305 m × 0.305 m, allowable working load = 338 kN, contribution of frictional resistance to working load = 280 kN, $E_p = 21 \times 10^6$ kN/m^2, $E_s = 30,000$ kN/m^2, $\mu_s = 0.3$, and $\xi = 0.62$ [Eq. (11.73)].

11.14 A 30-m long concrete pile is 305 mm × 305 mm in cross section and is fully embedded in a sand deposit. If $n_h = 9200$ kN/m^2, the moment at ground level, $M_g = 0$, the allowable displacement of pile head = 12 mm; $E_p = 21 \times 10^6$ kN/m^2; and $F_{Y(\text{pile})} = 21,000$ kN/m^2, calculate the allowable lateral load, Q_g, at the ground level. Use the elastic solution method.

11.15 Solve Problem 11.14 by Brom's method. Assume that the pile is flexible and free headed. Let the soil unit weight, $\gamma = 16$ kN/m^3; the soil friction angle, $\phi' = 30°$; and the yield stress of the pile material, $F_Y = 21$ MN/m^2.

11.16 A steel H-pile (section HP 330 × 149) is driven by a hammer. The maximum rated hammer energy is 54.23 kN-m, the weight of the ram is 53.4 kN, and the length of the pile is 27.44 m. Also, we have coefficient of restitution = 0.35, weight of the pile cap = 10.7 kN, hammer efficiency = 0.85, number of blows for the last inch of penetration = 10, and $E_p = 207 \times 10^6$ kN/m^2. Estimate the pile capacity using Eq. (11.106). Take FS = 6.

11.17 Solve Problem 11.16 using the modified EN formula. (See Table 11.16). Use FS = 4.

11.18 Solve Problem 11.16 using the Danish formula (See Table 11.16). Use FS = 3.

11.19 Figure 11.35a shows a pile. Let $L = 20$ m, D (pile diameter) = 450 mm, $H_f = 4$ m, $\gamma_{\text{fill}} = 17.5$ kN/m^3, and $\phi'_{\text{fill}} = 25°$. Determine the total downward drag force on the pile. Assume that the fill is located above the water table and that $\delta' = 0.5\phi'_{\text{fill}}$.

11.20 Redo Problem 11.19 assuming that the water table coincides with the top of the fill and that $\gamma_{\text{sat(fill)}} = 19.8$ kN/m^3. If the other quantities remain the same, what would be the downward drag force on the pile? Assume $\delta' = 0.5\phi'_{\text{fill}}$.

11.21 Refer to Figure 11.35b. Let $L = 15.24$ m, $\gamma_{\text{fill}} = 17.29$ kN/m^3, $\gamma_{\text{sat(clay)}} = 19.49$ kN/m^3, $\phi'_{\text{clay}} = 20°$, $H_f = 3.05$ m, and D (pile diameter) = 406 mm. The water table coincides with the top of the clay layer. Determine the total downward drag force on the pile. Assume $\delta' = 0.6\phi'_{\text{clay}}$.

Figure p11.23

Figure P11.25

11.22 A concrete pile measuring 0.406 m $\times$ 0.406 m in cross section is 18.3 m long. It is fully embedded in a layer of sand. The following is an approximation of the mechanical cone penetration resistance (q_c) and the friction ratio (F_r) for the sand layer. Estimate the allowable bearing capacity of the pile. Use FS $= 4$.

Depth below ground surface (m)	q_c (kN/m^2)	F_r (%)
0–6.1	2803	2.3
6.1–13.7	3747	2.7
13.7–19.8	8055	2.8

11.23 The plan of a group pile is shown in Figure P11.23. Assume that the piles are embedded in a saturated homogeneous clay having a $c_u = 86$ kN/m^2. Given: diameter of piles $(D) = 316$ mm, center-to-center spacing of piles $= 600$ mm, and length of piles $= 20$ m. Find the allowable load-carrying capacity of the pile group. Use FS $= 3$.

11.24 Redo Problem 11.23 with the following: center-to-center spacing of piles $= 762$ mm, length of piles $= 13.7$ m, $D = 305$ mm, $c_u = 41.2$ kN/m^2, $\gamma_{sat} = 19.24$ kN/m^3, and FS $= 3$.

11.25 The section of a 4×4 group pile in a layered saturated clay is shown in Figure P11.25. The piles are square in cross section (356 mm $\times$ 356 mm). The center-to-center spacing (d) of the piles is 1 m. Determine the allowable load-bearing capacity of the pile group. Use FS $= 3$.

11.26 Figure P11.26 shows a group pile in clay. Determine the consolidation settlement of the group. Use the 2:1 method of estimate the average effective stress in the clay layers.

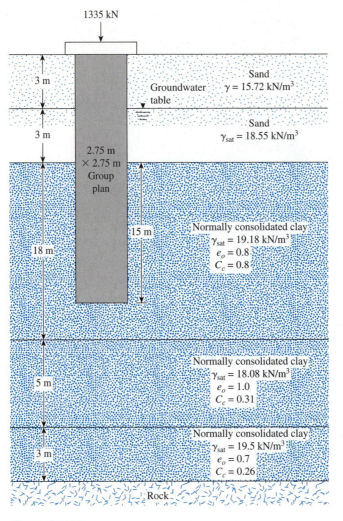

Figure P11.26

References

AMERICAN SOCIETY OF CIVIL ENGINEERS (1959). "Timber Piles and Construction Timbers," *Manual of Practice,* No. 17, American Society of Civil Engineers, New York.

AMERICAN SOCIETY OF CIVIL ENGINEERS (1993). *Design of Pile Foundations (Technical Engineering and Design Guides as Adapted from the U.S. Army Corps of Engineers,* No. 1), American Society of Civil Engineers, New York.

BALDI, G., BELLOTTI, R., GHIONNA, V., JAMIOLKOWSKI, M. and PASQUALINI, E. (1981). "Cone Resistance in Dry N.C. and O.C. Sands, Cone Penetration Testing and Experience," *Proceedings, ASCE Specialty Conference*, St. Louis, pp. 145–177.

BJERRUM, L., JOHANNESSEN, I. J., and EIDE, O. (1969). "Reduction of Skin Friction on Steel Piles to Rock," *Proceedings, Seventh International Conference on Soil Mechanics and Foundation Engineering,* Mexico City, Vol. 2, pp. 27–34.

BOWLES, J. E. (1982). *Foundation Analysis and Design,* McGraw-Hill, New York.

BOWLES, J. E. (1996). *Foundation Analysis and Design,* McGraw-Hill, New York.

BRIAUD, J. L., TUCKER, L., LYTTON, R. L., and COYLE, H. M. (1985). *Behavior of Piles and Pile Groups,* Report No. FHWA/RD-83/038, Federal Highway Administration, Washington, DC.

BROMS, B. B. (1965). "Design of Laterally Loaded Piles," *Journal of the Soil Mechanics and Foundations Division,* American Society of Civil Engineers, Vol. 91, No. SM3, pp. 79–99.

CHEN, Y. J., and KULHAWY, F. H. (1994). "Case History Evaluation of the Behavior of Drilled Shafts under Axial and Lateral Loading," *Final Report, Project 1493-04, EPRI TR-104601*, Geotechnical Group, Cornell University, Ithaca, NY, December.

COYLE, H. M., and CASTELLO, R. R. (1981). "New Design Correlations for Piles in Sand," *Journal of the Geotechnical Engineering Division,* American Society of Civil Engineers, Vol. 107, No. GT7, pp. 965–986.

DAVISSON, M. T. (1973). "High Capacity Piles" in *Innovations in Foundation Construction*, Proceedings of a Lecture Series, Illinois Section, American Society of Civil Engineers, Chicago.

DAVISSON, M. T. (1970). "BRD Vibratory Driving Formula," *Foundation Facts,* Vol. VI, No. 1, pp. 9–11.

DAVISSON, M. T., and GILL, H. L. (1963). "Laterally Loaded Piles in a Layered Soil System," *Journal of the Soil Mechanics and Foundations Division,* American Society of Civil Engineers, Vol. 89, No. SM3, pp. 63–94.

FENG, Z., and DESCHAMPS, R. J. (2000). "A Study of the Factors Influencing the Penetration and Capacity of Vibratory Driven Piles," *Soils and Foundations,* Vol. 40, No. 3, pp. 43–54.

GOODMAN, R. E. (1980). *Introduction to Rock Mechanics,* Wiley, New York.

GUANG-YU, Z. (1988). "Wave Equation Applications for Piles in Soft Ground," *Proceedings, Third International Conference on the Application of Stress-Wave Theory to Piles* (B. H. Fellenius, ed.), Ottawa, Ontario, Canada, pp. 831–836.

JANBU, N. (1953). *An Energy Analysis of Pile Driving with the Use of Dimensionless Parameters,* Norwegian Geotechnical Institute, Oslo, Publication No. 3.

KISHIDA, H., and MEYERHOF, G. G. (1965). "Bearing Capacity of Pile Groups under Eccentric Loads in Sand," *Proceedings, Sixth International Conference on Soil Mechanics and Foundation Engineering,* Montreal, Vol. 2, pp. 270–274.

LIU, J. L., YUAN, Z. I., and ZHANG, K. P. (1985). "Cap-Pile-Soil Interaction of Bored Pile Groups," *Proceedings, Eleventh International Conference on Soil Mechanics and Foundation Engineering*, San Francisco, Vol. 3, pp. 1433–1436.

MANSUR, C. I., and HUNTER, A. H. (1970). "Pile Tests—Arkansas River Project," *Journal of the Soil Mechanics and Foundations Division*, American Society of Civil Engineers, Vol. 96, No. SM6, pp. 1545–1582.

MATLOCK, H., and REESE, L. C. (1960). "Generalized Solution for Laterally Loaded Piles," *Journal of the Soil Mechanics and Foundations Division,* American Society of Civil Engineers, Vol. 86, No. SM5, Part I, pp. 63–91.

MEYERHOF, G. G. (1976). "Bearing Capacity and Settlement of Pile Foundations," *Journal of the Geotechnical Engineering Division,* American Society of Civil Engineers, Vol. 102, No. GT3, pp. 197–228.

NOTTINGHAM, L. C., and SCHMERTMANN, J. H. (1975). *An Investigation of Pile Capacity Design Procedures,* Research Report No. D629, Department of Civil Engineering, University of Florida, Gainesville, FL.

OLSON, R. E., and FLAATE, K. S. (1967). "Pile Driving Formulas for Friction Piles in Sand," *Journal of the Soil Mechanics and Foundations Division,* American Society of Civil Engineers, Vol. 93, No. SM6, pp. 279–296.

O'NEILL, M. W., and REESE, L. C. (1999). *Drilled Shafts: Construction Procedure and Design Methods*, FHWA Report No. IF-99-025.

SCHMERTMANN, J. H. (1978). *Guidelines for Cone Penetration Test: Performance and Design,* Report FHWA-TS-78-209, Federal Highway Administration, Washington, DC.

SEILER, J. F., and KEENEY, W. D. (1944). "The Efficiency of Piles in Groups," *Wood Preserving News,* Vol. 22, No. 11 (November).

SKOV, R., and DENVER, H. (1988). "Time Dependence of Bearing Capacity of Piles," *Proceedings, Third International Conference on Application of Stress Wave Theory to Piles,* Ottawa, Canada, pp. 879–889.

SLADEN, J. A. (1992). "The Adhesion Factor: Applications and Limitations," *Canadian Geotechnical Journal*, Vol. 29, No. 2, pp. 323–326.

SVINKIN, M. (1996). Discussion on "Setup and Relaxation in Glacial Sand," *Journal of Geo-technical Engineering,* ASCE, Vol. 22, pp. 319–321.

TERZAGHI, K., PECK, R. B., and MESRI, G. (1996). *Soil Mechanics in Engineering Practice*, John Wiley, NY.

VESIC, A. S. (1961). "Bending of Beams Resting on Isotropic Elastic Solids," *Journal of the Engineering Mechanics Division,* American Society of Civil Engineers, Vol. 87, No. EM2, pp. 35–53.

VESIC, A. S. (1969). *Experiments with Instrumented Pile Groups in Sand*, American Society for Testing and Materials, Special Technical Publication No. 444, pp. 177–222.

VESIC, A. S. (1970). "Tests on Instrumental Piles—Ogeechee River Site," *Journal of the Soil Mechanics and Foundations Division,* American Society of Civil Engineers, Vol. 96, No. SM2, pp. 561–584.

VESIC, A. S. (1977). *Design of Pile Foundations,* National Cooperative Highway Research Program Synthesis of Practice No. 42, Transportation Research Board, Washington, DC.

VIJAYVERGIYA, V. N., and FOCHT, J. A., JR. (1972). *A New Way to Predict Capacity of Piles in Clay,* Offshore Technology Conference Paper 1718, Fourth Offshore Technology Conference, Houston.

WONG, K. S., and TEH, C. I. (1995). "Negative Skin Friction on Piles in Layered Soil Deposit," *Journal of Geotechnical and Geoenvironmental Engineering,* American Society of Civil Engineers, Vol. 121, No. 6, pp. 457–465.

WOODWARD, R. J., GARDNER, W. S., and GREER, D. M. (1972). *Drilled Pier Foundations,* McGraw-Hill, New York.

12 Drilled-Shaft Foundations

12.1   Introduction

The terms *caisson, pier, drilled shaft,* and *drilled pier* are often used interchangeably in foundation engineering; all refer to a *cast-in-place pile generally having a diameter of about 750 mm* or more, with or without steel reinforcement and with or without an enlarged bottom. Sometimes the diameter can be as small as 305 mm.

To avoid confusion, we use the term *drilled shaft* for a hole drilled or excavated to the bottom of a structure's foundation and then filled with concrete. Depending on the soil conditions, casings may be used to prevent the soil around the hole from caving in during construction. The diameter of the shaft is usually large enough for a person to enter for inspection.

The use of drilled-shaft foundations has several advantages:

1. A single drilled shaft may be used instead of a group of piles and the pile cap.
2. Constructing drilled shafts in deposits of dense sand and gravel is easier than driving piles.
3. Drilled shafts may be constructed before grading operations are completed.
4. When piles are driven by a hammer, the ground vibration may cause damage to nearby structures. The use of drilled shafts avoids this problem.
5. Piles driven into clay soils may produce ground heaving and cause previously driven piles to move laterally. This does not occur during the construction of drilled shafts.
6. There is no hammer noise during the construction of drilled shafts; there is during pile driving.
7. Because the base of a drilled shaft can be enlarged, it provides great resistance to the uplifting load.
8. The surface over which the base of the drilled shaft is constructed can be visually inspected.
9. The construction of drilled shafts generally utilizes mobile equipment, which, under proper soil conditions, may prove to be more economical than methods of constructing pile foundations.
10. Drilled shafts have high resistance to lateral loads.

There are also a couple of drawbacks to the use of drilled-shaft construction. For one thing, the concreting operation may be delayed by bad weather and always needs close supervision. For another, as in the case of braced cuts, deep excavations for drilled shafts may induce substantial ground loss and damage to nearby structures.

12.2   Types of Drilled Shafts

Drilled shafts are classified according to the ways in which they are designed to transfer the structural load to the substratum. Figure 12.1a shows a drilled *straight shaft*. It extends through the upper layer(s) of poor soil, and its tip rests on a strong load-bearing soil layer or rock. The shaft can be cased with steel shell or pipe when required (as it is with cased, cast-in-place concrete piles; see Figure 11.4). For such shafts, the resistance to the applied load may develop from end bearing and also from side friction at the shaft perimeter and soil interface.

A *belled shaft* (see Figures 12.1b and c) consists of a straight shaft with a bell at the bottom, which rests on good bearing soil. The bell can be constructed in the shape of a dome (see Figure 12.1b), or it can be angled (see Figure 12.1c). For angled bells, the underreaming tools that are commercially available can make 30 to 45° angles with the vertical. For the majority of drilled shafts constructed in the United States, the entire load-carrying capacity is assigned to the end bearing only. However, under certain circumstances, the end-bearing capacity and the side friction are taken into account. In Europe, both the side frictional resistance and the end-bearing capacity are always taken into account.

Straight shafts can also be extended into an underlying rock layer. (See Figure 12.1d.) In the calculation of the load-bearing capacity of such shafts, the end bearing and the shear stress developed along the shaft perimeter and rock interface can be taken into account.

Figure 12.1 Types of drilled shaft: (a) straight shaft; (b) and (c) belled shaft; (d) straight shaft socketed into rock

12.3 ## Construction Procedures

The most common construction procedure used in the United States involves rotary drilling. There are three major types of construction methods: the dry method, the casing method, and the wet method.

Dry Method of Construction

This method is employed in soils and rocks that are above the water table and that will not cave in when the hole is drilled to its full depth. The sequence of construction, shown in Figure 12.2, is as follows:

Step 1. The excavation is completed (and belled if desired), using proper drilling tools, and the spoils from the hole are deposited nearby. (See Figure 12.2a.)

Step 2. Concrete is then poured into the cylindrical hole. (See Figure 12.2b.)

Step 3. If desired, a rebar cage is placed in the upper portion of the shaft. (See Figure 12.2c.)

Step 4. Concreting is then completed, and the drilled shaft will be as shown in Figure 12.2d.

Casing Method of Construction

This method is used in soils or rocks in which caving or excessive deformation is likely to occur when the borehole is excavated. The sequence of construction is shown in Figure 12.3 and may be explained as follows:

Step 1. The excavation procedure is initiated as in the case of the dry method of construction. (See Figure 12.3a.)

Step 2. When the caving soil is encountered, bentonite slurry is introduced into the borehole. (See Figure 12.3b.) Drilling is continued until the excavation goes past the caving soil and a layer of impermeable soil or rock is encountered.

Step 3. A casing is then introduced into the hole. (See Figure 12.3c.)

Step 4. The slurry is bailed out of the casing with a submersible pump. (See Figure 12.3d.)

Step 5. A smaller drill that can pass through the casing is introduced into the hole, and excavation continues. (See Figure 12.3e.)

Step 6. If needed, the base of the excavated hole can then be enlarged, using an underreamer. (See Figure 12.3f.)

Step 7. If reinforcing steel is needed, the rebar cage needs to extend the full length of the excavation. Concrete is then poured into the excavation and the casing is gradually pulled out. (See Figure 12.3g.)

Step 8. Figure 12.3h shows the completed drilled shaft.

Wet Method of Construction

This method is sometimes referred to as the *slurry displacement method.* Slurry is used to keep the borehole open during the entire depth of excavation. (See Figure 12.4.) Following are the steps involved in the wet method of construction:

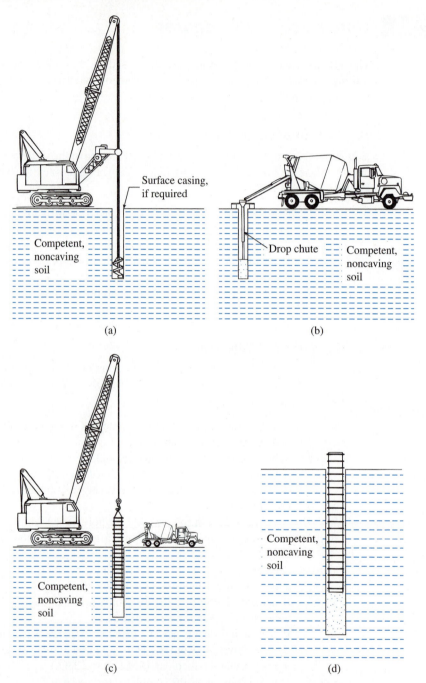

Figure 12.2 Dry method of construction: (a) initiating drilling; (b) starting concrete pour; (c) placing rebar cage; (d) completed shaft (After O'Neill and Reese, 1999)

Figure 12.3 Casing method of construction: (a) initiating drilling; (b) drilling with slurry; (c) introducing casing; (d) casing is sealed and slurry is being removed from interior of casing; (e) drilling below casing; (f) underreaming; (g) removing casing; (h) completed shaft (After O'Neill and Reese, 1999)

(e)

(f)

Level of
fluid concrete

Drifting fluid
forced from
space between
casing and soil

(g)

(h)

Figure 12.3 (*Continued*)

Figure 12.4 Slurry method of construction: (a) drilling to full depth with slurry; (b) placing rebar cage; (c) placing concrete; (d) completed shaft (After O'Neill and Reese, 1999)

Step 1. Excavation continues to full depth with slurry. (See Figure 12.4a.)

Step 2. If reinforcement is required, the rebar cage is placed in the slurry. (See Figure 12.4b.)

Step 3. Concrete that will displace the volume of slurry is then placed in the drill hole. (See Figure 12.4c.)

Step 4. Figure 12.4d shows the completed drilled shaft.

Figure 12.5 shows a drilled shaft under construction using the dry method. The construction of a drilled shaft using the wet method is shown in Figure 12.6. A typical auger, a reinforcement cage, and a typical clean-out bucket are shown in Figure 12.7.

Figure 12.5 Drilled shaft construction using the dry method (*Courtesy of Sanjeev Kumar, Southern Illinois University, Carbondale, Illinois*)

Figure 12.6 Drilled shaft construction using wet method (*Courtesy of Khaled Sobhan, Florida Atlantic Univetsity, Boca Raton, Florida*)

(a)　　　　　　　　　　　　　　　(b)

(c)

Figure 12.7 Drilled shaft construction: (a) A typical auger; (b) a reinforcement cage; (c) a clean-out bucket (*Courtesy of Khaled Sobhan, Florida Atlantic Univetsity, Boca Raton, Florida*)

12.4 *Other Design Considerations*

For the design of ordinary drilled shafts without casings, a minimum amount of vertical steel reinforcement is always desirable. Minimum reinforcement is 1% of the gross cross-sectional area of the shaft. For drilled shafts with nominal reinforcement, most building codes suggest using a design concrete strength, f_c, on the order of $f'_c/4$. Thus, the minimum shaft diameter becomes

$$f_c = 0.25f'_c = \frac{Q_w}{A_{\text{gs}}} = \frac{Q_w}{\frac{\pi}{4}D_s^2}$$

or

$$D_s = \sqrt{\frac{Q_w}{\left(\frac{\pi}{4}\right)(0.25)f_c'}} = 2.257\sqrt{\frac{Q_u}{f_c'}} \qquad (12.1)$$

where

D_s = diameter of the shaft
f_c' = 28-day concrete strength
Q_w = working load of the drilled shaft
A_{gs} = gross cross-sectional area of the shaft

If drilled shafts are likely to be subjected to tensile loads, reinforcement should be continued for the entire length of the shaft.

Concrete Mix Design

The concrete mix design for drilled shafts is not much different from that for any other concrete structure. When a reinforcing cage is used, consideration should be given to the ability of the concrete to flow through the reinforcement. In most cases, a concrete slump of about 15.0 mm (6 in.) is considered satisfactory. Also, the maximum size of the aggregate should be limited to about 20 mm (0.75 in.)

12.5 Load Transfer Mechanism

The load transfer mechanism from drilled shafts to soil is similar to that of piles, as described in Section 11.5. Figure 12.8 shows the load test results of a drilled shaft, conducted in a clay soil by Reese et al. (1976). The shaft (Figure 12.8a) had a diameter of 762 mm and a depth of penetration of 6.94 m. Figure 12.8b shows the load-settlement curves. It can be seen that the total load carried by the drilled shaft was 1246 kN. The load carried by side resistance was about 800 kN, and the rest was carried by point bearing. It is interesting to note that, with a downward movement of about 6 mm, full side resistance was mobilized. However, about 25 mm of downward movement was required for mobilization of full point resistance. This situation is similar to that observed in the case of piles. Figure 12.8c shows the average load-distribution curves for different stages of the loading.

12.6 Estimation of Load-Bearing Capacity

The ultimate load-bearing capacity of a drilled shaft (see Figure 12.9) is

$$Q_u = Q_p + Q_s \qquad (12.2)$$

Figure 12.8 Load test results of Reese et al. (1976) on a drilled shaft: (a) dimensions of the shaft; (b) plot of base, sides, and total load with mean settlement; (c) plot of load-distribution curve with depth

where

Q_u = ultimate load
Q_p = ultimate load-carrying capacity at the base
Q_s = frictional (skin) resistance

The ultimate base load Q_p can be expressed in a manner similar to the way it is expressed in the case of shallow foundations [Eq. (3.19)], or

$$Q_p = A_p\left(c'N_cF_{cs}F_{cd}F_{cc} + q'N_qF_{qs}F_{qd}F_{qc} + \frac{1}{2}\gamma'N_\gamma F_{\gamma s}F_{\gamma d}F_{\gamma c}\right) \qquad (12.3)$$

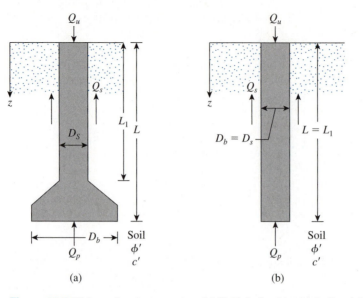

Figure 12.9 Ultimate bearing capacity of drilled shafts: (a) with bell and (b) straight shaft

where

$$
\begin{aligned}
c' &= \text{cohesion} \\
N_c, N_q, N_\gamma &= \text{bearing capacity factors} \\
F_{cs}, F_{qs}, F_{\gamma s} &= \text{shape factors} \\
F_{cd}, F_{qd}, F_{\gamma d} &= \text{depth factors} \\
F_{cc}, F_{qc}, F_{\gamma c} &= \text{compressibility factors} \\
\gamma' &= \text{effective unit weight of soil at the base of the shaft} \\
q' &= \text{effective vertical stress at the base of the shaft} \\
A_p &= \text{area of the base} = \frac{\pi}{4}D_b^2
\end{aligned}
$$

In most instances, the last term (the one containing N_γ) is neglected, except in the case of a relatively short drilled shaft. With this assumption, we can write

$$Q_u = A_p(c'N_cF_{cs}F_{cd}F_{cc} + qN_qF_{qs}F_{qd}F_{qc}) + Q_s \tag{12.4}$$

The procedure to estimate the ultimate capacity of drilled shafts in granular and cohesive soil is described in the following sections.

12.7 *Drilled Shafts in Granular Soil: Load-Bearing Capacity*

Estimation of Q_p

For a drilled shaft with its base located on a granular soil (that is, $c' = 0$), the *net ultimate load-carrying capacity* at the base can be obtained from Eq. (12.4) as

$$Q_{p(\text{net})} = A_p[q'(N_q - 1)F_{qs}F_{qd}F_{qc}] \tag{12.5}$$

The bearing capacity factor, N_q, for various soil friction angles (ϕ') can be taken from Table 3.3. It is also given in Table 12.1. Also,

$$F_{qs} = 1 + \tan \phi' \qquad (12.6)$$

$$F_{qd} = 1 + C \underbrace{\tan^{-1}\left(\frac{L}{D_b}\right)}_{\text{radian}} \qquad (12.7)$$

$$C = 2 \tan \phi'(1 - \sin \phi')^2 \qquad (12.8)$$

The variations of F_{qs} and C with ϕ' are given in Table 12.1.

According to Chen and Kulhawy (1994), F_{qc} can be calculated in the following manner.

Step 1. Calculate the critical rigidity index as

$$I_{cr} = 0.5 \exp\left[2.85 \cot\left(45 - \frac{\phi'}{2}\right)\right] \qquad (12.9)$$

where I_{cr} = critical rigidity index (see Table 12.1).

Table 12.1 Variation of N_q, F_{qs}, C, I_{cr}, μ_s, and n with ϕ'

Soil friction angle, ϕ' (deg)	N_q (Table 3.3)	F_{qs} [Eq. (12.6)]	C [Eq. (12.8)]	I_{cr} [Eq. (12.9)]	μ_s [Eq. (12.13)]	n [Eq. (12.15)]
25	10.66	1.466	0.311	43.84	0.100	0.00500
26	11.85	1.488	0.308	47.84	0.115	0.00475
27	13.20	1.510	0.304	52.33	0.130	0.00450
28	14.72	1.532	0.299	57.40	0.145	0.00425
29	16.44	1.554	0.294	63.13	0.160	0.00400
30	18.40	1.577	0.289	69.63	0.175	0.00375
31	20.63	1.601	0.283	77.03	0.190	0.00350
32	23.18	1.625	0.276	85.49	0.205	0.00325
33	26.09	1.649	0.269	95.19	0.220	0.00300
34	29.44	1.675	0.262	106.37	0.235	0.00275
35	33.30	1.700	0.255	119.30	0.250	0.00250
36	37.75	1.727	0.247	134.33	0.265	0.00225
37	42.92	1.754	0.239	151.88	0.280	0.00200
38	48.93	1.781	0.231	172.47	0.295	0.00175
39	55.96	1.810	0.223	196.76	0.310	0.00150
40	64.20	1.839	0.214	225.59	0.325	0.00125
41	73.90	1.869	0.206	259.98	0.340	0.00100
42	85.38	1.900	0.197	301.29	0.355	0.00075
43	99.02	1.933	0.189	351.22	0.370	0.00050
44	115.31	1.966	0.180	412.00	0.385	0.00025
45	134.88	2.000	0.172	486.56	0.400	0.00000

Step 2.   Calculate the reduced rigidity index as

$$I_{rr} = \frac{I_r}{1 + I_r \Delta} \tag{12.10}$$

where

$$I_r = \text{soil rididity index} = \frac{E_s}{2(1 + \mu_s)q' \tan \phi'} \tag{12.11}$$

in which

$$E_s = \text{drained modulus of elasticity of soil} = m p_a \tag{12.12}$$
$$p_a = \text{atmospheric pressure}\,(\approx 100\,\text{kN/m}^2)$$

$$m = \begin{cases} 100 \text{ to } 200 \text{ (loose soil)} \\ 200 \text{ to } 500 \text{ (medium dense soil)} \\ 500 \text{ to } 1000 \text{ (dense soil)} \end{cases}$$

$$\mu_s = \text{Poisson's ratio of soil} = 0.1 + 0.3\left(\frac{\phi' - 25}{20}\right)$$
$$(\text{for } 25° \le \phi' \le 45°)\,(\text{see Table 12.1}) \tag{12.13}$$

$$\Delta = n\frac{q'}{p_a} \tag{12.14}$$

$$n = 0.005\left(1 - \frac{\phi' - 25}{20}\right)(\text{see Table 12.1}) \tag{12.15}$$

Step 3.   If $I_{rr} \ge I_{cr}$, then

$$F_{qc} = 1 \tag{12.16}$$

However, if $I_{rr} < I_{cr}$, then

$$F_{qc} = \exp\left\{(-3.8 \tan \phi') + \left[\frac{(3.07 \sin \phi')(\log_{10} 2I_{rr})}{1 + \sin \phi'}\right]\right\} \tag{12.17}$$

The magnitude of $Q_{p(\text{net})}$ also can be reasonably estimated from a relationship based on the analysis of Berezantzev et al. (1961) that can be expressed as

$$Q_{p(\text{net})} = A_p q'(\omega N_q^* - 1) \tag{12.18}$$

where

$$N_q^* = \text{bearing capacity factor} = 0.21 e^{0.17\phi'} \text{ (See Table 12.2)} \tag{12.19}$$
$$\omega = \text{correction factor} = f(L/D_b)$$

In Eq. (12.19), ϕ' is in degrees. The variation of ω with L/D_b is given in Figure 12.10.

Table 12.2 Variation of N_q^* with ϕ' [Eq. (12.19)]

ϕ' (deg)	N_q^*
25	14.72
26	17.45
27	20.68
28	24.52
29	29.06
30	34.44
31	40.83
32	48.39
33	57.36
34	67.99
35	80.59
36	95.52
37	113.22
38	134.20
39	159.07
40	188.55
41	223.49
42	264.90
43	313.99
44	372.17
45	441.14

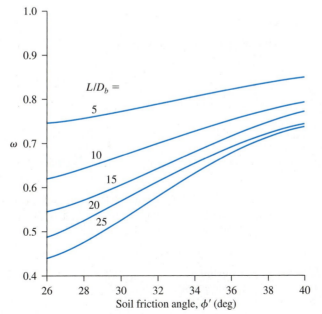

Figure 12.10 Variation of ω with ϕ' and L/D_b

Estimation of Q_s

The frictional resistance at ultimate load, Q_s, developed in a drilled shaft may be calculated as

$$Q_s = \int_0^{L_1} pf\, dz \tag{12.20}$$

where

p = shaft perimeter = πD_s

f = unit frictional (or skin) resistance = $K\sigma_o' \tan \delta'$ $\qquad$ (12.21)

K = earth pressure coefficient $\approx K_o = 1 - \sin \phi'$ $\qquad$ (12.22)

σ_o' = effective vertical stress at any depth z

Thus,

$$Q_s = \int_0^{L_1} pf\, dz = \pi D_s(1 - \sin \phi') \int_0^{L_1} \sigma_o' \tan \delta'\, dz \tag{12.23}$$

The value of σ_o' will increase to a depth of about $15 D_s$ and will remain constant thereafter, as shown in Figure 11.16.

For cast-in-pile concrete and good construction techniques, a rough interface develops and, hence, δ'/ϕ' may be taken to be one. With poor slurry construction, $\delta'/\phi' \approx 0.7$ to 0.8.

Allowable Net Load, $Q_{all (net)}$

An appropriate factor of safety should be applied to the ultimate load to obtain the net allowable load, or

$$Q_{all(net)} = \frac{Q_{p(net)} + Q_s}{\text{FS}} \tag{12.24}$$

12.8   *Load-Bearing Capacity Based on Settlement*

On the basis of a database of 41 loading tests, Reese and O'Neill (1989) proposed a method for calculating the load-bearing capacity of drilled shafts that is based on settlement. The method is applicable to the following ranges:

1. Shaft diameter: $D_s = 0.52$ to 1.2 m
2. Bell depth: $L = 4.7$ to 30.5 m
3. Field standard penetration resistance: $N_{60} = 5$ to 60
4. Concrete slump = 100 to 225 mm

Figure 12.11 Development of Eq. (12.25)

Reese and O'Neill's procedure (see Figure 12.11) gives

$$Q_{u(net)} = \sum_{i=1}^{N} f_i p \Delta L_i + q_p A_p \qquad (12.25)$$

where

f_i = ultimate unit shearing resistance in layer i
p = perimeter of the shaft = πD_s
q_p = unit point resistance
A_p = area of the base = $(\pi/4) D_b^2$

$$f_i = \beta_1 \sigma'_{ozi} < \beta_2 \qquad (12.26)$$

where σ'_{ozi} = vertical effective stress at the middle of layer i.

$$\beta_1 = \beta_3 - \beta_4 z_i^{0.5} \; (\text{for } 0.25 \leq \beta_1 \leq 1.2) \qquad (12.27)$$

The units for f_i, z_i, and σ'_{ozi} and the magnitude of β_2, β_3, and β_4 in the SI system are

Item	SI
f_i	kN/m^2
z_i	m
σ'_{ozi}	kN/m^2
β_2	192 kN/m^2
β_3	1.5
β_4	0.244

The point bearing capacity is

$$q_p = \beta_5 N_{60} \leq \beta_6 \quad [\text{for } D_b < 1.27\,\text{m}] \tag{12.28}$$

where N_{60} = field standard penetration number within a distance of $2D_b$ below the base of the drilled shaft.

The magnitudes of β_5 and β_6 and the unit of q_p in the SI system are given here.

Item	SI
β_5	57.5
β_6	4310 kN/m^2
q_p	kN/m^2

If D_b is equal to or greater than 1.27 m, excessive settlement may occur. In that case, q_p may be replaced by q_{pr} or

$$q_{pr} = \frac{1.27}{D_b(\text{m})} q_p \tag{12.29}$$

Based on the desired level of settlement, Figures 12.12 and 12.13 may now be used to calculate the allowable load, $Q_{\text{all(net)}}$. Note that the trend lines given in these figures is the average of all test results.

More recently, Rollins et al. (2005) have modified Eq. (12.27) for gravelly sands as follows:

For sand with 25 to 50% gravel,

$$\beta_1 = \beta_7 - \beta_8 z_i^{0.75} \quad (\text{for } 0.25 \leq \beta_1 \leq 1.8) \tag{12.30}$$

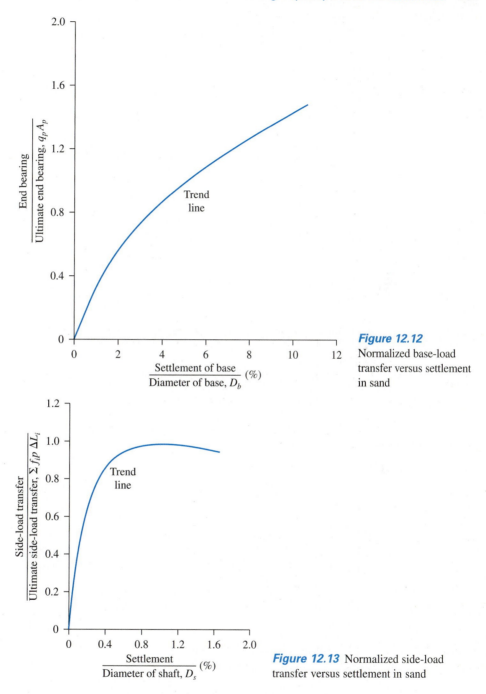

Figure 12.12 Normalized base-load transfer versus settlement in sand

Figure 12.13 Normalized side-load transfer versus settlement in sand

For sand with more than 50% gravel,

$$\beta_1 = \beta_9 e^{-\beta_{10} z_i} \ (\text{for } 0.25 \leq \beta_1 \leq 3.0) \tag{12.31}$$

The magnitudes of β_7, β_8, β_9, and β_{10} and the unit of z_i in the SI system are given here.

Item	SI
β_7	2.0
β_8	0.15
β_9	3.4
β_{10}	-0.085
z_i	m

Figure 12.14 provides the normalized side-load transfer trend based on the desired level of settlement for gravelly sand and gravel.

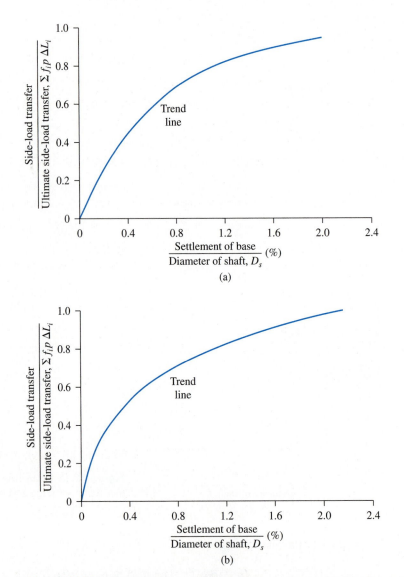

Figure 12.14 Normalized side-load transfer versus settlement: (a) gravelly sand (gravel 25–50%) and (b) gravel (more than 50%)

Example 12.1

A soil profile is shown in Figure 12.15. A point bearing drilled shaft with a bell is placed in a layer of dense sand and gravel. Determine the allowable load the drilled shaft could carry. Use Eq. (12.5) and a factor of safety of 4. Take $D_s = 1$ m and $D_b = 1.75$ m. For the dense sand layer, $\phi' = 36°$; $E_s = 500p_a$. Ignore the frictional resistance of the shaft.

Solution
We have

$$Q_{p(net)} = A_p[q'(N_q - 1)F_{qs}F_{qd}F_{qc}]$$

and

$$q' = (6)(16.2) + (2)(19.2) = 135.6 \text{ kN/m}^2$$

For $\phi' = 36°$, from Table 12.1, $N_q = 37.75$. Also,

$$F_{qs} = 1.727$$

and

$$F_{qd} = 1 + C \tan^{-1}\left(\frac{L}{D_b}\right)$$

$$= 1 + 0.247 \tan^{-1}\left(\frac{8}{1.75}\right) = 1.335$$

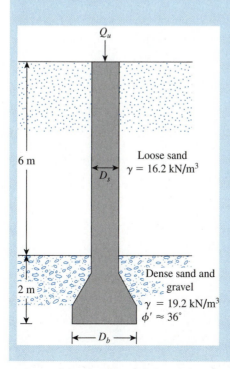

Q_u

6 m

D_s

Loose sand
$\gamma = 16.2$ kN/m³

2 m

Dense sand and gravel
$\gamma = 19.2$ kN/m³
$\phi' \approx 36°$

D_b

Figure 12.15 Allowable load of drilled shaft

From Eq. (12.9),

$$I_{cr} = 0.5 \exp\left[2.85 \cot\left(45 - \frac{\phi'}{2} \right) \right] = 134.3 \text{ (See Table 12.1)}$$

From Eq. (12.12), $E_s = mp_a$. With $m = 500$, we have

$$E_s = (500)(100) = 50,000 \text{ kN/m}^2$$

From Eq. (12.13) and Table 12.1,

$$\mu_s = 0.265$$

So

$$I_r = \frac{E_s}{2(1 + \mu_s)(q')(\tan \phi')} = \frac{50,000}{2(1 + 0.265)(135.6)(\tan 36)} = 200.6$$

From Eq. (12.10),

$$I_{rr} = \frac{I_r}{1 + I_r\Delta}$$

with

$$\Delta = n\frac{q'}{p_a} = 0.00225\left(\frac{135.6}{100} \right) = 0.0031$$

it follows that

$$I_{rr} = \frac{200.6}{1 + (200.6)(0.0031)} = 123.7$$

I_{rr} is less than I_{cr}. So, from Eq. (12.17),

$$F_{qc} = \exp\left\{ (-3.8 \tan \phi') + \left[\frac{(3.07 \sin \phi')(\log_{10} 2I_{rr})}{1 + \sin \phi'} \right] \right\}$$

$$= \exp\left\{ (-3.8 \tan 36) + \left[\frac{(3.07 \sin 36) \log(2 \times 123.7)}{1 + \sin 36} \right] \right\} = 0.958$$

Hence,

$$Q_{p(net)} = \left[\left(\frac{\pi}{4} \right)(1.75)^2 \right](135.6)(37.75 - 1)(1.727)(1.335)(0.958) = 26,474 \text{ kN}$$

and

$$Q_{p(all)} = \frac{Q_{p(net)}}{FS} = \frac{26,474}{4} \approx \mathbf{6619 \text{ kN}}$$

■

Example 12.2

Solve Example 12.1 using Eq. (12.18).

Solution

Equation (12.18) asserts that

$$Q_{p(\text{net})} = A_p q' (\omega N_q^* - 1)$$

We have (also see Table 12.2)

$$N_q^* = 0.21 e^{0.17\phi'} = 0.21 e^{(0.17)(36)} = 95.52$$

and

$$\frac{L}{D_b} = \frac{8}{1.75} = 4.57$$

From Figure 12.10, for $\phi' = 36°$ and $L/D_b = 4.57$, the value of ω is about 0.83. So

$$Q_{p(\text{net})} = \left[\left(\frac{\pi}{4} \right) (1.75)^2 \right] (135.6)[(0.83)(95.52) - 1] = 25{,}532 \text{ kN}$$

and

$$Q_{p(\text{all})} = \frac{25{,}532}{4} = \textbf{6383 kN} \qquad \blacksquare$$

Example 12.3

A drilled shaft is shown in Figure 12.16. The uncorrected average standard penetration number (N_{60}) within a distance of $2D_b$ below the base of the shaft is about 30. Determine

a. The ultimate load-carrying capacity
b. The load-carrying capacity for a settlement of 12mm. Use Eq. (12.30).

Solution

Part a

From Eqs. (12.26) and (12.27),

$$f_i = \beta_1 \sigma'_{ozi}$$

and

$$\beta_1 = 2.0 - 0.15 z^{0.75}$$

For this problem, $z_i = 6/2 = 3$ m, so

$$\beta = 2 - (0.15)(3)^{0.75} = 1.658$$

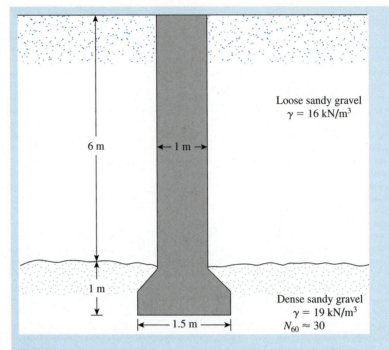

Figure 12.16 Drilled shaft supported by a dense layer of sandy gravel

and

$$\sigma'_{ozi} = \gamma z_i = (16)(3) = 48 \text{ kN/m}^2$$

Thus,

$$f_i = (48)(1.658) = 79.58 \text{ kN/m}^2$$

and

$$\Sigma f_i p \, \Delta L_i = (79.58)(\pi \times 1)(6) = 1500 \text{ kN}$$

From Eq. (12.28),

$$q_p = 57.5 N_{60} = (57.5)(30) = 1725 \text{ kN/m}^2$$

Note that D_b is greater than 1.27. So we will use Eq. (12.29a).

$$q_{pr} = \left(\frac{1.27}{D_b}\right) q_p = \left(\frac{1.27}{1.5}\right)(1725) \approx 1461 \text{ kN/m}^2$$

Now

$$q_{pr} A_p = (14.61)\left(\frac{\pi}{4} \times 1.5^2\right) \approx 2582 \text{ kN}$$

Hence,

$$Q_{\text{ult(net)}} = q_{pr} A_p + \Sigma f_i p \, \Delta L_i = 2582 + 1500 = \textbf{4082 kN}$$

Part b
We have

$$\frac{\text{Allowable settlement}}{D_s} = \frac{12}{(1.0)(1000)} = 0.12 = 1.2\%$$

The trend line in Figure 12.14a shows that, for a normalized settlement of 1.2%, the normalized load is about 0.8 . Thus, the side-load transfer is $(0.8)(1500) \approx 1200$ kN. Similarly,

$$\frac{\text{Allowable settlement}}{D_b} = \frac{12}{(1.5)(1000)} = 0.008 = 0.8\%$$

The trend line shown in Figure 12.12 indicates that, for a normalized settlement of 1.4%, the normalized base load is 0.317. So the base load is $(0.317)(2582) = 818.5$ kN. Hence, the total load is

$$Q = 1200 + 818.5 \approx \mathbf{2018.5 \ kN}$$ ∎

12.9 *Drilled Shafts in Clay: Load-Bearing Capacity*

For saturated clays with $\phi = 0$, the bearing capacity factor N_q in Eq. (12.4) is equal to unity. Thus, for this case,

$$Q_{p(\text{net})} \approx A_p c_u N_c F_{cs} F_{cd} F_{cc} \tag{12.32}$$

where c_u = undrained cohesion.

Assuming that $L \geq 3D_b$, we can rewrite Eq. (12.32) as

$$Q_{p(\text{net})} = A_p c_u N_c^* \tag{12.33}$$

where $N_c^* = N_c F_{cs} F_{cd} F_{cc} = 1.33[(\ln I_r) + 1]$ in which I_r = soil rigidity index.

$$\tag{12.34}$$

The soil rigidity index was defined in Eq. (12.11). For $\phi = 0$,

$$I_r = \frac{E_s}{3c_u} \tag{12.35}$$

O'Neill and Reese (1999) provided an approximate relationship between c_u and $E_s/3c_u$. This relationship is shown in Figure 12.17. For all practical purposes, if c_u/p_a is

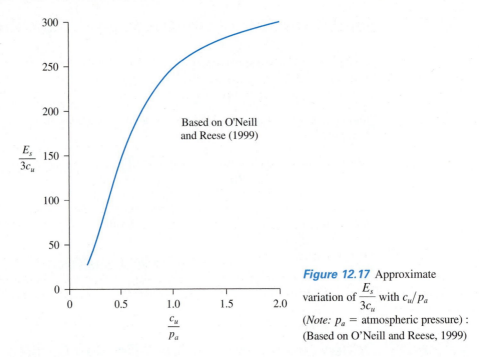

Figure 12.17 Approximate variation of $\dfrac{E_s}{3c_u}$ with c_u/p_a (*Note*: p_a = atmospheric pressure) : (Based on O'Neill and Reese, 1999)

equal to or greater than unity (p_a = atmospheric pressure $\approx 100 \text{ kN/m}^2$), then the magnitude of N_c^* can be taken to be 9.

Experiments by Whitaker and Cooke (1966) showed that, for belled shafts, the full value of $N_c^* = 9$ is realized with a base movement of about 10 to 15% of D_b. Similarly, for straight shafts ($D_b = D_s$), the full value of $N_c^* = 9$ is obtained with a base movement of about 20% of D_b.

The expression for the skin resistance of drilled shafts in clay is similar to Eq. (11.55), or

$$Q_s = \sum_{L=0}^{L=L_1} \alpha^* c_u p\, \Delta L \tag{12.36}$$

Kulhawy and Jackson (1989) reported the field-test result of 106 straight drilled shafts—65 in uplift and 41 in compression. The best correlation obtained from the results is

$$\alpha^* = 0.21 + 0.25\left(\frac{p_a}{c_u}\right) \le 1 \tag{12.37}$$

where p_a = atmospheric pressure $\approx 100 \text{ kN/m}^2$.

So, conservatively, we may assume that

$$\alpha^* = 0.4 \tag{12.38}$$

12.10 *Load-Bearing Capacity Based on Settlement*

Reese and O'Neill (1989) suggested a procedure for estimating the ultimate and allowable (based on settlement) bearing capacities for drilled shafts in clay. According to this procedure, we can use Eq. (12.25) for the net ultimate load, or

$$Q_{\text{ult(net)}} = \sum_{i=1}^{n} f_i p\, \Delta L_i + q_p A_p$$

The unit skin friction resistance can be given as

$$f_i = \alpha_i^* c_{u(i)} \tag{12.39}$$

The following values are recommended for α_i^*:

$\alpha_i^* = 0$ for the top 1.5 m (5 ft) and bottom 1 diameter, D_s, of the drilled shaft. (*Note:* If $D_b > D_s$, then $\alpha^* = 0$ for 1 diameter above the top of the bell and for the peripheral area of the bell itself.)
$\alpha_i^* = 0.55$ elsewhere.

The expression for q_p (point load per unit area) can be given as

$$q_p = 6c_{ub}\left(1 + 0.2\,\frac{L}{D_b}\right) \le 9c_{ub} \le 40p_a \tag{12.40}$$

where

c_{ub} = average undrained cohesion within a vertical distance of $2D_b$ below the base
p_a = atmospheric pressure

If D_b is large, excessive settlement will occur at the ultimate load per unit area, q_p, as given by Eq. (12.40). Thus, for $D_b > 1.91\,\text{m}$, q_p may be replaced by

$$q_{pr} = F_r q_p \tag{12.41}$$

where

$$F_r = \frac{2.5}{\psi_1 D_b + \psi_2} \le 1 \tag{12.42}$$

The relations for ψ_1 and ψ_2 along with the unit of D_b in the SI system are given in Table 12.3.

Figures 12.18 and 12.19 may now be used to evaluate the allowable load-bearing capacity, based on settlement. (Note that the ultimate bearing capacity in Figure 12.18 is q_p, not q_{pr}.) To do so,

Step 1. Select a value of settlement, s.

Step 2. Calculate $\sum\limits_{i=1}^{N} f_i p \,\Delta L_i$ and $q_p A_p$.

Step 3. Using Figures 12.18 and 12.19 and the calculated values in Step 2, determine the *side load* and the *end bearing load*.

Step 4. The sum of the side load and the end bearing load gives the total allowable load.

Table 12.3 Relationships for ψ_1 and ψ_2

Item	SI
ψ_1	$\psi_1 = 2.78 \times 10^{-4} + 8.26 \times 10^{-5}\left(\dfrac{L}{D_b}\right) \le 5.9 \times 10^{-4}$
ψ_2	$\psi_2 = 0.065[c_{ub}(\text{kN/m}^2)]^{0.5}$ $(0.5 \le \psi_2 \le 1.5)$
D_b	mm

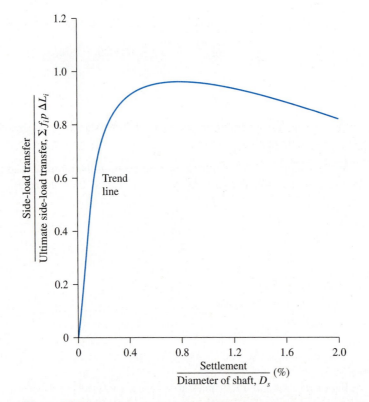

Figure 12.18 Normalized side-load transfer versus settlement in cohesive soil

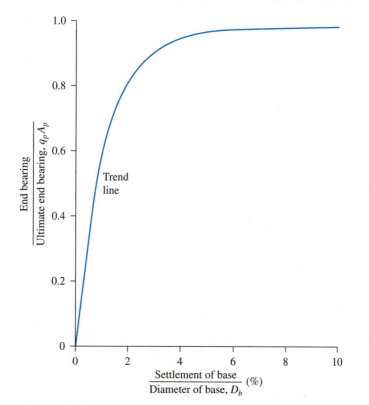

Figure 12.19 Normalized base-load transfer versus settlement in cohesive soil

Example 12.4

Figure 12.20 shows a drilled shaft without a bell. Here, $L_1 = 8.23$ m, $L_2 = 2.59$ m, $D_s = 1.0$ m, $c_{u(1)} = 50$ kN/m^2, and $c_{u(2)} = 108.75$ kN/m^2. Determine

  a. The net ultimate point bearing capacity
  b. The ultimate skin resistance
  c. The working load, Q_w (FS = 3)

Use Eqs. (12.33), (12.36), and (12.38).

Solution
Part a
From Eq. (12.33),

$$Q_{p(\text{net})} = A_p c_u N_c^* = A_p c_{u(2)} N_c^* = \left[\left(\frac{\pi}{4} \right)(1)^2 \right](108.75)(9) = \mathbf{768.7\ kN}$$

(*Note:* Since $c_{u(2)}/p_a > 1$, $N_c^* \approx 9$.)

Figure 12.20 A drill shaft without a bell

Part b

From Eq. (12.36),

$$Q_s = \Sigma \alpha^* c_u p \Delta L$$

From Eq. (12.38),

$$\alpha^* = 0.4$$

$$p = \pi D_s = (3.14)(1.0) = 3.14 \text{ m}$$

and

$$Q_s = (0.4)(3.14)[(50 \times 8.23) + (108.75 \times 2.59)] = \textbf{871 kN}$$

Part c

$$Q_w = \frac{Q_{p(\text{net})} + Q_s}{\text{FS}} = \frac{768.7 + 871}{3} = \textbf{546.6 kN} \quad \blacksquare$$

Example 12.5

A drilled shaft in a cohesive soil is shown in Figure 12.21. Use Reese and O'Neill's method to determine the following.

a. The ultimate load-carrying capacity.
b. The load-carrying capacity for an allowable settlement of 12 mm.

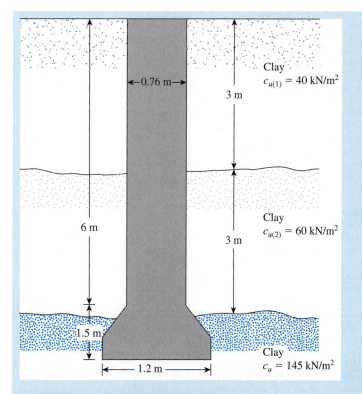

Figure 12.21 A drilled shaft in layered clay

Solution

Part a

From Eq. (12.39),

$$f_i = \alpha_i^* c_{u(i)}$$

From Figure 12.21,

$$\Delta L_1 = 3 - 1.5 = 1.5 \text{ m}$$

$$\Delta L_2 = (6 - 3) - D_s = (6 - 3) - 0.76 = 2.24 \text{ m}$$

$$c_{u(1)} = 40 \text{ kN/m}^2$$

and

$$c_{u(2)} = 60 \text{ kN/m}^2$$

Hence,

$$\Sigma f_i p \Delta L_i = \Sigma \alpha_i^* c_{u(i)} p \Delta L_i$$

$$= (0.55)(40)(\pi \times 0.76)(1.5) + (0.55)(60)(\pi \times 0.76)(2.24)$$

$$= 255.28 \text{ kN}$$

Again, from Eq. (12.40),

$$q_p = 6c_{ub}\left(1 + 0.2\frac{L}{D_b}\right) = (6)(145)\left[1 + 0.2\left(\frac{6 + 1.5}{1.2}\right)\right] = 1957.5 \text{ kN/m}^2$$

A check reveals that

$$q_p = 9c_{ub} = (9)(145) = 1305 \text{ kN/m}^2 < 1957.5 \text{ kN/m}^2$$

So we use $q_p = 1305 \text{ kN/m}^2$

$$q_pA_p = q_p\left(\frac{\pi}{4}D_b^2\right) = (1305)\left[\left(\frac{\pi}{4}\right)(1.2)^2\right] = 1475.9 \text{ kN}$$

Hence,

$$Q_{\text{ult}} = \Sigma\alpha_i^*c_{u(i)}p\Delta L_i + q_pA_p = 255.28 + 1475.9 \approx \mathbf{1731 \text{ kN}}$$

Part b
We have

$$\frac{\text{Allowable settlement}}{D_s} = \frac{12}{(0.76)(1000)} = 0.0158 = 1.58\%$$

The trend line shown in Figure 12.18 indicates that, for a normalized settlement of 1.58%, the normalized side load is about 0.9. Thus, the side load is

$$(0.9)(\Sigma f_ip\Delta L_i) = (0.9)(255.28) = 229.8 \text{ kN}$$

Again,

$$\frac{\text{Allowable settlement}}{D_b} = \frac{12}{(1.2)(1000)} = 0.01 = 1.0\%$$

The trend line shown in Figure 12.19 indicates that, for a normalized settlement of 1.0%, the normalized end bearing is about 0.63, so

$$\text{Base load} = (0.63)(q_pA_p) = (0.63)(1475.9) = 929.8 \text{ kN}$$

Thus, the total load is

$$Q = 229.8 + 929.8 = \mathbf{1159.6 \text{ kN}}$$ ∎

12.11 Settlement of Drilled Shafts at Working Load

The settlement of drilled shafts at working load is calculated in a manner similar to that outlined in Section 11.15. In many cases, the load carried by shaft resistance is small compared with the load carried at the base. In such cases, the contribution of s_3 may be ignored. (Note that in Eqs. (11.74) and (11.75) the term D should be replaced by D_b for drilled shafts.)

Example 12.6

Refer to Figure 12.20. Given: $L_1 = 8\,\text{m}$, $L_2 = 3\,\text{m}$, $D_s = 1.5\,\text{m}$, $c_{u(1)} = 50\,\text{kN/m}^2$, $c_{u(2)} = 105\,\text{kN/m}^2$, and working load $Q_w = 1005\,\text{kN}$. Estimate the elastic settlement at the working load. Use Eqs. (11.73), (11.75), and (11.76). Take $\xi = 0.65$, $E_p = 21 \times 10^6\,\text{kN/m}^2$, $E_s = 14{,}000\,\text{kN/m}^2$, $\mu_s = 0.3$, and $Q_{wp} = 250\,\text{kN}$.

Solution
From Eq. (11.73),

$$s_{e(1)} = \frac{(Q_{wp} + \xi Q_{ws})L}{A_p E_p}$$

Now,

$$Q_{ws} = 1005 - 250 = 755\,\text{kN}$$

so

$$s_{e(1)} = \frac{[250 + (0.65)(755)](11)}{\left(\dfrac{\pi}{4} \times 1.5^2\right)(21 \times 10^6)} = 0.00022\,\text{m} = 0.22\,\text{mm}$$

From Eq. (11.75),

$$s_{e(2)} = \frac{Q_{wp} C_p}{D_b q_p}$$

From Table 11.13, for stiff clay, $C_p \approx 0.04$; also,

$$q_p = c_{u(b)} N_c^* = (105)(9) = 945\,\text{kN/m}^2$$

Hence,

$$s_{e(2)} = \frac{(250)(0.04)}{(1.5)(945)} = 0.0071\,\text{m} = 7.1\,\text{mm}$$

Again, from Eqs. (11.76) and (11.77),

$$s_{e(3)} = \left(\frac{Q_{ws}}{pL}\right)\left(\frac{D_s}{E_s}\right)(1 - \mu_s^2)I_{ws}$$

where

$$I_{ws} = 2 + 0.35\sqrt{\frac{L}{D_s}} = 2 + 0.35\sqrt{\frac{11}{1.5}} = 2.95$$

$$s_{e(3)} = \left[\frac{755}{(\pi \times 1.5)(11)}\right]\left(\frac{1.5}{14{,}000}\right)(1 - 0.3^2)(2.95) = 0.0042\,\text{m} = 4.2\,\text{mm}$$

The total settlement is

$$s_e = s_{e(1)} + s_{e(2)} + s_{e(3)} = 0.22 + 7.1 + 4.2 \approx \textbf{11.52 mm}$$ ∎

12.12 Lateral Load-Carrying Capacity—Characteristic Load and Moment Method

Several methods for analyzing the lateral load-carrying capacity of piles, as well as the load-carrying capacity of drilled shafts, were presented in Section 11.16; therefore, they will not be repeated here. In 1994, Duncan et al. developed a *characteristic load method* for estimating the lateral load capacity for drilled shafts that is fairly simple to use. We describe this method next.

According to the characteristic load method, the *characteristic load* Q_c and *moment* M_c form the basis for the dimensionless relationship that can be given by the following correlations:

Characteristic Load

$$Q_c = 7.34D_s^2 \, (E_p R_I) \left(\frac{c_u}{E_p R_I} \right)^{0.68} \qquad \text{(for clay)} \qquad (12.43)$$

$$Q_c = 1.57D_s^2 \, (E_p R_I) \left(\frac{\gamma' D_s \phi' K_p}{E_p R_I} \right)^{0.57} \qquad \text{(for sand)} \qquad (12.44)$$

Characteristic Moment

$$M_c = 3.86D_s^3 \, (E_p R_I) \left(\frac{c_u}{E_p R_I} \right)^{0.46} \qquad \text{(for clay)} \qquad (12.45)$$

$$M_c = 1.33D_s^3 (E_p R_I) \left(\frac{\gamma' D_s \phi' K_p}{E_p R_I} \right)^{0.40} \qquad \text{(for sand)} \qquad (12.46)$$

In these equations,

D_s = diameter of drilled shafts
E_p = modulus of elasticity of drilled shafts
R_I = ratio of moment of inertia of drilled shaft section to moment of inertia of a solid section (*Note:* $R_I = 1$ for uncracked shaft without central void)
γ' = effective unit weight of sand
ϕ' = effective soil friction angle (degrees)
K_p = Rankine passive pressure coefficient = $\tan^2(45 + \phi'/2)$

Deflection Due to Load Q_g Applied at the Ground Line

Figures 12.22 and 12.23 give the plot of Q_g/Q_c versus x_o/D_s for drilled shafts in sand and clay due to the load Q_g applied at the ground surface. Note that x_o is the ground

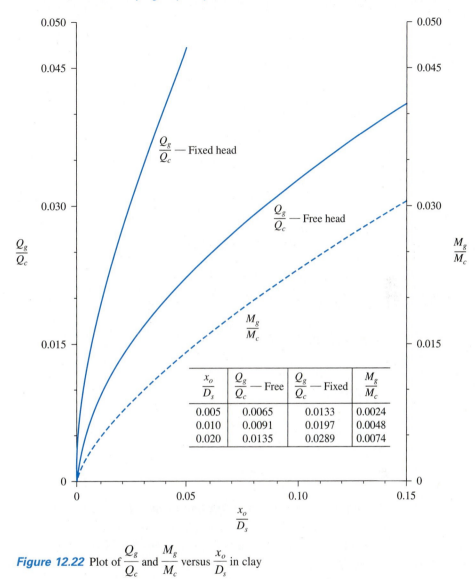

Figure 12.22 Plot of $\dfrac{Q_g}{Q_c}$ and $\dfrac{M_g}{M_c}$ versus $\dfrac{x_o}{D_s}$ in clay

line deflection. If the magnitudes of Q_g and Q_c are known, the ratio Q_g/Q_c can be calculated. The figure can then be used to estimate the corresponding value of x_o/D_s and, hence, x_o.

Deflection Due to Moment Applied at the Ground Line

Figures 12.22 and 12.23 give the variation plot of M_g/M_c with x_o/D_s for drilled shafts in sand and clay due to an applied moment M_g at the ground line. Again, x_o is the ground line deflection. If the magnitudes of M_g, M_c, and D_s are known, the value of x_o can be calculated with the use of the figure.

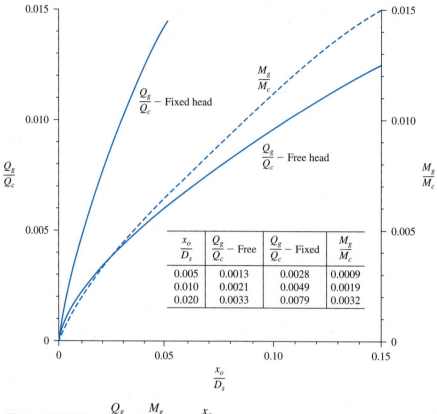

$\dfrac{x_o}{D_s}$	$\dfrac{Q_g}{Q_c}$ – Free	$\dfrac{Q_g}{Q_c}$ – Fixed	$\dfrac{M_g}{M_c}$
0.005	0.0013	0.0028	0.0009
0.010	0.0021	0.0049	0.0019
0.020	0.0033	0.0079	0.0032

Figure 12.23 Plot of $\dfrac{Q_g}{Q_c}$ and $\dfrac{M_g}{M_c}$ versus $\dfrac{x_o}{D_s}$ in sand

Deflection Due to Load Applied Above the Ground Line

When a load Q is applied above the ground line, it induces both a load $Q_g = Q$ and a moment $M_g = Qe$ at the ground line, as shown in Figure 12.24a. A superposition solution can now be used to obtain the ground line deflection. The step-by-step procedure is as follows (refer to Figure 12.24b):

Step 1. Calculate Q_g and M_g.

Step 2. Calculate the deflection x_{oQ} that would be caused by the load Q_g acting alone.

Step 3. Calculate the deflection x_{oM} that would be caused by the moment acting alone.

Step 4. Determine the value of a load Q_{gM} that would cause the same deflection as the moment (i.e., x_{oM}).

Step 5. Determine the value of a moment M_{gQ} that would cause the same deflection as the load (i.e., x_{oQ}).

Step 6. Calculate $(Q_g + Q_{gM})/Q_c$ and determine x_{oQM}/D_s.

Step 7. Calculate $(M_g + M_{gQ})/M_c$ and determine x_{oMQ}/D_s.

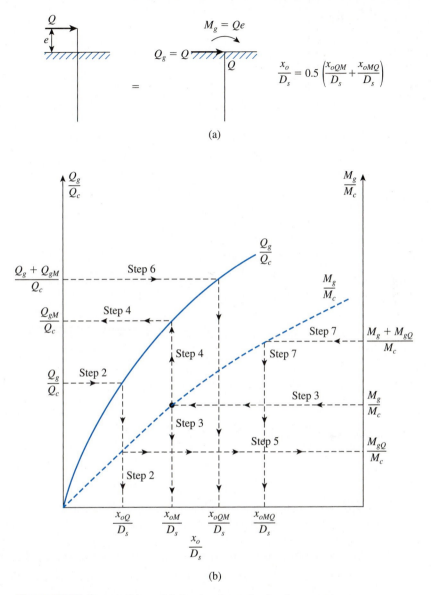

Figure 12.24 Superposition of deflection due to load and moment

Step 8. Calculate the combined deflection:

$$x_{o(\text{combined})} = 0.5(x_{oQM} + x_{oMQ}) \qquad (12.47)$$

Maximum Moment in Drilled Shaft Due to Ground Line Load Only

Figure 12.25 shows the plot of Q_g/Q_c with $M_{\max}/M_c$ for fixed- and free-headed drilled shafts due only to the application of a ground line load Q_g. For fixed-headed

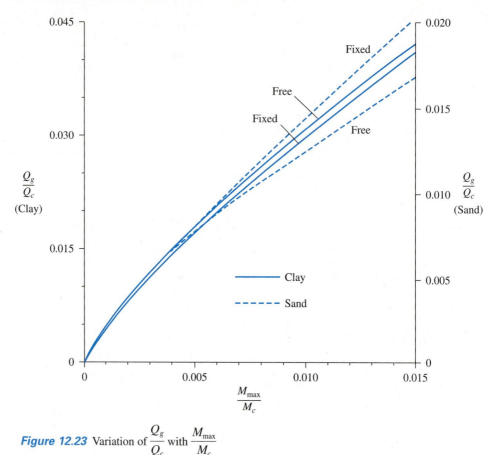

Figure 12.23 Variation of $\dfrac{Q_g}{Q_c}$ with $\dfrac{M_{\max}}{M_c}$

shafts, the maximum moment in the shaft, $M_{\max}$, occurs at the ground line. For this condition, if Q_c, M_c, and Q_g are known, the magnitude of $M_{\max}$ can be easily calculated.

Maximum Moment Due to Load and Moment at Ground Line

If a load Q_g and a moment M_g are applied at the ground line, the maximum moment in the drilled shaft can be determined in the following manner:

Step 1. Using the procedure described before, calculate $x_{o(\text{combined})}$ from Eq. (12.47).

Step 2. To solve for the characteristic length T, use the following equation:

$$x_{o(\text{combined})} = \frac{2.43Q_g}{E_p I_p} T^3 + \frac{1.62M_g}{E_p I_p} T^2 \tag{12.48}$$

Step 3. The moment in the shaft at a depth z below the ground surface can be calculated as

$$M_z = A_m Q_g T + B_m M_g \tag{12.49}$$

where A_m, B_m = dimensionless moment coefficients (Matlock and Reese, 1961); see Figure 12.26.

The value of the maximum moment M_{max} can be obtained by calculating M_z at various depths in the upper part of the drilled shaft.

The characteristic load method just described is valid only if L/D_s has a certain minimum value. If the actual L/D_s is less than $(L/D_s)_{min}$, then the ground line deflections will be underestimated and the moments will be overestimated. The values of $(L/D_s)_{min}$ for drilled shafts in sand and clay are given in the following table:

Clay		Sand	
$\dfrac{E_p R_I}{c_u}$	$(L/D_s)_{min}$	$\dfrac{E_p R_I}{\gamma' D_s \phi' K_p}$	$(L/D_s)_{min}$
1×10^5	6	1×10^4	8
3×10^5	10	4×10^4	11
1×10^6	14	2×10^5	14
3×10^6	18		

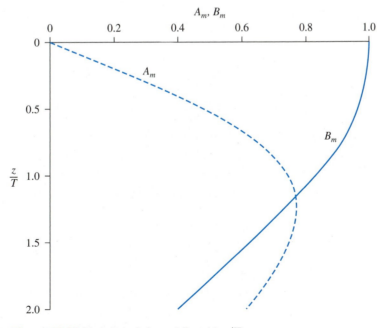

Figure 12.26 Variation of A_m and B_m with z/T

Example 12.7

A free-headed drilled shaft in clay is shown in Figure 12.27. Let $E_p = 22 \times 10^6$ kN/m^2. Determine

    a. The ground line deflection, $x_{o(\text{combined})}$
    b. The maximum bending moment in the drilled shaft
    c. The maximum tensile stress in the shaft
    d. The minimum penetration of the shaft needed for this analysis

Solution
We are given

$D_s = 1$ m
$c_u = 100$ kN/m^2
$R_I = 1$
$E_p = 22 \times 10^6$ kN/m^2

and

$$I_p = \frac{\pi D_s^4}{64} = \frac{(\pi)(1)^4}{64} = 0.049 \text{ m}^4$$

Part a
From Eq. (12.43),

$$Q_c = 7.34 D_s^2 \, (E_p R_I) \left(\frac{c_u}{E_p R_I} \right)^{0.68}$$

$$= (7.34)(1)^2 [(22 \times 10^6)(1)] \left[\frac{100}{(22 \times 10^6)(1)} \right]^{0.68}$$

$$= 37{,}607 \text{ kN}$$

$M_g = 200$ kN-m
$Q_g = 150$ kN

Clay
$c_u = 100$ kN/m^2

$D_s = $
1 m

Figure 12.27 Free-headed drilled shaft

From Eq. (12.45),

$$M_c = 3.86 D_s^3 (E_p R_I) \left(\frac{c_u}{E_p R_I} \right)^{0.46}$$

$$= (3.86)(1)^3 [(22 \times 10^6)(1)] \left[\frac{100}{(22 \times 10^6)(1)} \right]^{0.46}$$

$$= 296{,}139 \text{ kN-m}$$

Thus,

$$\frac{Q_g}{Q_c} = \frac{150}{37{,}607} = 0.004$$

From Figure 12.22, $x_{oQ} \approx (0.0025) D_s = 0.0025 \text{ m} = 2.5 \text{ mm}$. Also,

$$\frac{M_g}{M_c} = \frac{200}{296{,}139} = 0.000675$$

From Figure 12.22, $x_{oM} \approx (0.0014) D_s = 0.0014 \text{ m} = 1.4 \text{ mm}$, so

$$\frac{x_{oM}}{D_s} = \frac{0.0014}{1} = 0.0014$$

From Figure 12.22, for $x_{oM}/D_s = 0.0014$, the value of $Q_{gM}/Q_c \approx 0.002$. Hence,

$$\frac{x_{oQ}}{D_s} = \frac{0.0025}{1} = 0.0025$$

From Figure 12.22, for $x_{oQ}/D_s = 0.0025$, the value of $M_{gQ}/M_c \approx 0.0013$, so

$$\frac{Q_g}{Q_c} + \frac{Q_{gM}}{Q_c} = 0.004 + 0.002 = 0.006$$

From Figure 12.22, for $(Q_g + Q_{gM})/Q_c = 0.006$, the value of $x_{oQM}/D_s \approx 0.0046$. Hence,

$$x_{oQM} = (0.0046)(1) = 0.0046 \text{ m} = 4.6 \text{ mm}$$

Thus, we have

$$\frac{M_g}{M_c} + \frac{M_{gQ}}{M_c} = 0.000675 + 0.0013 \approx 0.00198$$

From Figure 12.22, for $(M_g + M_{gQ})/M_c = 0.00198$, the value of $x_{oMQ}/D_s \approx 0.0041$. Hence,

$$x_{oMQ} = (0.0041)(1) = 0.0041 \text{ m} = 4.1 \text{ mm}$$

Consequently,

$$x_{o \text{ (combined)}} = 0.5(x_{oQM} + x_{oMQ}) = (0.5)(4.6 + 4.1) = \textbf{4.35 mm}$$

Part b
From Eq. (12.48),

$$x_{o \text{ (combined)}} = \frac{2.43 Q_g}{E_p I_p} T^3 + \frac{1.62 M_g}{E_p I_p} T^2$$

so

$$0.00435 \text{ m} = \frac{(2.43)(150)}{(22 \times 10^6)(0.049)} T^3 + \frac{(1.62)(200)}{(22 \times 10^6)(0.049)} T^2$$

or

$$0.00435 \text{ m} = 338 \times 10^{-6} T^3 + 300.6 \times 10^{-6} T^2$$

and it follows that

$$T \approx 2.05 \text{ m}$$

From Eq. (12.49),

$$M_z = A_m Q_g T + B_m M_g = A_m(150)(2.05) + B_m(200) = 307.5 A_m + 200 B_m$$

Now the following table can be prepared:

$\dfrac{z}{T}$	A_m (Figure 12.26)	B_m (Figure 12.26)	M_z (kN-m)
0	0	1.0	200
0.4	0.36	0.98	306.7
0.6	0.52	0.95	349.9
0.8	0.63	0.9	373.7
1.0	0.75	0.845	399.6
1.1	0.765	0.8	395.2
1.25	0.75	0.73	376.6

So the maximum moment is 399.4 kN-m $\approx$ 400 kN-m and occurs at $z/T \approx 1$. Hence,

$$z = (1)(T) = (1)(2.05 \text{ m}) = \textbf{2.05 m}$$

Part c
The maximum tensile stress is

$$\sigma_{\text{tensile}} = \frac{M_{\text{max}}\left(\dfrac{D_s}{2}\right)}{I_p} = \frac{(400)\left(\dfrac{1}{2}\right)}{0.049} = \textbf{4081.6 kN/m}^2$$

Part d
We have

$$\frac{E_p R_I}{c_u} = \frac{(22 \times 10^6)(1)}{100} = 2.2 \times 10^5$$

By interpolation, for $(E_p R_I)/c_u = 2.2 \times 10^5$, the value of $(L/D_s)_{\min} \approx 8.5$. So

$$L \approx (8.5)(1) = \textbf{8.5 m}$$

■

12.13 *Drilled Shafts Extending into Rock*

In Section 12.1, we noted that drilled shafts can be extended into rock. Figure 12.28 shows a drilled shaft whose depth of embedment in rock is equal to L. When considering drilled shafts in rock, we can find several correlations between the end bearing capacity and the *unconfined compression strength of intact rocks, q_u.* It is important to recognize that, in the field, there are cracks, joints, and discontinuities in the rock, and the influence of those factors should be considered. Keeping this in mind, Zhang and Einstein (1998) analyzed a data base of 39 full-scale drilled shaft tests in which the

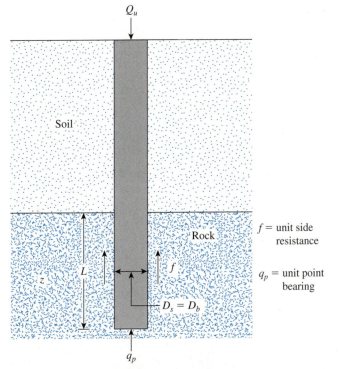

Figure 12.28 Drilled shaft socketed into rock

shaft bases were cast on or in generally *soft rock* with some degree of jointing. Based on these results, they proposed

$$Q_{u(\text{net})} = Q_p + Q_s = q_p A_p + f p L \tag{12.50}$$

where end bearing capacity Q_p can be expressed as

$$Q_p(\text{MN}) = q_p A_p = [4.83(q_u \text{ MN/m}^2)^{0.51}][A_p(\text{m}^2)] \tag{12.51}$$

Figure 12.29 shows the plot of $q_p(\text{MN/m}^2)$ versus $q_u(\text{MN/m}^2)$ obtained from the data on which Eq. (12.51) is based.

Also, the side resistance Q_s is

$$Q_s(\text{MN}) = f p L = [0.4(q_u \text{ MN/m}^2)^{0.5}][\pi D_s(\text{m})][L(\text{m})]$$
$$\text{(for smooth socket)} \tag{12.52}$$

and

$$Q_s(\text{MN}) = f p L = [0.8(q_u \text{ MN/m}^2)^{0.5}][\pi D_s(\text{m})][L(\text{m})]$$
$$\text{(for rough socket)} \tag{12.53}$$

Figure 12.29 Plot of q_p versus q_u (Adapted from Zhang and Einstein, 1998)

Example 12.8

Figure 12.30 shows a drilled shaft extending into a shale formation. For the intact rock cores, given $q_u = 4.2 \text{ MN/m}^2$. Estimate the allowable load-bearing capacity of the drilled shaft. Use a factor of safety (FS) = 3. Assume a smooth socket for side resistance.

Solution

From Eq. (12.51),

$$Q_p = A_p[4.83(q_u)^{0.51}] = \frac{\pi}{4}(1)^2[(4.83)(4.2)^{0.51}] = 7.89 \text{ MN}$$

Again, from Eq. (12.52),

$$Q_s = 0.4(q_u)^{0.5}(\pi D_s L) = 0.4(4.2)^{0.5}[(\pi)(1)(4)] = 10.3 \text{ MN}$$

Hence,

$$Q_{all} = \frac{Q_u}{\text{FS}} = \frac{Q_p + Q_s}{\text{FS}} = \frac{7.89 + 10.3}{3} = \mathbf{6.06MN}$$

Figure 12.30 Drilled shaft extending into rock

Problems

12.1 A drilled shaft is shown in Figure P12.1. Determine the net allowable point bearing capacity. Given

$D_b = 2 \text{ m}$ $\gamma_c = 15.6 \text{ kN/m}^3$
$D_s = 1.2 \text{ m}$ $\gamma_s = 17.6 \text{ kN/m}^3$
$L_1 = 6 \text{ m}$ $\phi' = 35°$
$L_2 = 3 \text{ m}$ $c_u = 35 \text{ kN/m}^2$
Factor of safety = 3

Use Eq. (12.18).

12.2 Redo Problem 12.1, this time using Eq. (12.15). Let $E_s = 600 p_a$.

12.3 For the drilled shaft described in Problem 12.1, what skin resistance would develop in the top 6 m, which are in clay? Use Eqs. (12.36) and (12.38).

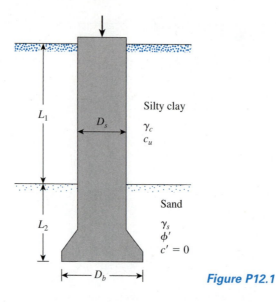

Silty clay

γ_c
c_u

Sand

γ_s
ϕ'
$c' = 0$

Figure P12.1

12.4 Redo Problem 12.1 with the following:

$D_b = 1.75$ m $\qquad \gamma_c = 17.8$ kN/m^3
$D_s = 1$ m $\qquad\quad \gamma_s = 18.2$ kN/m^3
$L_1 = 6.25$ m $\qquad \phi' = 32°$
$L_2 = 2.5$ m $\qquad\;\; c_u = 32$ kN/m^2
Factor of safety $= 4$

12.5 Redo Problem 12.4 using Eq. (12.5). Let $E_s = 400p_a$.

12.6 For the drilled shaft described in Problem 12.4, what skin friction would develop in the top 6.25 m?
a. Use Eqs. (12.36) and (12.38).
b. Use Eq. (12.39).

12.7 Figure P12.7 shows a drilled shaft without a bell. Assume the following values:

$L_1 = 6$ m $\qquad c_{u(1)} = 50$ kN/m^2
$L_2 = 7$ m $\qquad c_{u(2)} = 75$ kN/m^2
$D_s = 1.5$ m

Determine:

a. The net ultimate point bearing capacity [use Eqs. (12.33) and (12.34)]
b. The ultimate skin friction [use Eqs. (12.36) and (12.38)]
c. The working load Q_w (factor of safety $= 3$)

12.8 Repeat Problem 12.7 with the following data:

$L_1 = 6.1$ m $\qquad c_{u(1)} = 70$ kN/m^2
$L_2 = 3.05$ m $\quad c_{u(2)} = 120$ kN/m^2
$D_s = 0.91$ m

Use Eqs. (12.39) and (12.40).

12.9 A drilled shaft in a medium sand is shown in Figure P12.9. Using the method proposed by Reese and O'Neill, determine the following:

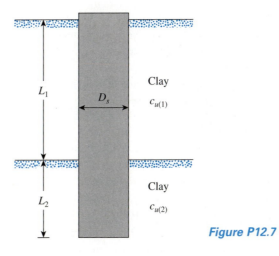

Clay
$c_{u(1)}$

Clay
$c_{u(2)}$

Figure P12.7

a. The net allowable point resistance for a base movement of 25 mm
b. The shaft frictional resistance for a base movement of 25 mm
c. The total load that can be carried by the drilled shaft for a total base movement of 25 mm

Assume the following values:

$$L = 12 \text{ m} \qquad \gamma = 18 \text{ kN/m}^3$$
$$L_1 = 11 \text{ m} \qquad \phi' = 38°$$
$$D_s = 1 \text{ m} \qquad D_r = 65\% \text{ (medium sand)}$$
$$D_b = 2 \text{ m}$$

Medium sand
γ
ϕ'

Average standard
penetration number (N_{60})
within $2D_b$ below the
drilled shaft = 19

Figure P12.9

12.10 In Figure P12.9, let $L = 7$ m, $L_1 = 6$ m, $D_s = 0.75$ m, $D_b = 1.25$ m, $\gamma = 18$ kN/m³, and $\phi' = 37°$. The average uncorrected standard penetration number (N_{60}) within $2D_b$ below the drilled shaft is 29. Determine
  a. The ultimate load-carrying capacity
  b. The load-carrying capacity for a settlement of 12 mm.
  The sand has 35% gravel. Use Eq. (12.30) and Figures 12.12 and 12.14.

12.11 For the drilled shaft described in Problem 12.7, determine
  a. The ultimate load-carrying capacity
  b. The load carrying capacity for a settlement of 25 mm
  Use the procedure outlined by Reese and O'Neill. (See Figures 12.18 and 12.19.)

12.12 For the drilled shaft described in Problem 12.7, estimate the total elastic settlement at working load. Use Eqs. (11.73), (11.75), and (11.76). Assume that $E_p = 20 \times 10^6$ kN/m², $C_p = 0.03$, $\xi = 0.65$, $\mu_s = 0.3$, $E_s = 12{,}000$ kN/m², and $Q_{ws} = 0.8Q_w$. Use the value of Q_w from Part (c) of Problem 12.7.

12.13 For the drilled shaft described in Problem 12.8, estimate the total elastic settlement at working load. Use Eqs. (11.73), (11.75), and (11.76). Assume that $E_p = 3 \times 10^6$ lb/in², $C_p = 0.03$, $\xi = 0.65$, $\mu_s = 0.3$, $E_s = 2000$ lb/in², and $Q_{ws} = 0.83Q_w$. Use the value of Q_w from Part (c) of Problem 12.8.

12.14 Figure P12.14 shows a drilled shaft extending into clay shale. Given: q_u(clay shale) $= 1.81$ MN/m². Considering the socket to be rough, estimate the allowable load-carrying capacity of the drilled shaft. Use FS $= 4$.

12.15 A free-headed drilled shaft is shown in Figure P12.15. Let $Q_g = 260$ kN, $M_g = 0$, $\gamma = 17.5$ kN/m³, $\phi' = 35°$, $c' = 0$, and $E_p = 22 \times 10^6$ kN/m². Determine
  a. The ground line deflection, x_o
  b. The maximum bending moment in the drilled shaft
  c. The maximum tensile stress in the shaft
  d. The minimum penetration of the shaft needed for this analysis

Figure P12.14   **Figure P12.15**

References

BEREZANTZEV, V. G., KHRISTOFOROV, V. S., and GOLUBKOV, V. N. (1961). "Load Bearing Capacity and Deformation of Piled Foundations," *Proceedings, Fifth International Conference on Soil Mechanics and Foundation Engineering,* Paris, Vol. 2, pp. 11–15.

CHEN, Y.-J., and KULHAWY, F. H. (1994). "Case History Evaluation of the Behavior of Drilled Shafts under Axial and Lateral Loading," *Final Report, Project 1493-04, EPRI TR-104601,* Geotechnical Group, Cornell University, Ithaca, NY, December.

DUNCAN, J. M., EVANS, L. T., JR., and OOI, P. S. K. (1994). "Lateral Load Analysis of Single Piles and Drilled Shafts," *Journal of Geotechnical Engineering,* ASCE, Vol. 120, No. 6, pp. 1018–1033.

KULHAWY, F. H., and JACKSON, C. S. (1989). "Some Observations on Undrained Side Resistance of Drilled Shafts," *Proceedings, Foundation Engineering: Current Principles and Practices,* American Society of Civil Engineers, Vol. 2, pp. 1011–1025.

MATLOCK, H., and REESE, L.C. (1961). "Foundation Analysis of Offshore Pile-Supported Structures," in *Proceedings, Fifth International Conference on Soil Mechanics and Foundation Engineering,* Vol. 2, Paris, pp. 91–97.

O'NEILL, M. W. (1997). Personal communication.

O'NEILL, M.W., and REESE, L.C. (1999). *Drilled Shafts: Construction Procedure and Design Methods,* FHWA Report No. IF-99-025.

REESE, L. C., and O'NEILL, M. W. (1988). *Drilled Shafts: Construction and Design,* FHWA, Publication No. HI-88-042.

REESE, L. C., and O'NEILL, M. W. (1989). "New Design Method for Drilled Shafts from Common Soil and Rock Tests," *Proceedings, Foundation Engineering: Current Principles and Practices,* American Society of Civil Engineers, Vol. 2, pp. 1026–1039.

REESE, L. C., TOUMA, F. T., and O'NEILL, M. W. (1976). "Behavior of Drilled Piers under Axial Loading," *Journal of Geotechnical Engineering Division,* American Society of Civil Engineers, Vol. 102, No. GT5, pp. 493–510.

ROLLINS, K. M., CLAYTON, R. J., MIKESELL, R. C., and BLAISE, B. C. (2005). "Drilled Shaft Side Friction in Gravelly Soils," *Journal of Geotechnical and Geoenvironmental Engineering,* American Society of Civil Engineers, Vol. 131, No. 8, pp. 987–1003.

WHITAKER, T., and COOKE, R. W. (1966). "An Investigation of the Shaft and Base Resistance of Large Bored Piles in London Clay," *Proceedings, Conference on Large Bored Piles,* Institute of Civil Engineers, London, pp. 7–49.

ZHANG, L., and EINSTEIN, H. H. (1998). "End Bearing Capacity of Drilled Shafts in Rock," *Journal of Geotechnical and Geoenvironmental Engineering,* American Society of Civil Engineers, Vol. 124, No. 7, pp. 574–584.

13 Foundations on Difficult Soils

13.1 Introduction

In many areas of the United States and other parts of the world, certain soils make the construction of foundations extremely difficult. For example, expansive or collapsible soils may cause high differential movements in structures through excessive heave or settlement. Similar problems can also arise when foundations are constructed over sanitary landfills. Foundation engineers must be able to identify difficult soils when they are encountered in the field. Although not all the problems caused by all soils can be solved, preventive measures can be taken to reduce the possibility of damage to structures built on them. This chapter outlines the fundamental properties of three major soil conditions—collapsible soils, expansive soils, and sanitary landfills—and methods of careful construction of foundations.

Collapsible Soil

13.2 Definition and Types of Collapsible Soil

Collapsible soils, which are sometimes referred to as *metastable soils,* are unsaturated soils that undergo a large change in volume upon saturation. The change may or may not be the result of the application of additional load. The behavior of collapsing soils under load is best explained by the typical void ratio effective pressure plot (e against $\log \sigma'$) for a collapsing soil, as shown in Figure 13.1. Branch ab is determined from the consolidation test on a specimen at its natural moisture content. At an effective pressure level of σ'_w, the equilibrium void ratio is e_1. However, if water is introduced into the specimen for saturation, the soil structure will collapse. After saturation, the equilibrium void ratio at the same effective pressure level σ'_w is e_2; cd is the branch of the e–$\log \sigma'$ curve under additional load after saturation. Foundations that are constructed on such soils may undergo large and sudden settlement if the soil under them becomes saturated with an unanticipated supply of moisture. The moisture may come from any of several sources, such as (a) broken water pipelines, (b) leaky sewers, (c) drainage

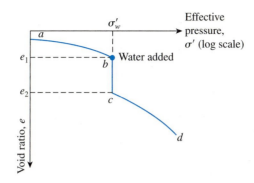

Figure 13.1 Nature of variation of void ratio with pressure for a collapsing soil

from reservoirs and swimming pools, (d) a slow increase in groundwater, and so on. This type of settlement generally causes considerable structural damage. Hence, identifying collapsing soils during field exploration is crucial.

The majority of naturally occurring collapsing soils are *aeolian*—that is, wind-deposited sands or silts, such as loess, aeolic beaches, and volcanic dust deposits. The deposits have high void ratios and low unit weights and are cohesionless or only slightly cohesive. *Loess* deposits have silt-sized particles. The cohesion in loess may be the result of clay coatings surrounding the silt-size particles. The coatings hold the particles in a rather stable condition in an unsaturated state. The cohesion also may be caused by the presence of chemical precipitates leached by rainwater. When the soil becomes saturated, the clay binders lose their strength and undergo a structural collapse. In the United States, large parts of the Midwest and the arid West have such types of deposit. Loess deposits are also found over 15 to 20% of Europe and over large parts of China. The thickness of loess deposits can range up to about 10 m in the central United States. In parts of China it can be up to 100 m. Figure 13.2 shows the extent of the loess deposits in the Mississippi River basin.

Many collapsing soils may be residual soils that are products of the weathering of parent rocks. Weathering produces soils with a large range of particle-size distribution. Soluble and colloidal materials are leached out by weathering, resulting in large void ratios and thus unstable structures. Many parts of South Africa and Zimbabwe have residual soils that are decomposed granites. Sometimes collapsing soil deposits may be left by flash floods and mudflows. These deposits dry out and are poorly consolidated. An excellent review of collapsing soils is that of Clemence and Finbarr (1981).

13.3 *Physical Parameters for Identification*

Several investigators have proposed various methods for evaluating the physical parameters of collapsing soils for identification. Some of these methods are discussed briefly in Table 13.1.

Jennings and Knight (1975) suggested a procedure for describing the *collapse potential* of a soil: An undisturbed soil specimen is taken at its natural moisture content in a consolidation ring. Step loads are applied to the specimen up to a pressure level σ'_w of 200 kN/m^2. (In Figure 13.1, this is σ'_w.) At that pressure, the specimen

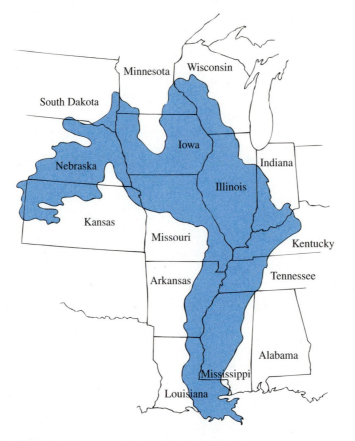

Figure 13.2 Loess deposit in Mississippi River basin

is flooded for saturation and left for 24 hours. This test provides the void ratios e_1 and e_2 before and after flooding, respectively. The collapse potential may now be calculated as

$$C_p = \Delta\varepsilon = \frac{e_1 - e_2}{1 + e_o} \tag{13.1}$$

where

e_o = natural void ratio of the soil
$\Delta\varepsilon$ = vertical strain

The severity of foundation problems associated with a collapsible soil have been correlated with the collapse potential C_p by Jennings and Knight (1975). They were summarized by Clemence and Finbarr (1981) and are given in Table 13.2.

Holtz and Hilf (1961) suggested that a loessial soil that has a void ratio large enough to allow its moisture content to exceed its liquid limit upon saturation is susceptible to collapse. So, for collapse,

$$w_{(saturated)} \geq LL \tag{13.2}$$

where LL = liquid limit.

However, for saturated soils,

$$e_o = wG_s \qquad (13.3)$$

where G_s = specific gravity of soil solids.

Table 13.1 Reported Criteria for Identification of Collapsing Soil[a]

Investigator	Year	Criteria
Denisov	1951	Coefficient of subsidence: $$K = \frac{\text{void ratio at liquid limit}}{\text{natural void ratio}}$$ $K = 0.5\text{--}0.75$: highly collapsible $K = 1.0$: noncollapsible loam $K = 1.5\text{--}2.0$: noncollapsible soils
Clevenger	1958	If dry unit weight is less than 12.6 kN/m^3, settlement will be large; if dry unit weight is greater than 14 kN/m^3 settlement will be small.
Priklonski	1952	$$K_D = \frac{\text{natural moisture content} - \text{plastic limit}}{\text{plasticity index}}$$ $K_D < 0$: highly collapsible soils $K_D > 0.5$: noncollapsible soils $K_D > 1.0$: swelling soils
Gibbs	1961	Collapse ratio, $R = \dfrac{\text{saturation moisture content}}{\text{liquid limit}}$ This was put into graph form.
Soviet Building Code	1962	$$L = \frac{e_o - e_L}{1 + e_o}$$ where e_o = natural void ratio and e_L = void ratio at liquid limit. For natural degree of saturation less than 60%, if $L > -0.1$, the soil is a collapsing soil.
Feda	1964	$$K_L = \frac{w_o}{S_r} - \frac{\text{PL}}{\text{PI}}$$ where w_o = natural water content, S_r = natural degree of saturation, PL = plastic limit, and PI = plasticity index. For $S_r < 100\%$, if $K_L > 0.85$, the soil is a subsident soil.
Benites	1968	A dispersion test in which 2 g of soil are dropped into 12 ml of distilled water and specimen is timed until dispersed; dispersion times of 20 to 30 s were obtained for collapsing Arizona soils.
Handy	1973	Iowa loess with clay (<0.002 mm) contents: $<16\%$: high probability of collapse $16\text{--}24\%$: probability of collapse $24\text{--}32\%$: less than 50% probability of collapse $>32\%$: usually safe from collapse

[a]Modified after Lutenegger and Saber (1988)

Table 13.2 Relation of Collapse Potential to the Severity of Foundation Problems[a]

C_p (%)	Severity of problem
0–1	No problem
1–5	Moderate trouble
5–10	Trouble
10–20	Severe trouble
20	Very severe trouble

[a]From Clemence, S. P., and Finbarr, A. O. (1981). "Design Considerations for Collapsible Soils," *Journal of the Geotechnical Engineering Division*, American Society of Civil Engineers, Vol. 107, GT3 pp. 305–317. With permission from ASCE.

Combining Eqs. (13.2) and (13.3) for collapsing soils yields

$$e_o \geqslant (\text{LL})(G_s) \tag{13.4}$$

The natural dry unit weight of the soil required for its collapse is

$$\gamma_d \leqslant \frac{G_s \gamma_w}{1 + e_o} = \frac{G_s \gamma_w}{1 + (\text{LL})(G_s)} \tag{13.5}$$

For an average value of $G_s = 2.65$, the limiting values of γ_d for various liquid limits may now be calculated from Eq. (13.5).

Figure 13.3 shows a plot of the preceding limiting values of dry unit weights against the corresponding liquid limits. For any soil, if the natural dry unit weight falls below the limiting line, the soil is likely to collapse.

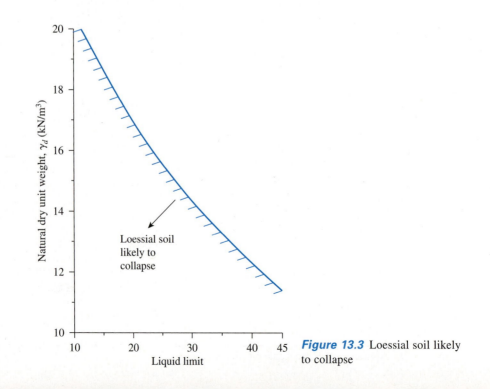

Figure 13.3 Loessial soil likely to collapse

Care should be taken to obtain undisturbed samples for determining the collapse potentials and dry unit weights—preferably block samples cut by hand. The reason is that samples obtained by thin-walled tubes may undergo some compression during the sampling process. However, if cut block samples are used, the boreholes should be made *without water.*

<h2>13.4   *Procedure for Calculating Collapse Settlement*</h2>

Jennings and Knight (1975) proposed the following laboratory procedure for determining the collapse settlement of structures upon saturation of soil:

Step 1.   Obtain *two* undisturbed soil specimens for tests in a standard consolidation test apparatus (oedometer).

Step 2.   Place the two specimens under 1 kN/m^2 pressure for 24 hours.

Step 3.   After 24 hours, saturate one specimen by flooding. Keep the other specimen at its natural moisture content.

Step 4.   After 24 hours of flooding, resume the consolidation test on both specimens by doubling the load (the same as in the standard consolidation test) to the desired pressure level.

Step 5.   Plot the e–$\log \sigma'$ graphs for both specimens (Figures 13.4a and b).

Step 6.   Calculate the *in situ* effective pressure, σ'_o. Draw a vertical line corresponding to the pressure σ'_o.

Step 7.   From the e–$\log \sigma'_o$ curve of the soaked specimen, determine the preconsolidation pressure, σ'_c. If $\sigma'_c/\sigma'_o = 0.8$–$1.5$, the soil is normally consolidated; however, if $\sigma'_c/\sigma'_o > 1.5$, the soil is preconsolidated.

Figure 13.4 Settlement calculation from double oedometer test: (a) normally consolidated soil; (b) overconsolidated soil

Step 8. Determine e_o', corresponding to σ_o' from the e–log σ_o' curve of the soaked specimen. (This procedure for normally consolidated and overconsolidated soils is shown in Figures 13.4a and b, respectively.)

Step 9. Through point (σ_o', e_o') draw a curve that is similar to the e–log σ_o' curve obtained from the specimen tested at its natural moisture content.

Step 10. Determine the incremental pressure, $\Delta\sigma'$, on the soil caused by the construction of the foundation. Draw a vertical line corresponding to the pressure of $\sigma_o' + \Delta\sigma'$ in the e–log σ' curve.

Step 11. Now, determine Δe_1 and Δe_2. The settlement of soil without change in the natural moisture content is

$$S_{c(1)} = \frac{\Delta e_1}{1 + e_o'} (H) \tag{13.6}$$

where H = thickness of soil susceptible to collapse.

Also, the settlement caused by collapse in the soil structure is

$$S_{c(2)} = \frac{\Delta e_2}{1 + e_o'} (H) \tag{13.7}$$

13.5 Foundation Design in Soils Not Susceptible to Wetting

For actual foundation design purposes, some standard field load tests may also be conducted. Figure 13.5 shows the relation of the nature of load per unit arca versus settlement in a field load test in a loessial deposit. Note that the load–settlement relationship is essentially linear up to a certain critical pressure, σ_{cr}', at which there is a breakdown of the soil structure and hence a large settlement. Sudden breakdown of soil structure is more common with soils having a high natural moisture content than with normally dry soils.

If enough precautions are taken in the field to prevent moisture from increasing under structures, spread foundations and mat foundations may be built on collapsible

Figure 13.5 Field load test in loessial soil: load per unit area versus settlement

soils. However, the foundations must be proportioned so that the critical stresses (see Figure 13.5) in the field are never exceeded. A factor of safety of about 2.5 to 3 should be used to calculate the allowable soil pressure, or

$$\sigma'_{all} = \frac{\sigma'_{cr}}{FS} \tag{13.8}$$

where

σ'_{all} = allowable soil pressure
FS = factor of safety

The differential and total settlements of these foundations should be similar to those of foundations designed for sandy soils.

Continuous foundations may be safer than isolated foundations over collapsible soils in that they can effectively minimize differential settlement. Figure 13.6 shows a typical procedure for constructing continuous foundations. This procedure uses footing beams and longitudinal load-bearing beams.

In the construction of heavy structures, such as grain elevators, over collapsible soils, settlements up to about 0.3 m are sometimes allowed (Peck, Hanson, and Thornburn, 1974). In this case, tilting of the foundation is not likely to occur, because there is no eccentric loading. The total expected settlement for such structures can be estimated from standard consolidation tests on specimens of field moisture content. Without eccentric loading, the foundations will exhibit uniform settlement over loessial deposits; however, if the soil is of residual or colluvial nature, settlement may not be uniform. The reason is the nonuniformity generally encountered in residual soils.

Extreme caution must be used in building heavy structures over collapsible soils. If large settlements are expected, drilled-shaft and pile foundations should be considered. These types of foundation can transfer the load to a stronger load-bearing stratum.

Figure 13.6 Continuous foundation with load-bearing beams (From Clemence, S. P., and Finbarr, A. O. (1981). "Design Considerations for Collapsible Soils," *Journal of the Geotechnical Engineering Division*, American Society of Civil Engineers, Vol. 107, GT3 pp. 305–317. With permission from ASCE.)

| 13.6 | *Foundation Design in Soils Susceptible to Wetting* |

If the upper layer of soil is likely to get wet and collapse sometime after construction of the foundation, several design techniques to avoid failure of the foundation may be considered. They are as follow:

Dynamic Compaction

If the expected depth of wetting is about 1.5 to 2 m from the ground surface, the soil may be moistened and recompacted by heavy rollers. Spread footings and mats may be constructed over the compacted soil. An alternative to recompaction by heavy rollers is *heavy tamping,* which is sometimes referred to as *dynamic compaction.* (See Chapter 14.) Heavy tamping consists primarily of dropping a heavy weight repeatedly on the ground. The height of the drop can vary from 8 to 30 m. The stress waves generated by the dropping weight help in the densification of the soil.

Lutenegger (1986) reported the use of dynamic compaction to stabilize a thick layer of friable loess before construction of a foundation in Russe, Bulgaria. During field exploration, the water table was not encountered to a depth of 10 m, and the natural moisture content was below the plastic limit. Initial density measurements made on undisturbed soil specimens indicated that the moisture content at saturation would exceed the liquid limit, a property usually encountered in collapsible loess. For dynamic compaction of the soil, the upper 1.7 m of crust material was excavated. A circular concrete weight of 133 kN was used as a hammer. At each grid point, compaction was achieved by dropping the hammer 7 to 12 times through a vertical distance of 2.5 m.

Figure 13.7 shows the dry unit weight of the soil before and after compaction. The increase in dry unit weight of the soil shows that dynamic compaction can be used effectively to stabilize collapsible soil.

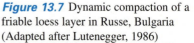

Figure 13.7 Dynamic compaction of a friable loess layer in Russe, Bulgaria (Adapted after Lutenegger, 1986)

Chemical Stabilization

If conditions are favorable, foundation trenches can be flooded with solutions of sodium silicate and calcium chloride to stabilize the soil chemically. The soil will then behave like a soft sandstone and resist collapse upon saturation. This method is successful only if the solutions can penetrate to the desired depth; thus, it is most applicable to fine sand deposits. Silicates are rather costly and are not generally used. However, in some parts of Denver, silicates have been used successfully.

The injection of a sodium silicate solution to stabilize collapsible soil deposits has been used extensively in the former Soviet Union and Bulgaria (Houston and Houston, 1989). This process, which is used for dry collapsible soils and for wet collapsible soils that are likely to compress under the added weight of the structure to be built, consists of three steps:

Step 1. Injection of carbon dioxide to remove any water that is present and for preliminary activation of the soil

Step 2. Injection of sodium silicate grout

Step 3. Injection of carbon dioxide to neutralize alkalies.

Vibrofloation and Ponding

When the soil layer is susceptible to wetting to a depth of about 10 m ($\approx$30 ft), several techniques may be used to cause collapse of the soil *before* construction of the foundation is begun. Two of these techniques are *vibroflotation* and *ponding* (also called *flooding*). Vibroflotation is used successfully in free-draining soil. (See Chapter 14.) Ponding—by constructing low dikes—is utilized at sites that have no impervious layers. However, even after saturation and collapse of the soil by ponding, some additional settlement of the soil may occur after construction of the foundation is begun. Additional settlement may also be caused by incomplete saturation of the soil at the time of construction. Ponding may be used successfully in the construction of earth dams.

Extending Foundation Beyond Zone of Wetting

If precollapsing the soil is not practical, foundations may be extended beyond the zone of possible wetting, although the technique may require drilled shafts and piles. The design of drilled shafts and piles must take into consideration the effect of negative skin friction resulting from the collapse of the soil structure and the associated settlement of the zone of subsequent wetting.

In some cases, a *rock-column type of foundation* (*vibroreplacement*) may be considered. Rock columns are built with large boulders that penetrate the collapsible soil layer. They act as piles in transferring the load to a more stable soil layer.

Expansive Soils

13.7 General Nature of Expansive Soils

Many plastic clays swell considerably when water is added to them and then shrink with the loss of water. Foundations constructed on such clays are subjected to large uplifting forces caused by the swelling. These forces induce heaving, cracking, and the breakup of both building foundations and slab-on-grade members. Figure 13.8 shows

Figure 13.8 Cracks in a wall due to heaving of an expansive clay (*Courtesy of Anand Puppala, University of Texas at Arlington, Arlington, Texas*)

the cracks in a wall due to excessive heaving. Expansive clays cover large parts of the United States, South America, Africa, Australia, and India. In the United States, these clays are predominant in Texas, Oklahoma, and the upper Missouri Valley. In general, expansive clays have liquid limits and plasticity indices greater than about 40 and 15, respectively.

As noted, the increase and decrease in moisture content causes clay to swell and shrink. Figure 13.9 shows shrinkage cracks on the ground surface of a clay weathered from the Eagle Ford shale formation in the Dallas-Fort Worth, Texas area. The depth in a soil to which periodic changes of moisture occur is usually referred to as the *active zone* (see Figure 13.10). The depth of the active zone varies, depending on the location of the site. Some typical active-zone depths in American cities are given in Table 13.3. In some clays and clay shales in the western United States, the depth of the active zone can be as much as 15 m. The active-zone depth can easily be determined by plotting the liquidity index against the depth of the soil profile over several seasons. Figure 13.11 shows such a plot for the Beaumont formation in the Houston area.

Figure 13.9 Shrinkage cracks on ground surface in a clay weathered from Eagle Ford shale formation in the Dallas-Fort Worth area (*Courtesy of Thomas M. Petry, Missouri University of Science and Technology, Rolla, Missouri*)

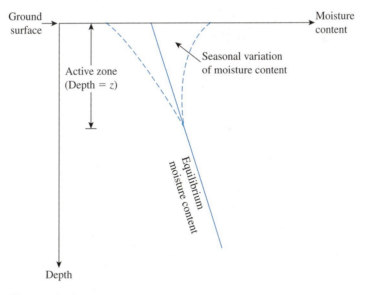

Figure 13.10 Definition of active zone

Table 13.3 Typical Active-Zone Depths in Some U.S. Cities[a]

City	Depth of active zone (m)
Houston	1.5 to 3
Dallas	2.1 to 4.6
San Antonio	3 to 9
Denver	3 to 4.6

[a] After O'Neill and Poormoayed (1980) (O'Neill, M. W., and Poormoayed, N.(1980). "Methodology for Foundations on Expansive Clays," *Journal of the Geotechnical Engineering Division*, American Society of Civil Engineers, Vol. 106, No. GT12, pp. 1345–1367. With permission from ASCE.)

Shrinkage cracks can extend deep into the active zone. Figure 13.12 shows interconnected shrinkage cracks extending from the ground surface into the active zone in an expansive clay.

To study the magnitude of possible swell in a clay, simple laboratory oedometer tests can be conducted on undisturbed specimens. Two common tests are the *unrestrained swell test* and the *swelling pressure test*. They are described in the following sections.

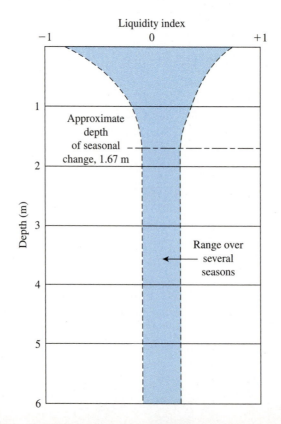

Figure 13.11 Active zone in Houston area, Beaumont formation (O'Neill, M. W. and Poormoayed, N. (1980). "Methodology for Foundations on Expansive Clays," *Journal of the Geotechnical Engineering Division*, American Society of Civil Engineers, Vol. 106, No. GT12, pp. 1345–1367. With permission from ASCE.)

Figure 13.12 Interconnected shrinkage cracks extended from the ground surface into the active zone (*Courtesy of Thomas M. Petry, Missouri University of Science and Technology, Rolla, Missouri*)

13.8   *Unrestrained Swell Test*

In the *unrestrained swell test,* the specimen is placed in an oedometer under a small surcharge of about 6.9 kN/m^2 (1 lb/in^2). Water is then added to the specimen, and the expansion of the volume of the specimen (i.e., its height; the area of cross section is constant) is measured until equilibrium is reached. The percentage of free swell may than be expressed as the ratio

$$s_{w(\text{free})}(\%) = \frac{\Delta H}{H}(100) \tag{13.9}$$

where

$s_{w(\text{free})}$ = free swell, as a percentage
ΔH = height of swell due to saturation
H = original height of the specimen

Vijayvergiya and Ghazzaly (1973) analyzed various soil test results obtained in this manner and prepared a correlation chart of the free swell, liquid limit, and natural moisture content, as shown in Figure 13.13. O'Neill and Poormoayed (1980) developed a relationship for calculating the free surface swell from this chart:

$$\Delta S_F = 0.0033 Z s_{w(\text{free})} \tag{13.10}$$

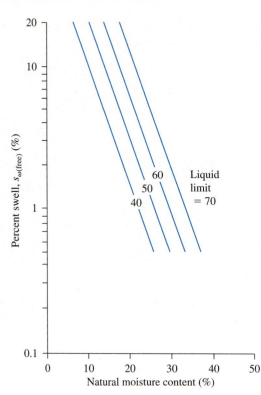

Figure 13.13 Relation between percentage of free swell, liquid limit, and natural moisture content (From Vijayvergiya, V. N. and Ghazzaly, O. I. (1973). "Prediction of Swelling Potential of Natural Clays," Proceedings, Third International Research and Engineering Conference on Expansive Clays, pp. 227–234. With permission from ASCE.)

where

$$\Delta S_F = \text{free surface swell}$$
$$Z = \text{depth of active zone}$$
$$s_{w(\text{free})} = \text{free swell, as a percentage}$$

Swelling Pressure Test

The swelling pressure can be determined from two different types of tests. They are

- Conventional consolidation test
- Constant volume test

The two methods are briefly described here.

Conventional Consolidation Test

In this type of test, the specimen is placed in a oedometer under a small surcharge of about 6.9 kN/m². Water is added to the specimen, allowing it to swell and reach an equilibrium position after some time. Subsequently, loads are added in convenient steps, and the specimen is consolidated. The plot of specimen deformation (δ) versus log σ' is shown in Figure 13.14. The δ versus log σ' plot crosses the horizontal line through the point of initial condition. The pressure corresponding to the point of intersection is the *zero swell pressure*, σ'_{sw}.

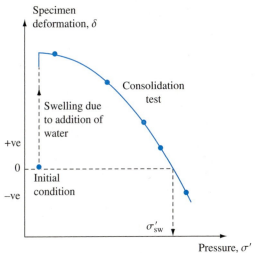

Figure 13.14 Zero swell pressure from conventional consolidation test

Constant Volume Test

The *constant volume test* can be conducted by taking a specimen in a consolidation ring and applying a pressure equal to the effective overburden pressure, σ'_o, plus the approximate anticipated surcharge caused by the foundation, σ'_s. Water is then added to the specimen. As the specimen starts to swell, pressure is applied in small increments to prevent swelling. Pressure is maintained until full swelling pressure is developed on the specimen, at which time the total pressure is

$$\sigma'_{sw} = \sigma'_o + \sigma'_s + \sigma'_1 \tag{13.11}$$

where

σ'_{sw} = total pressure applied to prevent swelling, or zero swell pressure
σ'_1 = additional pressure applied to prevent swelling after addition of water

Figure 13.15 shows the variation of the percentage of swell with effective pressure during a swelling pressure test. (For more information on this type of test, see Sridharan et al., 1986.)

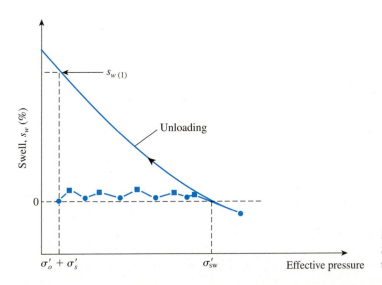

Figure 13.15
Swelling pressure test

A σ'_{sw} of about 20 to 30 kN/m^2 is considered to be low, and a σ'_{sw} of 1500 to 2000 kN/m^2 is considered to be high. After zero swell pressure is attained, the soil specimen can be unloaded in steps to the level of the effective overburden pressure, σ'_o. The unloading process will cause the specimen to swell. The equilibrium swell for each pressure level is also recorded. The variation of the swell, s_w in percent, and the applied pressure on the specimen will be like that shown in Figure 13.15.

The constant volume test can be used to determine the surface heave, ΔS, for a foundation (O'Neill and Poormoayed, 1980) as given by the formula

$$\Delta S = \sum_{i=1}^{n} [s_{w(1)} (\%)](H_i)(0.01) \qquad (13.12)$$

where

$s_{w(1)}(\%)$ = swell, in percent, for layer i under a pressure of $\sigma'_o + \sigma'_s$ (see Figure 13.15)
ΔH_i = thickness of layer i

It is important to point out that the zero swell pressure (σ'_{sw}) obtained from the conventional consolidation test and the constant volume test may not be the same. Table 13.4 summarizes some laboratory test results of Sridharan et al. (1986) to illustrate this point. It also was shown by Sridharan et al. (1986) that the zero swell pressure is a function of the dry unit weight of soil, but not of the initial moisture content (Figure 13.16).

Table 13.4 Comparison of Zero Swell Pressure Obtained from Conventional Consolidation Tests and Constant Volume Tests—Summary of Test Results of Sridharan et al. (1986)

Soil	Liquid limit	Plasticity index	Initial void ratio, e_i	σ'_{sw} (kN/m²)	
				Consolidation test	Constant volume test
BC-1	80	44	0.893	294.3	186.4
BC-4	98	57	1.002	382.6	251.8
BC-5	96	65	0.742	500.3	304.1
BC-7	96	66	0.572	1275.3	372.8
BC-8	94	62	0.656	147.2	68.7

Example 13.1

A soil profile has an active zone of expansive soil of 1.83 m. The liquid limit and the average natural moisture content during the construction season are 50% and 20%, respectively. Determine the free surface swell.

Figure 13.16 Plot of zero swell pressure with the dry unit weight of soil (After Sridharan et al., 1986. Copyright ASTM INTERNATIONAL. Reprinted with permission.)

Solution

From Figure 13.13 for LL = 50% and $w = 20\%$, $s_{w(\text{free})} = 3\%$. From Eq. (13.10),

$$\Delta S_F = 0.0033 Z s_{w(\text{free})}$$

Hence,

$$\Delta S_F = 0.0033(1.83)(3)(12) = \textbf{18.12 mm}$$ ■

Example 13.2

A soil profile's active-zone depth is 3.5 m. If a foundation is to be placed 0.5 m below the ground surface, what would be the estimated total swell? The following data were obtained from laboratory tests:

Depth (m)	Swell under effective overburden and estimated foundation surcharge pressure, $s_{w(1)}$(%)
0.5	2
1	1.5
2	0.75
3	0.25

Solution

The values of $s_{w(1)}$ (%) are plotted with depth in Figure 13.17a. The area of this diagram will be the total swell. The trapezoidal rule yields

$$\Delta S = \frac{1}{100}\left[\frac{1}{2}(1)(0+0.5) + \frac{1}{2}(1)(0.5+1.1) + \frac{1}{2}(1)(1.1+2)\right]$$

$$= 0.026 \text{ m} = \textbf{26 mm}$$

Figure 13.17 (a) Variation of $s_{w(1)}$ with depth; (b) variation of ΔS with depth

Example 13.3

In Example 13.2, if the allowable total swell is 10 mm, what would be the undercut necessary to reduce the total swell?

Solution

Using the procedure outlined in Example 13.2, we calculate the total swell at various depths below the foundation from Figure 13.17a as follows:

Depth (m)	Total swell, ΔS (mm)
3.5	0
3	$0 + \left[\frac{1}{2}(0.5)(0.25)\right]\frac{1}{100} = 0.000625 \text{ m} = 0.625 \text{ mm}$
2.5	$0.000625 + \frac{1}{100}\left[\frac{1}{2}(0.5)(0.25 + 0.5)\right] = 0.0025 \text{ m} = 2.5 \text{ mm}$
1.5	$0.0025 + \frac{1}{100}\left[\frac{1}{2}(1)(0.5 + 1.1)\right] = 0.0105 \text{ m} = 10.5 \text{ mm}$
0.5	26 mm

Plotted in Figure 13.17b, these total settlements show that a total swell of 10 mm corresponds to a depth of 1.6 m below the ground surface.

Hence, the undercut below the foundation is **$1.6 - 0.5 = 1.1$ m**. This soil should be excavated, replaced by nonswelling soil, and recompacted.

Classification of Expansive Soil on the Basis of Index Tests

Classification systems for expansive soils are based on the problems they create in the construction of foundations (potential swell). Most of the classifications contained in the literature are summarized in Figure 13.18 and Table 13.5. However, the classification system developed by the U.S. Army Waterways Experiment Station (Snethen et al., 1977) is the one most widely used in the United States. It has also been summarized by O'Neill and Poormoayed (1980); see Table 13.6. Sridharan (2005) proposed an index called the *free swell ratio* to predict the clay type, potential swell classification, and dominant clay minerals present

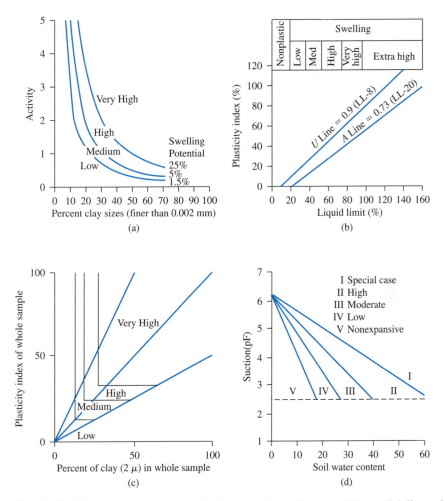

Figure 13.18 Commonly used criteria for determining swell potential (From Abduljauwad, S. N. and Al-Sulaimani, G. J. (1993). "Determination of Swell Potential of Al-Qatif Clay," *Geotechnical Testing Journal*, American Society for Testing and Materials, vol. 16, No. 4, pp. 469–484. Copyright ASTM INTERNATIONAL. Reprinted with permission.)

Table 13.5 Summary of Some Criteria for Identifying Swell Potential (From Abduljauwad, S. N. and Al-Sulaimani, G. J. (1993). "Determination of Swell Potential of Al-Qatif Clay," Geotechnical Testing Journal, American Society for Testing and Materials, vol. 16, No. 4, pp. 469–484. Copyright ASTM INTERNATIONAL. Reprinted with permission.)

Reference	Criteria	Remarks
Holtz (1959)	CC > 28, PI > 35, and SL < 11 (very high) $20 \leqslant$ CC $\leqslant 31$, $25 \leqslant$ PI $\leqslant 41$, and $7 \leqslant$ SL $\leqslant 12$ (high) $13 \leqslant$ CC $\leqslant 23$, $15 \leqslant$ PI $\leqslant 28$, and $10 \leqslant$ SL $\leqslant 16$ (medium) CC $\leqslant 15$, PI $\leqslant 18$, and SL $\geqslant 15$ (low)	Based on CC, PI, and SL
Seed et al. (1962)	See Figure 13.18a	Based on oedometer test using compacted specimen, percentage of clay <2 μm, and activity
Altmeyer (1955)	LS < 5, SI > 12, and PS < 0.5 (noncritical) $5 \leqslant$ LS $\leqslant 8$, $10 \leqslant$ SL $\leqslant 12$, and $0.5 \leqslant$ PS $\leqslant 1.5$ (marginal) LS > 8, SL < 10, and PS > 1.5 (critical)	Based on LS, SL, and PS Remolded sample ($\rho_{d(max)}$ and w_{opt}) soaked under 6.9 kPa surcharge
Dakshanamanthy and Raman (1973)	See Figure 13.18b	Based on plasticity chart
Raman (1967)	PI > 32 and SI > 40 (very high) $23 \leqslant$ PI $\leqslant 32$ and $30 \leqslant$ SI $\leqslant 40$ (high) $12 \leqslant$ PI $\leqslant 23$ and $15 \leqslant$ SI $\leqslant 30$ (medium) PI < 12 and SI < 15 (low)	Based on PI and SI
Sowers and Sowers (1970)	SL < 10 and PI > 30 (high) $10 \leqslant$ SL $\leqslant 12$ and $15 \leqslant$ PI $\leqslant 30$ (moderate) SL > 12 and PI < 15 (low)	Little swell will occur when w_o results in LI of 0.25
Van Der Merwe (1964)	See Figure 13.18c	Based on PI, percentage of clay <2 μm, and activity
Uniform Building Code, 1968	EI > 130 (very high) and $91 \leqslant$ EI $\leqslant 130$ (high) $51 \leqslant$ EI $\leqslant 90$ (medium) and $21 \leqslant$ EI $\leqslant 50$ (low) $0 \leqslant$ EI $\leqslant 20$ (very low)	Based on oedometer test on compacted specimen with degree of saturation close to 50% and surcharge of 6.9 kPa
Snethen (1984)	LL > 60, PI > 35, τ_{nat} > 4, and SP > 1.5 (high) $30 \leqslant$ LL $\leqslant 60$, $25 \leqslant$ PI $\leqslant 35$, $1.5 \leqslant \tau_{nat} \leqslant 4$, and $0.5 \leqslant$ SP $\leqslant 1.5$ (medium) LL < 30, PI < 25, τ_{nat} < 1.5, and SP < 0.5 (low)	PS is representative for field condition and can be used without τ_{nat}, but accuracy will be reduced
Chen (1988)	PI ≥ 35 (very high) and $20 \leqslant$ PI $\leqslant 55$ (high) $10 \leqslant$ PI $\leqslant 35$ (medium) and PI $\leqslant 15$ (low)	Based on PI
McKeen (1992)	Figure 13.18d	Based on measurements of soil water content, suction, and change in volume on drying
Vijayvergiya and Ghazzaly (1973)	$\log SP = (1/12)(0.44 \, LL - w_o + 5.5)$	Empirical equations
Nayak and Christensen (1974)	$SP = (0.00229 \, PI)(1.45C)/w_o + 6.38$	Empirical equations
Weston (1980)	$SP = 0.00411(LL_w)^{4.17} q^{-3.86} w_o^{-2.33}$	Empirical equations

Table 13.5 (*continued*)

Note: C = clay, %

CC = colloidal content, %

EI = Expansion index = $100 \times$ percent swell $\times$ fraction passing No. 4 sieve

LI = liquidity index, %

LL = liquid limit, %

LL_w = weighted liquid limit, %

LS = linear shrinkage, %

PI = plasticity index, %

PS = probable swell, %

q = surcharge

SI = shrinkage index = LL − SL, %

SL = shrinkage limit, %

SP = swell potential, %

w_o = natural soil moisture

w_{opt} = optimum moisture content, %

τ_{nat} = natural soil suction in tsf

$\rho_{d(max)}$ = maximum dry density

Table 13.6 Expansive Soil Classification System[a]

Liquid limit	Plasticity index	Potential swell (%)	Potential swell classification
<50	<25	<0.5	Low
50–60	25–35	0.5–1.5	Marginal
>60	>35	>1.5	High

Potential swell = vertical swell under a pressure equal to overburden pressure

[a]Compiled from O'Neill and Poormoayed (1980)

in a given soil. The free swell ratio can be determined by finding the equilibrium sediment volumes of 10 grams of an oven-dried specimen passing No. 40 U.S. sieve (0.425 mm opening) in distilled water (V_d) and in CCl_4 or kerosene (V_K). The free swell ratio (FSR) is defined as

$$\text{FSR} = \frac{V_d}{V_K} \tag{13.13}$$

Table 13.7 gives the expansive soil classification based on free swell ratio. Also, Figure 13.19 shows the classification of soil based on the free swell ratio.

Table 13.7 Expansive Soil Classification Based on Free Swell Ratio

Free swell ratio	Clay type	Potential swell classification	Dominant clay mineral
≤ 1.0	Non-swelling	Negligible	Kaolinite
1.0–1.5	Mixture of swelling and non-swelling	Low	Kaolinite and montmorillonite
1.5–2.0	Swelling	Moderate	Montmorillonite
2.0–4.0	Swelling	High	Montmorillonite
> 4.0	Swelling	Very High	Montmorillonite

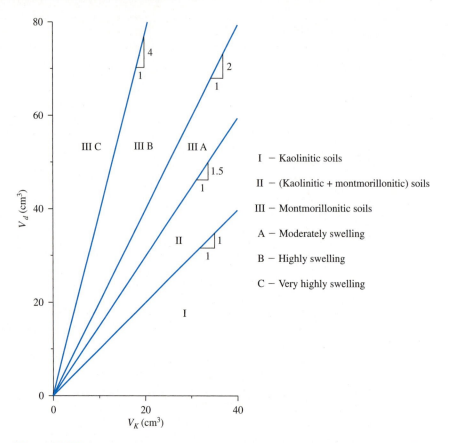

Figure 13.19 Classification based on free swell ratio (Adapted from Sridharan, 2005)

13.11   *Foundation Considerations for Expansive Soils*

If a soil has a low swell potential, standard construction practices may be followed. However, if the soil possesses a marginal or high swell potential, precautions need to be taken, which may entail

1. Replacing the expansive soil under the foundation
2. Changing the nature of the expansive soil by compaction control, prewetting, installation of moisture barriers, or chemical stabilization
3. Strengthening the structures to withstand heave, constructing structures that are flexible enough to withstand the differential soil heave without failure, or constructing isolated deep foundations below the depth of the active zone

One particular method may not be sufficient in all situations. Combining several techniques may be necessary, and local construction experience should always be considered. Following are some details regarding the commonly used techniques for dealing with expansive soils.

Replacement of Expansive Soil

When shallow, moderately expansive soils are present at the surface, they can be removed and replaced by less expansive soils and then compacted properly.

Changing the Nature of Expansive Soil

1. *Compaction:* The heave of expansive soils decreases substantially when the soil is compacted to a lower unit weight on the high side of the optimum moisture content (possibly 3 to 4% above the optimum moisture content). Even under such conditions, a slab-on-ground type of construction should not be considered when the total probable heave is expected to be about 38 mm or more.

2. *Prewetting:* One technique for increasing the moisture content of the soil is ponding and hence achieving most of the heave before construction. However, this technique may be time consuming because the seepage of water through highly plastic clays is slow. After ponding, 4 to 5% of hydrated lime may be added to the top layer of the soil to make it less plastic and more workable (Gromko, 1974).

3. *Installation of moisture barriers:* The long-term effect of the differential heave can be reduced by controlling the moisture variation in the soil. This is achieved by providing vertical moisture barriers about 1.5 m deep around the perimeter of slabs for the slab-on-grade type of construction. These moisture barriers may be constructed in trenches filled with gravel, lean concrete, or impervious membranes.

4. *Stabilization of soil:* Chemical stabilization with the aid of lime and cement has often proved useful. A mix containing about 5% lime is sufficient in most cases. The effect of lime in stabilizing expansive soils, thereby reducing the shrinking and swelling characteristics, can be demonstrated with the aid of Figure 13.20. For this, expansive clay weathered from the Eagle Ford shale formation in the Dallas-Fort Worth, Texas area was taken. Some of it was mixed with water to about its liquid limit. It was placed in two molds that were about 152 mm long and 12.7 mm × 12.7 mm in cross section. Figure 13.20a shows the shrinkage of the soil specimens in the mold in a *dry condition*. The same soil also was mixed with 6% lime (by dry weight) and then with a similar amount of water and placed in six similar molds. Figures 13.20b shows the shrinkage of the lime-stabilized specimens in a dry condition, which was practically negligible compared to that seen in Figure 13.20a. Lime or cement and water are mixed with the top layer of soil and compacted. The addition

(a)

Figure 13.20 Shrinkage of expansive clay (Eagle Ford soil) mixed with water to about its liquid limit in molds of 152 mm × 12.7 mm × 12.7 mm: (a) without addition of lime;

(b)

Figure 13.20 *(continued)* (b) with addition of 6% lime by weight *(Courtesy of Thomas M. Petry, Missouri University of Science and Technology, Rolla, Missouri)*

of lime or cement will decrease the liquid limit, the plasticity index, and the swell characteristics of the soil. This type of stabilization work can be done to a depth of 1 to 1.5 m. Hydrated high-calcium lime and dolomite lime are generally used for lime stabilization.

Another method of stabilization of expansive soil is the *pressure injection* of lime slurry or lime–fly-ash slurry into the soil, usually to a depth of 4 to 5 m or and occasionally deeper to cover the active zone. Further details of the pressure injection technique are presented in Chapter 14. Depending on the soil conditions at a site, single or multiple injections can be planned, as shown in Figure 13.21. Figure 13.22 shows the slurry pressure injection work for a building pad. The stakes that are marked are the planned injection points. Figure 13.23 shows lime–fly-ash stabilization by pressure injection of the bank of a canal that had experienced sloughs and slides.

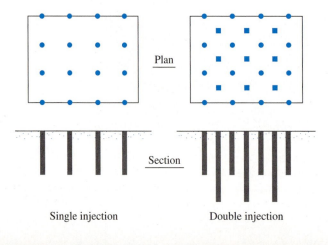

Plan

Section

Single injection   Double injection

Figure 13.21 Multiple lime slurry injection planning for a building pad

Figure 13.22 Pressure injection of lime slurry for a building pad (*Courtesy of Hayward Baker Inc., Odenton, Maryland.*)

Figure 13.23 Slope stabilization of a canal bank by pressure injection of lime–fly-ash slurry (*Courtesy of Hayward Baker Inc., Odenton, Maryland.*)

13.12 *Construction on Expansive Soils*

Care must be exercised in choosing the type of foundation to be used on expansive soils. Table 13.8 shows some recommended construction procedures based on the total predicted heave, ΔS, and the length-to-height ratio of the wall panels. For example, the table proposes the use of waffle slabs as an alternative in designing rigid buildings that are capable of tolerating movement. Figure 13.24 shows a schematic diagram of a waffle slab. In this type of construction, the ribs hold the structural load. The waffle voids allow the expansion of soil.

Table 13.8 Construction Procedures for Expansive Clay Soils[a]

Total predicted heave (mm)		Recommended construction	Method	Remarks
L/H = 1.25	L/H = 2.5			
0 to 6.35	12.7	No precaution		
6.35 to 12.7	12.7 to 50.8	Rigid building tolerating movement (steel reinforcement as necessary)	*Foundations:* Pads Strip footings mat (waffle)	Footings should be small and deep, consistent with the soil-bearing capacity. Mats should resist bending.
			Floor slabs: Waffle Tile	Slabs should be designed to resist bending and should be independent of grade beams.
			Walls:	Walls on a mat should be as flexible as the mat. There should be no vertical rigid connections. Brickwork should be strengthened with tie bars or bands.
12.7 to 50.8	50.8 to 101.6	Building damping movement	*Joints:* Clear Flexible	Contacts between structural units should be avoided, or flexible, waterproof material may be inserted in the joints.
			Walls: Flexible Unit construction Steel frame	Walls or rectangular building units should heave as a unit.
			Foundations: Three point Cellular Jacks	Cellular foundations allow slight soil expansion to reduce swelling pressure. Adjustable jacks can be inconvenient to owners. Three-point loading allows motion without duress.
>50.8	>101.6	Building independent of movement	*Foundation drilled shaft:* Straight shaft Bell bottom	Smallest-diameter and widely spaced shafts compatible with the load should be placed. Clearance should be allowed under grade beams.
			Suspended floor:	Floor should be suspended on grade beams 305 to 460 mm above the soil.

[a]Gromko, G. J., (1974). "Review of Expansive Soils," *Journal of the Geotechnical Engineering Division*, American Society of Civil Engineers, Vol. 100, No. GT6, pp. 667–687. With permission from ASCE.

Figure 13.24 Waffle slab

Figure 13.25 (a) Construction of drilled shafts with bells and grade beam; (b) definition of parameters in Eq. (13.14)

Table 13.8 also suggests the use of a drilled shaft foundation with a suspended floor slab when structures are constructed independently of movement of the soil. Figure 13.25a shows a schematic diagram of such an arrangement. The bottom of the shafts should be placed below the active zone of the expansive soil. For the design of the shafts, the uplifting force, U, may be estimated (see Figure 13.25b) from the equation

$$U = \pi D_s Z \sigma'_{sw} \tan \phi'_{ps} \tag{13.14}$$

where

D_s = diameter of the shaft
Z = depth of the active zone
ϕ'_{ps} = effective angle of plinth–soil friction
σ'_{sw} = pressure for zero swell (see Figures 13.14 and 13.15;
$\qquad \sigma'_{sw} = \sigma'_o + \sigma'_s + \sigma'_1$)

In most cases, the value of ϕ'_{ps} varies between 10 and 20°. An average value of the zero horizontal swell pressure must be determined in the laboratory. In the absence of laboratory results, $\sigma'_{sw} \tan \phi'_{ps}$ may be considered equal to the undrained shear strength of clay, c_u, in the active zone.

The belled portion of the drilled shaft will act as an anchor to resist the uplifting force. Ignoring the weight of the drilled shaft, we have

$$Q_{net} = U - D \qquad (13.15)$$

where

Q_{net} = net uplift load
D = dead load

Now,

$$Q_{net} \approx \frac{c_u N_c}{\text{FS}} \left(\frac{\pi}{4} \right) (D_b^2 - D_s^2) \qquad (13.16)$$

where

c_u = undrained cohesion of the clay in which the bell is located
N_c = bearing capacity factor
FS = factor of safety
D_b = diameter of the bell of the drilled shaft

Combining Eqs. (13.15) and (13.16) gives

$$U - D = \frac{c_u N_c}{\text{FS}} \left(\frac{\pi}{4} \right) (D_b^2 - D_s^2) \qquad (13.17)$$

Conservatively, from Tables 3.3 and 3.4,

$$N_c \approx N_{c(\text{strip})} F_{cs} = N_{c(\text{strip})} \left(1 + \frac{N_q B}{N_c L} \right) \approx 5.14 \left(1 + \frac{1}{5.14} \right) = 6.14$$

A drilled-shaft design is examined in Example 13.4.

Example 13.4

Figure 13.26 shows a drilled shaft with a bell. The depth of the active zone is 5 m. The zero swell pressure of the swelling clay (σ'_{sw}) is 450 kN/m². For the drilled shaft, the dead load (D) is 600 kN and the live load is 300 kN. Assume $\phi'_{ps} = 12°$.
a. Determine the diameter of the bell, D_b.
b. Check the bearing capacity of the drilled shaft assuming zero uplift force.

Solution
Part a: Determining the Bell Diameter, D_b
The uplift force, Eq. (13.14), is

$$U = \pi D_s Z \sigma'_{sw} \tan \phi'_{ps}$$

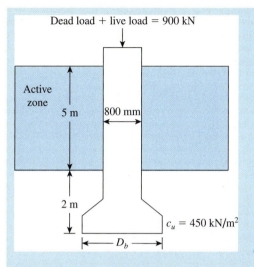

Dead load + live load = 900 kN

Active zone

5 m

800 mm

2 m

$c_u = 450 \text{ kN/m}^2$

D_b

Figure 13.26 Drilled shaft in a swelling clay

Given: $Z = 5$ m and $\sigma'_{sw} = 450 \text{ kN/m}^2$. Then

$$U = \pi(0.8)(5)(450)\tan 12° = 1202 \text{ kN}$$

Assume the dead load and live load to be zero, and FS in Eq. (13.17) to be 1.25. So, from Eq. (13.17),

$$U = \frac{c_u N_c}{\text{FS}}\left(\frac{\pi}{4}\right)(D_b^2 - D_s^2)$$

$$1202 = \frac{(450)(6.14)}{1.25}\left(\frac{\pi}{4}\right)(D_b^2 - 0.8^2); \quad D_b = \mathbf{1.15 \text{ m}}$$

The factor of safety against uplift with the dead load also should be checked. A factor of safety of at least 2 is desirable. So, from Eq. (13.17)

$$\text{FS} = \frac{c_u N_c \left(\frac{\pi}{4}\right)(D_b^2 - D_s^2)}{U - D}$$

$$= \frac{(450)(6.14)\left(\frac{\pi}{4}\right)(1.15^2 - 0.8^2)}{1202 - 600} = \mathbf{2.46 > 2—OK}$$

Part b: Check for Bearing Capacity
Assume that $U = 0$. Then

$$\text{Dead load + live load} = 600 + 300 = 900 \text{ kN}$$

$$\text{Downward load per unit area} = \frac{900}{\left(\frac{\pi}{4}\right)(D_b^2)} = \frac{900}{\left(\frac{\pi}{4}\right)(1.15^2)} = 866.5 \text{ kN/m}^2$$

Net bearing capacity of the soil under the bell $= q_{u(net)} = c_u N_c$

$$= (450)(6.14) = 2763 \text{ kN/m}^2$$

Hence, the factor of safety against bearing capacity failure is

$$\text{FS} = \frac{2763}{866.5} = \textbf{3.19} > \textbf{3—OK}$$ ∎

Sanitary Landfills

13.13 General Nature of Sanitary Landfills

Sanitary landfills provide a way to dispose of refuse on land without endangering public health. Sanitary landfills are used in almost all countries, to varying degrees of success. The refuse disposed of in sanitary landfills may contain organic, wood, paper, and fibrous wastes, or demolition wastes such as bricks and stones. The refuse is dumped and compacted at frequent intervals and is then covered with a layer of soil, as shown in Figure 13.27. In the compacted state, the average unit weight of the refuse may vary between 5 and 10 kN/m^3. A typical city in the United States, with a population of 1 million, generates about 3.8×10^6 m^3 of compacted landfill material per year.

As property values continue to increase in densely populated areas, constructing structures over sanitary landfills becomes more and more tempting. In some instances, a visual site inspection may not be enough to detect an old sanitary landfill. However, construction of foundations over sanitary landfills is generally problematic because of poisonous gases (e.g., methane), excessive settlement, and low inherent bearing capacity.

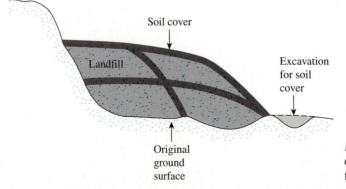

Figure 13.27 Schematic diagram of a sanitary landfill in progress

13.14   *Settlement of Sanitary Landfills*

Sanitary landfills undergo large continuous settlements over a long time. Yen and Scanlon (1975) documented the settlement of several landfill sites in California. After completion of the landfill, the settlement rate (Figure 13.28) may be expressed as

$$m = \frac{\Delta H_f \,(\text{m})}{\Delta t \,(\text{month})} \tag{13.18}$$

where

m = settlement rate
H_f = maximum height of the sanitary landfill

On the basis of several field observations, Yen and Scanlon determined the following empirical correlations for the settlement rate:

$$m = a - b \log t_1 \qquad [\text{for fill heights ranging from 12 to 24 m}] \tag{13.19}$$

$$m = c - d \log t_1 \qquad [\text{for fill heights ranging from 24 to 30 m}] \tag{13.20}$$

$$m = e - f \log t_1 \qquad [\text{for fill heights larger than 30 m}] \tag{13.21}$$

In these equations,

m is in m/mo (ft/mo).
t_1 is the median fill age, in months

In SI units, the values of a, b, c, d, e, and f given in Eqs. (13.19) through (13.21) are

Item	SI
a	0.0268
b	0.0116
c	0.038
d	0.0155
e	0.0435
f	0.0183

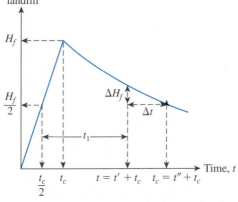

Figure 13.28 Settlement of sanitary landfills

The median fill age may be defined from Figure 13.28 as

$$t_1 = t - \frac{t_c}{2} \tag{13.22}$$

where

t = time from the beginning of the landfill
t_c = time for completion of the landfill

Equations (13.19), (13.20), and (13.21) are based on field data from landfills for which t_c varied from 70 to 82 months. To get an idea of the approximate length of time required for a sanitary landfill to undergo complete settlement, consider Eq. (13.19). For a fill 12 m high and for $t_c = 72$ months,

$$m = 0.0268 - 0.0116 \log t_1$$

so

$$\log t_1 = \frac{0.0268 - m}{0.0116}$$

If $m = 0$ (zero settlement rate), $\log t_1 = 2.31$, or $t_1 \approx 200$ months. Thus, settlement will continue for $t_1 - t_c/2 = 200 - 36 = 164$ months (≈ 14 years) after completion of the fill—a fairly long time. This calculation emphasizes the need to pay close attention to the settlement of foundations constructed on sanitary landfills.

A comparison of Eqs. (13.19) through (13.21) for rates of settlement shows that the value of m increases with the height of the fill. However, for fill heights greater than about 30 m, the rate of settlement should not be much different from that obtained from Eq. (13.21). The reason is that the decomposition of organic matter close to the surface is mainly the result of an anaerobic environment. For deeper fills, the decomposition is slower. Hence, for fill heights greater than about 30 m, the rate of settlement does not exceed that for fills that are about 30 m in height.

Sowers (1973) also proposed a formula for calculating the settlement of a sanitary landfill, namely,

$$\Delta H_f = \frac{\alpha H_f}{1 + e} \log \left(\frac{t''}{t'} \right) \tag{13.23}$$

where

H_f = height of the fill
e = void ratio
α = a coefficient for settlement
t', t'' = times (see Figure 13.28)
ΔH_f = settlement between times t' and t''

The coefficients α fall between

$$\alpha = 0.09e \quad \text{(for conditions favorable to decomposition)} \tag{13.24}$$

and

$$\alpha = 0.03e \quad \text{(for conditions unfavorable to decomposition)} \qquad (13.25)$$

Equation (13.23) is similar to the equation for secondary consolidation settlement.

Problems

13.1 For leossial soil, given $G_s = 2.69$. Plot a graph of γ_d (kN/m^3) versus liquid limit to identify the zone in which the soil is likely to collapse on saturation. If a soil has a liquid limit of 33, $G_s = 2.69$, and $\gamma_d = 13.5$ kN/m^3, is collapse likely to occur?

13.2 A collapsible soil layer in the field has a thickness of 3 m. The average effective overburden pressure on the soil layer is 62 kN/m^2. An undisturbed specimen of this soil was subjected to a double oedometer test. The preconsolidation pressure of the specimen as determined from the soaked specimen was 84 kN/m^2. Is the soil in the field normally consolidated or preconsolidated?

13.3 An expansive soil has an active-zone thickness of 8 m. The natural moisture content of the soil is 20% and its liquid limit is 50. Calculate the free surface swell of the expansive soil upon saturation.

13.4 An expansive soil profile has an active-zone thickness of 5.2 m. A shallow foundation is to be constructed at a depth of 1.2 m below the ground surface. Based on the swelling pressure test, the following are given.

Depth from ground surface (m)	Swell under overburden and estimated foundation surcharge pressure, $s_{w(1)}$ (%)
1.2	3.0
2.2	2.0
3.2	1.2
4.2	0.55
5.2	0.0

Estimate the total possible swell under the foundation.

13.5 Refer to Problem 13.4. If the allowable total swell is 15 mm., what would be the necessary undercut?

13.6 Repeat Problem 13.4 with the following: active zone thickness = 6 m, depth of shallow foundation = 1.5 m.

Depth from ground surface (m)	Swell under overburden and estimated foundation surcharge pressure, $s_{w(1)}$ (%)
1.5	5.5
2.0	3.1
3.0	1.5
4.0	0.75
5.0	0.4
6.0	0.0

13.7 Refer to Problem 13.6. If the allowable total swell is 30 mm, what would be the necessary undercut?

13.8 Refer to Figure 13.25b. For the drilled shaft with bell, given:

Thickness of active zone, $Z = 9.15$ m
Dead load $= 1334$ kN
Live load $= 267$ kN
Diameter of the shaft, $D_s = 1.07$ m
Zero swell pressure for the clay in the active zone $= 574.6$ kN/m^2
Average angle of plinth-soil friction, $\phi'_{ps} = 15°$
Average undrained cohesion of the clay around the bell $= 144.6$ kN/m^2

Determine the diameter of the bell, D_b. A factor of safety of 3 against uplift is required with the assumption that dead load plus live load is equal to zero.

13.9 Refer to Problem 13.8. If an additional requirement is that the factor of safety against uplift is at least 3 with the dead load on (live load $= 0$), what should be the diameter of the bell?

References

ABDULJAUWAD, S. N., and AL-SULAIMANI, G. J. (1993). "Determination of Swell Potential of Al-Qatif Clay," *Geotechnical Testing Journal*, American Society for Testing and Materials, Vol. 16, No. 4, pp. 469–484.

ALTMEYER, W. T. (1955). "Discussion of Engineering Properties of Expansive Clays," *Journal of the Soil Mechanics and Foundations Division*, American Society of Civil Engineers, Vol. 81, No. SM2, pp. 17–19.

BENITES, L. A. (1968). "Geotechnical Properties of the Soils Affected by Piping near the Benson Area, Cochise County, Arizona," M. S. Thesis, University of Arizona, Tucson.

CHEN, F. H. (1988). *Foundations on Expansive Soils*, Elsevier, Amsterdam.

CLEMENCE, S. P., and FINBARR, A. O. (1981). "Design Considerations for Collapsible Soils," *Journal of the Geotechnical Engineering Division*, American Society of Civil Engineers, Vol. 107, No. GT3, pp. 305–317.

CLEVENGER, W. (1958). "Experience with Loess as Foundation Material," *Transactions*, American Society of Civil Engineers, Vol. 123, pp. 151–170.

DAKSHANAMANTHY, V., and RAMAN, V. (1973). "A Simple Method of Identifying an Expansive Soil," *Soils and Foundations*, Vol. 13, No. 1, pp. 97–104.

DENISOV, N. Y. (1951). *The Engineering Properties of Loess and Loess Loams*, Gosstroiizdat, Moscow.

FEDA, J. (1964). "Colloidal Activity, Shrinking and Swelling of Some Clays," *Proceedings, Soil Mechanics Seminar*, Loda, Illinois, pp. 531–546.

GIBBS, H. J. (1961). *Properties Which Divide Loose and Dense Uncemented Soils*, Earth Laboratory Report EM-658, Bureau of Reclamation, U.S. Department of the Interior, Washington, DC.

GROMKO, G. J. (1974). "Review of Expansive Soils," *Journal of the Geotechnical Engineering Division*, American Society of Civil Engineers, Vol. 100, No. GT6, pp. 667–687.

HANDY, R. L. (1973). "Collapsible Loess in Iowa," *Proceedings, Soil Science Society of America*, Vol. 37, pp. 281–284.

HOLTZ, W. G. (1959). "Expansive Clays—Properties and Problems," *Journal of the Colorado School of Mines*, Vol. 54, No. 4, pp. 89–125.

HOLTZ, W. G., and HILF, J. W. (1961). "Settlement of Soil Foundations Due to Saturation," *Proceedings, Fifth International Conference on Soil Mechanics and Foundation Engineering*, Paris, Vol. 1, 1961, pp. 673–679.

HOUSTON, W. N., and HOUSTON, S. L. (1989). "State-of-the-Practice Mitigation Measures for Collapsible Soil Sites," *Proceedings, Foundation Engineering: Current Principles and Practices,* American Society of Civil Engineers, Vol. 1, pp. 161–175.

JENNINGS, J. E., and KNIGHT, K. (1975). "A Guide to Construction on or with Materials Exhibiting Additional Settlements Due to 'Collapse' of Grain Structure," *Proceedings, Sixth Regional Conference for Africa on Soil Mechanics and Foundation Engineering,* Johannesburg, pp. 99–105.

LUTENEGGER, A. J. (1986). "Dynamic Compaction in Friable Loess," *Journal of Geotechnical Engineering,* American Society of Civil Engineers, Vol. 112, No. GT6, pp. 663–667.

LUTENEGGER, A. J., and SABER, R. T. (1988). "Determination of Collapse Potential of Soils," *Geotechnical Testing Journal,* American Society for Testing and Materials, Vol. 11. No. 3, pp. 173–178.

McKEEN, R. G. (1992). "A Model for Predicting Expansive Soil Behavior," *Proceedings, Seventh International Conference on Expansive Soils,* Dallas, Vol. 1, pp. 1–6.

NAYAK, N. V., and CHRISTENSEN, R. W. (1974). "Swell Characteristics of Compacted Expansive Soils," *Clay and Clay Minerals,* Vol. 19, pp. 251–261.

O'NEILL, M. W., and POORMOAYED, N. (1980). "Methodology for Foundations on Expansive Clays," *Journal of the Geotechnical Engineering Division,* American Society of Civil Engineers, Vol. 106, No. GT12, pp. 1345–1367.

PECK, R. B., HANSON, W. E., and THORNBURN, T. B. (1974). *Foundation Engineering,* Wiley, New York.

PRIKLONSKII, V. A. (1952). *Gruntovedenic, Vtoraya Chast',* Gosgeolzdat, Moscow.

RAMAN, V. (1967). "Identification of Expansive Soils from the Plasticity Index and the Shrinkage Index Data," *The Indian Engineer,* Vol. 11, No. 1, pp. 17–22.

SEED, H. B., WOODWARD, R. J., JR., and LUNDGREN, R. (1962). "Prediction of Swelling Potential for Compacted Clays," *Journal of the Soil Mechanics and Foundations Division,* American Society of Civil Engineers, Vol. 88, No. SM3, pp. 53–87.

SEMKIN, V. V., ERMOSHIN, V. M., and OKISHEV, N. D. (1986). "Chemical Stabilization of Loess Soils in Uzbekistan," *Soil Mechanics and Foundation Engineering* (trans. from Russian), Vol. 23, No. 5, pp. 196–199.

SNETHEN, D. R. (1984). "Evaluation of Expedient Methods for Identification and Classification of Potentially Expansive Soils," *Proceedings, Fifth International Conference on Expansive Soils,* Adelaide, pp. 22–26.

SNETHEN, D. R., JOHNSON, L. D., and PATRICK, D. M. (1977). *An Evaluation of Expedient Methodology for Identification of Potentially Expansive Soils,* Report No. FHWA-RD-77-94, U.S. Army Engineers Waterways Experiment Station, Vicksburg, MS.

SOWERS, G. F. (1973). "Settlement of Waste Disposal Fills," *Proceedings, Eighth International Conference on Soil Mechanics and Foundation Engineering,* Moscow, pp. 207–210.

SOWERS, G. B., and SOWERS, G. F. (1970). *Introductory Soil Mechanics and Foundations,* 3d ed. Macmillan, New York.

SRIDHARAN, A. (2005). "On Swelling Behaviour of Clays," *Proceedings, International Conference on Problematic Soils,* North Cyprus, Vol. 2, pp. 499–516.

SRIDHARAN, A., RAO, A. S., and SIVAPULLAIAH, P. V. (1986), "Swelling Pressure of Clays," *Geotechnical Testing Journal,* American Society for Testing and Materials, Vol. 9, No. 1, pp. 24–33.

Uniform Building Code (1968). *UBC Standard No. 29-2.*

VAN DER MERWE, D. H. (1964), "The Prediction of Heave from the Plasticity Index and Percentage Clay Fraction of Soils," *Civil Engineer in South Africa,* Vol. 6, No. 6, pp. 103–106.

VIJAYVERGIYA, V. N., and GHAZZALY, O. I. (1973). "Prediction of Swelling Potential of Natural Clays," *Proceedings, Third International Research and Engineering Conference on Expansive Clays,* pp. 227–234.

WESTON, D. J. (1980). "Expansive Roadbed Treatment for Southern Africa," *Proceedings, Fourth International Conference on Expansive Soils,* Vol. 1, pp. 339–360.

YEN, B. C., and SCANLON, B. (1975). "Sanitary Landfill Settlement Rates," *Journal of the Geotechnical Engineering Division,* American Society of Civil Engineers, Vol. 101, No. GT5, pp. 475–487.

14 Soil Improvement and Ground Modification

14.1   *Introduction*

The soil at a construction site may not always be totally suitable for supporting structures such as buildings, bridges, highways, and dams. For example, in granular soil deposits, the *in situ* soil may be very loose and indicate a large elastic settlement. In such a case, the soil needs to be densified to increase its unit weight and thus its shear strength.

Sometimes the top layers of soil are undesirable and must be removed and replaced with better soil on which the structural foundation can be built. The soil used as fill should be well compacted to sustain the desired structural load. Compacted fills may also be required in low-lying areas to raise the ground elevation for construction of the foundation.

Soft saturated clay layers are often encountered at shallow depths below foundations. Depending on the structural load and the depth of the layers, unusually large consolidation settlement may occur. Special soil-improvement techniques are required to minimize settlement.

In Chapter 13, we mentioned that the properties of expansive soils could be altered substantially by adding stabilizing agents such as lime. Improving *in situ* soils by using additives is usually referred to as *stabilization*.

Various techniques are used to

1. Reduce the settlement of structures
2. Improve the shear strength of soil and thus increase the bearing capacity of shallow foundations
3. Increase the factor of safety against possible slope failure of embankments and earth dams
4. Reduce the shrinkage and swelling of soils

This chapter discusses some of the general principles of soil improvement, such as compaction, vibroflotation, precompression, sand drains, wick drains, stabilization by admixtures, jet grouting, and deep mixing, as well as the use of stone columns and sand compaction piles in weak clay to construct foundations.

14.2   *General Principles of Compaction*

If a small amount of water is added to a soil that is then compacted, the soil will have a certain unit weight. If the moisture content of the same soil is gradually increased and the energy of compaction is the same, the dry unit weight of the soil will gradually increase. The reason is that water acts as a lubricant between the soil particles, and under compaction it helps rearrange the solid particles into a denser state. The increase in dry unit weight with increase of moisture content for a soil will reach a limiting value beyond which the further addition of water to the soil will result in a *reduction* in dry unit weight. The moisture content at which the *maximum dry unit weight* is obtained is referred to as the *optimum moisture content.*

The standard laboratory tests used to evaluate maximum dry unit weights and optimum moisture contents for various soils are

- The Standard Proctor test (ASTM designation D-698)
- The Modified Proctor test (ASTM designation D-1557)

The soil is compacted in a mold in several layers by a hammer. The moisture content of the soil, w, is changed, and the dry unit weight, γ_d, of compaction for each test is determined. The maximum dry unit weight of compaction and the corresponding optimum moisture content are determined by plotting a graph of γ_d against w (%). The standard specifications for the two types of Proctor test are given in Tables 14.1 and 14.2.

Table 14.1 Specifications for Standard Proctor Test (Based on ASTM Designation 698)

Item	Method A	Method B	Method C
Diameter of mold	101.6 mm	101.6 mm	152.4 mm
Volume of mold	944 cm^3	944 cm^3	2124 cm^3
Mass of hammer	2.5 kg	2.5 kg	2.5 kg
Height of hammer drop	304.8 mm	304.8 mm	304.8 mm
Number of hammer blows per layer of soil	25	25	56
Number of layers of compaction	3	3	3
Energy of compaction	600 kN·m/m^3	600 kN·m/m^3	600 kN·m/m^3
Soil to be used	Portion passing No. 4 4.57-mm sieve. May be used if 20% *or less* by weight of material is retained on No. 4 sieve.	Portion passing 9.5-mm sieve. May be used if soil retained on No. 4 sieve *is more* than 20% and 20% *or less* by weight is retained on 9.5-mm ($\frac{3}{8}$-in.) sieve.	Portion passing 19.0-mm sieve. May be used if *more than* 20% by weight of material is retained on 9.5 mm sieve and *less than* 30% by weight is retained on 19.00-mm sieve.

Table 14.2 Specifications for Modified Proctor Test (Based on ASTM Designation 1557)

Item	Method A	Method B	Method C
Diameter of mold	101.6 mm	101.6 mm	152.4 mm
Volume of mold	944 cm^3	944 cm^3	2124 cm^3
Mass of hammer	4.54 kg	4.54 kg	4.54 kg
Height of hammer drop	457.2 mm	457.2 mm	457.2 mm
Number of hammer blows per layer of soil	25	25	56
Number of layers of compaction	5	5	5
Energy of compaction	2700 kN·m/m^3	2700 kN·m/m^3	2700 kN·m/m^3
Soil to be used	Portion passing No. 4 (4.57-mm) sieve. May be used if 20% *or less* by weight of material is retained on No. 4 sieve.	Portion passing 9.5-mm sieve. May be used if soil retained on No. 4 sieve *is more* than 20% and 20% *or less* by weight is retained on 9.5-mm sieve.	Portion passing 19.0-mm ($\frac{3}{4}$-in.) sieve. May be used if *more than* 20% by weight of material is retained on 9.5-mm sieve and *less than* 30% by weight is retained on 19-mm sieve.

Figure 14.1 shows a plot of γ_d against w (%) for a clayey silt obtained from standard and modified Proctor tests (method A). The following conclusions may be drawn:

1. The maximum dry unit weight and the optimum moisture content depend on the degree of compaction.
2. The higher the energy of compaction, the higher is the maximum dry unit weight.
3. The higher the energy of compaction, the lower is the optimum moisture content.
4. No portion of the compaction curve can lie to the right of the zero-air-void line. The zero-air-void dry unit weight, γ_{zav}, at a given moisture content is the theoretical maximum value of γ_d, which means that all the void spaces of the compacted soil are filled with water, or

$$\gamma_{zav} = \frac{\gamma_w}{\dfrac{1}{G_s} + w} \tag{14.1}$$

where

γ_w = unit weight of water
G_s = specific gravity of the soil solids
w = moisture content of the soil

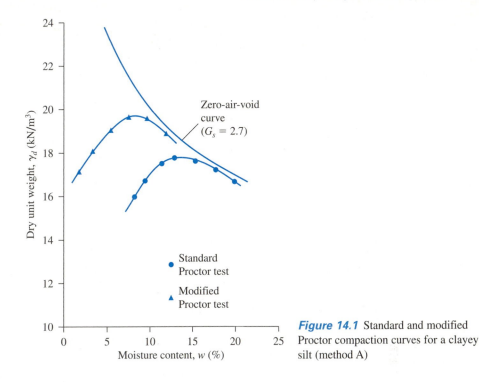

Figure 14.1 Standard and modified Proctor compaction curves for a clayey silt (method A)

5. The maximum dry unit weight of compaction and the corresponding optimum moisture content will vary from soil to soil.

   Using the results of laboratory compaction (γ_d versus w), specifications may be written for the compaction of a given soil in the field. In most cases, the contractor is required to achieve a relative compaction of 90% or more on the basis of a specific laboratory test (either the standard or the modified Proctor compaction test). The relative compaction is defined as

$$\text{RC} = \frac{\gamma_{d(\text{field})}}{\gamma_{d(\text{max})}} \qquad (14.2)$$

   Chapter 1 introduced the concept of relative density (for the compaction of granular soils), defined as

$$D_r = \left[\frac{\gamma_d - \gamma_{d(\text{min})}}{\gamma_{d(\text{max})} - \gamma_{d(\text{min})}} \right] \frac{\gamma_{d(\text{max})}}{\gamma_d}$$

where

$$\gamma_d = \text{dry unit weight of compaction in the field}$$
$$\gamma_{d(\text{max})} = \text{maximum dry unit weight of compaction as determined in the laboratory}$$
$$\gamma_{d(\text{min})} = \text{minimum dry unit weight of compaction as determined in the laboratory}$$

For granular soils in the field, the degree of compaction obtained is often measured in terms of relative density. Comparing the expressions for relative density and relative compaction reveals that

$$RC = \frac{A}{1 - D_r(1 - A)} \tag{14.3}$$

where $A = \dfrac{\gamma_{d(\text{min})}}{\gamma_{d(\text{max})}}$.

Omar, et al. (2003) recently presented the results of modified Proctor compaction tests on 311 soil samples. Of these samples, 45 were gravelly soil (GP, GP-GM, GW, GW-GM, and GM), 264 were sandy soil (SP, SP-SM, SW-SM, SW, SC-SM, SC, and SM), and two were clay with low plasticity (CL). All compaction tests were conducted using ASTM 1557 method C to avoid over-size correction. Based on the tests, the following correlations were developed.

$$\rho_{d(\text{max})}(\text{kg/m}^3) = [4{,}804{,}574G_S - 195.55(LL)^2 + 156{,}971(\text{R\# }4)^{0.5}$$
$$- 9{,}527{,}830]^{0.5} \tag{14.4}$$

$$\ln(w_{\text{opt}}) = 1.195 \times 10^{-4}(LL)^2 - 1.964G_s - 6.617 \times 10^{-5}(\text{R\# }4)$$
$$+ 7.651 \tag{14.5}$$

where

$$\rho_d(\text{max}) = \text{maximum dry density}$$
$$w_{\text{opt}} = \text{optimum moisture content}$$
$$G_s = \text{specific gravity of soil solids}$$
$$LL = \text{liquid limit, in percent}$$
$$\text{R \# }4 = \text{percent retained on No. 4 sieve}$$

It needs to be pointed out that Eqs. (14.4 and 14.5) contain the term for liquid limit. This is because the soils that were considered included silty and clayey sands.

Osman et al. (2008) analyzed a number of laboratory compaction-test results on fine-grained (cohesive) soil. Based on this study, the following correlations were developed:

$$w_{\text{opt}} = (1.99 - 0.165 \ln E)(\text{PI}) \tag{14.6}$$

and

$$\gamma_{d(\text{max})} = L - Mw_{\text{opt}} \tag{14.7}$$

where

$$L = 14.34 + 1.195 \ln E \qquad (14.8)$$
$$M = -0.19 + 0.073 \ln E \qquad (14.9)$$

w_{opt} = optimum moisture content (%)
PI = plasticity index (%)
$\gamma_{d(\text{max})}$ = maximum dry unit weight (kN/m^3)
E = compaction energy (kN-m/m^3)

14.3   *Field Compaction*

Ordinary compaction in the field is done by rollers. Of the several types of roller used, the most common are

1. Smooth-wheel rollers (or smooth drum rollers)
2. Pneumatic rubber-tired rollers
3. Sheepsfoot rollers
4. Vibratory rollers

Figure 14.2 shows a *smooth-wheel roller* that can also create vertical vibration during compaction. Smooth-wheel rollers are suitable for proof-rolling subgrades and for finishing the construction of fills with sandy or clayey soils. They provide 100% coverage under the wheels, and the contact pressure can be as high as

Figure 14.2 Vibratory smooth-wheel rollers (*Courtesy of Tampo Manufacturing Co., Inc., San Antonio, Texas*)

Figure 14.3 Pneumatic rubber-tired roller (*Courtesy of Tampo Manufacturing Co., Inc., San Antonio, Texas*)

300 to 400 kN/m². However, they do not produce a uniform unit weight of compaction when used on thick layers.

Pneumatic rubber-tired rollers (Figure 14.3) are better in many respects than smooth-wheel rollers. Pneumatic rollers, which may weigh as much as 2000 kN, consist of a heavily loaded wagon with several rows of tires. The tires are closely spaced—four to six in a row. The contact pressure under the tires may range up to 600 to 700 kN/m², and they give about 70 to 80% coverage. Pneumatic rollers, which can be used for sandy and clayey soil compaction, produce a combination of pressure and kneading action.

Sheepsfoot rollers (Figure 14.4) consist basically of drums with large numbers of projections. The area of each of the projections may be 25 to 90 cm². These rollers are *most effective in compacting cohesive soils.* The contact pressure under the projections may range from 1500 to 7500 kN/m². During compaction in the field, the initial passes compact the lower portion of a lift. Later, the middle and top of the lift are compacted.

Vibratory rollers are efficient in compacting granular soils. Vibrators can be attached to smooth-wheel, pneumatic rubber-tired or sheepsfoot rollers to send vibrations into the soil being compacted. Figures 14.2 and 14.4 show vibratory smooth-wheel rollers and a vibratory sheepsfoot roller, respectively.

In general, compaction in the field depends on several factors, such as the type of compactor, type of soil, moisture content, lift thickness, towing speed of the compactor, and number of passes the roller makes.

Figure 14.5 shows the variation of the unit weight of compaction with depth for a poorly graded dune sand compacted by a vibratory drum roller. Vibration was

Figure 14.4 Vibratory sheepsfoot roller (*Courtesy of Tampo Manufacturing Co., Inc., San Antonio, Texas*)

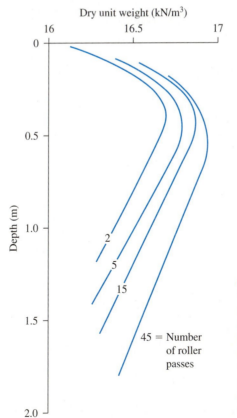

Figure 14.5 Vibratory compaction of a sand: Variation of dry unit weight with depth and number of roller passes; lift thickness = 2.44 m (After D'Appolonia et al., 1969) (After D'Appolonia, D. J., Whitman, R. V. and D'Appolonia, E. (1969). "Sand Compaction with Vibratory Rollers," Journal of the Soil Mechanics and Foundations Division, American Society of Civil Engineers, Vol. 95, N. SM1, pp. 263–284. With permission from ASCE.)

produced by mounting an eccentric weight on a single rotating shaft within the drum cylinder. The weight of the roller used for this compaction was 55.7 kN, and the drum diameter was 1.19 m. The lifts were kept at 2.44 m. Note that, at any depth, the dry unit weight of compaction increases with the number of passes the roller makes. However, the rate of increase in unit weight gradually decreases after about 15 passes. Note also the variation of dry unit weight with depth by the number of roller passes. The dry unit weight and hence the relative density, D_r, reach maximum values at a depth of about 0.5 m and then gradually decrease as the depth increases. The reason is the lack of confining pressure toward the surface. Once the relation between depth and relative density (or dry unit weight) for a soil for a given number of passes is determined, for satisfactory compaction based on a given specification, the approximate thickness of each lift can be easily estimated.

Hand-held vibrating plates can be used for effective compaction of granular soils over a limited area. Vibrating plates are also gang-mounted on machines. These can be used in less restricted areas.

14.4 Compaction Control for Clay Hydraulic Barriers

Compacted clays are commonly used as hydraulic barriers in cores of earth dams, liners and covers of landfills, and liners of surface impoundments. Since the primary purpose of a barrier is to minimize flow, the hydraulic conductivity, k, is the controlling factor. In many cases, it is desired that the hydraulic conductivity be less than 10^{-7}cm/s. This can be achieved by controlling the minimum degree of saturation during compaction, a relation that can be explained by referring to the compaction characteristics of three soils described in Table 14.3 (Othman and Luettich, 1994).

Figures 14.6, 14.7, and 14.8 show the standard and modified Proctor test results and the hydraulic conductivities of compacted specimens. Note that the solid symbols represent specimens with hydraulic conductivities of 10^{-7}cm/s or less. As can be seen from these figures, the data points plot generally parallel to the line of full saturation. Figure 14.9 shows the effect of the degree of saturation during compaction on the hydraulic conductivity of the three soils. It is evident from the figure that, if it is desired that the maximum hydraulic conductivity be 10^{-7}cm/s, then all soils should be compacted at a minimum degree of saturation of 88%.

Table 14.3 Characteristics of Soils Reported in Figures 14.6, 14.7, and 14.8

Soil	Classification	Liquid limit	Plasticity index	Percent finer than No. 200 sieve (0.075 mm)
Wisconsin A	CL	34	16	85
Wisconsin B	CL	42	19	99
Wisconsin C	CH	84	60	71

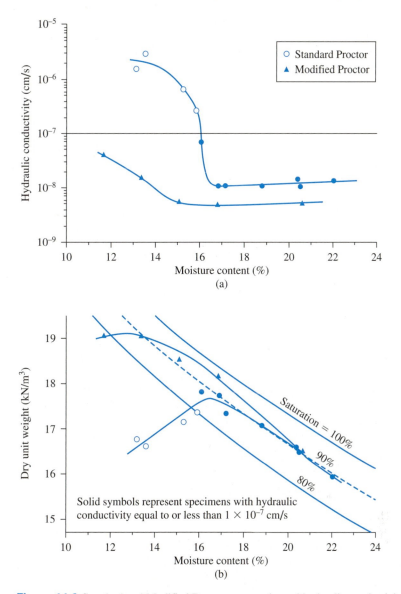

Figure 14.6 Standard and Modified Proctor test results and hydraulic conductivity of Wisconsin A soil (After Othman and Luettich, 1994) (From Othman, M. A., and S. M. Luettich. *Compaction Control Criteria for Clay Hydraulic Barriers*. In Transportation Research Record 1462, Transportation Research Board, National Research Council, Washington, D.C., 1994, Figures 4 and 5, p. 32, and Figures 6 and 7, p. 33. Reproduced with permission of the Transportation Research Board.)

In field compaction at a given site, soils of various composition may be encountered. Small changes in the content of fines will change the magnitude of hydraulic conductivity. Hence, considering the various soils likely to be encountered at a given site the procedure just described aids in developing a minimum-degree-of-saturation criterion for compaction to construct hydraulic barriers.

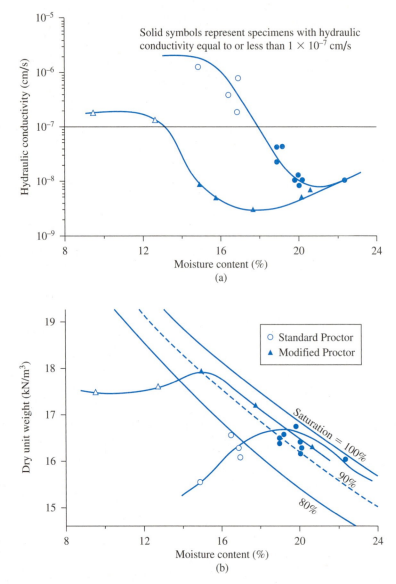

Figure 14.7 Standard and Modified Proctor test results and hydraulic conductivity of Wisconsin B soil (After Othman and Luettich, 1994) (From Othman, M. A., and S. M. Luettich. Compaction Control Criteria for Clay Hydraulic Barriers. In Transportation Research Record 1462, Transportation Research Board, National Research Council, Washington, D.C., 1994, Figures 4 and 5, p. 32, and Figures 6 and 7, p. 33. Reproduced with permission of the Transportation Research Board.)

14.5 *Vibroflotation*

Vibroflotation is a technique developed in Germany in the 1930s for *in situ* densification of thick layers of loose granular soil deposits. Vibroflotation was first used in the United States about 10 years later. The process involves the use of a *vibroflot* (called the *vibrating unit*), as shown in Figure 14.10. The device is about 2 m in length. This vibrating unit has an eccentric weight inside it and can develop a centrifugal force. The weight enables the unit to vibrate horizontally. Openings at the bottom and top of the unit are for water jets. The vibrating unit is attached to a follow-up pipe. The figure shows the vibroflotation equipment necessary for compaction in the field.

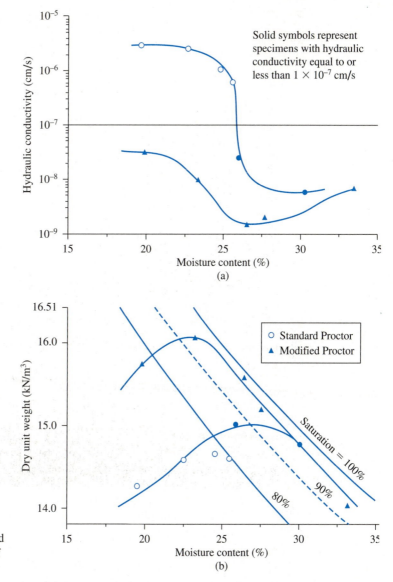

Figure 14.8 Standard and Modified Proctor test results and hydraulic conductivity of Wisconsin C soil (After Othman and Luettich, 1994) (From Othman, M. A., and S. M. Luettich. Compaction Control Criteria for Clay Hydraulic Barriers. In Transportation Research Record 1462, Transportation Research Board, National Research Council, Washington, D.C., 1994, Figures 4 and 5, p. 32, and Figures 6 and 7, p. 33. Reproduced with permission of the Transportation Research Board.)

The entire compaction process can be divided into four steps (see Figure 14.11):

Step 1. The jet at the bottom of the vibroflot is turned on, and the vibroflot is lowered into the ground.

Step 2. The water jet creates a quick condition in the soil, which allows the vibrating unit to sink.

Step 3. Granular material is poured into the top of the hole. The water from the lower jet is transferred to the jet at the top of the vibrating unit. This water carries the granular material down the hole.

Step 4. The vibrating unit is gradually raised in about 0.3-m lifts and is held vibrating for about 30 seconds at a time. This process compacts the soil to the desired unit weight.

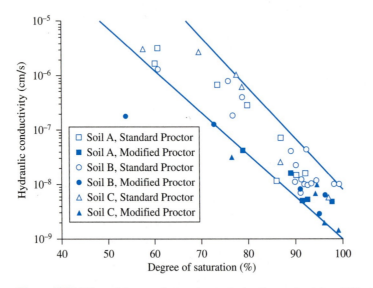

Figure 14.9 Effect of degree of saturation on hydraulic conductivity of Wisconsin A, B, and C soils (After Othman and Luettich, 1994) (From Othman, M. A., and S. M. Luettich. Compaction Control Criteria for Clay Hydraulic Barriers. In Transportation Research Record 1462, Transportation Research Board, National Research Council, Washington, D.C., 1994, Figures 4 and 5, p. 32, and Figures 6 and 7, p. 33. Reproduced with permission of the Transportation Research Board.)

Table 14.4 gives the details of various types of vibroflot unit used in the United States. The 23 kW electric units have been used since the latter part of the 1940s. The 100-HP units were introduced in the early 1970s. The zone of compaction around a single probe will vary according to the type of vibroflot used. The cylindrical zone of compaction will have a radius of about 2 m for a 23 kW unit and about 3 m for a 75 kW unit. Compaction by vibroflotation involves various probe spacings, depending on the zone of compaction. (See Figure 14.12.) Mitchell (1970) and Brown (1977) reported several successful cases of foundation design that used vibroflotation.

The success of densification of *in situ* soil depends on several factors, the most important of which are the grain-size distribution of the soil and the nature of the backfill used to fill the holes during the withdrawal period of the vibroflot. The range of the grain-size distribution of *in situ* soil marked Zone 1 in Figure 14.13 is most suitable for compaction by vibroflotation. Soils that contain excessive amounts of fine sand and silt-size particles are difficult to compact; for such soils, considerable effort is needed to reach the proper relative density of compaction. Zone 2 in Figure 14.13 is the approximate lower limit of grain-size distribution for compaction by vibroflotation. Soil deposits whose grain-size distribution falls into Zone 3 contain appreciable amounts of gravel. For these soils, the rate of probe penetration may be rather slow, so compaction by vibroflotation might prove to be uneconomical in the long run.

The grain-size distribution of the backfill material is one of the factors that control the rate of densification. Brown (1977) defined a quantity called *suitability number* for rating a backfill material. The suitability number is given by the formula

$$S_N = 1.7\sqrt{\frac{3}{(D_{50})^2} + \frac{1}{(D_{20})^2} + \frac{1}{(D_{10})^2}} \qquad (14.10)$$

Power
supply

Water
pump

Follow-up
pipe

Vibrating
unit

A
Cylinder of compacted
material, added from the
suface to compensate
for the loss of volume
caused by the increase in
density of the compacted
soil

B
Cylinder of compacted
material, produced by a
single vibroflot compaction

A

B

Figure 14.10 Vibroflotation unit (After Brown, 1977. With permission from ASCE.)

where D_{50}, D_{20}, and D_{10} are the diameters (in mm) through which 50%, 20%, and 10%, respectively, of the material is passing. The smaller the value of S_N, more desirable is the backfill material. Following is a backfill rating system proposed by Brown (1977):

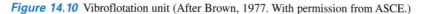

Range of S_N	Rating as backfill
0–10	Excellent
10–20	Good
20–30	Fair
30–50	Poor
>50	Unsuitable

Step 1	Step 2	Step 3	Step 4

Figure 14.11 Compaction by the vibroflotation process (After Brown, 1977. With permission from ASCE.)

Table 14.4 Types of Vibrating Units[a]

	75 kW electric and hydraulic motors	23 kW electric motors
(a) Vibrating tip		
Length	2.1 m	1.86 m
Diameter	406 mm	381 mm
Weight	18 kN	18 kN
Maximum movement when free	12.5 mm	7.6 mm
Centrifugal force	160 kN	90 kN
(b) Eccentric		
Weight	1.16 kN	0.76 kN
Offset	38 mm	32 mm
Length	610 mm	387 mm
Speed	1800 rpm	1800 rpm
(c) Pump		
Operating flow rate	0–1.6 m^3/min	0–0.6 m^3/min
Pressure	690–1035 kN/m^2	690–1035 kN/m^2
(d) Lower follow-up pipe and extensions		
Diameter	305 mm	305 mm
Weight	3.65 kN/m	3.65 kN/m

[a] After Brown, R. E. (1977), "Vibroflotation Compaction of Cohesionless Soils," *Journal of the Geotechnical Engineering Division*, Vol. 103, No. GT12. With permission from ASCE.

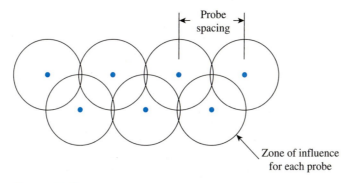

Figure 14.12 Nature of probe spacing for vibroflotation

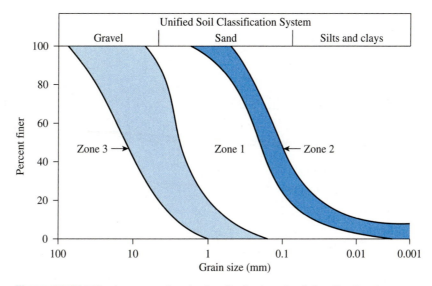

Figure 14.13 Effective range of grain-size distribution of soil for vibroflotation

An excellent case study that evaluated the benefits of vibroflotation was presented by Basore and Boitano (1969). Densification of granular subsoil was necessary for the construction of a three-story office building at the Treasure Island Naval Station in San Francisco, California. The top 9 m of soil at the site was loose to medium-dense sand fill that had to be compacted. Figure 14.14a shows the nature of the layout of the vibroflotation points. Sixteen compaction points were arranged in groups of four, with 1.22 m, 1.52 m, 1.83 m, and 2.44 m spacing. Prior to compaction, standard penetration tests were conducted at the centers of groups of three compaction points. After the completion of compaction by vibroflotation, the variation of the standard penetration resistance with depth was determined at the same points.

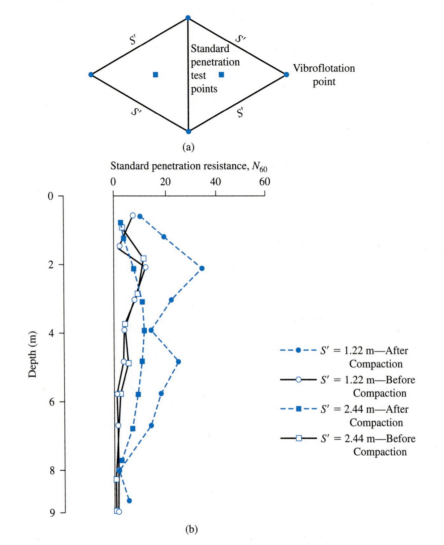

Figure 14.14 (a) Layout of vibroflotation compaction points; (b) variation of standard penetration resistance (N_{60}) before and after compaction (Basore, C. E. and Boitano, J. D. (1969). "Sand Densification by Piles and Vibrofloation," *Journal of Soil Mechanics and Foundation Engineering Division*, American Society of Civil Engineers, Vol. 95, No. 6, pp. 1301–1323, Figure 16. With permission from ASCE.)

Figure 14.14b shows the variation of standard penetration resistance, N_{60}, with depth before and after compaction for vibroflotation point spacings $S' = 1.22$ m and 2.44 m. From this figure, the following general conclusions can be drawn:

- For any given S', the magnitude of N_{60} after compaction decreases with an increase in depth.
- An increase in N_{60} indicates an increase in the relative density of sand.

- The degree of compaction decreases with the increase in S'. At $S' = 1.22$ m, the degree of compaction at any depth is the largest. However, at $S' = 2.44$ m, the vibroflotation had practically no effect in compacting soil.

During the past 30 to 35 years, the vibroflotation technique has been used successfully on large projects to compact granular subsoils, thereby controlling structural settlement.

<h2>14.6   Blasting</h2>

Blasting is a technique that has been used successfully in many projects (Mitchell, 1970) for the densification of granular soils. The general soil grain sizes suitable for compaction by blasting are the same as those for compaction by vibroflotation. The process involves the detonation of explosive charges such as 60% dynamite at a certain depth below the ground surface in saturated soil. The lateral spacing of the charges varies from about 3 to 9 m. Three to five successful detonations are usually necessary to achieve the desired compaction. Compaction (up to a relative density of about 80%) up to a depth of about 18 m over a large area can easily be achieved by using this process. Usually, the explosive charges are placed at a depth of about two-thirds of the thickness of the soil layer desired to be compacted. The sphere of influence of compaction by a 60% dynamite charge can be given as follows (Mitchell, 1970):

$$r = \sqrt{\frac{W_{EX}}{C}} \tag{14.11}$$

where r = sphere of influence
W_{EX} = weight of explosive -60% dynamite
$C = 0.0122$ when W_{EX} is in kg and r is in m

Figure 14.15 shows the test results of soil densification by blasting in an area measuring 15 m by 9 m (Mitchell, 1970). For these tests, twenty 2.09-kg charges of Gelamite No. 1 (Hercules Powder Company, Wilmington, Delaware) were used.

<h2>14.7   Precompression</h2>

When highly compressible, normally consolidated clayey soil layers lie at a limited depth and large consolidation settlements are expected as the result of the construction of large buildings, highway embankments, or earth dams, precompression of soil may be used to minimize postconstruction settlement. The principles of precompression are best explained

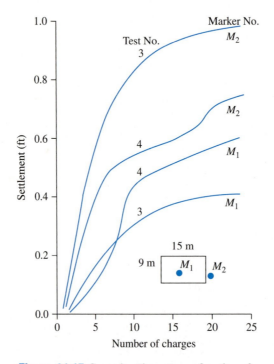

Figure 14.15 Ground settlement as a function of number of explosive charges

by reference to Figure 14.16. Here, the proposed structural load per unit area is $\Delta\sigma'_{(p)}$, and the thickness of the clay layer undergoing consolidation is H_c. The maximum primary consolidation settlement caused by the structural load is then

$$S_{c(p)} = \frac{C_c H_c}{1 + e_o} \log \frac{\sigma'_o + \Delta\sigma'_{(p)}}{\sigma'_o} \tag{14.12}$$

The settlement–time relationship under the structural load will be like that shown in Figure 14.16b. However, if a surcharge of $\Delta\sigma'_{(p)} + \Delta\sigma'_{(f)}$ is placed on the ground, the primary consolidation settlement will be

$$S_{c(p+f)} = \frac{C_c H_c}{1 + e_o} \log \frac{\sigma'_o + [\Delta\sigma'_{(p)} + \Delta\sigma'_{(f)}]}{\sigma'_o} \tag{14.13}$$

The settlement–time relationship under a surcharge of $\Delta\sigma'_{(p)} + \Delta\sigma'_{(f)}$ is also shown in Figure 14.16b. Note that a total settlement of $S_{c(p)}$ would occur at time t_2, which is much

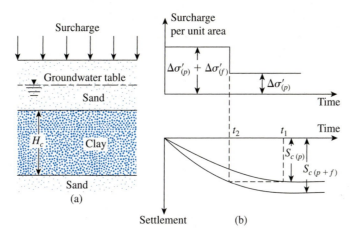

Figure 14.16 Principles of precompression

shorter than t_1. So, if a temporary total surcharge of $\Delta\sigma'_{(p)} + \Delta\sigma'_{(f)}$ is applied on the ground surface for time t_2, the settlement will equal $S_{c(p)}$. At that time, if the surcharge is removed and a structure with a permanent load per unit area of $\Delta\sigma'_{(p)}$ is built, no appreciable settlement will occur. The procedure just described is called *precompression.* The total surcharge $\Delta\sigma'_{(p)} + \Delta\sigma'_{(f)}$ can be applied by means of temporary fills.

Derivation of Equations for Obtaining $\Delta\sigma'_{(f)}$ and t_2

Figure 14.16b shows that, under a surcharge of $\Delta\sigma'_{(p)} + \Delta\sigma'_{(f)}$, the degree of consolidation at time t_2 after the application of the load is

$$U = \frac{S_{c(p)}}{S_{c(p+f)}} \tag{14.14}$$

Substitution of Eqs. (14.12) and (14.13) into Eq. (14.14) yields

$$U = \frac{\log\left[\dfrac{\sigma'_o + \Delta\sigma'_{(p)}}{\sigma'_o}\right]}{\log\left[\dfrac{\sigma'_o + \Delta\sigma'_{(p)} + \Delta\sigma'_{(f)}}{\sigma'_o}\right]} = \frac{\log\left[1 + \dfrac{\Delta\sigma'_{(p)}}{\sigma'_o}\right]}{\log\left\{1 + \dfrac{\Delta\sigma'_{(p)}}{\sigma'_o}\left[1 + \dfrac{\Delta\sigma'_{(f)}}{\Delta\sigma'_{(p)}}\right]\right\}} \tag{14.15}$$

Figure 14.17 gives magnitudes of U for various combinations of $\Delta\sigma'_{(p)}/\sigma'_o$, and $\Delta\sigma'_{(f)}/\Delta\sigma'_{(p)}$. The degree of consolidation referred to in Eq. (14.15) is actually the average degree of consolidation at time t_2, as shown in Figure 14.17b. However, if the average degree

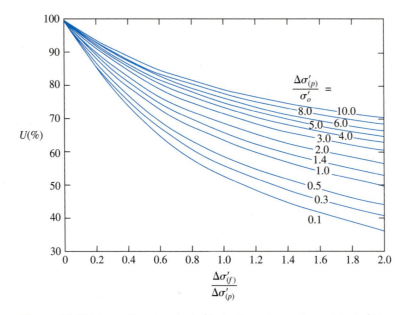

Figure 14.17 Plot of U against $\Delta\sigma'_{(f)}/\Delta\sigma'_{(p)}$ for various values of $\Delta\sigma'_{(p)}/\sigma'_o$— Eq. (14.15)

of consolidation is used to determine t_2, some construction problems might occur. The reason is that, after the removal of the surcharge and placement of the structural load, the portion of clay close to the drainage surface will continue to swell, and the soil close to the midplane will continue to settle. (See Figure 14.18.) In some cases, net continuous settlement might result. A conservative approach may solve the problem; that is, assume that U in Eq. (14.15) is the midplane degree of consolidation (Johnson, 1970a). Now, from Eq. (1.73),

$$U = f(T_v) \tag{1.73}$$

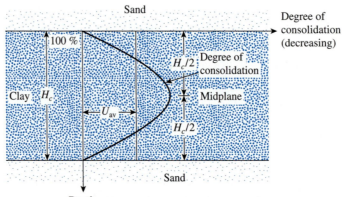

Figure 14.18

where

T_v = time factor = $C_v t_2 / H^2$
C_v = coefficient of consolidation
t_2 = time
H = maximum drainage path ($= H_c/2$ for two-way drainage and H_c for one-way drainage)

The variation of U (the midplane degree of consolidation) with T_v is given in Figure 14.19.

Procedure for Obtaining Precompression Parameters

Two problems may be encountered by engineers during precompression work in the field:

1. The value of $\Delta\sigma'_{(f)}$ is known, but t_2 must be obtained. In such a case, obtain σ'_o, $\Delta\sigma_{(p)}$, and solve for U, using Eq. (14.15) or Figure 14.17. For this value of U, obtain T_v from Figure 14.19. Then

$$t_2 = \frac{T_v H^2}{C_v} \tag{14.16}$$

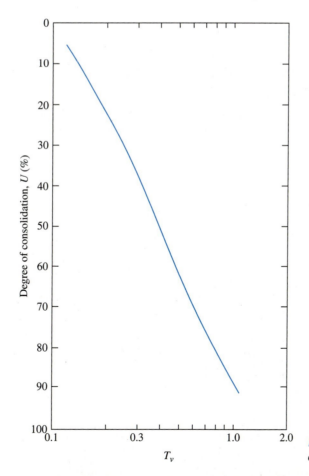

Figure 14.19 Plot of midplane degree of consolidation against T_v

2. For a specified value of t_2, $\Delta\sigma'_{(f)}$ must be obtained. In such a case, calculate T_v. Then use Figure 14.19 to obtain the midplane degree of consolidation, U. With the estimated value of U, go to Figure 14.17 to get the required value of $\Delta\sigma'_{(f)}/\Delta\sigma'_{(p)}$, and then calculate $\Delta\sigma'_{(f)}$.

Several case histories on the successful use of precompression techniques for improving foundation soil are available in the literature (for example, Johnson, 1970a).

Example 14.1

Examine Figure 14.16. During the construction of a highway bridge, the average permanent load on the clay layer is expected to increase by about 115 kN/m². The average effective overburden pressure at the middle of the clay layer is 210 kN/m². Here, $H_c = 6$ m, $C_c = 0.28$, $e_o = 0.9$, and $C_v = 0.36$ m²/mo. The clay is normally consolidated. Determine

    a. The total primary consolidation settlement of the bridge without precompression
    b. The surcharge, $\Delta\sigma'_{(f)}$, needed to eliminate the entire primary consolidation settlement in nine months by precompression.

Solution

Part a
The total primary consolidation settlement may be calculated from Eq. (14.12):

$$S_{c(p)} = \frac{C_c H_c}{1 + e_o} \log\left[\frac{\sigma'_o + \Delta\sigma'_{(p)}}{\sigma'_o}\right] = \frac{(0.28)(6)}{1 + 0.9} \log\left[\frac{210 + 115}{210}\right]$$
$$= 0.1677 \text{ m} = \mathbf{167.7 \text{ mm}}$$

Part b
We have

$$T_v = \frac{C_v t_2}{H^2}$$
$$C_v = 0.36 \text{ m}^2/\text{mo.}$$
$$H = 3 \text{ m (two-way drainage)}$$
$$t_2 = 9 \text{ mo.}$$

Hence,

$$T_v = \frac{(0.36)(9)}{3^2} = 0.36$$

According to Figure 14.19, for $T_v = 0.36$, the value of U is 47%. Now,

$$\Delta\sigma'_{(p)} = 115 \text{ kN/m}^2$$

and

$$\sigma'_o = 210 \text{ kN/m}^2$$

so

$$\frac{\Delta\sigma'_{(p)}}{\sigma'_o} = \frac{115}{210} = 0.548$$

According to Figure 14.17, for $U = 47\%$ and $\Delta\sigma'_{(p)}/\sigma'_o = 0.548$, $\Delta\sigma'_{(f)}/\Delta\sigma'_{(p)} \approx 1.8$; thus,

$$\Delta\sigma'_{(f)} = (1.8)(115) = \textbf{207 kN/m}^2 \qquad \blacksquare$$

14.8 *Sand Drains*

The use of sand drains is another way to accelerate the consolidation settlement of soft, normally consolidated clay layers and achieve precompression before the construction of a desired foundation. Sand drains are constructed by drilling holes through the clay layer(s) in the field at regular intervals. The holes are then backfilled with sand. This can be achieved by several means, such as (a) rotary drilling and then backfilling with sand; (b) drilling by continuous-flight auger with a hollow stem and backfilling with sand (through the hollow steam); and (c) driving hollow steel piles. The soil inside the pile is then jetted out, after which backfilling with sand is done. Figure 14.20 shows a schematic diagram of sand drains. After backfilling the drill holes with sand, a surcharge is applied at the ground surface. The surcharge will increase the pore water pressure in the clay. The excess pore water pressure in the clay will be dissipated by drainage—both vertically and radially to the sand drains—thereby accelerating settlement of the clay layer. In Figure 14.20a, note that the radius of the sand drains is r_w. Figure 14.20b shows the plan of the layout of the sand drains. The effective zone from which the radial drainage will be directed toward a given sand drain is approximately cylindrical, with a diameter of d_e.

To determine the surcharge that needs to be applied at the ground surface and the length of time that it has to be maintained, see Figure 14.16 and use the corresponding equation, Eq. (14.15):

$$U_{v,r} = \frac{\log\left[1 + \dfrac{\Delta\sigma'_{(p)}}{\sigma'_o}\right]}{\log\left\{1 + \dfrac{\Delta\sigma'_{(p)}}{\sigma'_o}\left[1 + \dfrac{\Delta\sigma'_{(f)}}{\Delta\sigma'_{(p)}}\right]\right\}} \qquad (14.17)$$

The notations $\Delta\sigma'_{(p)}$, σ'_o, and $\Delta\sigma'_{(f)}$ are the same as those in Eq. (14.15); however, the left-hand side of Eq. (14.17) is the *average degree* of consolidation instead of the degree of

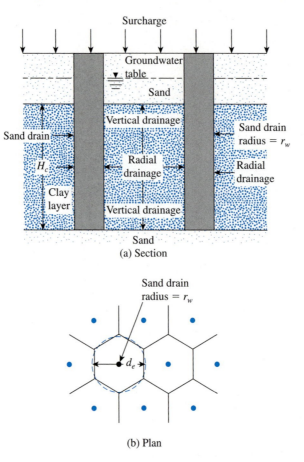

(a) Section

(b) Plan

Figure 14.20 Sand drains

consolidation at midplane. Both *radial* and *vertical* drainage contribute to the average degree of consolidation. If $U_{v,r}$ can be determined for any time t_2 (see Figure 14.16b), the total surcharge $\Delta\sigma'_{(f)} + \Delta\sigma'_{(p)}$ may be obtained easily from Figure 14.17. The procedure for determining the average degree of consolidation $(U_{v,r})$ follows:

For a given surcharge and duration, t_2, the average degree of consolidation due to drainage in the vertical and radial directions is

$$U_{v,r} = 1 - (1 - U_r)(1 - U_v) \tag{14.18}$$

where

U_r = average degree of consolidation with radial drainage only
U_v = average degree of consolidation with vertical drainage only

The successful use of sand drains has been described in detail by Johnson (1970b). As with precompression, constant field settlement observations may be necessary during the period the surcharge is applied.

Average Degree of Consolidation Due to Radial Drainage Only

Figure 14.21 shows a schematic diagram of a sand drain. In the figure, r_w is the radius of the sand drain and $r_e = d_e/2$ is the radius of the effective zone of drainage. It is also important to realize that, during the installation of sand drains, a certain zone of clay surrounding them is smeared, thereby changing the hydraulic conductivity of the clay. In the figure, r_s is the radial distance from the center of the sand drain to the farthest point of the smeared zone. Now, for the average-degree-of-consolidation relationship, we will use the *theory of equal strain*. Two cases may arise that relate to the nature of the application of surcharge, and they are shown in Figure 14.22. (See the notations shown in Figure 14.16). Either (a) the entire surcharge is applied instantaneously (see Figure 14.22a), or (b) the surcharge is applied in the form of a ramp load (see Figure 14.22b). When the entire surcharge is applied instantaneously (Barron, 1948),

$$U_r = 1 - \exp\left(\frac{-8T_r}{m}\right) \tag{14.19}$$

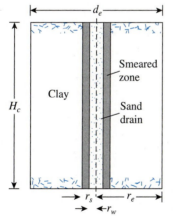

Figure 14.21 Schematic diagram of a sand drain

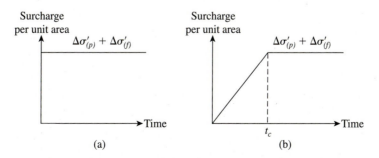

Figure 14.22 Nature of application of surcharge

where

$$m = \frac{n^2}{n^2 - S^2} \ln\left(\frac{n}{S}\right) - \frac{3}{4} + \frac{S^2}{4n^2} + \frac{k_h}{k_s}\left(\frac{n^2 - S^2}{n^2}\right) \ln S \tag{14.20}$$

in which

$$n = \frac{d_e}{2r_w} = \frac{r_e}{r_w} \tag{14.21}$$

$$S = \frac{r_s}{r_w} \tag{14.22}$$

and

k_h = hydraulic conductivity of clay in the horizontal direction in the unsmeared zone

k_s = horizontal hydraulic conductivity in the smeared zone

T_r = nondimensional time factor for radial drainage only $= \dfrac{C_{vr}t_2}{d_e^2}$ $\tag{14.23}$

C_{vr} = coefficient of consolidation for radial drainage

$$= \frac{k_h}{\left[\dfrac{\Delta e}{\Delta\sigma'(1 + e_{av})}\right]\gamma_w} \tag{14.24}$$

For a *no-smear case*, $r_s = r_w$ and $k_h = k_s$, so $S = 1$ and Eq. (14.20) becomes

$$m = \left(\frac{n^2}{n^2 - 1}\right) \ln(n) - \frac{3n^2 - 1}{4n^2} \tag{14.25}$$

Table 14.5 gives the values of U_r for various values of T_r and n.

If the surcharge is applied in the form of a *ramp* and *there is no smear,* then (Olson, 1977)

$$U_r = \frac{T_r - \dfrac{1}{A}[1 - \exp(-AT_r)]}{T_{rc}} \qquad (\text{for } T_r \leq T_{rc}) \tag{14.26}$$

and

$$U_r = 1 - \frac{1}{AT_{rc}}[\exp(AT_{rc}) - 1]\exp(-AT_{rc}) \qquad (\text{for } T_r \geq T_{rc}) \tag{14.27}$$

Table 14.5 Variation of U_r for Various Values of T_r and n, No-Smear Case [Eqs. (14.19) and (14.25)]

Degree of consolidation U_r (%)	Time factor T_r for value of $n\ (=r_e/r_w)$				
	5	10	15	20	25
0	0	0	0	0	0
1	0.0012	0.0020	0.0025	0.0028	0.0031
2	0.0024	0.0040	0.0050	0.0057	0.0063
3	0.0036	0.0060	0.0075	0.0086	0.0094
4	0.0048	0.0081	0.0101	0.0115	0.0126
5	0.0060	0.0101	0.0126	0.0145	0.0159
6	0.0072	0.0122	0.0153	0.0174	0.0191
7	0.0085	0.0143	0.0179	0.0205	0.0225
8	0.0098	0.0165	0.0206	0.0235	0.0258
9	0.0110	0.0186	0.0232	0.0266	0.0292
10	0.0123	0.0208	0.0260	0.0297	0.0326
11	0.0136	0.0230	0.0287	0.0328	0.0360
12	0.0150	0.0252	0.0315	0.0360	0.0395
13	0.0163	0.0275	0.0343	0.0392	0.0431
14	0.0177	0.0298	0.0372	0.0425	0.0467
15	0.0190	0.0321	0.0401	0.0458	0.0503
16	0.0204	0.0344	0.0430	0.0491	0.0539
17	0.0218	0.0368	0.0459	0.0525	0.0576
18	0.0232	0.0392	0.0489	0.0559	0.0614
19	0.0247	0.0416	0.0519	0.0594	0.0652
20	0.0261	0.0440	0.0550	0.0629	0.0690
21	0.0276	0.0465	0.0581	0.0664	0.0729
22	0.0291	0.0490	0.0612	0.0700	0.0769
23	0.0306	0.0516	0.0644	0.0736	0.0808
24	0.0321	0.0541	0.0676	0.0773	0.0849
25	0.0337	0.0568	0.0709	0.0811	0.0890
26	0.0353	0.0594	0.0742	0.0848	0.0931
27	0.0368	0.0621	0.0776	0.0887	0.0973
28	0.0385	0.0648	0.0810	0.0926	0.1016
29	0.0401	0.0676	0.0844	0.0965	0.1059
30	0.0418	0.0704	0.0879	0.1005	0.1103
31	0.0434	0.0732	0.0914	0.1045	0.1148
32	0.0452	0.0761	0.0950	0.1087	0.1193
33	0.0469	0.0790	0.0987	0.1128	0.1239
34	0.0486	0.0820	0.1024	0.1171	0.1285
35	0.0504	0.0850	0.1062	0.1214	0.1332
36	0.0522	0.0881	0.1100	0.1257	0.1380
37	0.0541	0.0912	0.1139	0.1302	0.1429
38	0.0560	0.0943	0.1178	0.1347	0.1479
39	0.0579	0.0975	0.1218	0.1393	0.1529
40	0.0598	0.1008	0.1259	0.1439	0.1580
41	0.0618	0.1041	0.1300	0.1487	0.1632
42	0.0638	0.1075	0.1342	0.1535	0.1685
43	0.0658	0.1109	0.1385	0.1584	0.1739
44	0.0679	0.1144	0.1429	0.1634	0.1793

(continued)

Table 14.5 (*continued*)

Degree of consolidation U_r (%)	Time factor T_r for value of n (=r_e/r_w)				
	5	**10**	**15**	**20**	**25**
45	0.0700	0.1180	0.1473	0.1684	0.1849
46	0.0721	0.1216	0.1518	0.1736	0.1906
47	0.0743	0.1253	0.1564	0.1789	0.1964
48	0.0766	0.1290	0.1611	0.1842	0.2023
49	0.0788	0.1329	0.1659	0.1897	0.2083
50	0.0811	0.1368	0.1708	0.1953	0.2144
51	0.0835	0.1407	0.1758	0.2020	0.2206
52	0.0859	0.1448	0.1809	0.2068	0.2270
53	0.0884	0.1490	0.1860	0.2127	0.2335
54	0.0909	0.1532	0.1913	0.2188	0.2402
55	0.0935	0.1575	0.1968	0.2250	0.2470
56	0.0961	0.1620	0.2023	0.2313	0.2539
57	0.0988	0.1665	0.2080	0.2378	0.2610
58	0.1016	0.1712	0.2138	0.2444	0.2683
59	0.1044	0.1759	0.2197	0.2512	0.2758
60	0.1073	0.1808	0.2258	0.2582	0.2834
61	0.1102	0.1858	0.2320	0.2653	0.2912
62	0.1133	0.1909	0.2384	0.2726	0.2993
63	0.1164	0.1962	0.2450	0.2801	0.3075
64	0.1196	0.2016	0.2517	0.2878	0.3160
65	0.1229	0.2071	0.2587	0.2958	0.3247
66	0.1263	0.2128	0.2658	0.3039	0.3337
67	0.1298	0.2187	0.2732	0.3124	0.3429
68	0.1334	0.2248	0.2808	0.3210	0.3524
69	0.1371	0.2311	0.2886	0.3300	0.3623
70	0.1409	0.2375	0.2967	0.3392	0.3724
71	0.1449	0.2442	0.3050	0.3488	0.3829
72	0.1490	0.2512	0.3134	0.3586	0.3937
73	0.1533	0.2583	0.3226	0.3689	0.4050
74	0.1577	0.2658	0.3319	0.3795	0.4167
75	0.1623	0.2735	0.3416	0.3906	0.4288
76	0.1671	0.2816	0.3517	0.4021	0.4414
77	0.1720	0.2900	0.3621	0.4141	0.4546
78	0.1773	0.2988	0.3731	0.4266	0.4683
79	0.1827	0.3079	0.3846	0.4397	0.4827
80	0.1884	0.3175	0.3966	0.4534	0.4978
81	0.1944	0.3277	0.4090	0.4679	0.5137
82	0.2007	0.3383	0.4225	0.4831	0.5304
83	0.2074	0.3496	0.4366	0.4992	0.5481
84	0.2146	0.3616	0.4516	0.5163	0.5668
85	0.2221	0.3743	0.4675	0.5345	0.5868
86	0.2302	0.3879	0.4845	0.5539	0.6081
87	0.2388	0.4025	0.5027	0.5748	0.6311
88	0.2482	0.4183	0.5225	0.5974	0.6558
89	0.2584	0.4355	0.5439	0.6219	0.6827
90	0.2696	0.4543	0.5674	0.6487	0.7122

Table 14.5 (*continued*)

Degree of consolidation U_r (%)	Time factor T_r for value of n ($=r_e/r_w$)				
	5	10	15	20	25
91	0.2819	0.4751	0.5933	0.6784	0.7448
92	0.2957	0.4983	0.6224	0.7116	0.7812
93	0.3113	0.5247	0.6553	0.7492	0.8225
94	0.3293	0.5551	0.6932	0.7927	0.8702
95	0.3507	0.5910	0.7382	0.8440	0.9266
96	0.3768	0.6351	0.7932	0.9069	0.9956
97	0.4105	0.6918	0.8640	0.9879	1.0846
98	0.4580	0.7718	0.9640	1.1022	1.2100
99	0.5391	0.9086	1.1347	1.2974	1.4244

where

$$T_{rc} = \frac{C_{vr} t_c}{d_e^2} \text{ (see Figure 14.22b for the definition of } t_c) \tag{14.28}$$

and

$$A = \frac{2}{m} \tag{14.29}$$

Average Degree of Consolidation Due to Vertical Drainage Only

Using Figure 14.22a, for instantaneous application of a surcharge, we may obtain the average degree of consolidation due to vertical drainage only from Eqs. (1.74) and (1.75). We have

$$T_v = \frac{\pi}{4}\left[\frac{U_v(\%)}{100}\right]^2 \quad \text{(for } U_v = 0 \text{ to } 60\%\text{)} \qquad \text{[Eq. (1.74)]}$$

and

$$T_v = 1.781 - 0.933 \log\left[100 - U_v(\%)\right] \quad \text{(for } U_v > 60\%\text{)} \qquad \text{[Eq. (1.75)]}$$

where U_v = average degree of consolidation due to vertical drainage only, and

$$T_v = \frac{C_v t_2}{H^2} \qquad \text{[Eq. (1.69)]}$$

where C_v = coefficient of consolidation for vertical drainage.

For the case of ramp loading, as shown in Figure 14.22b, the variation of U_v with T_v can be expressed as (Olson, 1977):

For $T_v \leq T_c$:

$$U_v = \frac{T_v}{T_c} \left\{ 1 - \frac{2}{T_v} \Sigma \frac{1}{M^4} [1 - \exp(-M^2 T_v)] \right\} \tag{14.30}$$

For $T_v \leq T_c$:

$$U_v = 1 - \frac{2}{T_c} \Sigma \frac{1}{M^4} [\exp(-M^2 T_c) - 1] \exp(-M^2 T_v) \tag{14.31}$$

where

$M = \dfrac{\pi}{2}(2m' + 1)$

$m' = 0, 1, 2, \ldots$

$$T_c = \frac{C_v t_c}{H^2} \tag{14.32}$$

where H = length of maximum vertical drainage path. Figure 14.23 shows the variation of $U_v(\%)$ with T_c and T_v.

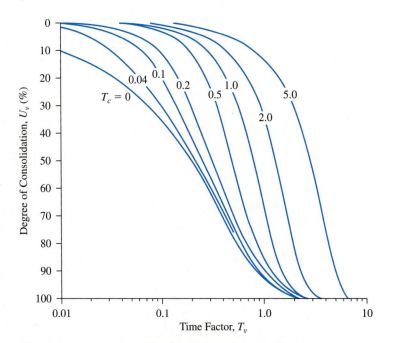

Figure 14.23 Variation of U_v with T_v and T_c [Eqs. (14.30) and (14.31)]

Example 14.2

Redo Example 14.1, with the addition of some sand drains. Assume that $r_w = 0.1$ m, $d_e = 3$ m, $C_v = C_{vr}$, and the surcharge is applied instantaneously. (See Figure 14.22a.) Also assume that this is a no-smear case.

Solution
Part a
The total primary consolidation settlement will be 167.7 mm, as before.

Part b
From Example 14.1, $T_v = 0.36$. Using Eq. (1.74), we obtain

$$T_v = \frac{\pi}{4}\left[\frac{U_v(\%)}{100}\right]^2$$

or

$$U_v = \sqrt{\frac{4T_v}{\pi}} \times 100 = \sqrt{\frac{(4)(0.36)}{\pi}} \times 100 = 67.7\%$$

Also,

$$n = \frac{d_e}{2r_w} = \frac{3}{2 \times 0.1} = 15$$

Again,

$$T_r = \frac{C_{vr}t_2}{d_e^2} = \frac{(0.36)(9)}{(3)^2} = 0.36$$

From Table 14.5 for $n = 15$ and $T_r = 0.36$, the value of U_r is about 77%, Hence,

$$U_{v,r} = 1 - (1 - U_v)(1 - U_r) = 1 - (1 - 0.67)(1 - 0.77)$$

$$= 0.924 = 92.4\%$$

Now, from Figure 14.17, for $\Delta\sigma_p'/\sigma_o' = 0.548$ and $U_{v,r} = 92.4\%$, the value of $\Delta\sigma_f'/\Delta\sigma_p' \approx 0.12$. Hence,

$$\Delta\sigma_{(f)}' = (115)(0.12) = \textbf{13.8 kN/m}^2$$ ∎

Example 14.3

Suppose that, for the sand drain project of Figure 14.20, the clay is normally consolidated. We are given the following data:

$$\text{Clay:} \quad H_c = 4.57 \text{ m (two-way drainage)}$$
$$C_c = 0.31$$
$$e_o = 1.1$$

Effective overburden pressure at the middle of the clay layer

$$= 47.92 \text{ kN/m}^2$$
$$C_v = 106.15 \times 10^{-4} \text{ m}^2/\text{day}$$

$$\text{Sand drain:} \quad r_w = 0.091 \text{ m}$$
$$d_e = 1.83 \text{ m}$$
$$C_v = C_{vr}$$

A surcharge is applied as shown in Figure 14.24. Assume this to be a no-smear case. Calculate the degree of consolidation 30 days after the surcharge is first applied. Also, determine the consolidation settlement at that time due to the surcharge.

Solution

From Eq. (14.32),

$$T_c = \frac{C_v t_c}{H^2} = \frac{(106.15 \times 10^{-4} \text{ m}^2/\text{day})(60)}{\left(\dfrac{4.57}{2}\right)^2} = 0.122$$

and

$$T_v = \frac{C_v t_2}{H^2} = \frac{(106.15 \times 10^{-4})(30)}{\left(\dfrac{4.57}{2}\right)^2} = 0.061$$

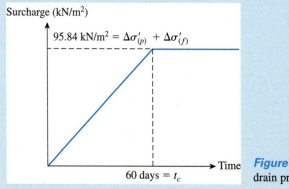

Surcharge (kN/m²)

$95.84 \text{ kN/m}^2 = \Delta\sigma'_{(p)} + \Delta\sigma'_{(f)}$

60 days $= t_c$

Time

Figure 14.24 Ramp load for a sand drain project

Using Figure 14.23 for $T_c = 0.123$ and $T_v = 0.061$, we have $U_v \approx 9\%$. For the sand drain,

$$n = \frac{d_e}{2r_w} = \frac{1.83}{(2)(0.091)} = 10$$

From Eq. (14.28),

$$T_{rc} = \frac{C_{vr}t_c}{d_e^2} = \frac{(106.15 \times 10^{-4})(60)}{(1.83)^2} = 0.19$$

and

$$T_r = \frac{C_{vr}t_2}{d_e^2} = \frac{(106.15 \times 10^{-4})(30)}{(1.83)^2} = 0.095$$

Again, from Eq. (14.26),

$$U_r = \frac{T_r - \frac{1}{A}[1 - \exp(-AT_r)]}{T_{rc}}$$

Also, for the no-smear case,

$$m = \frac{n^2}{n^2 - 1} \ln (n) - \frac{3n^2 - 1}{4n^2} = \frac{10^2}{10^2 - 1} \ln (10) - \frac{3(10)^2 - 1}{4(10)^2} = 1.578$$

and

$$A = \frac{2}{m} = \frac{2}{1.578} = 1.267$$

so

$$U_r = \frac{0.096 - \frac{1}{1.267}[1 - \exp(-1.267 \times 0.095)]}{0.19} = 0.03 = 3\%$$

From Eq. (14.18),

$$U_{v,r} = 1 - (1 - U_r)(1 - U_v) = 1 - (1 - 0.03)(1 - 0.09) = 0.117 = \textbf{11.7\%}$$

The total primary settlement is thus

$$S_{c(p)} = \frac{C_c H_c}{1 + e_o} \log \left[\frac{\sigma_o' + \Delta\sigma_{(p)}' + \Delta\sigma_f'}{\sigma_o'} \right]$$

$$= \frac{(0.31)(4.57)}{1 + 1.1} \log \left(\frac{47.92 + 95.84}{47.92} \right) = 0.332 \text{ m}$$

and the settlement after 30 days is

$$S_{c(p)}U_{v,r} = (0.322)(0.117)(1000) = \textbf{37.67 mm}$$

∎

14.9   *Prefabricated Vertical Drains*

Prefabricated vertical drains (PVDs), also referred to as *wick* or *strip drains,* were originally developed as a substitute for the commonly used sand drain. With the advent of materials science, these drains began to be manufactured from synthetic polymers such as polypropylene and high-density polyethylene. PVDs are normally manufactured with a corrugated or channeled synthetic core enclosed by a geotextile filter, as shown schematically in Figure 14.25. Installation rates reported in the literature are on the order of 0.1 to 0.3 m/s, excluding equipment mobilization and setup time. PVDs have been used extensively in the past for expedient consolidation of low-permeability soils under surface surcharge. The main advantage of PVDs over sand drains is that they do not require drilling; thus, installation is much faster. Figures 14.26a and b are photographs of the installation of PVDs in the field.

Design of PVDs

The relationships for the average degree of consolidation due to radial drainage into sand drains are given in Eqs. (14.19) through (14.24) for equal-strain cases. Yeung (1997) used these relationships to develop design curves for PVDs. The theoretical developments used by Yeung are given next.

Figure 14.27 shows the layout of a square-grid pattern of prefabricated vertical drains. See also Figure 14.25 for the definition of *a* and *b*). The equivalent diameter of a PVD can be given as

$$d_w = \frac{2(a + b)}{\pi} \tag{14.33}$$

Now, Eq. (14.19) can be rewritten as

$$U_r = 1 - \exp\left(-\frac{8C_{vr}t}{d_w^2}\frac{d_w^2}{d_e^2m}\right) = 1 - \exp\left(-\frac{8T_r'}{\alpha'}\right) \tag{14.34}$$

Polypropylene core

Geotextile fabric

Figure 14.25 Prefabricated vertical drain (PVD)

(a)

(b)

Figure 14.26 Installation of PVDs in the field (*Note:* (b) is a closeup view of (a)) (*Courtesy of E. C. Shin, University of Incheon, Korea*)

Figure 14.27 Square-grid pattern layout of prefabricated vertical drains

where d_e = diameter of the effective zone of drainage = $2r_e$. Also,

$$T'_r = \frac{C_{vr}t}{d_w^2} \qquad (14.35)$$

$$\alpha' = n^2 m = \frac{n^4}{n^2 - S^2} \ln\left(\frac{n}{S}\right) - \left(\frac{3n^2 - S^2}{4}\right) + \frac{k_h}{k_s}(n^2 - S^2)\ln S \qquad (14.36)$$

and

$$n = \frac{d_e}{d_w} \qquad (14.37)$$

From Eq. (14.34),

$$T'_r = -\frac{\alpha'}{8}\ln(1 - U_r)$$

or

$$(T'_r)_1 = \frac{T'_r}{\alpha'} = -\frac{\ln(1 - U_r)}{8} \qquad (14.38)$$

Table 14.6 gives the variation of $(T'_r)_1$ with U_r. Also, Figure 14.28 shows plots of α' versus n for $k_h/k_s = 5$ and 10 and $S = 2$ and 3.

Table 14.6 Variation of $(T_r')_1$ with U_r [Eq. (14.38)]

$U_r(\%)$	$(T_r')_1$
0	0
5	0.006
10	0.013
15	0.020
20	0.028
25	0.036
30	0.045
35	0.054
40	0.064
45	0.075
50	0.087
55	0.100
60	0.115
65	0.131
70	0.150
75	0.173
80	0.201
85	0.237
90	0.288
95	0.374

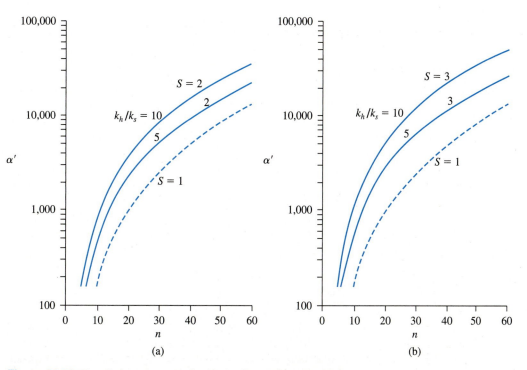

Figure 14.28 Plot of α' versus n: (a) $S = 2$: (b) $S = 3$ [Eq. (14.36)]

Following is a step-by-step procedure for the design of prefabricated vertical drains:

Step 1. Determine time t_2 available for the consolidation process and the $U_{v,r}$ required therefore [Eq. (14.17)]

Step 2. Determine U_v at time t_2 due to vertical drainage. From Eq. (14.18)

$$U_r = 1 - \frac{1 - U_{v,r}}{1 - U_v} \tag{14.39}$$

Step 3. For the PVD that is to be used, calculate d_w from Eq. (14.33).

Step 4. Determine $(T'_r)_1$ from Eqs. (14.38) and (14.39).

Step 5. Determine T'_r from Eq. (14.35).

Step 6. Determine

$$\alpha' = \frac{T'_r}{(T'_r)_1}.$$

Step 7. Using Figure 14.28 and α' determined from Step 6, determine n.

Step 8. From Eq. (14.37),

$$d_e = n \qquad d_w$$
$$\uparrow \qquad \uparrow$$
$$\text{Step 7} \quad \text{Step 3}$$

Step 9. Choose the drain spacing:

$$d = \frac{d_e}{1.05} \qquad (\text{for triangular pattern})$$

$$d = \frac{d_e}{1.128} \qquad (\text{for square pattern})$$

A Case History

The installation of PVDs combined with preloading is an efficient way to gain strength in soft clays for construction of foundations. An example of a field study can be found in the works of Shibuya and Hanh (2001) which describes a full-scale test embankment 40 m × 40 m in plan constructed over a soft clay layer located at Nong Ngu Hao, Thailand. PVDs were installed in the soft clay layer in a triangular pattern (Figure 14.29a). Figure 14.29b shows the pattern of preloading at the site along with the settlement-time plot at the ground surface below the center of the test embankment. Maximum settlement was reached after about four months. The variation of the undrained shear strength (c_u) with depth in the soft clay layer before and after the soil improvement is shown in Figure 14.29c. The variation of c_u with depth is based on field vane shear tests. The undrained shear strength increases by about 50 to 100% at various depths.

14.10   *Lime Stabilization*

As mentioned in Section 14.1, admixtures are occasionally used to stabilize soils in the field—particularly fine-grained soils. The most common admixtures are lime, cement, and lime–fly ash. The main purposes of stabilizing the soil are to (a) modify the soil, (b) expedite construction, and (c) improve the strength and durability of the soil.

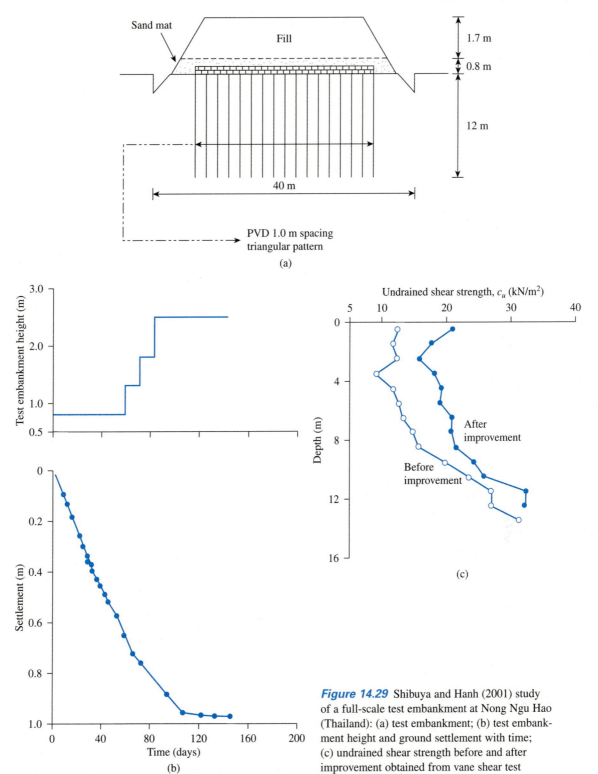

Figure 14.29 Shibuya and Hanh (2001) study of a full-scale test embankment at Nong Ngu Hao (Thailand): (a) test embankment; (b) test embankment height and ground settlement with time; (c) undrained shear strength before and after improvement obtained from vane shear test

The types of *lime* commonly used to stabilize fine-grained soils are hydrated high-calcium lime [$Ca(OH)_2$], calcitic quicklime (CaO), monohydrated dolomitic lime [$Ca(OH)_2 \cdot MgO$], and dolomitic quicklime. The quantity of lime used to stabilize most soils usually is in the range from 5 to 10%. When lime is added to clayey soils, two *pozzolanic* chemical reactions occur: *cation exchange* and *flocculation–agglomeration*. In the cation exchange and flocculation–agglomeration reactions, the *monovalent* cations generally associated with clays are replaced by the *divalent* calcium ions. The cations can be arranged in a series based on their affinity for exchange:

$$Al^{3+} > Ca^{2+} > Mg^{2+} > NH_4^+ > K^+ > Na^+ > Li^+$$

Any cation can replace the ions to its right. For example, calcium ions can replace potassium and sodium ions from a clay. Flocculation–agglomeration produces a change in the texture of clay soils. The clay particles tend to clump together to form larger particles, thereby (a) decreasing the liquid limit, (b) increasing the plastic limit, (c) decreasing the plasticity index, (d) increasing the shrinkage limit, (e) increasing the workability, and (f) improving the strength and deformation properties of a soil. Some examples in which lime influences the plasticity of clayey soils are given in Table 14.7.

Pozzolanic reaction between soil and lime involves a reaction between lime and the silica and alumina of the soil to form cementing material. One such reaction is

$$Ca(OH)_2 + SiO_2 \rightarrow CSH$$
$$\uparrow$$
$$\text{Clay silica}$$

where

$C = CaO$
$S = SiO_2$
$H = H_2O$

The pozzolanic reaction may continue for a long time.

The first 2 to 3% lime (on the dry-weight basis) substantially influences the workability and the property (such as plasticity) of the soil. The addition of lime to clayey soils also affects their compaction characteristics.

Table 14.7 Influence of Lime on Plasticity of Clay (Compiled from Thompson, 1967)

Soil	AASHTO Classification	0% Lime		5% Lime	
		Liquid limit	Plasticity index	Liquid limit	Plasticity index
Bryce B	A-7-6(18)	53	29	NP	NP
Cowden B	A-7-6(19)	54	33	NP	NP
Drummer B	A-7-6(19)	54	31	NP	NP
Huey B	A-7-6(17)	46	29	NP	NP

Note: NP—Non-plastic

Properties of Cured Lime-Stabilized Soils

The unconfined compression strength (q_u) of fine-grained soils compacted at optimum moisture content may range from 170 kN/m² to 2100 kN/m², depending upon the nature of the soil. With about 3 to 5% addition of lime and a curing period of 28 days, the unconfined compression strength may increase by 700 kN/m² or more.

The tensile strength (σ_T) of cured fine-grained soils also increases with lime stabilization. Tullock, Hudson, and Kennedy (1970) gave the following relationship between σ_T and q_u:

$$\sigma_T \text{ (kN/m}^2) = 47.54 + 50.6 q_u \text{ (MN/m}^2) \tag{14.40}$$

where σ_T is the indirect tensile strength.

Thompson (1966) provided the following relationship to estimate the modulus of elasticity (E_s) of lime-stabilized soils:

$$E_s \text{ (MN/m}^2) = 68.86 + 0.124 q_u \text{ (kN/m}^2) \tag{14.41}$$

Poisson's ratio (μ_s) of cured stabilized soils with about 5% lime varies between 0.08 to 0.12 (with an average of 0.11) at a stress level of 25% or less of the ultimate compressive strength. It increases to about 0.27 to 0.37 (with an average of 0.31) at a stress level greater than 50% to 75% of the ultimate compression strength (Transportation Research Board, 1987).

Lime Stabilization in the Field

Lime stabilization in the field can be done in three ways. They are

1. The *in situ* material or the borrowed material can be mixed with the proper amount of lime at the site and then compacted after the addition of moisture.
2. The soil can be mixed with the proper amount of lime and water at a plant and then hauled back to the site for compaction.
3. Lime slurry can be pressure injected into the soil to a depth of 4 to 5 m. Figure 14.30 shows a vehicle used for pressure injection of lime slurry.

The slurry-injection mechanical unit is mounted to the injection vehicle. A common injection unit is a hydraulic-lift mast with crossbeams that contain the injection rods. The rods are pushed into the ground by the action of the lift mast beams. The slurry is generally mixed in a batching tank about 3 m in diameter and 12 m long and is pumped at high pressure to the injection rods. Figure 14.31 is a photograph of the lime slurry pressure-injection process. The ratio typically specified for the preparation of lime slurry is 1.13 kg of dry lime to a gallon of water.

Because the addition of hydrated lime to soft clayey soils immediately increases the plastic limit, thus changing the soil from plastic to solid and making it appear to "dry up," limited amounts of the lime can be thrown on muddy and troublesome construction sites. This action improves trafficability and may save money and time. Quicklimes have also been successfully used in drill holes having diameters of 100 to 150 mm for stabilization of subgrades and slopes. For this type of work, holes are drilled in a grid pattern and then filled with quicklime.

Figure 14.30 Equipment for pressure injection of lime slurry (*Courtesy of Hayward Baker Inc., Odenton, Maryland.*)

14.11 *Cement Stabilization*

Cement is being increasingly used as a stabilizing material for soil, particularly in the construction of highways and earth dams. The first controlled soil–cement construction in the United States was carried out near Johnsonville, South Carolina, in 1935. Cement can be used to stabilize sandy and clayey soils. As in the case of lime, cement helps decrease the liquid limit and increase the plasticity index and workability of clayey soils. Cement stabilization is effective for clayey soils when the liquid limit is less than 45 to 50 and the plasticity index is less than about 25. The optimum requirements of cement by volume for effective stabilization of various types of soil are given in Table 14.8.

Like lime, cement helps increase the strength of soils, and strength increases with curing time. Table 14.9 presents some typical values of the unconfined compressive strength of various types of untreated soil and of soil–cement mixtures made with approximately 10% cement by weight.

Granular soils and clayey soils with low plasticity obviously are most suitable for cement stabilization. Calcium clays are more easily stabilized by the addition of cement, whereas sodium and hydrogen clays, which are expansive in nature, respond better to lime stabilization. For these reasons, proper care should be given in the selection of the stabilizing material.

For field compaction, the proper amount of cement can be mixed with soil either at the site or at a mixing plant. If the latter approach is adopted, the mixture can then be carried to the site. The soil is compacted to the required unit weight with a predetermined amount of water.

Figure 14.31 Pressure injection of lime slurry (*Courtesy of Hayward Baker Inc., Odenton, Maryland.*)

Table 14.8 Cement Requirement by Volume for Effective Stabilization of Various Soils[a]

Soil type		Percent cement by volume
AASHTO classification	**Unified classification**	
A-2 and A-3	GP, SP, and SW	6–10
A-4 and A-5	CL, ML, and MH	8–12
A-6 and A-7	CL, CH	10–14

[a] After Mitchell, J. K. and Freitag, D. R. (1959). "A Review and Evaluation of Soil-Cement Pavements," Journal of the Soil Mechanics and Foundations Division, American Society of Civil Engineers, Vol. 85, No. SM6, pp. 49–73. With permission from ASCE.

Table 14.9 Typical Compressive Strengths of Soils and Soil–Cement Mixtures[a]

Material	Unconfined compressive strength range kN/m²
Untreated soil:	
Clay, peat	Less than 350
Well-compacted sandy clay	70–280
Well-compacted gravel, sand, and clay mixtures	280–700
Soil–cement (10% cement by weight):	
Clay, organic soils	Less than 350
Silts, silty clays, very poorly graded sands, slightly organic soils	350–1050
Silty clays, sandy clays, very poorly graded sands, and gravels	700–1730
Silty sands, sandy clays, sands, and gravels	1730–3460
Well-graded sand–clay or gravel–sand–clay mixtures and sands and gravels	3460–10,350

[a] After Mitchell, J. K. and Freitag, D. R. (1959). "A Review and Evaluation of Soil-Cement Pavements," *Journal of the Soil Mechanics and Foundations Division*, American Society of Civil Engineers, Vol. 85, No. SM6, pp. 49–73. With permission from ASCE.

Similar to lime injection, cement slurry made of portland cement and water (in a water–cement ratio of 0.5:5) can be used for pressure grouting of poor soils under foundations of buildings and other structures. Grouting decreases the hydraulic conductivity of soils and increases their strength and load-bearing capacity. For the design of low-frequency machine foundations subjected to vibrating forces, stiffening the foundation soil by grouting and thereby increasing the resonant frequency is sometimes necessary.

14.12   *Fly-Ash Stabilization*

Fly ash is a by-product of the pulverized coal combustion process usually associated with electric power-generating plants. It is a fine-grained dust and is composed primarily of silica, alumina, and various oxides and alkalies. Fly ash is pozzolanic in nature and can react with hydrated lime to produce cementitious products. For that reason, lime–fly-ash mixtures can be used to stabilize highway bases and subbases. Effective mixes can be prepared with 10 to 35% fly ash and 2 to 10% lime. Soil–lime–fly-ash mixes are compacted under controlled conditions, with proper amounts of moisture to obtain stabilized soil layers.

A certain type of fly ash, referred to as "Type C" fly ash, is obtained from the burning of coal primarily from the western United States. This type of fly ash contains a fairly large proportion (up to about 25%) of free lime that, with the addition of water, will react with other fly-ash compounds to form cementitious products. Its use may eliminate the need to add manufactured lime.

Stone Columns

A method now being used to increase the load-bearing capacity of shallow foundations on soft clay layers is the construction of stone columns. This generally consists of water-jetting a vibroflot (see Section 14.5) into the soft clay layer to make a circular hole that extends through the clay to firmer soil. The hole is then filled with an imported gravel. The gravel in the hole is gradually compacted as the vibrator is withdrawn. The gravel used for the stone column has a size range of 6 to 40 mm. Stone columns usually have diameters of 0.5 to 0.75 m and are spaced at about 1.5 to 3 m center to center. Figure 14.32 shows the construction of a stone column.

Figure 14.32 Construction of a stone column (*Courtesy of The Reinforced Earth Company, Vienna, Virginia and Menard Soil Treatment Inc., Orange, California*)

After stone columns are constructed, a fill material should always be placed over the ground surface and compacted before the foundation is constructed. The stone columns tend to reduce the settlement of foundations at allowable loads. Several case histories of construction projects using stone columns are presented in Hughes and Withers (1974), Hughes et al. (1975), Mitchell and Huber (1985), and other works.

Stone columns work more effectively when they are used to stabilize a large area where the undrained shear strength of the subsoil is in the range of 10 to 50 kN/m² than to improve the bearing capacity of structural foundations (Bachus and Barksdale, 1989). Subsoils weaker than that may not provide sufficient lateral support for the columns. For large-site improvement, stone columns are most effective to a depth of 6 to 10 m. However, they have been constructed to a depth of 31 m. Bachus and Barksdale provided the following general guidelines for the design of stone columns to stabilize large areas.

Figure 14.33a shows the plan view of several stone columns. The area replacement ratio for the stone columns may be expressed as

$$a_s = \frac{A_s}{A} \tag{14.42}$$

where

A_s = area of the stone column
A = total area within the unit cell

For an *equilateral triangular pattern* of stone columns,

$$a_s = 0.907\left(\frac{D}{s}\right)^2 \tag{14.43}$$

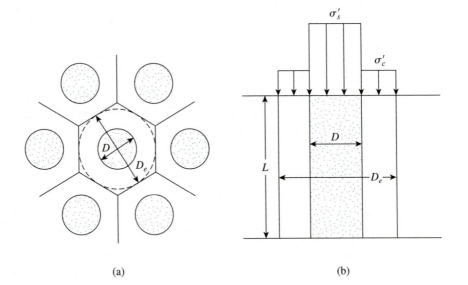

(a)   (b)

Figure 14.33 (a) Stone columns in a triangular pattern; (b) stress concentration due to change in stiffness

where

D = diameter of the stone column
s = spacing between the columns

When a uniform stress by means of a fill operation is applied to an area with stone columns to induce consolidation, a stress concentration occurs due to the change in the stiffness between the stone columns and the surrounding soil. (See Figure 14.33b.) The stress concentration factor is defined as

$$n' = \frac{\sigma'_s}{\sigma'_c} \tag{14.44}$$

where

σ'_s = effective stress in the stone column
σ'_c = effective stress in the subgrade soil

The relationships for σ'_s and σ'_c are

$$\sigma'_s = \sigma'\left[\frac{n'}{1 + (n' - 1)a_s}\right] = \mu_s\sigma' \tag{14.45}$$

and

$$\sigma'_c = \sigma'\left[\frac{1}{1 + (n' - 1)a_s}\right] = \mu_c\sigma' \tag{14.46}$$

where

σ' = average effective vertical stress
μ_s, μ_c = stress concentration coefficients

The improvement in the soil owing to the stone columns may be expressed as

$$\frac{S_{e(t)}}{S_e} = \mu_c \tag{14.47}$$

where

$S_{e(t)}$ = settlement of the treated soil
S_e = total settlement of the untreated soil

Load-Bearing Capacity of Stone Columns

If a foundation is constructed over a stone column as shown in Figure 14.34, failure will occur by bulging of the column at ultimate load. The bulging will occur within a length of 2.5D to 3D, measured from the top of the stone column, where D is the diameter of the column.

Hughes et al. (1975) provided an approximate relationship for the ultimate bearing capacity of stone columns, which can be given as

$$q_u = \tan^2\left(45 + \frac{\phi'}{2}\right)(4c_u + \sigma'_r) \tag{14.48}$$

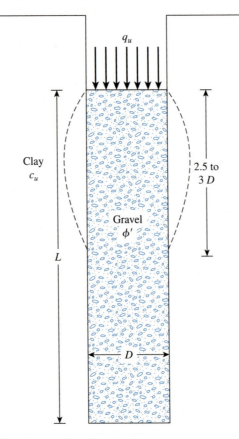

Figure 14.34 Bearing capacity of stone column

where

c_u = undrained shear strength of the clay
σ_r' = effective radial stress as measured by a pressuremeter ($\approx 2c_u$)
ϕ' = effective stress friction angle of the stone column material

    Thus, assuming that the stone column carries the entire load of the foundation, the ultimate load can be given as

$$Q_u = \frac{\pi}{4} D^2 \tan^2\left(45 + \frac{\phi'}{2}\right)(4c_u + \sigma_r') \tag{14.49}$$

On the basis of large-scale-model tests, Christoulas et al. (2000) suggested that

$$Q_u = \pi D L c_u \tag{14.50}$$

In the opinion of the author, the lower of the two values of Q_u obtained from Eqs. (14.49) and (14.50) should be used for actual design purposes. The allowable load can then be given as

$$Q_{all} = \frac{Q_u}{FS} \qquad (14.51)$$

where FS = factor of safety ($\approx$1.5 to 2).

Christoulas et al. (2000) also suggested a relationship between the load Q and the elastic settlement S_e for stone columns that can be expressed as

$$S_e = \left(\frac{Q}{LE_{clay}}\right)I_d \qquad (\text{for } Q \leq Q_1) \qquad (14.52)$$

and

$$S_e = \left(\frac{Q_1}{LE_{clay}}\right)I_d + \left(\frac{Q - Q_1}{4LE_{clay}}\right)I_d \qquad (\text{for } Q_1 \leq Q \leq Q_u) \qquad (14.53)$$

where

$$Q_1 = \frac{0.1DLE_{clay}}{I_d} \qquad (14.54)$$

in which

E_{clay} = modulus of elasticity of clay
I_d = influence factor (Mattes and Poulos, 1969)

The influence factor proposed by Mattes and Poulos is a function of three quantities:

1. $K = \dfrac{E_{col}}{E_{clay}}$

   where E_{col} = modulus of elasticity of the stone column material.

2. $\dfrac{L}{D}$

3. Poisson's ratio of clay, μ_{clay}. A value of $\mu_{clay} = 0.5$ will give a conservative result.

The variation of I_d (for $\mu_{clay} = 0.5$) with K is shown in Figure 14.35.

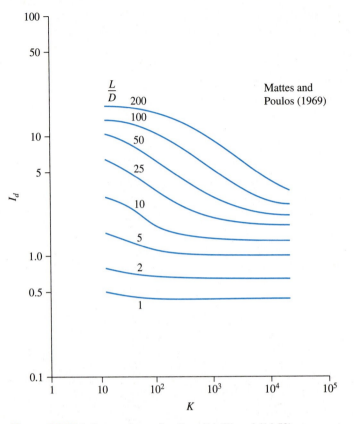

Figure 14.35 Influence factor I_d—Eqs. (14.52) and (14.53)

14.14   *Sand Compaction Piles*

Sand compaction piles are similar to stone columns, and they can be used in marginal sites to improve stability, control liquefaction, and reduce the settlement of various structures. Built in soft clay, these piles can significantly accelerate the pore water pressure-dissipation process and hence the time for consolidation.

Sand piles were first constructed in Japan between 1930 and 1950 (Ichimoto, 1981). Large-diameter compacted sand columns were constructed in 1955, using the Compozer technique (Aboshi et al., 1979). The Vibro-Compozer method of sand pile construction was developed by Murayama in Japan in 1958 (Murayama, 1962).

Sand compaction piles are constructed by driving a hollow mandrel with its bottom closed during driving. On partial withdrawal of the mandrel, the bottom doors open. Sand is poured from the top of the mandrel and is compacted in steps by applying air pressure as the mandrel is withdrawn. The piles are usually 0.46 to 0.76 m in diameter and are placed at about 1.5 to 3 m center to center. The pattern of layout of sand compaction piles

Figure 14.36 Construction of sand compaction pile in Yokohama, Japan Harbor (*Courtesy of E. C. Shin, University of Incheon, Korea*)

is the same as for stone columns. Figure 14.36 shows the construction of sand compaction piles in the harbor of Yokohama, Japan.

Basore and Boitano (1969) reported a case history on the densification of a granular subsoil having a thickness of about 9 m at the Treasure Island Naval Station in San Francisco, California, using sand compaction piles. The sand piles had diameters of 356 mm. Figure 14.37a shows the layout of the sand piles. The spacing, S', between the piles was varied. The standard penetration resistance, N_{60}, before and after the construction of piles are shown in Figure 14.37b (see location of SPT test in Figure 14.37a). From this figure, it can be seen that the effect of densification at any given depth decreases with the increase in S' (or S'/D). These tests show that when S'/D exceeds about 4 to 5, the effect of densification is practically negligible.

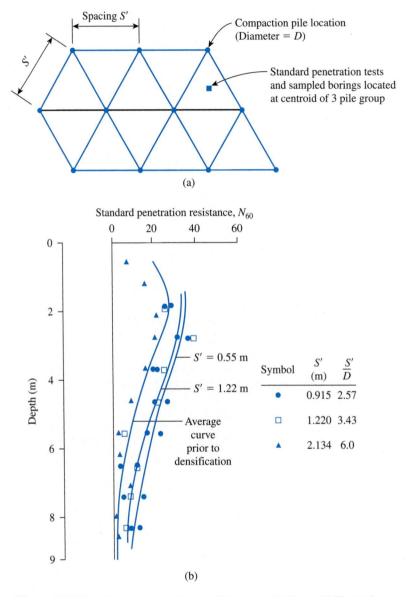

Figure 14.37 Sand compaction pile test of Basore and Boitano (1969): (a) Layout of the compaction piles; (b) Standard penetration resistance variation with depth and S'

14.15 *Dynamic Compaction*

Dynamic compaction is a technique that is beginning to gain popularity in the United States for densification of granular soil deposits. The process primarily involves dropping a heavy weight repeatedly on the ground at regular intervals. The weight of the hammer used varies from 8 to 35 metric tons, and the height of the hammer drop varies between 7.5 and 30.5 m.

The stress waves generated by the hammer drops help in the densification. The degree of compaction achieved depends on

- The weight of the hammer
- The height of the drop
- The spacing of the locations at which the hammer is dropped

Leonards et al. (1980) suggested that the significant depth of influence for compaction is approximately

$$\text{DI} \simeq \tfrac{1}{2}\sqrt{W_H h} \tag{14.55}$$

where

DI = significant depth of densification (m)
W_H = dropping weight (metric ton)
h = height of drop (m)

Partos et al. (1989) provided several case histories of site improvement that used dynamic compaction. In 1992, Poran and Rodriguez suggested a rational method for conducting dynamic compaction for granular soils in the field. According to their method, for a hammer of width D having a weight W_H and a drop h, the approximate shape of the densified area will be of the type shown in Figure 14.38 (i.e., a semiprolate spheroid). Note that in this figure $b = \text{DI}$. Figure 14.39 gives the design chart for a/D and b/D versus $NW_H h/Ab$ (D = width of the hammer if not circular in cross section; A = area of cross section of the hammer; N = number of required hammer drops). The method uses the following steps:

Step 1. Determine the required significant depth of densification, $\text{DI}(=b)$.
Step 2. Determine the hammer weight (W_H), height of drop (h), dimensions of the cross section, and thus the area A and the width D.
Step 3. Determine $\text{DI}/D = b/D$.
Step 4. Use Figure 14.39 and determine the magnitude of $NW_H h/Ab$ for the value of b/D obtained in Step 3.
Step 5. Since the magnitudes of W_H, h, A, and b are known (or assumed) from Step 2, the number of hammer drops can be estimated from the value of $NW_H h/Ab$ obtained from Step 4.
Step 6. With known values of $NW_H h/Ab$, determine a/D and thus a from Figure 14.39.
Step 7. The grid spacing, S_g, for dynamic compaction may now be assumed to be equal to or somewhat less than a. (See Figure 14.40.)

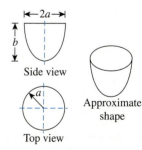

Side view

Approximate shape

Top view

Figure 14.38 Approximate shape of the densified area due to dynamic compaction (Poran, C. J. and Rodriguez, J. A. (1992). "Design of Dynamic Compaction," *Canadian Geotechnical Journal*, Vol. 29, No. 5, pp. 796–802. © 2008 NRC Canada or its licensors. Reproduced with permission.)

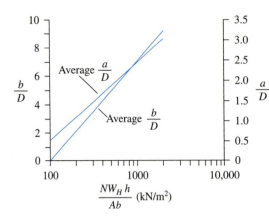

Figure 14.39 Plot of a/D and b/D versus $NW_H h/Ab$ (After Poran and Rodriguez, 1992) (Poran, C. J. and Rodriguez, J. A. (1992). "Design of Dynamic Compaction," *Canadian Geotechnical Journal*, Vol. 29, No. 5, pp. 796–802. © 2008 NRC Canada or its licensors. Reproduced with permission.)

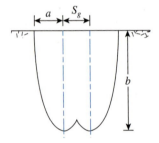

Figure 14.40 Approximate grid spacing for dynamic compaction

14.16 *Jet Grouting*

Jet grouting is a soil stabilization process whereby cement slurry in injected into soil at a high velocity to form a soil–concrete matrix. Conceptually, the process of jet grouting was first developed in the 1960s. Most of the research work after that was conducted in Japan (Ohta and Shibazaki, 1982). The technique was introduced into Europe in the late 1970s, whereas the process was first used in the United States in the early 1980s (Welsh, Rubright, and Coomber, 1986).

Three basic systems of jet grouting have been developed—single, double, and triple rod systems. In all cases, hydraulic rotary drilling is used to reach the design depth at which the soil has to be stabilized. Figure 14.41a shows the *single rod system* in which a cement slurry is injected at a high velocity to form a soil–cement matrix. In the *double rod system* (Figure 14.41b), the cement slurry is injected at a high velocity sheathed in a cone of air at an equally high velocity to erode and mix the soil well. The *triple rod system* (Figure 14.41c) uses high-pressure water shielded in a cone of air to erode the soil. The void created in this process is then filled with a pre-engineering cement slurry.

The effectiveness of the jet grouting is very much influenced by the nature of erodibility of soil. Gravelly soil and clean sand are highly erodible, whereas highly

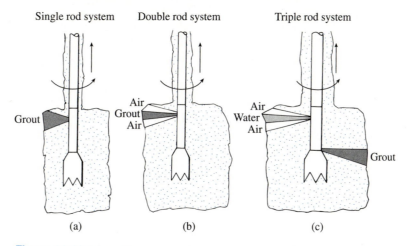

Figure 14.41 Jet grouting

plastic clays are difficult to erode. A summary of the range of parameters generally encountered for the three systems above follows (Welsh and Burke, 1991; Burke, 2004):

Single Rod System:

A. Grout slurry

Pressure . . . . 0.4–0.7 MN/m²
Volume . . . . 100–300 l/min
Specific gravity 1.25–1.6
Number of nozzles 1–6

B. Lift

Step height . . . . 5–600 mm
Step time . . . . 4–30 sec

C. Rotation . . . . 7–20 rpm
D. Stabilized soil column diameter

Soft clay . . . . 0.4–0.9 m
Silt . . . . 0.6–1.1 m
Sand . . . . 0.8–1.2 m

Double Rod System:

A. Grout slurry

Pressure . . . . 0.3–0.7 MN/m²
Volume . . . . 100 –600 l/min
Specific gravity 1.25–1.8
Number of nozzles 1–2

B. Air

Pressure . . . . 700–1500 kN/m²
Volume . . . . 8–30 m³/min

C. Lift

Step height .... 25–400 mm
Step time .... 4–30 sec

D. Rotation .... 7–15 rpm
E. Stabilized soil column diameter

Soft clay .... 0.9–1.8 m
Silt .... 0.9–1.8 m
Sand .... 1.2–2.1 m

Triple Rod System:

A. Grout slurry

Pressure .... 700 kN/m^2–1 MN/m^2
Volume .... 120–200 l/min
Specific gravity 1.5–2.0
Number of nozzles 1–3

B. Air

Pressure .... 700–1500 kN/m^2
Volume .... 4–15 m^3/min

C. Water

Pressure .... 0.3–0.4 MN/m^2
Volume .... 80–200 l/min

D. Lift

Step height .... 20–50 mm
Step time .... 4–20 sec

E. Rotation .... 7–15 rpm
F. Stabilized soil column diameter

Soft clay .... 0.9–1.2 m
Silt .... 0.9–1.4 m
Sand .... 0.9–2.5 m

Problems

14.1 A sandy soil has maximum and minimum dry unit weights of 18.08 kN/m^3 and 14.46 kN/m^3 respectively, and a dry unit weight of compaction in the field of 16.35 kN/m^3. Estimate the following:
  a. The relative compaction in the field
  b. The relative density in the field

14.2 A silty clay soil has a plastic limit (PL) of 18. Estimate the optimum moisture content and the maximum dry unit weight of the soil when compacted using the procedure of:
  a. Standard Proctor test
  b. Modified Proctor test
  Use Eqs. (14.6) and (14.7).

14.3 The following are given for a natural soil deposit:
Moist unit weight, $\gamma = 17.8$ kN/m^3
Moisture content, $w = 14\%$
$G_s = 2.7$
This soil is to be excavated and transported to a construction site for use in a compacted fill. If the specification calls for the soil to be compacted to a minimum dry unit weight of 18.4 kN/m^3 at the same moisture content of 14%, how many cubic meters of soil from the excavation site are needed to produce 20,000 m^3 of compacted fill?

14.4 A proposed embankment fill required 10,000 m^3 of compacted soil. The void ratio of the compacted fill is specified to be 0.65. Four available borrow pits are shown below along with the void ratios of the soil and the cost per cubic meter for moving the soil to the proposed construction site.

Borrow pit	Void ratio	Cost ($/m^3)
A	0.8	6
B	1.1	5
C	0.95	8
D	0.75	10

Make the necessary calculations to select the pit from which the soil should be brought to minimize the cost. Assume G_s to be the same for all borrow-pit soil.

14.5 For a vibroflotation work, the backfill to be used has the following characteristics:

$D_{50} = 2$ mm
$D_{20} = 0.7$ mm
$D_{10} = 0.65$ mm

Determine the suitability number of the backfill. How would you rate the material?

14.6 Repeat Problem 14.5 with the following:

$D_{50} = 1.8$ mm
$D_{20} = 0.72$ mm
$D_{10} = 0.25$ mm

14.7 Refer to Figure 14.16. For a large fill operation, the average permanent load $[\Delta\sigma'_{(p)}]$ on the clay layer will increase by about 75 kN/m^2. The average effective overburden pressure on the clay layer before the fill operation is 110 kN/m^2. For the clay layer, which is normally consolidated and drained at top and bottom, given: $H_c = 8$ m, $C_c = 0.27$, $e_o = 1.02$, $C_v = 0.52$ m^2/month. Determine the following:

a. The primary consolidation settlement of the clay layer caused by the addition of the permanent load $\Delta\sigma'_{(p)}$

b. The time required for 80% of primary consolidation settlement under the additional permanent load only

c. The temporary surcharge, $\Delta\sigma'_{(f)}$, that will be required to eliminate the entire primary consolidation settlement in 12 months by the precompression technique

14.8   Repeat Problem 14.7 with the following: $\Delta\sigma'_{(p)} = 57.5$ kN/m², average effective overburden pressure on the clay layer $= 71.88$ kN/m², $H_c = 4.57$ m, $C_c = 0.3$, $e_o = 1.0$, and $C_v = 9.68 \times 10^{-2}$ cm²/min.

14.9   The diagram of a sand drain is shown in Figures 14.21 and 14.22. Given: $r_w = 0.2$ m, $r_s = 0.3$ m, $d_e = 5$ m, $C_v = C_{vr} = 0.3$ m²/month, $k_h/k_s = 2$, and $H = 6$ m. Determine:

    a. The degree of consolidation for the clay layer caused only by the sand drains after six months of surcharge application

    b. The degree of consolidation for the clay layer that is caused by the combination of vertical drainage (drained on top and bottom) and radial drainage after six months of the application of surcharge. Assume that the surcharge is applied instantaneously.

14.10   A 3.05 m thick clay layer is drained at the top and bottom. Its characteristics are $C_{vr} = C_v$ (for vertical drainage) $= 39.02$ cm²/day, $r_w = 203$ mm, and $d_e = 1.83$ m. Estimate the degree of consolidation of the clay layer caused by the combination of vertical and radial drainage at $t = 0.2, 0.4, 0.8,$ and 1 year. Assume that the surcharge is applied instantaneously, and there is no smear.

14.11   For a sand drain project (Figure 14.20), the following are given:

    Clay: Normally consolidated
                  $H_c = 5.5$ m (one-way drainage)
                  $C_c = 0.3$
                  $e_o = 0.76$
                  $C_v = 0.015$ m²/day
                  Effective overburden pressure at the middle of clay
                  layer $= 80$ kN/m²
    Sand drain: $r_w = 0.07$ m
                  $r_w = r_s$
                  $d_e = 2.5$ m
                  $C_v = C_{vr}$

    A surcharge is applied as shown in Figure P14.11. Calculate the degree of consolidation and the consolidation settlement 50 days after the beginning of the surcharge application.

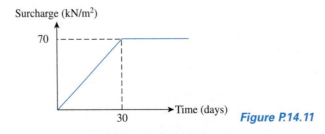

Surcharge (kN/m²)

70

30

Time (days)

Figure P.14.11

References

ABOSHI, H., ICHIMOTO, E., and HARADA, K. (1979). "The Compozer—a Method to Improve Characteristics of Soft Clay by Inclusion of Large Diameter Sand Column," *Proceedings, International Conference on Soil Reinforcement, Reinforced Earth and Other Techniques,* Vol. 1, Paris, pp. 211–216.

AMERICAN SOCIETY FOR TESTING and MATERIALS (2007). *Annual Book of Standards,* Vol. 04.08, West Conshohocken, PA.

BACHUS, R. C., and BARKSDALE, R. D. (1989). "Design Methodology for Foundations on Stone Columns," *Proceedings, Foundation Engineering: Current Principles and Practices* American Society of Civil Engineers, Vol. 1, pp. 244–257.

BARRON, R. A. (1948). "Consolidation of Fine-Grained Soils by Drain Wells," *Transactions,* American Society of Civil Engineers, Vol. 113, pp. 718–754.

BASORE, C. E., and BOITANO, J. D. (1969). "Sand Densification by Piles and Vibroflotation," *Journal of the Soil Mechanics and Foundations Division,* American Society of Civil Engineers, Vol. 95, No. SM6, pp. 1303–1323.

BROWN, R. E. (1977). "Vibroflotation Compaction of Cohesionless Soils," *Journal of the Geotechnical Engineering Division,* American Society of Civil Engineers, Vol. 103, No. GT12, pp. 1437–1451.

BURKE, G. K. (2004). "Jet Grouting Systems: Advantages and Disadvantages," *Proceedings, GeoSupport 2004: Drilled Shafts, Micropiling, Deep Mixing, Remedial Methods, and Special Foundation Systems*, American Society of Civil Engineers, pp. 875–886.

CHRISTOULAS, S., BOUCKOVALAS, G., and GIANNAROS, C. (2000). "An Experimental Study on Model Stone Columns," *Soils and Foundations,* Vol. 40, No. 6, pp. 11–22.

D'APPOLONIA, D. J., WHITMAN, R. V., and D'APPOLONIA, E. (1969). "Sand Compaction with Vibratory Rollers," *Journal of the Soil Mechanics and Foundations Division,* American Society of Civil Engineers, Vol. 95, No. SM1, pp. 263–284.

HUGHES, J. M. O., and WITHERS, N. J. (1974). "Reinforcing of Soft Cohesive Soil with Stone Columns," *Ground Engineering,* Vol. 7, pp. 42–49.

HUGHES, J. M. O., WITHERS, N. J., and GREENWOOD, D. A. (1975). "A Field Trial of Reinforcing Effects of Stone Columns in Soil," *Geotechnique,* Vol. 25, No. 1, pp. 31–34.

ICHIMOTO, A. (1981). "Construction and Design of Sand Compaction Piles," *Soil Improvement, General Civil Engineering Laboratory* (in Japanese), Vol. 5. pp. 37–45.

JOHNSON, S. J. (1970a). "Precompression for Improving Foundation Soils," *Journal of the Soil Mechanics and Foundations Division,* American Society of Civil Engineers. Vol. 96, No. SM1, pp. 114–144.

JOHNSON, S. J. (1970b). "Foundation Precompression with Vertical Sand Drains," *Journal of the Soil Mechanics and Foundations Division,* American Society of Civil Engineers. Vol. 96, No. SM1, pp. 145–175.

LEONARDS, G. A., CUTTER, W. A., and HOLTZ, R. D. (1980). "Dynamic Compaction of Granular Soils," *Journal of Geotechnical Engineering Division,* ASCE, Vol. 96, No. GT1, pp. 73–110.

MATTES, N. S., and POULOS, H. G. (1969). "Settlement of Single Compressible Pile," *Journal of the Soil Mechanics and Foundations Division,* ASCE, Vol. 95, No. SM1, pp. 189–208.

MITCHELL, J. K. (1970). "In-Place Treatment of Foundation Soils," *Journal of the Soil Mechanics and Foundations Division,* American Society of Civil Engineers, Vol. 96, No. SM1, pp. 73–110.

MITCHELL, J. K., and FREITAG, D. R. (1959). "A Review and Evaluation of Soil–Cement Pavements," *Journal of the Soil Mechanics and Foundations Division,* American Society of Civil Engineers, Vol. 85, No. SM6, pp. 49–73.

MITCHELL, J. K., and HUBER, T. R. (1985). "Performance of a Stone Column Foundation," *Journal of Geotechnical Engineering,* American Society of Civil Engineers, Vol. 111, No. GT2, pp. 205–223.

MURAYAMA, S. (1962). "An Analysis of Vibro-Compozer Method on Cohesive Soils," *Construction in Mechanization* (in Japanese), No. 150, pp. 10–15.

OHTA, S., and SHIBAZAKI, M. (1982). "A Unique Underpinning of Soil Specification Utilizing Super-High Pressure Liquid Jet," *Proceedings, Conference on Grouting in Geotechnical Engineering*, New Orleans, Louisiana.

OLSON, R. E. (1977). "Consolidation under Time-Dependent Loading," *Journal of Geotechnical Engineering Division, ASCE*, Vol. 102, No. GT1, pp. 55–60.

OMAR, M., ABDALLAH, S., BASMA, A., and BARAKAT, S. (2003). "Compaction Characteristics of Granular Soils in the United Arab Emirates," *Geotechnical and Geological Engineering*, Vol. 21, No. 3, pp. 283–295.

OSMAN, S., TOGROL, E., and KAYADELEN, C. (2008). "Estimating Compaction Behavior of Fine-Grained Soils Based on Compaction Energy," *Canadian Geotechnical Journal*, Vol. 45, No. 6, pp. 877–887.

OTHMAN, M. A., and LUETTICH, S. M. (1994). "Compaction Control Criteria for Clay Hydraulic Barriers," *Transportation Research Record*, No. 1462, National Research Council, Washington, DC, pp. 28–35.

PARTOS, A., WELSH, J. P., KAZANIWSKY, P. W., and SANDER, E. (1989). "Case Histories of Shallow Foundation on Improved Soil," *Proceedings, Foundation Engineering: Current Principles and Practices*, American Society of Civil Engineers, Vol. 1, pp. 313–327.

PORAN, C. J., and RODRIGUEZ, J. A. (1992). "Design of Dynamic Compaction," *Canadian Geotechnical Journal*, Vol. 2, No. 5, pp. 796–802.

SHIBUYA, S., and HANH, L. T. (2001). "Estimating Undrained Shear Strength of Soft Clay Ground Improved by Preloading with PVD—Case History in Bangkok," *Soils and Foundations*, Vol. 41, No. 4, pp. 95–101.

THOMPSON, M. R. (1967). *Bulletin 492, Factors Influencing the Plasticity and Strength of Lime-Soil Mixtures*, Engineering Experiment Station, University of Illinois.

THOMPSON, M. R. (1966). "Shear Strength and Elastic Properties of Lime-Soil Mixtures," *Highway Research Record 139*, National Research Council, Washington, D.C., pp. 1–14.

TRANSPORTATION RESEARCH BOARD (1987). *Lime Stabilization: Reactions, Properties, Design and Construction*, National Research Council, Washington, D.C.

TULLOCK, W. S., II, HUDSON, W. R., and KENNEDY, T. W. (1970). *Evaluation and Prediction of the Tensile Properties of Lime-Treated Materials*, Research Report 98-5, Center for Highway Research, University of Texas, Austin, Texas.

WELSH, J. P., and BURKE, G. K. (1991). "Jet Grouting–Uses for Soil Improvement," *Proceedings, Geotechnical Engineering Congress*, American Society of Civil Engineers, Vol. 1, pp. 334–345.

WELSH, J. P., RUBRIGHT, R. M., and COOMBER, D. B. (1986). "Jet Grouting for support of Structures," presented at the Spring Convention of the American Society of Civil Engineers, Seattle, Washington.

YEUNG, A. T. (1997). "Design Curves for Prefabricated Vertical Drains," *Journal of Geotechnical and Geoenvironmental Engineering*, Vol. 123, No. 8, pp. 755–759.

Answers to Selected Problems

Chapter 1

1.1 **a.** 0.39
   b. 58%
   c. 16.05 kN/m^3
1.3 **a.** 0.55
   b. 0.355
   c. 57.8%
   d. 106.7 lb/ft^3
1.5 $\gamma_d = 16.07$ kN/m^3; $\gamma = 17.68$ kN/m^3
1.7 Soil A: SM, silty sand
   Soil B: SM, silty sand
   Soil C: MH, elastic silt with sand
   Soil D: ML, sandy silt
   Soil E: SM, silty sand
   Soil F: CL, sandy lean clay
1.9 **a.** 0.01 cm/sec
   b. 0.034 cm/sec

1.11

Point	kN/m^2		
	σ	u	σ'
A	0	0	0
B	50.52	0	50.52
C	81.74	14.72	67.02
D	174.49	63.77	110.72

1.13 25.56 mm
1.15 **a.** 0.299
   b. 105.74 mm
1.17 10.9 days
1.19 $S_c = 7.5$ mm @ $t = 30$ days; $S_c = 40.5$ mm @ $t = 120$ days
1.21 **a.** 30.7°
   b. 33.67°

1.23 $\phi' = 28°$
$c' = 30 \text{ kN/m}^2$

1.25 $\sigma_1 = 302.6 \text{ kN/m}^2$
$u = 61.2 \text{ kN/m}^2$

Chapter 2

2.1   8.96%

2.3

Depth (m)	$(N_1)_{60}$
1.5	12
3.0	11
4.5	10
6.0	8
7.5	12
9.0	12

2.5 ϕ' (average) = 34°

2.7

Depth (m)	D_r (%)
1.5	52.9
3.0	55.5
4.5	51.1
6.0	50.2
7.5	42.3
9.0	44.3

Average $D_r \approx 49.4\%$

2.9 $15,000 \text{ kN/m}^2$
2.11 51.4 kN/m^2
2.13   **a.** 35.00 kN/m^2
          b. 31.86 kN/m^2
2.15   42°
2.17 $c_u = 45.6 \text{ kN/m}^2$; OCR = 3.37
2.19   **a.**   0.65
          b.   1.37
          c. 2131 kN/m^2
2.21 3125 kN/m^2

Chapter 3

3.1   **a.** 252.6 kN/m^2
          b. 176.8 kN/m^2
          c. 280 kN/m^2
3.3   **a.** 267.6 kN/m^2
          b. 184.7 kN/m^2
          c. 368 kN/m^2

3.5	5760 kN
3.7	825 kN/m^2
3.9	287.37 kN
3.11	1066 kN/m
3.13	455.9 kN

Chapter 4

4.1	1711.6 kN
4.3	997 kN
4.5	77.1 kN/m^2
4.7	1282.5 kN
4.9	509.5 kN/m^2
4.11	356 kN/m^2
4.13	589 kN

Chapter 5

5.1	a.	21.9 kN/m^2
	b.	14.07 kN/m^2
5.3	18.78 kN/m^2	
5.5	69.9 kN/m^2	
5.7	a.	@ A—160.5 kN/m^2
	b.	@ B—153 kN/m^2
	c.	@ C—14.45 kN/m^2
5.9	34.8 mm	
5.11	10.9 mm	
5.13	13.6 mm	
5.15	12.48 mm	
5.17	216.8 kN/m^2	
5.19	4000 kN	
5.21	32.4 mm	

Chapter 6

6.1	771 kN/m^2
6.3	181.4 kN/m^2
6.5	3.39 m
6.7	0.193 m
6.9	3260 kN/m^3

Chapter 7

7.1	$P_o = 97.63$ kN/m; $\bar{z} = 1.39$ m	
7.3	b.	3.4 m
	c.	79.89 kN/m
7.5	$P_a = 118.6$ kN/m; $\bar{z} = 1.67$ m	
7.7	81.57 kN/m	
7.9	62.96 kN/m	

7.11 $P_{ae} = 107.7$ kN/m; $\bar{z} = 2.35$ m

7.13

z (m)	$\sigma_a'(z)$ (kN/m²)
1.5	12.01
3.0	18.30
4.5	21.23
6.0	22.32

7.15   390.72 kN/m

Chapter 8

8.1 $FS_{(overturning)} = 3.41$
$FS_{(sliding)} = 1.5$
$FS_{(bearing)} = 5.4$

8.3 $FS_{(overturning)} = 8.28$
$FS_{(sliding)} = 2.79$

8.5   **a.**   903.8 kN
b.   369.8 kN

8.7

z (m)	$\sigma_a'(z)$ (kN/m²)
2	24.15
4	25.54
6	30.79
8	38.48

8.9   **a.**   23.2
b.   4.37
c.   11.68

8.11 $FS_{(overturning)} = 3.43$
$FS_{(sliding)} = 1.35$

Chapter 9

9.1   **a.**   13.31 m
b.   29.3 m
c.   2762 kN-m/m

9.3 $D_{theory} = 3.18$ m; $M_{max} = 59.8$ kN-m/m

9.5 $D = 1.6$ m; $M_{max} = 51.32$ kN-m/m

9.7   PZ 35

9.9 $D = 5.9$ m
$F = 232.8$ kN/m
$M_{max} = 51.91$ kN-m/m

9.11   100.6 kN

9.13

B (m)	P_u (kN)
0.3	15.37
0.6	21.48
0.9	28.00

Chapter 10

10.1   A $\rightarrow$ 169.72 kN
            B $\rightarrow$ 150.68 KN
            C $\rightarrow$ 233.77 kN
10.3   A $\rightarrow$ 148.5 kN
            B $\rightarrow$ 78.4 kN
            C $\rightarrow$ 202 kN
10.5   **a.** $\gamma_{av} = 17.08$ kN/m^3
                    $c_{av} = 19.58$ kN/m^3
            b. $\sigma_a = 30.74$ kN/m^2
10.7   A $\rightarrow$ 306.5 kN
            B $\rightarrow$ 405.55 kN
            C $\rightarrow$ 413.45 kN
10.9   A $\rightarrow$ 306.5 kN
            B $\rightarrow$ 439.35 kN
            C $\rightarrow$ 218.9 kN
10.11 3.57

Chapter 11

11.1   **a.**   2995.5 kN
            b.   2358 kN
            c.   2661 kN
11.3   793 kN
11.5   175 kN
11.7   389 kN
11.9   448.4 kN
11.11 493.9 kN
11.13 5.26 mm
11.15 32.5 kN
11.17 1298 kN
11.19 25.3 kN
11.21 171.2 kN
11.23 2846 kN
11.25 4362 kN

Chapter 12

12.1   9911 kN
12.3   316.7 kN
12.5   5064 kN

12.7   894 kN

12.9   3752 kN

12.11 2356 kN

12.13 6.25 mm

12.15 **a.**   3.13 mm

      b.   594.9 kN-m

      c.   3104 kN/m^2

      d.   7.5 m

Chapter 13

13.1

LL	γ_d below which collapse will occur (kN/m^3)
10	20.8
15	18.8
20	17.16
25	15.78
30	14.60
35	13.59
40	12.71

Collapse will occur @ LL = 30%

13.3   79.2 mm

13.5   1.71 m below the bottom of the foundation

13.7   1 m below the bottom of the foundation

13.9   3.97 m

Chapter 14

14.1   **a.**   90.4%

      b.   57.5%

14.3   23,573 m^3

14.5 $S_N = 3.86$; Excellent

14.7   **a.**   0.241 m

      b.   17.45 months

      c.   108.4 kN/m^2

14.9   **a.**   23%

      b.   61.9%

14.11 $U_{v,r} = 17.8\%$; Settlement = 45.6 mm

Index